Business Data Networks and Telecommunications

Eighth Edition

Business Data Networks and Telecommunications

International Edition

Raymond R. Panko
University of Hawai`i at Mānoa

Julia L. Panko
University of California, Santa Barbara

Boston Columbus Indianapolis New York San Francisco Upper Saddle River
Amsterdam Cape Town Dubai London Madrid Milan Munich Paris Montréal Toronto
Delhi Mexico City São Paulo Sydney Hong Kong Seoul Singapore Taipei Tokyo

Editorial Director: Sally Yagen
Editor in Chief: Eric Svendsen
Executive Editor: Bob Horan
Editorial Project Manager: Kelly Loftus
Editorial Assistant: Jason Calcano
Director of Marketing: Patrice Lumumba Jones
Senior Marketing Manager: Anne Fahlgren
Marketing Assistant: Melinda Jensen
Senior Managing Editor: Judy Leale
Production Project Manager: Kerri Tomasso
Senior Operations Supervisor: Arnold Vila
Operations Specialist: Ilene Kahn
Creative Director: Jayne Conte
Full-Service Project Management: Elm Street Publishing Services
Composition: Integra Software Services Pvt. Ltd.
Printer/Binder: Edwards Brothers
Cover Printer: Lehigh-Phoenix Color/ Hagerstown
Text Font: Palatino 10

Credits and acknowledgments borrowed from other sources and reproduced, with permission, in this textbook appear on appropriate page within text.

10 9 8 7 6 5 4 3 2 1

ISBN 10: 0-13-255245-0
ISBN 13: 978-0-13-255245-5

BRIEF CONTENTS

Basic Concepts and Principles

Applying Concepts and Principles: Up Through the Layers

CONTENTS

PREFACE FOR STUDENTS

PERSPECTIVE

Initially, information systems (IS) graduates had a single career track: programmer–analyst–database administrator–manager. Today, however, many IS graduates find themselves on the networking/security career track—often to their surprise. This course is an introduction to the networking/security track.

The network track is very attractive, especially if you take a security course as well as a networking course. In the most recent U.S. Bureau of Labor Statistics employment projections for 821 job categories, network analyst was the *top category of all* in terms of percentage growth.

Even programmers today need a strong understanding of networking. In the past, programmers wrote stand-alone programs that ran on a single computer. Today, however, most programmers write networked applications that work cooperatively with other programs on other computers.

Whichever IT career path you choose, learn as much as you can from each course. Employers today do not just look at transcripts. If you make it past initial job interviews, expect to be grilled for hours in subsequent interviews to see what you really know. Employers do not expect perfection, but they demand a high level of knowledge and the retention of that knowledge. Study for mastery and retention.

WHAT'S NEW IN THIS EDITION?

- This edition is a major redesign of the book. This redesign was undertaken on the basis of discussions with adopters, discussions with networking professionals, the examination of statistical market trends in networking, and my experiences in teaching the course. The redesign gives students better preparation for dealing with specific network standards, and it gives tighter integration to material on specific chapters.
- The book begins with four introductory chapters dealing with networking in general, networks standards, network security, and network design and other network management matters. After these four chapters, the student is able to apply these concepts to a series of progressively more complex types of networks.
- The security chapter is enhanced, with a greater focus on security management through the plan-protect-respond security management process and policy-based security. More security is added throughout the book, for instance, 802.1X in Chapter 6; the 802.11 core security protocols and advanced WLAN security topics in Chapter 8; TCP/IP security in Chapter 10; and application security in Chapter 11.
- This edition tightens the book's up-through-the-layers flow for the post-introductory chapters. There are two chapters on wired switched networks, two chapters on wireless networks, two chapters on TCP/IP internetworking, and a final chapter on applications.
- The wireless material is significantly enhanced and streamlined, with the de-emphasizing of older 802.11 standards and a sharper focus on security, on 3G, 4G, and WiMAX services, and on security.

- The networked applications chapter now has a strong section on cloud computing, a rewritten section on service-oriented architectures and Web services, and an updated section on peer-to-peer computing with information on BitTorrent and Skype.
- The material on TCP/IP has been expanded somewhat as a result of discussions with networking managers and professionals. More is covered in the first chapter, allowing the standards chapter to be more streamlined yet also add some new material, especially on port numbers. Chapters 9 and 10 provide a more focused and streamlined treatment of advanced TCP/IP core, supervisory, and security standards.
- In switched wired networks, wireless networks, and TCP/IP, LAN and WAN technology are now covered together.

HOW TO LEARN NETWORKING

Networking Is Difficult

Networking is an exciting topic. It is also a difficult topic. In programming, the focus is on creating and running programs. In networking, the critical skills are design, product selection, and troubleshooting. These rather abstract skills require a broad and deep knowledge of many concepts. Many IS students have a difficult time adjusting to these cerebral skill requirements. Security is even more of a "head game" because we are pitted against intelligent adversaries, not simply difficult choices and complex situations.

Employers Are Growing More Demanding

In the past, many teachers tried to deal with the complexity of networking by selecting what was in essence a "network appreciation" book—a feel-good book that lacked the detailed knowledge needed for actual networking jobs.

Today, however, employers demand—and get—strong job readiness from new graduates. If you want to get a job in the IS field, you will need to have a competitive level of knowledge in every IS subject that you study. Even applicants for database jobs are grilled in networking knowledge (and networking applicants are grilled in database and other areas).

How to Study the Book

There are several keys to studying this book:

- Reading chapters once will not be enough. You will need to really study the chapters.
- Slow down for the hard parts. Some sections will be fairly easy, others difficult. We all have a tendency to go at the same speed through all sections. Also, if you don't get it the first time, work at it. You may be able to get Bs and Cs not knowing the hard stuff, but you won't have the knowledge you need in the workforce.
- When you finish reading a section, immediately do the Test Your Understanding questions. If you don't get one of the questions, go back over the text. The understanding of networking is strongly cumulative, and if you skim over one section, you will have problems with other sections later. Multiple-choice questions in the test item file are taken entirely from the Test Your Understanding questions and the end-of-chapter questions.

- If you can, get together with other students to go over your answers to the Test Your Understanding questions to see if everybody agrees.
- Study the figures. Nearly every key point in each chapter is covered in the figures. If there is something in a figure you don't understand, you need to study the corresponding section in the chapter. Pretend you are trying to explain the figure to someone else. Also, study the "study figures" that summarize information.
- Network design is about selecting among alternatives. If several concepts are presented in a section or chapter, do not just study them individually. You need to know which one to use in a particular situation, and that requires comparing/contrasting knowledge. Study the figures that compare concepts, and make your own figures and lists or charts of features if the book does not provide them. Comparing different technologies in order to select the best one for corporate needs is a critical skill for all IT professionals.
- Study the synopsis section at the end of each chapter. The synopsis summarizes the core concepts in the chapter. Be very sure that you know them well. You might even study them before beginning the chapter, to get a broad understanding of the material.

Hands-On

One way to make networking less abstract to you is to do as many hands-on activities as possible.

A NETWORKING CAREER

If you like the networking course and think you want a networking career, there are a number of steps you should take before graduation, even if your school does not have advanced networking courses.

- Most importantly, do a networking internship. Employers really want workers with job experience—often preferring it to an unwarranted degree over academic preparation.
- Learn systems administration (the management of servers). Learn the essentials of UNIX and Windows Server. You can download a server version of Linux and install it on your home computer in order to play with UNIX commands and network management functions.
- Learn about security. Security and networking are now inextricably intertwined.
- Consider getting one or more industry certifications. In networking, the low-level CompTIA Network + certification should be obtainable with just a bit more study after you take your core networking course. However, this certification is not highly valued. Cisco's more valuable CCNA (Cisco Certified Network Associate) certification, which focuses on switching and routing, will require substantially more study. You also have to learn hands-on commands. This will require you to take a class or buy a network simulation program designed for this certification. Microsoft network certifications are also valuable, but they focus more on clients and servers. Employers like certifications, but they know that certifications are no substitutes for job experience.

ABOUT THE AUTHORS

Ray Panko is a professor of IT management at the University of Hawai`i's Shidler College of Business. His main courses are networking and security. Before coming to the university, he was a project manager at Stanford Research Institute (now SRI International), where he worked for Doug Englebart (the inventor of the mouse). He received his B.S. in Physics and his M.B.A. from Seattle University. He received his doctorate from Stanford University, where his dissertation was conducted under contract to the Office of the President of the United States. He has been awarded the Shidler College of Business's Dennis Ching award as the outstanding teacher among senior faculty. He is also a Shidler Fellow.

Julia Panko is a Ph.D. candidate in English at the University of California, Santa Barbara. Her research interests include the twentieth- and twenty-first-century novel, the history and theory of information technology, and the digital humanities. Her dissertation focuses on the relationship between information culture and the modern and contemporary novel. Before this edition, she was a technical editor for the book.

Business Data Networks and Telecommunications

CHAPTER 1

Networking: How We Got Here

LEARNING OBJECTIVES

By the end of this chapter, you should be able to:

- Describe basic network terminology, including the definition of *networking*, networked applications, hosts, network addresses, the network core, access links, the evolution of networked applications, and speed.
- Describe packet switching as a way to reduce long-distance transmission costs and digital signaling, modems, The ARPANET.
- Describe physical links and data links.
- Describe the origins of Internet standards, including e-mail.
- Explain why internetworking was needed; Kahn and Cerf's basic concept for internetworking; and packets, packet switches, and other duplicated network concepts.
- Describe the internet and transport layers.
- Explain the five basic layers of standards in the TCP/IP-OSI Hybrid Standards Architecture.
- Describe TCP/IP standards.
- Explain the evolution of the Internet from research network to commercial network.
- List TCP/IP supervisory standards.
- Explain distinctions between LANs and WANs.
- Describe a small home network.

THE CANDIDATE'S BLACKBERRY

During the 2008 presidential election campaign, Senator Barack Obama nearly always carried his BlackBerry mobile smart phone on his hip. Whenever he had a few minutes free, he opened it, read some messages, and sent messages of his own. The news media picked up on the BlackBerry as a sign that the candidate was "tech savvy." In contrast, Senator McCain only used his mobile phone occasionally, and only for voice calling.

Test Your Understanding

1. Note: These questions will require you to use your imagination. a) How do you think Senator Obama may have used his BlackBerry for e-mail during the campaign? Try to come up with concrete examples. b) What non-e-mail applications may he have run on his smart phone? How would he have used each? c) What did Senator Obama have to know to use his BlackBerry? d) What did he have to understand about how cellular telephony worked? e) How do you think he controlled who could call him? f) What would have happened if his mobile phone number had been published on a website? g) After Senator Obama became president, he was told he could not use his BlackBerry any more. Why do you think that was the case? h) How do you think he got around this problem?

BASIC NETWORK TERMINOLOGY

Networks, Hosts, and Applications

In a book on networking, it makes sense to begin by defining what a network is. As a working definition, we will define a **network** as a system that permits networked applications on different hosts to work together. Figure 1-1 illustrates this definition graphically. (In Chapter 2, we will extend this working definition.)

A network is a system that permits networked applications on different hosts to work together.

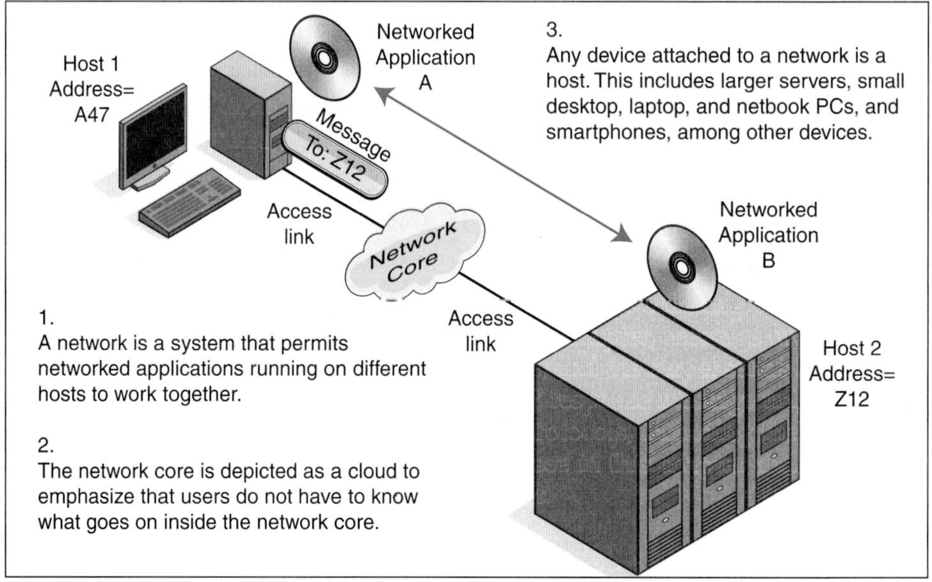

FIGURE 1-1 Basic Network Terminology

NETWORKED APPLICATIONS Note that the definition focuses on *applications*. Although you will spend this course learning the details of how networks work, our preliminary definition takes the *user's* point of view. Users need to do their jobs, and they need applications to help them. Networks are only valuable if they connect the applications that users need to have connected. For Senator Obama, the BlackBerry's e-mail application was a tool for getting his job done. He had no desire to know how the network worked, and he had no need to know. He simply wanted to use his e-mail as painlessly as he could. Applications are the only things users care about and should have to know about.

Computer applications existed long before networking. However, as networks began to connect different computers and to do so with greater reliability and speed, there was an explosion in **networked applications,** which require networks to work.

Networked applications are applications that require networks to work.

The 1970s brought the first networked applications, including e-mail and the File Transfer Protocol (FTP) for moving large files. Networked applications continued to grow steadily, but there was an explosion of applications in the 1990s, as the rapid growth and commercialization of the Internet brought the World Wide Web, e-commerce, and many of the other networked applications we use so extensively today.

This century has brought **Web 2.0** applications, in which users provide the content. In early (Web 1.0) web applications, an information provider created the content. This was expensive and gave information providers control of what appeared on the World Wide Web. With Web 2.0, these barriers are gone. Wikipedia is an excellent example of a Web 2.0 application. Although Wikipedia has some problems, this "crowd-sourced" application is by far the most comprehensive encyclopedia the world has ever known. On a less august level, there is YouTube.

In Web 2.0 applications, users provide the content.

Included in Web 2.0 are **social media applications,** such as Facebook and dating services, which are designed to facilitate relationships. E-mail, mobile telephony, and other communication applications have long been useful in building relationships, but social media sites have developed specialized tools to enhance social networking. Facebook works on the individual level, connecting an individual with several of his or her friends.

Social media applications are designed to facilitate group relationships.

HOSTS AND ADDRESSES In the definition of networking, it was noted that applications run on devices called *hosts*. We use the term *host* rather than *computer* because not all devices connected to networks are traditional computers. The BlackBerry smart phone is a good example of this today. In the future, toasters and coffee pots will be connected to networks within individual homes and across the Internet.[1] Formally, we will define a **host** as any device connected to a network.

[1] In fact, there already is an *Internet Coffee Pot Control Protocol* (RFC 2324) for remotely managing coffee brewers. The standard was created on April 1, 1998. In America, the first of April is April Fool's Day, which is a traditional day for verbal pranks. Although the Internet Coffee Pot Control Protocol was a joke standard, the future will hold many similar but serious protocols.

A host is any device attached to a network.

Each host computer needs a unique address. When a source host sends a message to a destination host, the source host places the address of the destination host in the message. Based on this address, the network delivers the message to the destination host. The source host does not have to know the path the message takes to the destination host—just as you simply dial a number and the telephone network connects you to the other telephone without you knowing how your voice travels to the other phone.

THE NETWORK CORE For users, a network is simply there. Consequently, Figure 1-1 depicts the **network core**—the central part of the network—as a cloud. Just as you cannot see the inside of a cloud, users do not have to look inside the network.

The network core is the central part of the network.

Of course, you will spend this course looking *deeply* inside various types of network technologies ranging from the network you may have in your home to the global Internet. Before my grandfather came to America, he was told that the streets of America were paved with gold. When he got here, he found that they were not even paved. Guess who was going to pave them?

Fortunately for you, the network "streets" in corporations today are already reasonably well paved. Your job will be to turn them into true super highways and to help your organization assimilate the ever growing number of networked applications that new network capabilities will make possible.

ACCESS LINKS Users connect to networks through **access links,** which may use copper wire, optical fiber, or radio transmission. Users may need to know a little more about access links than about the network core. For example, they may have to plug in access link technology, configure it, and troubleshoot simple problems. However, required knowledge should be reduced as far as possible.

Test Your Understanding

2. a) Give the book's definition of *network.* b) What is a networked application? c) What are Web 2.0 applications? d) What are social media applications? e) What is a host? f) Is your laptop PC or desktop PC a host? g) Is a mobile smart phone a host? h) Why is the network core shown as a cloud? i) Why may the user need to know more about his or her access link than about the network cloud?

The Evolution of Networked Applications

TERMINAL–HOST COMMUNICATION In the early years of networked computing, the user had a **dumb terminal,** which had only enough intelligence to communicate with the central host and to show characters on the screen. (This was before microprocessors, so placing intelligence on the desktop would have been prohibitively expensive.) The

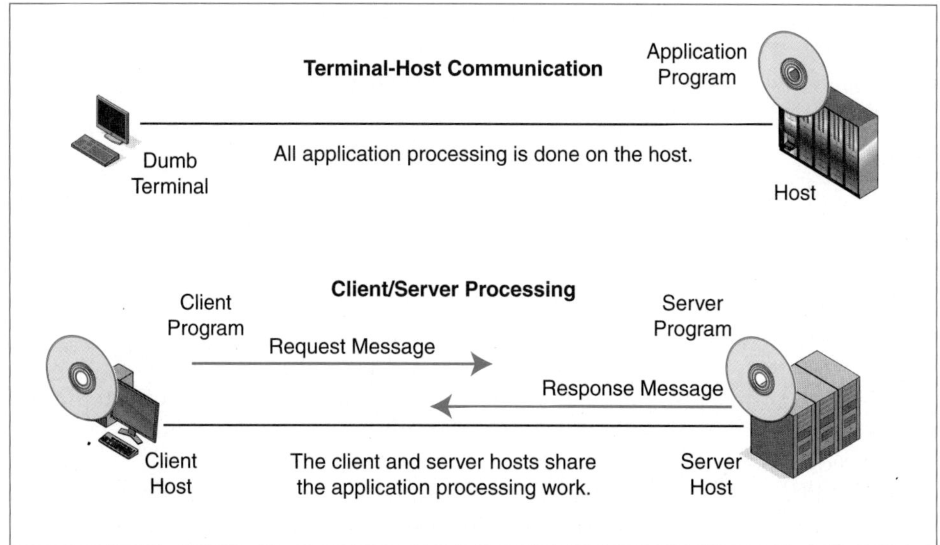

FIGURE 1-2 Terminal–Host Computing and Client/Server Computing

application software ran on a large host computer. Here we are not using the term *host* to refer to any device on the network. We are using it in its original sense, as a large central computer serving many terminal users. Figure 1-2 shows that the application software ran entirely on the central host computer.[2]

CLIENT/SERVER PROCESSING When PCs appeared, the user's computer could handle some of the processing work. Networked applications today usually require an application program on one host to work with an application program on another host. (Yes, we are using the term *host* in its generic meaning again.) Often, this involves **client/server processing,** in which a **server** program on a **server host** provides services to a **client** program on a **client host.**

> In client/server processing, a server program on a server host provides services to a client program on a client host.

Figure 1-2 illustrates how client/server processing works when you use the World Wide Web. Here the application protocol is the Hypertext Transfer Protocol (HTTP). Your PC is the client host, and the webserver is the server host. The webserver host runs webserver application software, while your PC runs a client application program, namely your browser. The webserver application provides service to your browser by sending you the webpages you request.

[2] Initially, only these large computers connected to networks. Consequently, standards for advanced networks used the term *host* when they referred to computers connected to networks. Later, as PCs began to connect to data networks, the only alternatives were to extend the term *host* to PCs and anything else connected to networks or to rewrite all of the standards. Standards developers opted for personal energy conservation.

Although browser/webserver applications are very common, the server is not always a webserver, and the client program is not always a browser. In database processing, of course, the server is a database server. The client PC, in turns, often runs a dedicated database client program.

Client/server processing typically uses a **request–response cycle.** The client sends a request message to the server. The server sends back a response message. For instance, when you type a URL or click on a link, your browser sends a request message to the webserver. This request message specifies a file to be downloaded. The webserver sends back a response message containing the specified file (or an error message). If the user downloads another webpage a few seconds or minutes later, this is a separate request–response cycle.

Client/server computing is a challenge for programmers. Instead of writing stand-alone programs, programmers today usually have to write application programs that must work with application programs on other computers. This can be difficult and often requires a good understanding of how networks work.

Today, PCs are very powerful, and we are seeing the rise of **peer-to-peer (P2P)** applications, in which client hosts provide to services directly to other client hosts. Although P2P processing has often been used for illegal purposes, we will see in the final chapter that P2P applications are likely to have a big role in corporations tomorrow.

In peer-to-peer (P2P) applications, client hosts provided service directly to other client hosts.

Another popular term today is **cloud computing,** in which the servers for internal corporate applications are managed by an outside company. With cloud computing, corporations do not have to maintain their own servers and may not have to maintain their own applications. Outsourcing is nothing new, but cloud computing is bringing a radically new level of outsourcing for internal applications. We will also look at cloud computing in the final chapter.

Test Your Understanding

3. a) Where is processing done in terminal–host communication? b) In client/server processing? c) In P2P applications? d) What is the client/server request–response cycle? e) What is cloud computing?

Speed

The first question people ask about a new-born baby is "Is it a boy or girl?" The first question people ask about a network is "How fast is it?" Network speeds[3] are measured in **bits per second (bps).** Note that this is *bits* per second, not *bytes* per second. Occasionally, when you download files, your software will tell you how many bytes per second are

---

[3] Strictly speaking, speed is a poor description because it implies velocity. Signals always travel at the speed of light through propagation media, regardless of how many bits are sent. We really should say rate, which is like the amount of water flowing through a pipe. Looked at another way, velocity is like *running* faster. Bits per second is like *talking* faster.

Transmission Speed Base Units		
Bits per second (bps)		
Usually not bytes per second (Bps)		
Metric Prefixes		
Metric and Base Unit	**Spoken**	**Meaning**
kbps (lower-case k)	kilobits per second	1,000 bits per second (not 1,024 bps)
Mbps	megabits per second	1,000 kbps 1,000,000 bps
Gbps	gigabits per second	1,000 Mbps 1,000,000,000 bps
Tbps	Terabits per second	1,000 Gbps 1,000,000,000,000 bps

FIGURE 1-3 Transmission Speed

being downloaded. However, that is an exception. In such cases, the B usually is capitalized, so that *bytes per second* is written as **Bps.**

Figure 1-3 shows that, in increasing factors of one thousand, there are **kilobits per second (kbps), megabits per second (Mbps), gigabits per second (Gbps),** and **terabits per second (Tbps).**

In the metric system, numbers are expressed in **base units,** such as bits per second, preceded by a **metric prefix** that multiplies the base unit. Without a metric prefix, five million bits per second would be 5,000,000 bps. With the metric prefix for mega (M), it is 5 Mbps. To give another example, 7.2 kbps without its metric prefix is 7,200 bps.

Note that metric prefixes for speeds increase in factors of 1,000—not 1,024. Factors of 1,024 are used for computer memory and storage. In networking, the bit is the basic unit rather than the byte. Straight metric units are fine for speed.

Note also that the abbreviation for kilobits per second is *kbps*, not *Kbps*. In the metric system, kilo is a lowercase k. (Capital K is for Kelvins, which is a measure of temperature.) Networking people are educated and know the metric system. Database people and computer scientists, who tend to use a capital K, apparently are not.

How fast does a network *need* to be? The answer depends upon the sizes of messages transmitted. Figure 1-4 shows download times at various transmission speeds for some applications you might use yourself.

- An e-mail message of 250 words will be downloaded instantly regardless of the network transmission speed.
- A JPEG photograph is about 2 megabytes. At 100 kbps, it will take 2.7 minutes to download the photograph. This is totally unacceptable. Even at 500 kbps, it will take a half minute to download the photograph. This is better but still painful. Unless the network speed is at least a few megabits per second, the download will not seem fast to the user.

	100 kbps	0.5 Mbps	1 Mbps	5 Mbps	10 Mbps	100 Mbps
E-mail Message (250 words)	0.02 sec	0 sec	0 sec	0 sec	0 sec	0 sec
Photograph (2 MB)	2.7 min	.5 min	16 sec	3 sec	2 sec	0 sec
MP3 Song (3 min)	2.9 min	.5 min	17 sec	3 sec	2 sec	0 sec
Limited-Quality TV (1.25 Mbps)	13 hrs	3 hrs	1 hr	15 min	8 min	1 min

FIGURE 1-4 Download Times for Various Applications

- For a three-minute MP3 file, the situation is almost identical to that of a JPEG photograph.
- TV requires even faster download speeds. Even with low-quality television, which requires 1.25 Mbps for speed for real-time streaming, even 1 Mbps is too slow. If an entire one-hour show is to be downloaded, this will take eight minutes even at 10 Mbps.

Test Your Understanding

4. a) Are network speeds usually measured in bits per second or bytes per second? b) How many bits per second (without a metric prefix) is 20 kbps? Use commas. c) How many bits per second (without a metric prefix) is 7 Mbps? Use commas. d) How many bits per second (without a metric prefix) is 320 kbps? Use commas. e) Is the metric prefix for kilo k or K?

PACKET SWITCHING AND THE ARPANET

So far, we have been looking at networks today. In this section, we will begin looking at how we got here.

Larry Roberts Has a Burstiness Problem

During the 1960s, the U.S. Department of Defense's **Advanced Research Projects Agency (ARPA)**[4] funded a great deal of basic research across the United States. While some of this research had direct military application, most funding was for basic science and technology.

CONNECTING TO REMOTE HOSTS Dr. Larry Roberts was head of ARPA's Information Processing Technology Office (IPTO). IPTO funded a great deal of software research. In those days, software usually was not portable from one computer to another. If

---

[4] Is it ARPA or DARPA? It depends on the year. It was born ARPA in 1958. In 1972, it became DARPA to emphasize its status as a Department of Defense agency. In 1993, it went back to ARPA. Then it went back to DARPA in 1996. DARPA, "ARPA-DARPA: The Name Chronicles," undated. http://www.darpa.gov/arpa-darpa.html. Last viewed August 2009.

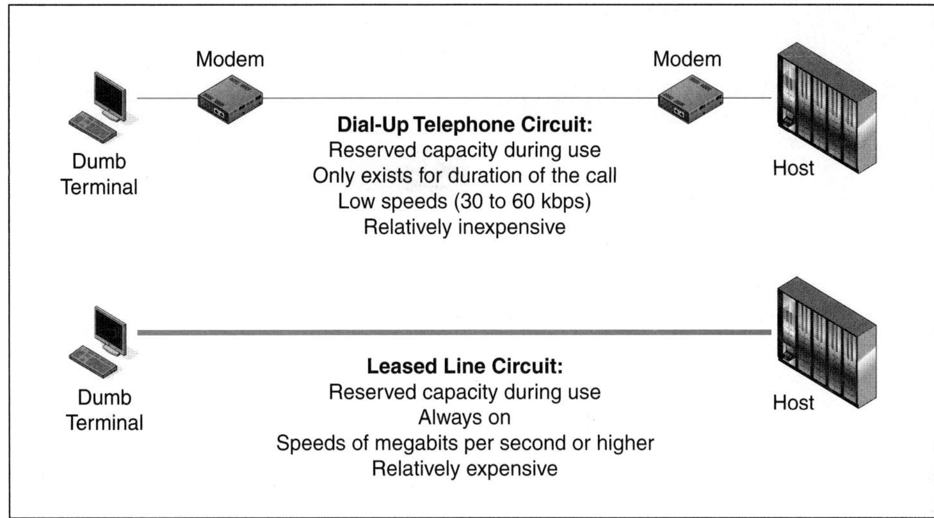

FIGURE 1-5 Telephone Connections

someone needed to use a piece of software, he or she would need a computer terminal and a transmission line to the host computer running the software. Figure 1-5 illustrates this situation.

The figure shows two ways to connect terminals and hosts over long distances. The first is a **dial-up** circuit. When you place a normal telephone call, you are given a circuit to the called party. A **circuit** is a link that gives you reserved capacity all the way between you and the other party. In reality, your voice travels over multiple transmission lines and telephone switches. However, it appears that you are connected by a single direct line. Consequently, this service is called **circuit switching.** In dial-up circuits, the circuit is terminated after the two parties hang up.

A circuit gives you reserved capacity all the way between you and the other party.

You only pay for a dial-up circuit when you are actually using it. However, in the 1960s, per-minute charges were very high. In addition, Figure 1-5 shows that you needed a device called a *modem* to convert between your computer's ones and zeros and the smoothly varying electrical signals carried by the telephone line going to your house. Modems were expensive in the 1960s. If you have ever connected to the Internet with a telephone modem, you also know two other problems with dial-up circuits. First, you cannot transmit and receive at high speeds. Webpages of any complexity take forever to download. Second, you cannot use your phone for voice conversations when you are connected to the Internet. This inevitably creates family discord.

The fact that dial-up circuits give reserved capacity means that no matter how many other people want to use the telephone system, your data transmission speed will never slow down. It is like getting season tickets for a sports team or the opera. You will always have your seat. The bad thing about reserved capacity is that

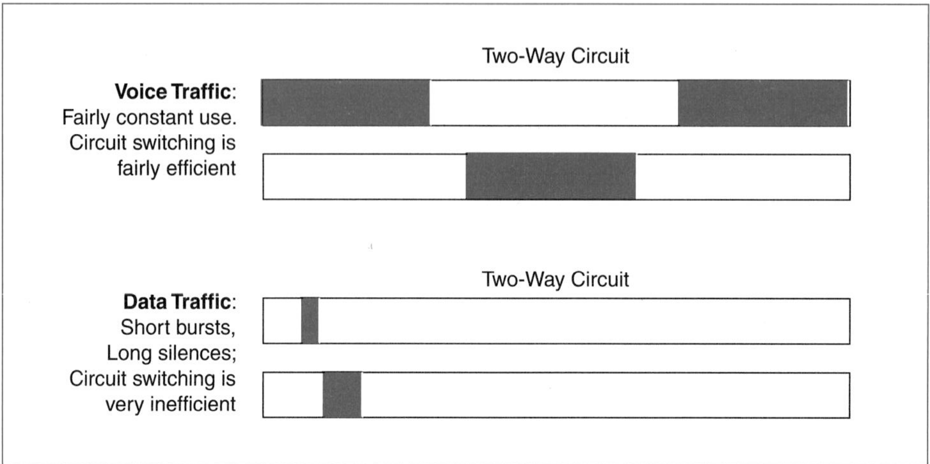

FIGURE 1-6 Data Burstiness and Reserved Circuit Capacity

you pay for it whether you use it or not. If your team has a lousy year or your opera company has a lousy tenor, you don't get any money back for the time you are not using your seat.

Figure 1-6 shows that when two people are on the telephone, someone is speaking most of the time. Ideally, only one person speaks at a time, so you only use your two-way circuit to about half of its capacity. In addition, there will always be some silences, so a third of capacity is probably more typical. That is somewhat wasteful, but it's not too bad.

In contrast to telephone conversations, data transmission is **bursty,** which means that there are bursts of traffic followed by long silences. To see this, consider what happens when you visit a website. To download a webpage, you send a request and get back a response containing the webpage. All of this usually takes a second or less. Now, you probably will look at the webpage for about 30 seconds. (Count it some time!). This means that you are only using one-thirtieth of your circuit—about 3 percent. So while reserved capacity is only somewhat wasteful for voice, it is *very* wasteful for data.

Data transmission is bursty, which means that there are bursts of traffic separated by long silences. This is very wasteful if you are using reserved-capacity circuits.

Figure 1-5 also shows a second way to connect remotely to a host computer. This is a **leased line circuit.** Leased line circuits are good for two reasons. First, they can provide much faster speeds than telephone modem service. Even today, telephone modems can only send at 33.6 kbps and receive at 56 kbps. Leased lines usually transmit at a megabit per second or more. Second, leased line circuits are **always-on connections.** You are connected 24 hours a day, every single day. Of course, this much reserved capacity for this much time is extremely expensive. The burstiness problem is far worse than it is for telephone modem connections.

Dr. Roberts funded many of the calls and leased line circuits that connected user terminals to remote computers that ran ARPA-funded software. He needed a way to reduce the cost of circuit switching for bursty computer traffic.

Test Your Understanding

5. a) In telephony, what is a circuit? b) What two types of circuits can corporations use to link terminals with hosts? c) Compare and contrast these two types of circuits. d) What is good about reserved capacity? e) What is bad about reserved capacity? f) What does it mean that data traffic is bursty? g) Why is circuit switching very inefficient for bursty data traffic?

Binary Signaling, Digital Signaling, and Modems

ANALOG SIGNALING Before looking at how to solve the problem of burstiness, we will step back and take a closer look at how telephone modems work. Figure 1-7 shows how the telephone system works in general. It shows that the human voice rises and falls rapidly but smoothly in loudness. A microphone in the handset translates loudness changes into electrical intensity. When voice loudness rises, the electrical signal rises in intensity. When voice loudness falls, the electrical signal falls. In other words, the electrical signal is *analogous* (similar) to the original voice signal. For this reason we call this an **analog signal.**

An analog signal rises and falls smoothly in intensity.

BINARY TRANSMISSION For computer signals, analog signaling creates a problem. Computers do not transmit sounds. They transmit ones and zeros. Speaking very loosely, we can transmit bits with two **states** (conditions), such as two voltage levels. A

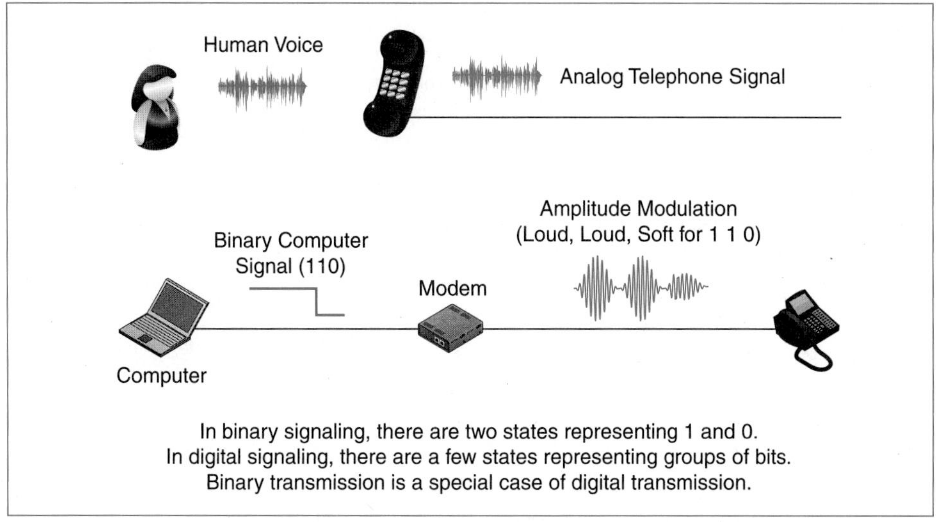

FIGURE 1-7 Telephone Modems

one might be a high voltage, and a zero might be a low voltage. There are only two states, so this type of transmission is called **binary transmission.**

DIGITAL TRANSMISSION Sometimes, there are *a few* states instead of two. For instance, if you have four voltage levels, you can represent two bits at a time. The lowest state might represent 00. The next-lowest might be 01. The two remaining states might be 10 and 11. When there are *a few* states (two or more, but not many), this is called **digital signaling.** Binary signaling, then, is a special case of digital signaling.

Transmission with two states is binary signaling.

Transmission with a few states (two or more, but not many) is digital signaling. Binary signaling is a special case of digital signaling.

THE ADVANTAGE OF DIGITAL SIGNALING In Chapter 5, we will see that signals degrade as they propagate from the sender to the receiver. For example, they attenuate (get weaker). The received signal will be different from the transmitted signal. In other words, it will be incorrect. In analog signaling, if the signal attenuates by 10 percent, then there will be a 10 percent error in transmission. Analog signaling is very sensitive to propagation attenuation and other degradations.

Digital signaling, in contrast, has some resistance to errors, as Figure 1-8 indicates. In this case, we have binary signaling. A 0 is defined as anything from 3 to 15 volts. A zero, in contrast, is defined as anything from negative 3 to negative 15 volts.[5] Any other voltage is undefined and is treated as an error.

Suppose that the transmitted signal is 12 volts. This is a zero. During propagation, the strength of the signal falls to 6 volts. This is a propagation error of 50 percent. Yet the

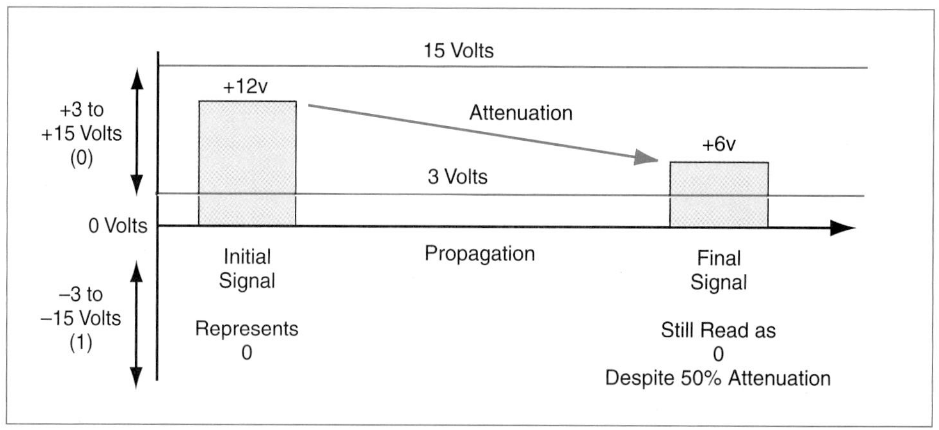

FIGURE 1-8 Resistance to Errors in Digital Signaling

[5] This is the way things were done in RS-232 serial ports that used to be found on computers. The operation of newer ports, such as USB and HDMI ports, is too complicated to be used in examples.

receiver correctly reads it as a 0. So binary transmission (and digital transmission in general) has *resistance to propagation error*. This is not complete *immunity*, however. If the signal had dropped by 90 percent, it would be only 1.2 volts, and this would be an error.

MODEMS The telephone network expects analog signals to come from the telephone subscriber's premises. However, computers send digital signals. The purpose of a **modem,** then, is to translate between digital computer signals and analog telephone line signals. **Modulation** is translating digital computer signals into analog transmission line signals; **demodulation** is translating analog transmission line signals into digital computer signals. Mercifully, *modulator/demodulator* is always shortened to *modem*.

Modems translate between digital computer signals and analog telephone line signals.

Figure 1-7 illustrates a simple type of modulation called **amplitude modulation.** In amplitude modulation, a one is represented by a loud signal; a zero is represented by a soft signal. So if a computer sends 1-1-0-1, the modem transmits LOUD, LOUD, soft, LOUD. Again, this is modulation.

When signals come down the telephone line from the other computer, they come in as loud and soft sounds. The modem translates them back into 1s and 0s. Again, this is demodulation.

What happens if two ones are sent in a row? There will be no change in the loudness level. How can the receiver tell if the sender has transmitted two 1s or only a single 1? The answer is that the sender and receiver divide time into fixed-length **clock cycles.** For instance, a clock cycle might be 1/2400 of a second. So if a binary signal remains high for 2/2400 of a second, this is two 1s rather than a single 1. In digital signaling, then, the signal stays in one state throughout the clock cycle. At the end of the clock cycle, the signal can stay in the same state for the next clock cycle, or it can jump to the other state.

In digital signaling, the signal stays in one state throughout the clock cycle. At the end of the clock cycle, the signal can stay in the same state for the next clock cycle, or it can jump to the other state.

Test Your Understanding

6. a) How did analog telephone line signaling get its name? b) Distinguish between analog and binary signaling. c) What is a state? d) How many states are there in binary signaling? e) How many states are there in digital signaling? f) Is all binary signaling digital signaling? g) Is all digital signaling binary signaling? h) Why does digital signaling give resistance to error? i) What does a modem do? j) Describe amplitude modulation briefly. k) Why are clock cycles necessary in signaling?

Packet Switching Presents a Possible Solution

PACKET SWITCHING During the 1960s, several researchers identified a solution for the inefficiency of reserved circuits for bursty data transmission. This solution was called **packet switching.** Figure 1-9 shows how packet switching works.

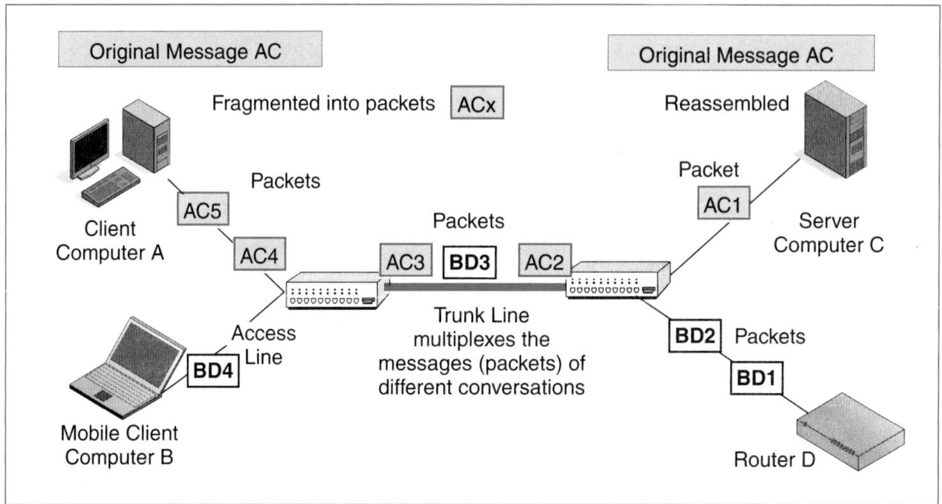

FIGURE 1-9 Packet Switching and Multiplexing

In the figure, an application on Host A wishes to send an application message (Original Message AC) to an application on Host C. The figure shows that Host A fragments the message into many smaller segments and sends each in a separate message called a **packet.** A typical packet is about 100 bytes long. Even sending very brief e-mail messages may require two or three packets. Longer documents, graphics files, audio messages, and video files may be sent in hundreds or thousands of packets. However many packets are sent, the network delivers them to Host C. The destination host reassembles the packets and passes them to the destination application program.

In packet switching the source host fragments the application messages into many smaller pieces called packets; the network delivers the packets to the destination host, which reassembles the application message.

This seems needlessly complex, but it has two major virtues. The first is easier error correction. If there is a transmission error that destroys a packet, only the lost packet needs to be resent. Early transmission lines had substantial error rates, and when whole messages were sent, messages might have to be resent several times before being received correctly.

More fundamentally, Figure 1-9 shows that the packets of multiple conversations can be **multiplexed** (mixed) on long-distance circuits. If each conversation between hosts is using only 3 percent of capacity, then, roughly speaking, about 30 conversations can be multiplexed onto the circuit. Packet switching, then, saves money by multiplexing multiple conversations over expensive circuits. Bursty traffic only has to pay for the capacity it actually uses.

Packet switching saves money by multiplexing multiple conversations over expensive circuits.

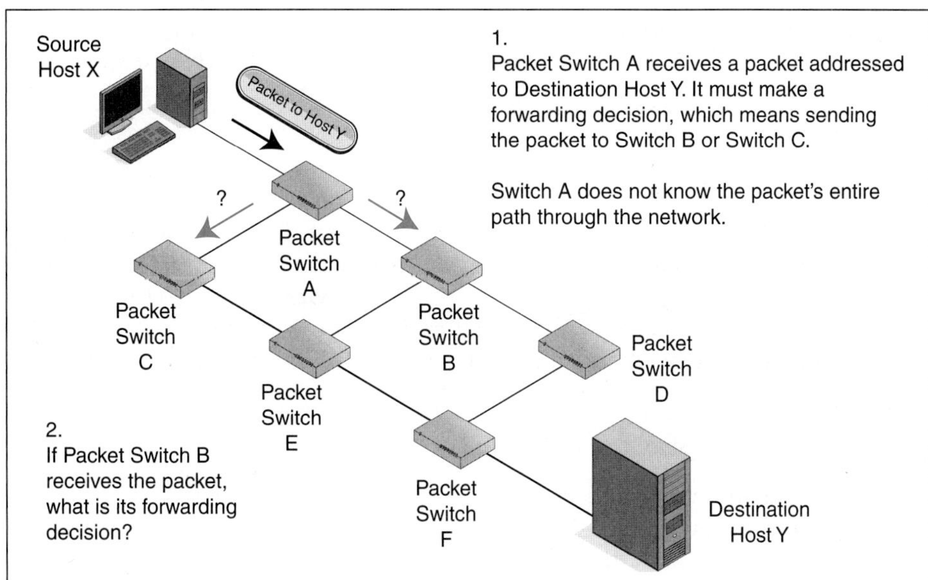

FIGURE 1-10 Sequential Switch Forwarding Decisions

PACKET SWITCH FORWARDING DECISIONS Figure 1-10 looks at the packet switches that make this work. Here, there are six packet switches, imaginatively labeled A through F. The source host transmits a packet to Packet Switch A—the switch to which the source host connects directly. Packet Switch A has to make a **forwarding decision.** In other words, it must decide where to send the packet next. It can either forward the packet to Packet Switch B or send it to Packet Switch C. Packet Switch A thinks that Packet Switch B is a better choice for getting the packet to Destination Host Y, so Switch A will forward the packet to Switch B.

> Each packet switch makes a forwarding decision, which means deciding where to send the packet next. This forwarding decision is based on the packet's destination address.

Now the packet is at Switch B. Packet Switch B has to make its own forwarding decision. It can forward the packet to Switch D or to Switch E. In this case, it decides to forward the packet to Switch D. Again, it made this decision only knowing about the switches to which it directly connected—Switch D and Switch E.

Note that neither Packet Switch A nor Packet Switch B knows the entire path the packet takes through the network. Each makes a local decision; it only decides where to send the packet *next*. It does not matter how many hops there are across packet switches from the source host to the destination host. Each packet switch along the way will make a local forwarding decision—despite the fact that the packet switches do not know the entire path the packet will take.

Individual packet switches do not know the packet's entire path through the network. They only make a local decision in which they decide where to send the packet next.

Test Your Understanding

7. a) In packet switching, what does the source host do? b) About how long is a packet? c) Why is fragmentation done? d) Where is reassembly done? e) What is the benefit of multiplexing? f) Why is packet switching good for bursty data traffic? g) When a packet switch receives a packet, what decision does it make? h) Do packet switches know a packet's entire path through a network?

Physical Links and Data Links

In a packet-switched network, there are two types of links or transmission paths. First, there are **physical links** between hosts and their switches and between switches. These physical links along the path the packet takes between the two hosts may use different technologies.

There also needed to be a name for the path across switches that a packet takes between the source host and the destination host. Figure 1-11 shows that these paths are called **data links.** When one host transmits a packet to another host, there may be multiple physical links along the way, but there is only one data link.

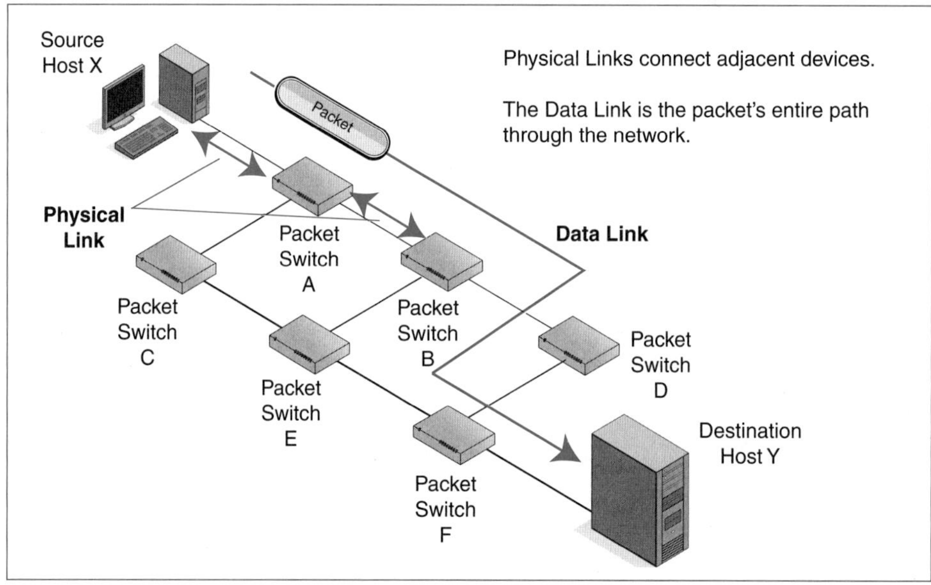

FIGURE 1-11 Physical Links and Data Links

Test Your Understanding

8. a) In Figure 1-11, how many physical links are there between the source host and the destination host along the indicated data link? b) How many data links are there between the source host and the destination host? c) If a packet passes through eight switches between the source and destination hosts, how many physical links would there be? (Careful!) d) How many data links will there be?

Larry Roberts Builds a Solution

Larry Roberts saw packet switching as a way to reduce the cost of connecting remote terminals to the hosts that ran software funded by ARPA. He also saw it as a research project. Packet switching seemed to have great potential. To see if that potential was real, someone had to build a packet-switched network.

Roberts funded the creation of a packet-switched network called the **ARPANET,** which Figure 1-12 illustrates. Most of the network chores were handled by minicomputers called **interface message processors (IMPs).** An IMP received a message from a host attached to it, broke the message into small pieces, and placed each piece in its own packet. The IMPs then acted as packet switches, forwarding the packet to the IMP of the destination host. That final IMP reassembled the original message and passed the message to the destination host. Note in the figure that an IMP can serve multiple hosts. A small team at Bolt Beranek and Newman programmed the IMPs.

Each host ran software called the **Network Control Program (NCP).** This handled details of host-to-host interactions above the level of packetization, delivery, and reassembly.

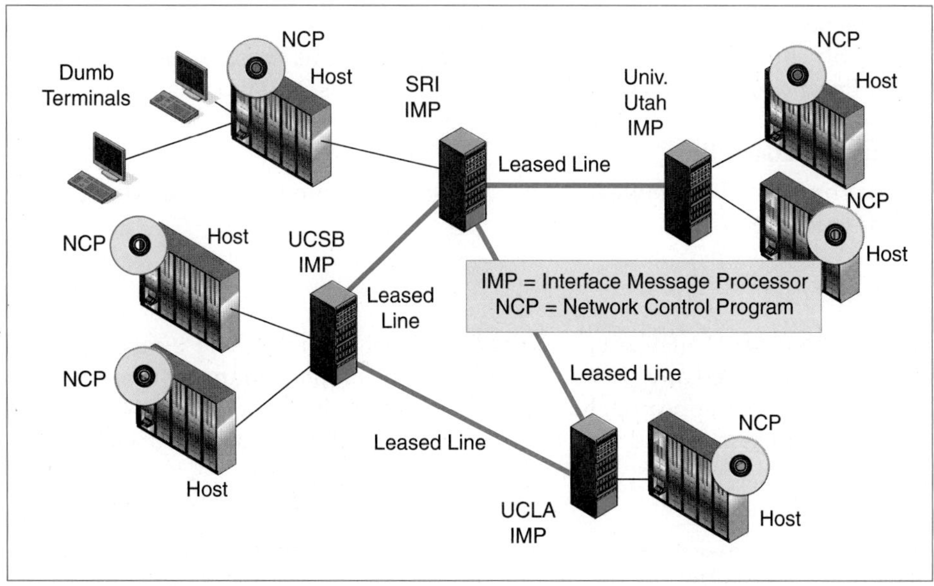

FIGURE 1-12 The ARPANET

For instance, if the source host transmitted too rapidly for the destination host to process, the NCP on the destination host could tell the NCP on the source host to slow down.

In 1969, the first two IMPs were installed at UCLA in Los Angeles and at Stanford Research Institute (now called SRI International).[6] As soon as that link was established, the University of California at Santa Barbara and the University of Utah were added. When it was clear that NCP and the IMPs were working as intended on these four sites, other sites were added, and some sites began to connect multiple computers to their IMPs.

Although the ARPANET was officially "born" in 1969, its performance was flakey until the mid-1970s. More fundamentally, there were no applications at all beyond those that ran on hosts for terminal users. Even in 1973, when the first author used the ARPANET, reliability and services were quite limited. Over time, however, reliability improved and more applications appeared. Users stopped calling the ARPANET the "notwork."

Test Your Understanding

9. a) On the ARPANET, explain the functions of IMPs. b) Explain the functions of NCPs.

The Network Working Group

Each of the first four sites had to devise its own way to connect its hosts to its IMP. This had to change. To make the new ARPANET useful, fundamental standards were needed for host–IMP connections. There also needed to be standards for applications, such as e-mail.

Nobody was filling this vacuum for standards, so graduate students from the four initial schools got together to create an informal **Network Working Group (NWG)**. Led by Steve Crocker at UCLA, the NWG began to create necessary standards. The group had no charter to develop standards, so the NWG members decided not to call their designs *standards* but rather **Requests for Comment (RFCs)**. Steve Crocker wrote RFC 1 to describe needs for software on two hosts to communicate. Over time, the NWG grew considerably, but it kept its informal flavor. Today, this work is carried on by the **Internet Engineering Task Force (IETF)**, which still calls its standards RFCs.

Test Your Understanding

10. a) Why was the Network Working Group created? b) What did it call its standards? c) Why?

E-Mail

Even before the ARPANET, users on the same host computer had e-mail. When you logged in, you received any messages that someone on that computer had left for you. When the ARPANET appeared, Ray Tomlinson at Bolt, Beranek, and Newman saw the opportunity for users to send mail to other users on different machines. He decided to

---

[6] Protests during the Vietnam War caused Stanford University to sever ties with Stanford Research Institute. The Institute decided to call itself SRI, but there was already a small consultancy with that name, so Stanford Research Institute became SRI International. However, faculty and graduate students continued to move frequently between the university and the institute. (In one case, a PhD student got a contract worth about $200,000 in today's dollars to do his dissertation.)

adapt the local SNDMSG program on his local host for message delivery over the ARPANET. To send a message, you would type SNDMSG and hit Enter. You would then enter a value for To: and then type the body.

On a single computer, your e-mail address was simply your username. Tomlinson realized that there would be a need to add the name of the host to the name of the mailbox. Now he needed a way to separate the username from the host name. He looked at his keyboard and saw a character that was hardly ever used. This was the "at" sign, @. So to reach Ray Panko at the Office1 computer at Stanford Research Institute, you sent the mail to Ra3y@Office1.

Now, Tomlinson had to code the e-mail program for host computers. He was not being paid to do this, so he did it on his own time, over a weekend. Networked e-mail was born. However, SNDMSG was only a message *sending* program. The receiver could only read incoming messages one at a time, in order. One intense early e-mail user was Larry Roberts. He often received dozens of messages per day. Initially, he had to go through his messages one at a time. On his own time, he wrote the first useful e-mail *reading* program, which he imaginatively called RD. Soon, other ARPANET users created better e-mail reading programs, such as bananard. (This was the 1970s.) This kind of do-it-yourself spirit and an absence of any profit motive made the ARPANET a great development environment.

ARPANET e-mail was also a great social networking tool. ARPA funded a great deal of computer science research in the 1960s and early 1970s, and people in different organizations funded by ARPA often knew each other well. Individual e-mail messages and computer conferencing message groups increased the cohesion of this research community. For example, the Message Service Group was a mailing list for people who engaged in discussions about issues in e-mail.

In 1974, a young researcher was visiting ARPA. In a meeting, he asked about the cost of an e-mail message. Nobody gave an answer. Later that day, a senior official pulled the visitor into an office and told him to stop asking about the price of e-mail because e-mail's cost was estimated to be about $60 per message, and e-mail made up 75 percent of all ARPANET traffic. Government auditors would scream if they knew this, and ARPA would soon have to come down hard against e-mail use. The researcher said that that $60 per message was impossible. They sat down and poured over the data for several hours. It turned out that the average message cost about as much as a postage stamp to deliver across the network. In the next few months, ARPA's concerns about e-mail relaxed.

Test Your Understanding

11. a) How did Ray Tomlinson extend e-mail? b) Why did he need the @ sign? c) Why did Larry Roberts have to write a mail reading program? d) Why was ARPA initially concerned about e-mail?

THE INTERNET

Bob Kahn Has a Problem

By the early 1970s, the ARPANET was (reasonably) stable. In addition, packet switching was proving itself in wireless environments. The ALOHANET project at the

University of Hawai`i demonstrated that packet switching could be done over satellite circuits, despite the long time lags in upward and downward transmission. Terrestrial (earth-bound) packet radio projects also began to appear, using backpack radios.

Dr. Bob Kahn, who was by now in charge of IPTO at ARPA, was happy to see packet switching blossoming. However, he also saw a deep problem. Users on the ARPANET, the PRNET packet radio network, and the SATNET packet-switched satellite network could not communicate with each other. Packet switched networking was becoming a Tower of Babel.

Test Your Understanding

12. What problem that Bob Kahn faced led to the Internet?

Bob Kahn and Vint Cerf Find a Solution

Kahn discussed the problem with a young Stanford professor, Vint Cerf. Together, Kahn and Cerf explored various solutions to the problem of internetworking. Their solution, which Figure 1-13 illustrates, was to use special devices called **gateways** to connect different networks together into an "internet." Today, we call these devices **routers.**

Routers connect different networks together into an internet.

Test Your Understanding

13. a) What device connects different networks into an internet? b) What is the old name for this device?

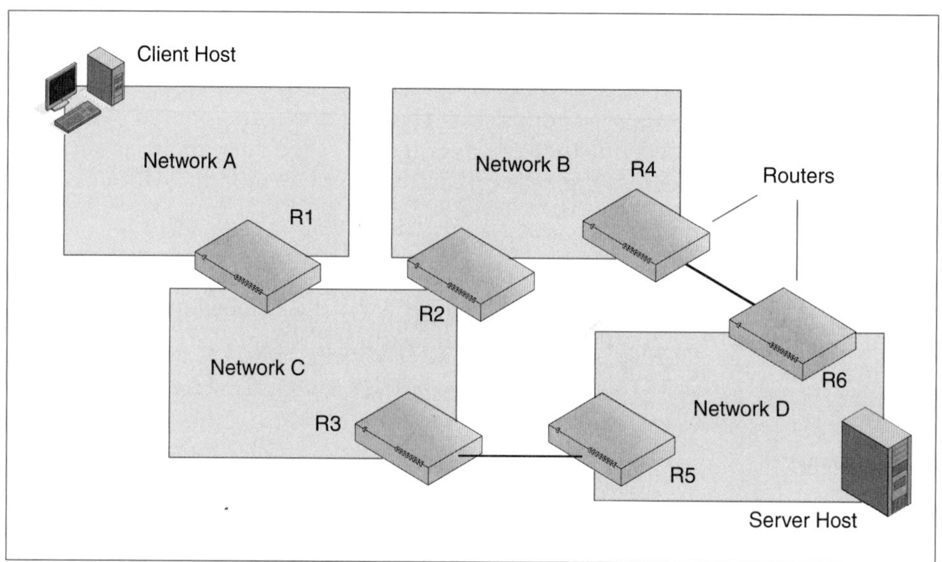

FIGURE 1-13 Internetworking

Component	Single Switched or Wireless Networks	Internets
Addresses	Vary by network technology	32-bit IP Addresses
Packets are called	Frames	Packets
Packet switches are called	Switches	Routers
End-to-end routes are called	Data links	Routes

FIGURE 1-14 Two Layers of Networking

A Second Layer of Networking

CAUTION: THE FOLLOWING MATERIAL IS DIFFICULT BECAUSE THE CREATION OF A SECOND LEVEL OF NETWORKING CREATES MANY PAIRS OF CONCEPTS THAT ARE SIMILAR BUT DIFFERENT.

One problem in connecting different types of networks is that they use incompatible data link technologies. They organized packets in different ways, and they had different and deeply incompatible addressing systems. To handle such problems, Kahn and Cerf essentially created a higher layer of networking over all of these individual incompatible networks technologies. We will call this higher layer the **internet layer.** Figure 1-14 shows that their solution needed to do more than create a second layer of networking. It also had to duplicate the concept of addresses, packet switches, packets, and paths. Unfortunately, this causes a great deal of confusion for students.

POINT OF TERMINOLOGY "Internet" is both a general concept and a name for the largest internet of all, the global Internet. To make distinctions, capitalization will be reserved for the global *Internet*, except in titles and at the start of sentences. An *internet* (in lowercase) is any internet. Lowercase is also used to refer to the *internet layer*.

A SECOND LEVEL OF ADDRESSES Different network technologies use different syntaxes for addresses. For example, Ethernet uses 48-bit MAC addresses, while Frame Relay uses 10-bit DLCI addresses. In addition, these addresses were not always assigned uniquely. Nearly all Frame Relay networks, for example, had a host represented by DLCI 1.

For the second layer of networking, Kahn and Cerf created a new globally unique address, the Internet Protocol (IP) address. This address was 32 bits long. For human reading, it usually is expressed in **dotted decimal notation.** In this notation, the IP address is broken into four 8-bit segments. Each segment's bits are treated as a binary number and are converted into a decimal number. The four segment decimal numbers are written out with dots (periods) between them. So 128.171.17.13 is a typical IP address in dotted decimal notation. Computers have no problems dealing with 32-bit strings, so only inferior biological entities use dotted decimal notation as a memory aid.

Computers have no problems dealing with 32-bit strings, so only inferior biological entities use dotted decimal notation as a memory aid.

In a later innovation, a system was given for dividing up the billions of possible IP addresses in a way that would make each IP address globally unique. For example, the University of Hawaii was given control over all IP addresses beginning with 128.171. This is 16 bits. No other organization could use IP addresses beginning with these bits. The University of Hawaii then assigned the remaining 16 bits internally in a way that preserved uniqueness. For example, one host is 128.171.17.13.

If you think about it, telephone numbers are assigned this way. Each country gets to assign its internal telephone numbers any way it chooses, as long as each phone has a phone number and no two phones have the same number.

A SECOND LEVEL OF PACKETS Adding a second layer of networking creates two of several things besides addresses. We now had two levels of packets—one at the data link layer and one at the internet layer. In time, terminology emerged to handle these differences. Packets at the data link layer are called **frames,** and packets at the internet layer are called packets.

> Packets at the data link layer are called frames, and packets at the internet layer are called packets.

THE RELATIONSHIP BETWEEN PACKETS AND FRAMES Figure 1-15 shows the relationship between packets and frames. Here, there are three networks connected by routers. The packet sent by the source host travels all the way to the destination host. The packet is addressed to the IP address of the destination host. Within each network, however, the packet travels in a frame specific to that network. In other words, packets always travel inside frames. In this case, there is one packet, but there are three frames

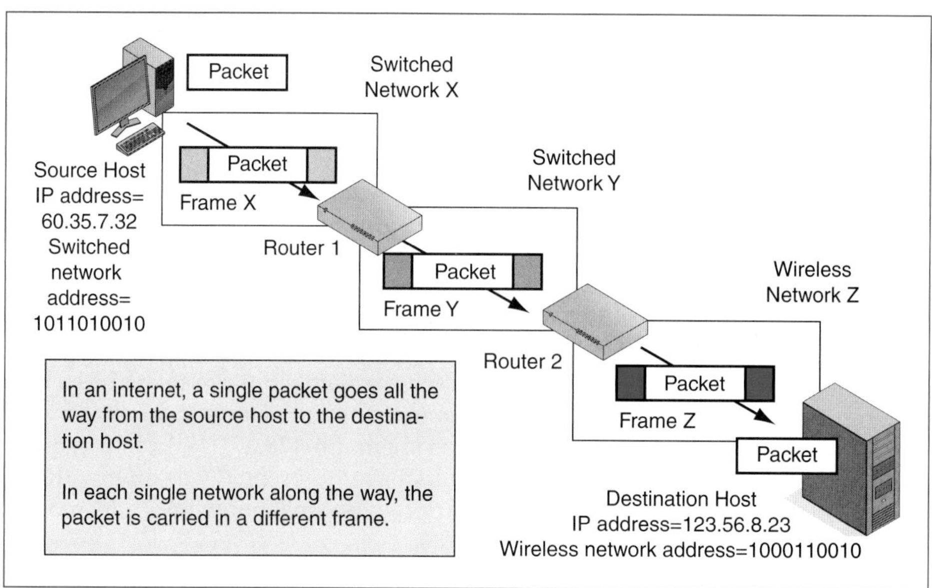

FIGURE 1-15 Packets and Frames

along the way, one for each network. If 10 networks separated the source host from the destination host, there would have been one packet traveling in 10 different frames.

SWITCHES AND ROUTERS Packet switches are needed at both the network and internet levels. At the network level, packet switches are simply called switches. At the internet layer, packet switches are called routers or gateways.

PHYSICAL LINKS, DATA LINKS, AND ROUTES We also need two types of network links above the physical layer. We saw that the path a packet (now called a frame) takes across its single network is called its data link. Figure 1-16 shows that the path that a

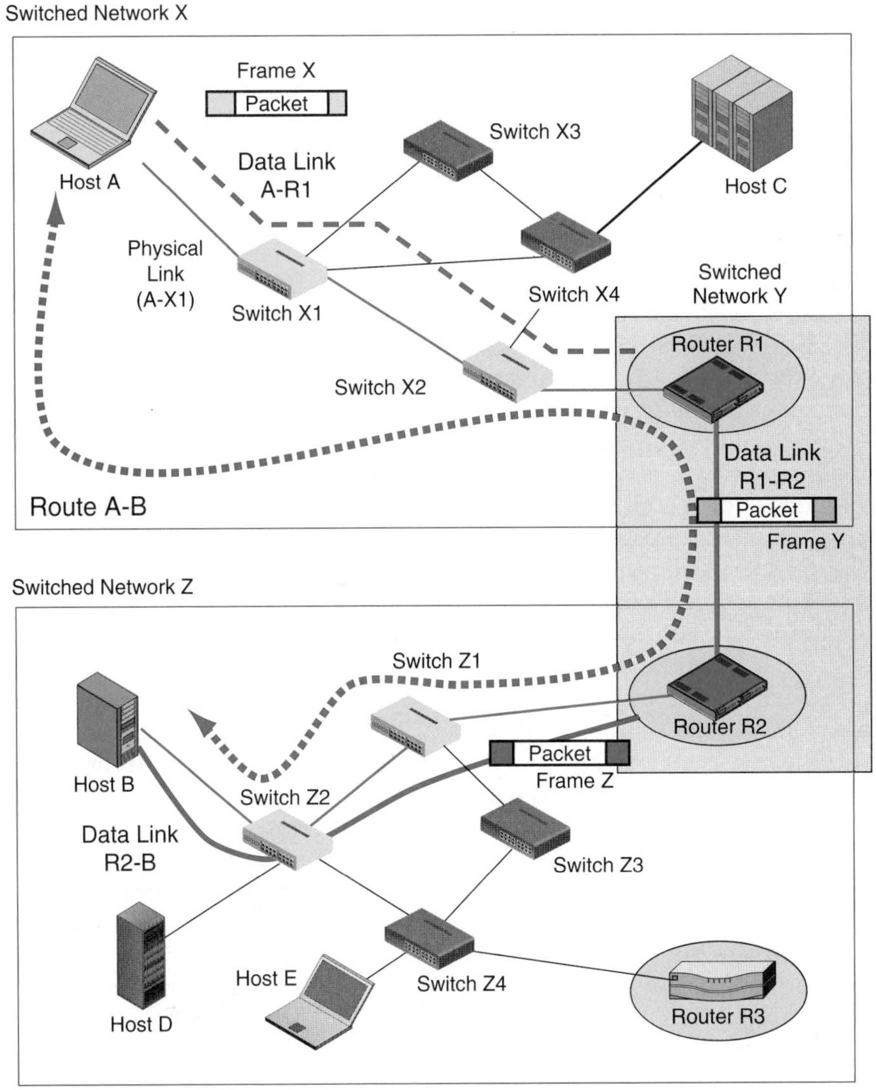

FIGURE 1-16 Physical Links, Data Links, and Routes

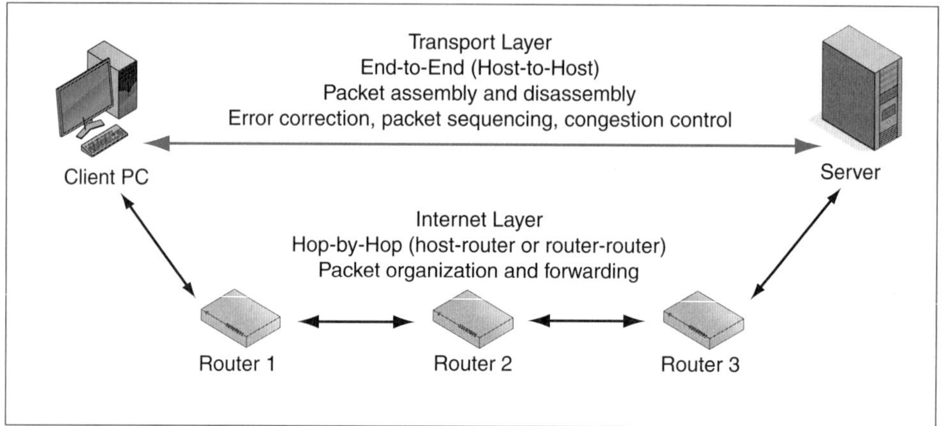

FIGURE 1-17 The Internet and Transport Layers

packet takes at the *internet* layer across the entire internet is called its **route.** If there are 10 networks between the source host and the destination host, there will be 10 data links along the way but only one route.

> The path that a packet takes at the internet layer across the entire internet is called its route.

THE TRANSPORT LAYER Actually, internetworking also required the creation of a *fourth* level of standards, the **transport layer.** Figure 1-17 shows relationship between the internet layer and the transport layer.

The internet layer is concerned with moving packets across an internet, across a series of routers. It governs what happens on each router along the way. It also governs packet formats and addressing.

The transport layer, which is above the internet layer, is only concerned with what happens on the source host and destination host. On the source host, the transport layer takes application messages and divides them into small units called segments (usually). It passes each segment down to the internet layer, which places each segment into a single packet and sends the packet on its way. On the destination host, the internet layer program removes the segment and passes the segment up to the transport layer program. The transport layer program reassembles the segments into the original application message. It passes the message to the application program.

The transport layer has a number of other functions. It typically provides error correction, which means that packets that are damaged or lost in transmission are retransmitted. The transport layer also causes the source host to reduce its transmission rate if there is congestion. Overall, the internet layer is a best-effort service that tries to get packets through but may fail to do so. It may even deliver packets out of sequence. The transport layer usually is a fix-up layer that supplies the functionality that the internet layer lacks.

THE APPLICATION LAYER The four lowest layers get a packet to the destination host, possibly with error correction and other services. The fifth layer is the **application layer.**

Layer Number	Layer Name	Specific Purpose	Broad Purpose
5	Application	Communication between application programs.	Application Communication
4	Transport	Application message fragmentation and reassembly. Ordering of packets, error correction, and congestion reduction.	Internet Transmission
3	Internet	Connection across an internet. Defines packet formats and router operation.	
2	Data Link	Connection across a single network. Defines frame formats and switch operation.	Single-Network Transmission
1	Physical	Physical connections between adjacent devices.	

FIGURE 1-18 Networking Layers

This layer controls communication between the two application programs that are communicating. For example, when browsers talk to webservers, this requires application layer standards (HTTP) to specify the communication. There are many applications, and many of them have their own standards. Consequently, there are more application layer standards than there are standards at other layers.

FIVE LAYERS Overall, network functionality is described fairly well by thinking of the five layers we have seen in this chapter. Figure 1-18 shows that these are the physical layer, data link layer, internet layer, transport layer, and application layer. Each layer provides services to the layer above it. The bottom two layers provide transmission through single networks. The internet and transport layer provide host-to-host transmission through an internet. Finally, the application layer provides application–application communication.

Test Your Understanding

14. a) Distinguish between *internet* with a lowercase i and *Internet* with an uppercase I. b) Why are many networking concepts duplicated in switched networks and internets? c) What are the two levels of addresses? d) How long are IP addresses? e) How are IP addresses usually expressed for humans? f) Distinguish between packets and frames. g) A host transmits a packet that travels through 47 networks. How many packets will there be along the route? h) How many frames will there be along the route? i) Distinguish between switches and routers. j) Distinguish between physical links, data links, and routes. k) Distinguish between what happens at the internet and transport layers. l) Why are application layer standards needed? m) List the numbers and names of the five layers. n) Are frames carried inside packets?

THE INTERNET EVOLVES

The TCP/IP Standards

Kahn and Cerf realized that their new approach would need new standards. They sketched out the details of core protocol needed for hosts to transmit packets. In place of the Network Core Program used on ARPANET hosts, they proposed a new protocol, which they called the Transmission Control Protocol. They published the first version of this protocol in 1973.

Originally, there was only one standard for the internet and transport layers, the Transmission Control Protocol. However, while every application needs packet delivery across an internet, not all applications need high functionality at the transport layer. Consequently, Figure 1-19 shows that the functionality for the internet layer and transport layer was split into two layers of standards.

For the functionality for moving packets across an internet, a single protocol was created, the **Internet Protocol (IP).** This layer deals with addresses, so addresses on an internet are called IP addresses.

At the transport layer, TCP/IP has two alternatives. For a transport layer protocol with high functionality, the transport standard retained the name **Transmission Control Protocol (TCP).** TCP fragments application messages. Each fragment is carried in a TCP message, which is called a TCP segment. TCP also puts information in packets in order if the packets arrive out of order, correct errors, and reduce the likelihood of congestion.

The second alternative is for applications that do not need or cannot use high transport layer functionality. This was the **User Datagram Protocol (UDP).** UDP messages are called **datagrams.** UDP does *not* do fragmentation, so the application message must be able to fit inside a single UDP datagram. UDP does no error correction or congestion control. It does not need to worry about placing packets in order because the entire transmission is done in a single packet.

Why would an application use UDP? Many applications do not need error correction or cannot wait for errors to be corrected. For instance, Voice over IP (VoIP) cannot wait for packet retransmission because voice must flow in real time. Consequently, VoIP uses UDP at the transport layer instead of TCP.

The original standards specified three protocols—IP, TCP, and UDP. However, they became known collectively as the **TCP/IP standards.** Later, many more standards were created to add functionality missing from TCP, UDP, and IP. However, this growing family of standards is still called **TCP/IP.**

TCP/IP is a family of standards including IP, TCP, UDP, and many other standards.

Layer	Standard(s)	
Transport Layer	Transmission Control Protocol (TCP)	User Datagram Protocol (UDP)
Internet Layer	Internet Protocol (IP)	

FIGURE 1-19 Core TCP/IP Standards

Test Your Understanding

15. a) What are the roles of the Internet Protocol? b) What are the roles of the Transmission Control Protocol? c) When would the User Datagram Protocol be used at the transport layer?

The First Test

In 1977, the standards were ready for their first tests. Bob Kahn had originally been concerned with bridging three networks with very different technologies. One was the ARPANET. Another was the ARPA-funded packet radio network at Stanford Research Institute, PRNET. The third was the ARPA-funded Atlantic packet satellite network, SATNET, which spanned the Atlantic. The test would do precisely that.

A host computer was placed in a van traveling on the Bayshore Freeway south of San Francisco. Using Bold Beranek and Newman routers, the host's packets traveled through the PRNET, the SATNET to England and back, and the ARPANET to the destination host. Amazingly, this lash-up worked perfectly.

Test Your Understanding

16. What three networks were involved in the first test of the TCP/IP standards?

The Internet Is Born—Slowly

When did the Internet start? The answer is surprisingly difficult because the Internet did not appear all at once. It emerged very slowly, in parallel with traditional ARPANET protocols. Rather than create a sharp continuity, hosts were allowed to continue using NCP or to implement TCP/IP. The two standards continued in parallel for several years. Finally, on January 1, 1983, hosts were required to stop using NCP. Although there were a few loud protests, most hosts had long since transitioned TCP/IP. In that sense, the Internet was born on January 1, 1983. At the same time, it had effectively existed for several years.

Test Your Understanding

17. a) In what sense is January 1, 1983, the birthday of the Internet? b) In what sense is it not?

E-Mail Interconnection

On both the ARPANET and the new Internet, it was assumed that the user would work at a terminal and connect to a particular host computer. On this base computer, the user could run e-mail software and other software. This application software on the base computer could use services on other computers.

E-mail was a major driving force for the growth of the Internet. However, Internet users could only talk to other Internet users, and to be an Internet user, you generally had to have an ARPA contract. Consequently, faculty members and researchers who did not have ARPA contracts began to create their own networks. Computer scientists created CSNET, and behavioral scientists and business schools created BITNET. Commercial computer networks also appeared with their own e-mail applications. There was an e-mail Tower of Babel.

Slowly, during the 1970s, mail systems on these other networks were connected to Internet e-mail. At first, the integration was awkward, but it soon became invisible to users on and off the Internet. Later, a major precedent was set when ARPA allowed MCI Communications to connect its commercial e-mail service to the Internet. Soon, other commercial networks connected their e-mail nets to the Internet.

Test Your Understanding

18. When did Internet e-mail begin to interconnect with e-mail on other networks?

The Internet Goes Commercial

Initially, ARPA funded ARPANET and Internet transmission. In 1986, the National Science Foundation created NSFNET, which two years later became the backbone of the Internet. NSFNET brought higher speeds to the Internet backbone. NSFNET also brought the **Acceptable Use Policy (AUP)** to the Internet. Basically, the AUP explicitly forbade the use of the Internet for commercial purposes such as buying, selling, and advertising. The Internet was to be a pure research network.

On April 30, 1995, the NSF discontinued the NSFNET as the core of the Internet. Before this, a number of companies called Internet Service Providers (ISPs) had formed to connect users to the Internet. These ISPs were also connected to one another. Consequently, when NSF pulled the plug on NSFNET support for the Internet, the Internet continued without any visible change to users.

Figure 1-20 shows the Internet today. Technologically, the Internet is simply a large collection of routers. However, these routers are owned by different ISPs. The ISPs are interconnected at **Network Access Points (NAPs),** which allow the ISPs to exchange packets.

To use the Internet today, you *must* connect to it via an ISP. The ISP sends your packets into the Internet, to destination hosts. Your ISP delivers reply packets to you.

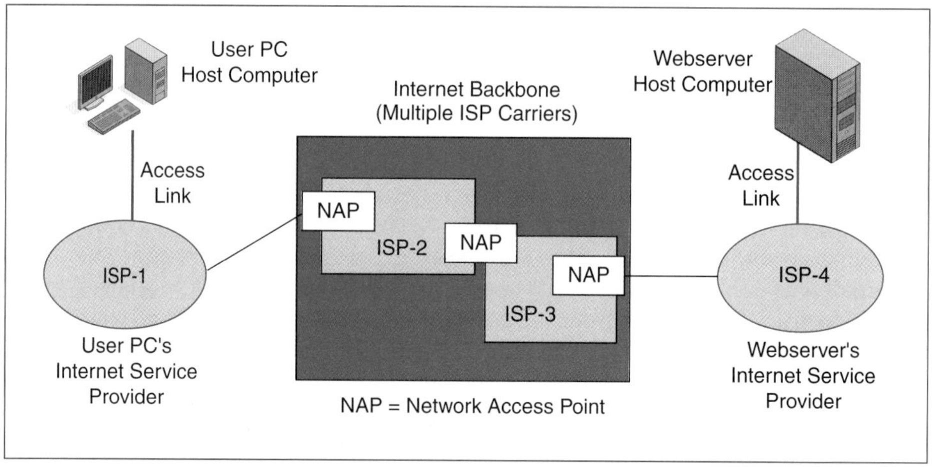

FIGURE 1-20 The Internet Today

You need an access line from your home or place of business to your ISP's nearest office. Corporations, which have far larger volumes of Internet traffic, need much faster access lines.

Who pays for Internet transmission? The answer is that *you* do. You pay money each month to your ISP. Corporations also pay money to their ISPs. While you pay ten to a hundred dollars a month, large corporations pay tens of thousands of dollars per month for service. Given the number of people and corporations on the Internet, there is enough money to pay for all of the Internet's transmission volume.

Although Internet service did not change when the Internet backbone became entirely commercial, the vanishing of NSF support led to the vanishing of NSF's acceptable use policy. From that point on, commercial activities were perfectly fine.

The termination of the NSFNET did not come as a surprise to anyone. NSF announced the change far ahead of time, and companies were ready to offer commercial service. Immediately after the NSFNET disconnected itself, companies began to offer e-commerce service massively. In a few months, it was hard to believe that e-commerce was not several years old. Many Internet-only companies began to sell stock through initial public offerings. A vast land rush of dot-com companies appeared, many based on absurd business models. In 2000 and 2001, the bottom dropped out of dot-com stocks. However, e-commerce did not collapse. After one year of stagnant growth in the middle of a recession, e-commerce continued to grow very rapidly.

Most e-commerce activity relied upon the World Wide Web, which was invented at CERN by Tim Berners-Lee. Although he created the HTML standard in 1991, it was just beginning to grow explosively on the Internet when commercial activity became possible.

Test Your Understanding

19. a) What was the Acceptable Use Policy in place on the Internet before 1995? b) Why did commercial activities on the Internet become acceptable in 1995? c) What are the carriers that provide Internet service? d) Why do they need to be interconnected? e) At what locations do ISPs interconnect? f) Did e-commerce collapse after the dot-com failures of 2000 and 2001?

Supervisory Applications

TCP, IP, and UDP are sufficient for delivering packets over an internet, but the TCP/IP family of standards today is much larger than this. Many TCP/IP standards are user application standards, such as standards for e-mail and the Web. Many of these standards, however, are **supervisory standards** that keep an internet working. We will look at two TCP/IP supervisory applications in this section. We will see many more throughout the book.

THE DYNAMIC HOST CONFIGURATION PROTOCOL (DHCP) Servers must have permanent IP addresses. Otherwise, clients could not find them. (Imagine what it would be like if your favorite store kept changing its street address.) Unchanging addresses are called **static IP addresses.**

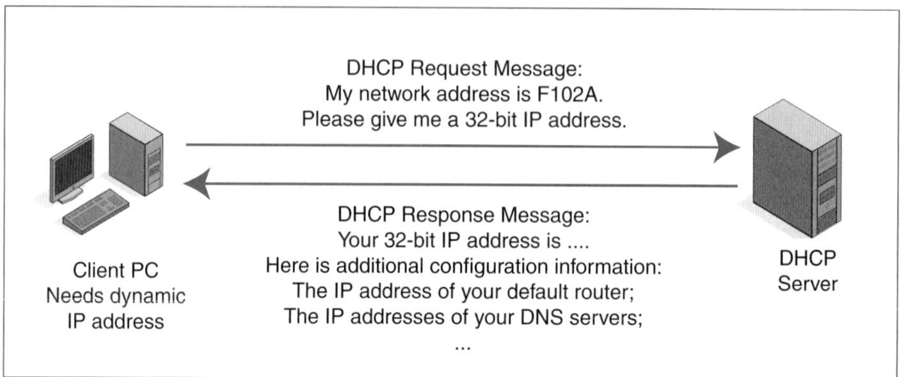

FIGURE 1-21 The Dynamic Host Configuration Protocol (DHCP)

Client PCs normally get their IP addresses a different way. When a client PC boots up, it realizes that it has no IP address. Figure 1-21 shows that the PC broadcasts[7] a **Dynamic Host Configuration Protocol (DHCP)** request message. This message asks for an IP address.

Because this message is broadcast, all nearby hosts receive it. However, hosts that are not DHCP servers simply ignore it. When the DHCP server receives the message, the server picks an available IP address from its address database. It then sends this IP address in a DHCP response message. The client PC uses that IP address as its IP address. This is called a **dynamic IP address.**

The figure indicates that DHCP does more than give the client PC an IP address. As "Configuration" in the name suggests, it sends general configuration information.

- This includes the IP address of a default router. When the PC needs to send a packet to a host that is not nearby, it sends the packet to this default router.
- Configuration information also includes the IP addresses of local Domain Name System servers, which we will see next.
- It includes other information such as a subnet address mask, which we will see later in this book.

It would be possible to simply enter this information in every corporate PC just once, manually. Afterward, the client PC would not have to use DHCP. However, consider what would happen if the firm later changed the IP addresses of its Domain Name System servers. The firm would have to reconfigure every client PC manually. That would be painfully expensive. With DHCP, every client PC automatically gets hot fresh information every time it boots up. If configuration information changes, all client PCs will be updated automatically.

---

[7] In Ethernet, if a frame has a destination address that is all 1s, switches will broadcast the frame to all hosts on the network.

THE DOMAIN NAME SYSTEM (DNS) To send a packet to another host, a source host obviously needs to know the IP address of the destination host. However, people are not good at memorizing 32-bit IP addresses. Consequently, servers are often given host names, such as Voyager.shidler.hawaii.edu. Host names are much easier to memorize than IP addresses, so when we type URLs, we use host names instead of IP addresses. (In fact, you probably didn't even know that you could use IP addresses instead of host names in URLs.)

For the ARPANET, there was a Network Information Center (NIC) at Stanford Research Institute that maintained a file containing the host names and host ARPANET address of every named host on the ARPANET. The NIC was run by a single person, Elizabeth (Jake) Feinler. When the ARPANET grew, she eventually got an assistant.

As the ARPANET and then the Internet grew, this centralized manual approach to maintaining host names and associated addresses became impossible to continue. Consequently, the IETF created the **Domain Name System (DNS)** in 1984. In this system, each organization with a second-level domain name, such as Panko.com or Hawaii.edu, must maintain one or more DNS servers. ISPs that provide service directly to customers also need to maintain DNS servers for their customers.

Figure 1-22 shows that when a source host needs to know the IP address of a destination host, the source host sends a DNS request message to its local DNS server. The DNS request message contains the host name of the target host.

As the figure shows, the DNS server looks up the IP address associated with the host name. It then sends back a DNS response message that contains the IP address.

Now the source host knows the IP address of the target host. It has no more need for the DNS server. The source host simply sends packets to the IP address of the target host.

What happens if the local DNS server does not know the IP address of the host name? In that case, the local DNS server will contact other DNS servers until it finds the correct IP address (or until it gives up).

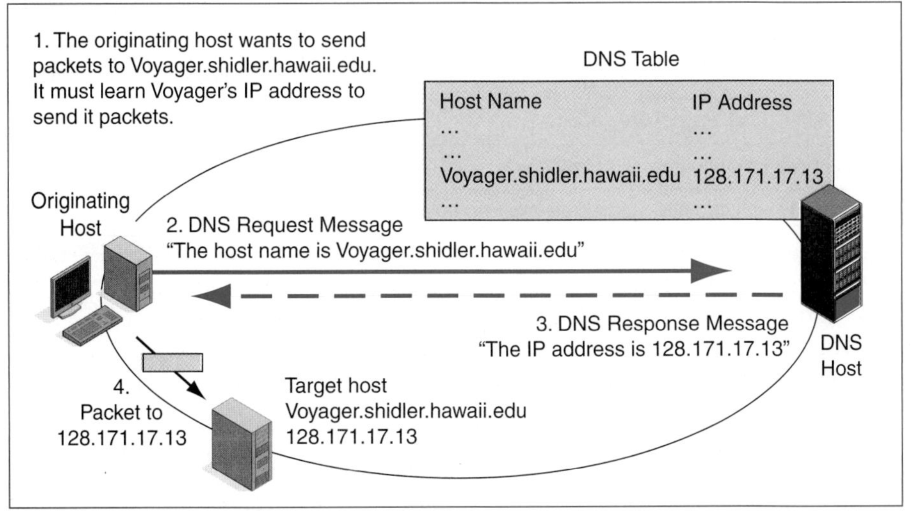

FIGURE 1-22 The Domain Name System (DNS)

Test Your Understanding

20. a) Distinguish between static and dynamic IP addresses. b) What protocol provides a client PC with its dynamic IP address? c) What other configuration information does it provide? d) Why should PCs get their configuration information dynamically instead of manually?

21. a) To send packets to a target host, what must the source host know? b) If the source host knows the host name of the target host, how can it get the target host's IP address?

22. Both DHCP servers and DNS servers send a host an IP address. These are the IP addresses of what hosts?

LANS AND WANS

One of the fundamental distinctions in networking is the one between local area networks (LANs) and wide area networks (WANs). Figure 1-23 shows how these two types of networks differ.

ON AND OFF THE CUSTOMER PREMISES Some authors base the difference between LANs and WANs on physical distance. For instance, some say that the dividing line between LANs and WANs is one mile or one kilometer. However, the real distinction appears to be that **local area networks (LANs)** exist within a company's site, while

Category	Local Area Networks	Wide Area Networks
Abbreviation	LAN	WAN
Service Area	Customer premises (apartment, office, building, campus, etc.)	Between sites within a corporation or between different corporations
Implementation	Self	Carrier with rights of way
Ability to choose technologies	High	Low
Need to manage technologies	High	Low
Cost per bit transmitted	Low	High with Arbitrary Changes
Therefore, typical transmission speed	Usually 100 Mbps to 10 Gbps	About 256 kbps to 50 Mbps
Can use switched network technology?	Yes	Yes
Can use routed network technology?	Yes, especially in large LANs	Yes, in fact, that is what the Internet is

FIGURE 1-23 LANs versus WANs

wide area networks (WANs) connect different sites within an organization or between organizations.

Local area networks (LANs) exist within a company's site, while wide area networks (WANs) connect different sites within an organization or between organizations.

For LANs, then, the company owns the property and can do anything it wants. It can choose any LAN technology it wishes, and it can implement it any way it wishes.

There is no such freedom for WANs. A company cannot legally lay wires between two of its sites. (Consider how your neighbors would feel if you started laying wires across their yards.) The government gives certain companies called **carriers**[8] permission (**rights of way**) to lay wires in public areas and offer service to customers. In return, carriers are subject to government regulation.

When you deal with carriers, you can only get the services they offer, and you must pay their prices. Although there may be multiple carriers in an area, the total number of service choices is likely to be quite limited.

On the positive side, you don't need to hire and maintain a large staff to deal with WANs because carriers handle nearly all of the details. In contrast, if you install a LAN, you also have to maintain it. As the old saying goes, anything you own ends up owning you.

ECONOMICS Another fundamental difference between LANs and WANs stems from economics. You know that if you place a long-distance call, this will cost more than a local call. An international call will cost even more. As distance increases, the price of transmission increases. The cost per bit transmitted therefore is higher in WANs than in LANs.

You know from basic economics that as unit price increases, fewer units are demanded. Or, in normal English, when the price for an item increases, you usually buy less of it. Consequently, companies tend to purchase lower-speed WAN links than LAN links. Typically, LANs bring 100 Mbps to 1 Gbps to each desktop. WAN speeds more typically vary from 256 kbps to about 50 Mbps.

In addition, companies spend more time optimizing their expensive WAN traffic than their relatively inexpensive LAN traffic. For example, companies may be somewhat tolerant of looking at YouTube videos on LANs, but they almost always clamp down on this type of information on their WAN links. They also tend to compress data before sending across a WAN so that it can be handled with a lower-capacity WAN link.

SINGLE NETWORKS OR INTERNETS? Some people think that LANs are single networks, while WANs are internets. However, that need not be the case. Small LANs usually will be single networks, but a larger LAN, such as one on a university campus, is likely to be a local internet.

---

[8] Carriers were originally called *common carriers*. The name reflected the fact that these carriers were required by law to provide service to anyone or any organization requesting services. Regulation was originally instituted in the railroad industry because many companies that owned railroads also owned other companies and refused to provide services to competitors of these other companies.

For WANs, there can also be single networks or internets. Of course, the Global Internet is a WAN internet. However, companies also use Frame Relay networks and Metropolitan Area Ethernet networks for WAN communication. These are large networks, but they are still switched single networks.

Test Your Understanding

23. a) Distinguish between LANs and WANs. b) Why do you have more flexibility with LAN service than with WAN service? Why? c) What are rights of way? d) What are carriers? e) What is the advantage of using carriers?

24. Why are typical WAN speeds slower than typical LAN speeds? Give a clear and complete argument.

25. a) Are LANs single networks or internets? b) Are WANs single networks or internets?

A SMALL HOME NETWORK

We have been looking at networking principles so far. We will close this chapter by looking at a real, although very small network—a network in a residential home. This is a network on the family's premises, so by definition, it is a local area network. Although this is a small network, it has most of the elements you have studied in this chapter.

COMPONENTS Figure 1-24 illustrates the basic hardware devices in a typical home computer network.

- The heart of the network is a wireless access router. We will look at this device in more detail a little later.

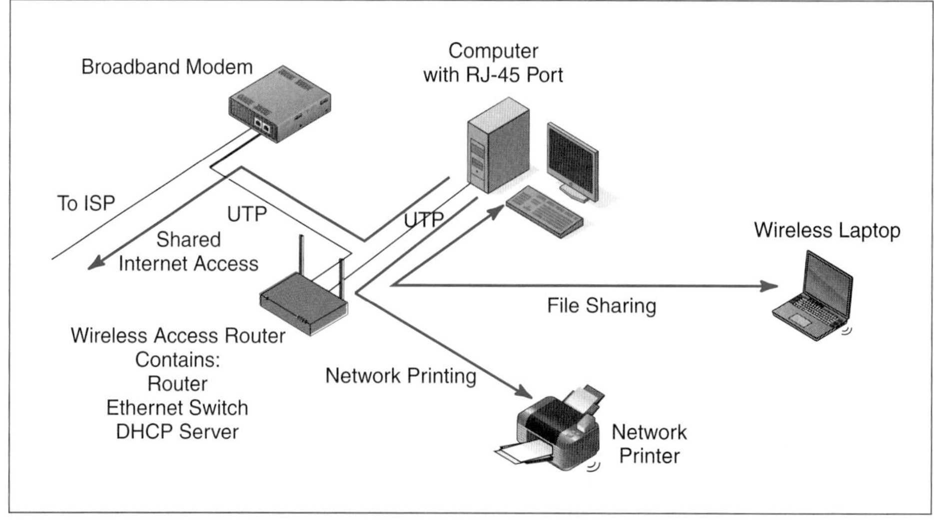

FIGURE 1-24 A Small Home Network

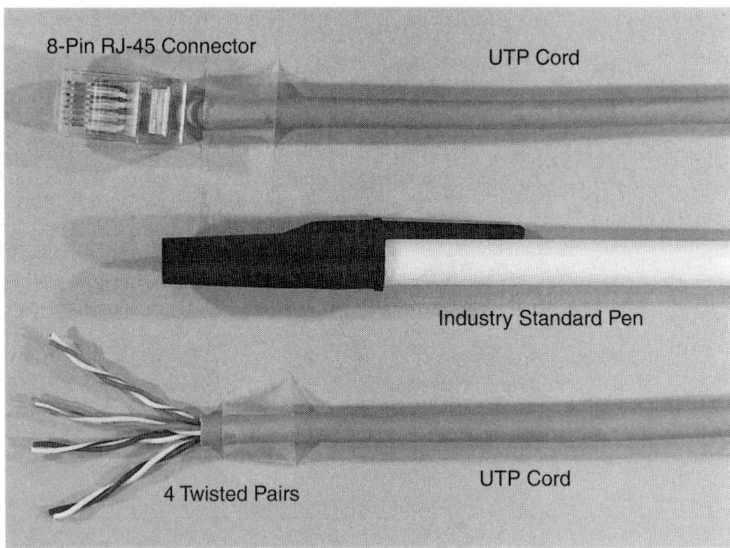

FIGURE 1-25 Unshielded Twisted Pair (UTP) Wiring

- A broadband modem connects the home network to an Internet service provider via a wired connection or wireless connection. We will look at wired ISP connections in Chapter 6 and wireless ISP connections in Chapter 8.
- There are two client PCs. One is connected to the wireless access router using a 4-pair unshielded twisted pair (UTP) cable. Figure 1-25 shows that 4-pair UTP consists of eight copper wires arranged in pairs, with each pair twisted around each other several times per inch. 4-pair UTP looks like a fat home telephone wire. It terminates in an RJ-45 connector. We will see more about 4-pair UTP in Chapter 6.
- The other client PC is a laptop computer with wireless capability. It connects to the wireless access point via radio signals. With wireless connections, there is no need to buy UTP cables and run them to each computer. However, as we will see in Chapter 7, wireless transmission is not always reliable or as fast as UTP transmission. We will also see in Chapter 7 that the main standard for wireless LAN (WLAN) transmission is 802.11.
- The final element is a wireless **network printer.** This printer also communicates with the wireless access router via 802.11. An increasing number of printers are network printers, which communicate with the access router via UTP, 802.11, or both.

THE WIRELESS ACCESS ROUTER The **wireless access router** deserves special attention because it contains several important hardware functions.

- First, the access router really is a router. Routers connect two networks. The access router connects the home network to the network of the ISP that provides Internet access.

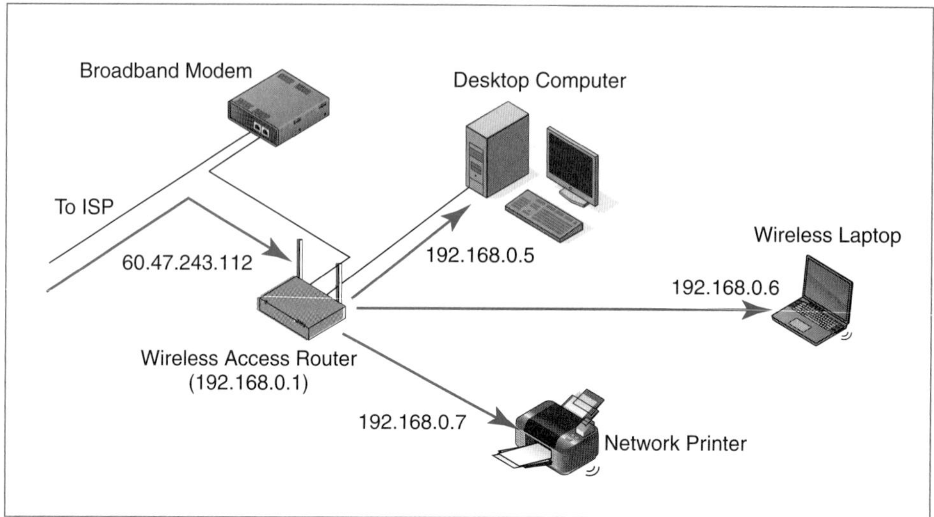

FIGURE 1-26 DHCP in a Small Home Network

- Second, the router has a built-in Ethernet switch. Most home access routers have at least four RJ-45 ports for UTP connections. It has another RJ-45 port to connect to the broadband modem.
- The router has a wireless access point to connect wirelessly to the access point. Access routers that lack this capability are called, simply, access routers.
- In this chapter, we saw that client PCs get their IP addresses from DHCP servers. Somewhat amazingly, the wireless access router has a built-in DHCP server. Figure 1-26 shows that the DHCP server gives IP addresses to the two client PCs (192.168.0.5 and 192.168.0.6) and to the network printer (192.168.0.7). The wireless access router's DHCP server also gives the wireless access router its own IP address (192.168.0.1).
- The wireless access router also provides **network address translation (NAT),** which translates between the internal IP addresses and the single IP address the ISP gives to the household (60.47.243.112). As Figure 1-27 illustrates, the ISP's DHCP server only gives the household a single IP address. When an internal device transmits, NAT converts the IP source addresses in its packets to the ISP's single allocated IP address. It then sends the packet on to the ISP. When packets arrive from the ISP, all have the IP address provided by the ISP as their destination addresses. The NAT function in the wireless access router places the internal IP address of the internal PC or the network printer into the packet's destination IP address field.

SERVICES Once the network is set up, the users can focus on the services their home network provides. Three of these services dominate today.

- **Shared Internet access** allows the two client PCs to use the Internet simultaneously, as if each was plugged directly into the broadband modem.

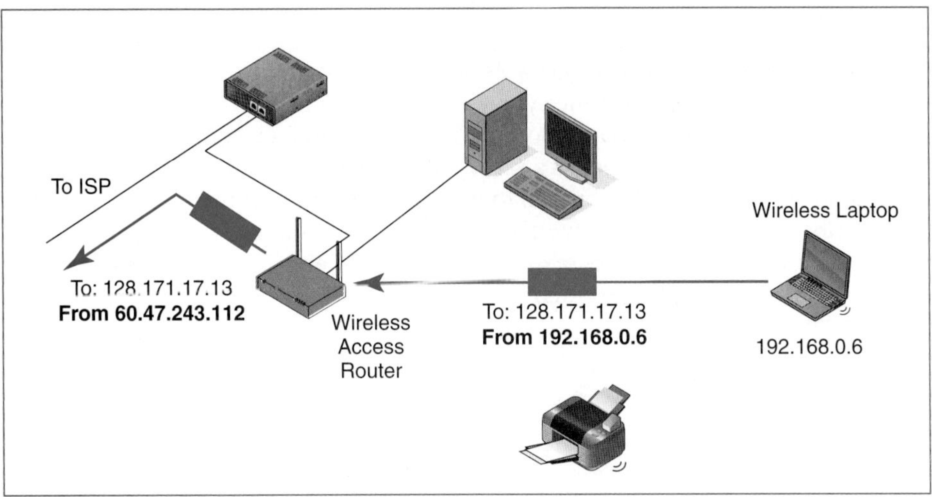

FIGURE 1-27 Network Address Translation (NAT)

- **File sharing** allows the two client PCs to share files on each other's hard drives. For example, one PC may have the family budget. A user at the other PC can access the budget file at any time.[9]
- **Printer sharing** allows either client PC to print to the printer. Both can even print at the same time. If they do, the printer will store one job in the print queue until it finishes printing the other.

CONFIGURATION Shared Internet access is completely automatic and requires no setup. However, file sharing and printer sharing require some setup on each of the client PCs. The process varies between operating systems and between versions of each operating system (for instance, Vista versus Windows 7).

The wireless printer and wireless access point also have to be configured. They do not have displays and keyboards to allow configuration. Consequently, they are configured from one of the PCs. They are first connected to the network. The PC doing the configuration finds them and does the configuration work.

Test Your Understanding

26. a) List the hardware elements in the small home network described in this section. b) For wired connections, what transmission medium is used? c) What is its connector standard? d) What is the standard for wireless PCs and printers to connect to a wireless access point? e) What are the five hardware functions in a wireless access router? f) Why is the DHCP function necessary? g) Why is NAT necessary? h) Which devices need to be configured? (List them.)

---

[9] Network hard drives can be attached to the network like a network printer. Networked PCs can access them directly.

THIS BOOK

This is the first of four introductory chapters. This chapter introduced basic concepts in networking. Chapter 2 will introduce you to principles in network standards. Chapter 3 will introduce important security concepts. Chapter 4 will focus on network design.

Using the first four chapters as a knowledge base, subsequent chapters will focus in detail in important areas. In general, they "move up through the layers." Chapters 5 and 6 describe single switched networks operating at Layers 1 and 2. Chapters 7 and 8 describe single wireless networks operating at Layers 1 and 2. Chapters 9 and 10 focus on internets and TCP/IP management at Layers 3 and 4. Finally, Chapter 11 discusses application layer standards.

CONCLUSION

Synopsis

After a brief look at a particular network user, we looked at basic network terminology. We defined a network as a system that permits applications running on different hosts to work together. Networked applications, quite simply, are applications that require a network to work (for instance, e-mail). A host, in turn, is any device attached to a network. A host can be a client PC, a server, a mobile smart phone, or any other networked device. The network core is the central part of the network. Access lines connect hosts to the network core.

We then looked at traditional networked computing, including terminal–host computing and client/server computing. In terminal–host communication, the user had a dumb terminal that connected to a large central host computer. In client/server computing, a client program on a client computer receives services from a server program on a server computer. Client/server computing works through a request–response cycle, in which the client sends a request to the server, and the server sends back a response. Peer-to-peer (P2P) computing, in which client PCs provide service to each other, is growing rapidly. Cloud computing, in which processing is outsourced, is also beginning to grow.

We noted that speeds are measured in *bits* per second, not *bytes* per second. Speeds are expressed in the metric system, with kilobits per second (kbps), megabits per second (Mbps), and gigabits per second (Gbps). Metric measurements increase in factors of 1,000, not 1,024.

Terminals and hosts communicate over long-distance telephone circuits. Telephone circuits provide reserved capacity. Dial-up circuits provide reserved capacity for the duration of a single call. Leased line circuits are always-on, high-speed circuits. For data transmission over dial-up circuits, the terminal user needs a device called a modem. Computers transmit in binary (with two possible states representing 1 and 0) or in digital communication (in which there are two or more but still few) possible states. A modem converts digital computer signals into analog signals that can travel over dial-up telephone lines.

Data transmission is bursty, with short and intense bursts of data separated by long silences. For reserved circuit capacity, this is very wasteful. Packet switching addresses this inefficiency by dividing application messages into small segments and sends each segment in an individual packet. The packets of multiple conversations can share leased line circuits. This is called multiplexing. It uses reserved capacity efficiently.

When a packet arrives at a packet switch, the switch may have several alternatives for forwarding the packet to another packet switch. The switch makes a decision and sends the packet on to the next packet switch. That switch makes its own forwarding decision to another packet switch. These sequential decisions eventually get the packet all the way to the destination host. Links between hosts and packet switches and between pairs of packet switches are called physical links. The path a packet takes across a packet switched network is called a data link.

The ARPANET was the first major packet switched network. The first ARPANET standards were created by an informal Network Working Group. This group of graduate students had no formal permission to create standards, so they called their standards requests for comments (RFCs). One of the first applications on the ARPANET was e-mail, but there were soon many more. Later, the job of creating standards for the successor of the ARPANET—the Internet—was taken over by the Internet Engineering Task Force (IETF).

As more networks began to appear, there was a need to interconnect them. Bob Kahn and Vint Cerf came up with an approach for doing this. Their idea was to create internets by connecting individual networks together with routers. They essentially created a second layer of networking. Consequently, internet transmission involves two types of addresses, two types of packet switches (switches and routers), two types of packets (frames and packets), and two types of paths—data links across individual switched networks and routes all the way from the source to the destination address.

A new set of standards was created for internetworking. These are generically called the TCP/IP standards. The basic work of internetworking was divided into two layers. The internet layer governs packet organization and how packet switches forward packets to their destination. The main standard at this layer is the Internet Protocol (IP). IP addresses are 32 bits long and are often represented as dotted decimal notation, in which there are four numbers separated by dots. An example would be 128.171.17.13.

The transport layer lies above the internet layer. The TCP standard at the transport layer provides application message fragmentation and reassembly, placing packets in their correct sequence, and handling error correction, among other things. The UDP standard at the transport layer does none of these things. UDP is good when an application cannot use the services of TCP.

In addition to these three core protocols, TCP/IP has many application protocols. It also has many supervisory protocols to keep an internet operating (including the global Internet). We briefly looked at two supervisory protocols. DHCP provides your PC with an IP address every time you connect to the Internet, as well as other configuration information. DNS allows your computer to learn another computer's IP address if you only know its host name.

The global Internet grew out of the original ARPANET. In 1995, the National Science Foundation stopped paying for the operation of the Internet backbone. Its role was taken over by commercial Internet service providers (ISPs), which interconnect at network access points. Once government money was no longer used, the Internet could be used for commercial activities. E-commerce was born immediately and grew meteorically.

Single networks and internets can be local area networks (LANs) or wide area networks (WANs). A LANs operates on a corporate site, so the company can use any LAN technology it wishes. WANs connect sites. For WAN transmission, companies must use commercial carriers that have rights of way to lay transmission facilities through public areas. WAN transmission is more expensive than LAN transmission per bit transmitted, just as international telephone calls cost more per minute than local telephone calls. Consequently, companies must live with slower WAN transmission speeds than LAN transmission speeds.

The chapter closed by looking at a small PC network in a home. These networks are familiar to most students. Although they are small, they encompass most major concepts in networking.

END-OF-CHAPTER QUESTIONS

Thought Questions

1. a) In Figure 1-11, when Host X transmits a packet along the data link shown, how many physical links are there along the way along the data link shown? b) How many data links?
2. a) In Figure 1-15, when the client host sends a packet to the server host, how many data links will there be along the way? (Assume that the packet will take the minimum number of router hops.) b) How many routes? c) How many packets? d) How many frames?
3. a) In Figure 1-16, how many physical links, data links, and routes are there along the way when Host A sends a packet to Host B? b) When Host E sends to Host C? (Assume that hops will be minimized across switches and routers.) c) When Host D sends to Host E? (Assume that hops will be minimized across switches and routers.)
4. A host sends out a light flash in each clock cycle to represent data. Light flashes can be off, red, green, or blue. a) Is this digital signaling? b) Is it binary signaling?
5. There are eight routers between Host R and Host S. a) How many data links will there be along the way when Host R transmits a packet to Host S? (Hint: Draw a picture.) b) How many routes? c) How many frames?
6. Why does it make sense to make only the transport layer reliable? This is not a simple question.
7. a) What does it mean that data transmission is bursty? b) Why is burstiness bad in circuit switching?
8. What layer fragments application messages so that each fragment can fit inside an individual packet?

Case Study

A friend of yours wishes to open a small business. She will sell microwave slow cookers. She wishes to operate out of her house in a nice residential area. She is thinking of using a wireless LAN to connect her four PCs. What problems is she likely to run into? Explain each as well as you can. Your explanation should be directed to her, not to your teacher.

Perspective Questions

1. What was the most surprising thing you learned in this chapter?
2. What was the most difficult part of this chapter for you?

CHAPTER 1a

Hands-On Windows Networking

LEARNING OBJECTIVES

By the end of this chapter, you should be able to:

- Use basic networking commands in Microsoft Windows.
- Discuss network concepts in Chapter 1 with better understanding.

HANDS-ON NETWORKING TOOLS

This chapter introduces you to basic hands-on networking tools in Microsoft Windows. It assumes that you have read Chapter 1.

Binary and Decimal Conversions Using the Microsoft Windows Calculator

It is relatively easy to convert 32-bit IP addresses into dotted decimal notation if you use the Microsoft Windows Calculator. As shown in Figure 1a-1, go to the *Start* button, then to *Programs* or *All Programs*, then to *Accessories*, and then click on *Calculator*. The Windows Calculator will pop up. Initially, it is a very simple calculator. Choose *View* and click on *Scientific* to make Calculator an advanced scientific calculator.

BINARY TO DECIMAL To convert eight binary bits to decimal, first divide the 32 bits into four 8-bit segments. Click on the *Bin* (binary) radio button and type in the 8-bit binary sequence you wish to convert. Then click on the *Dec* (decimal) radio button. The decimal value for that segment will appear.

Note that you cannot convert the whole 32-bit IP address at one time. You have to do it in four 8-bit segments.

Once you have the four decimal segment values, write them in order with dots between them. It will look something like 128.171.17.13. You have now converted the 32-bit IP address to dotted decimal notation.

DECIMAL TO BINARY To convert decimal to binary, go to *View* and choose *Scientific* if you have not already done so. Click on *Dec* to indicate that you are entering a decimal number. Type the number. Now click on *Bin* to convert this number to binary.

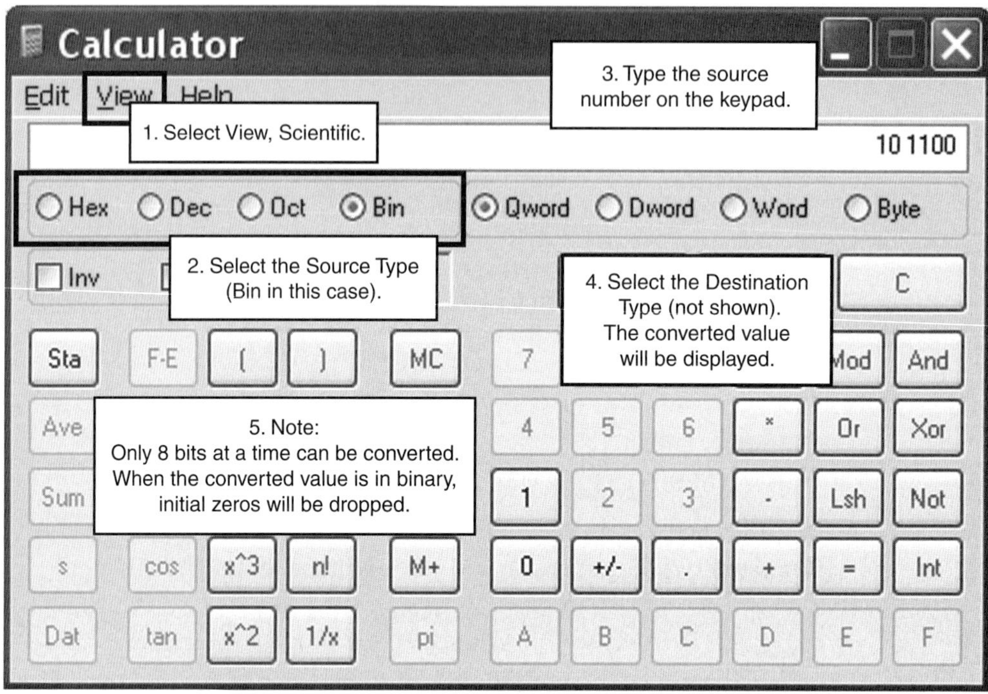

FIGURE 1a-1 Windows Calculator

One subtlety is that Calculator drops initial zeros. So if you convert 17, you get 10001. You must add three initial zeros to make this an 8-bit segment: 00010001.

Another subtlety is that you can only convert one 8-bit segment at a time.

1. a) What is 11001010 in decimal? b) Express the following IP address in binary: 128.171.17.13. (*Hint:* 128 is 10000000. Put spaces between each group of 8 bits.) c) Convert the following address in binary to dotted decimal notation: 11110000 10101010 00001111 11100011. (Spaces are added between bytes to make reading easier.) (*Hint:* 11110000 is 240 in decimal.)

Test Your Download Speed

How fast is your Internet connection? Test your download speed at two sites. The following is a list of sites offering free download scans. If you can, test your bandwidth during periods of light and heavy use.

http://www.speakeasy.net/speedtest/

http://reviews.cnet.com/internet-speed-test/

http://www.speedtest.net/

2. a) What kind of connection do you have (telephone modem, cable modem, LAN, etc.)? b) What site did you use for your first test? c) What did you learn? d) What site did you use for your second test? e) Did you get different results?

Working with the Windows Command Line

Windows offers a number of tools from its command line prompt. Network professionals need to learn to work with these commands.

GETTING TO THE COMMAND LINE To get to the command line, click on the *Start* button and choose *Run*. Type either *cmd* or *command*, depending on your version of Windows, and then hit *OK*.

COMMAND LINE RULES At the command line, you need to type carefully because even a single-letter error will ruin the command. You also need to hit *Enter* at the end of each line. You can clear the command line screen by typing **cls** and then pressing *Enter*.

3. Go to the command line. Clear the screen.

SEE YOUR CONFIGURATION In Windows, you can find information about your own computer by typing **ipconfig/all** *Enter* at the command line. This will give you your IP address, your physical address (your Ethernet address), and other information.

4. Use ipconfig/all or winipconfig. a) What is your computer's IP address? b) What is its Ethernet address? c) What is your default router (gateway)? d) What are the IP addresses of your DNS hosts? e) What is the IP address of your DHCP server? f) When you get a dynamic IP address, you are given a lease specifying how long you may use it. What is the starting time and ending time of your lease?

Ping and Tracert

PING To find out whether you can reach a host and to see how much latency there is when you contact a host, use the **ping** command. You ping an IP address or host name much as a submarine pings a target to see whether it exists and how far away it is. To use the command, type **ping *hostname* and press *Enter*,** or type **ping *IPaddress* and press *Enter*.** Ping may not work if the host is behind a firewall, because firewalls typically block pings.

5. Ping a host whose name you know and that you use frequently. What is the latency? If this process does not work because the host is behind a firewall, try pinging other hosts until you succeed.

PING 127.0.0.1 (PC, CALL HOME) Ping the address 127.0.0.1. This is your computer's **loopback address**. In effect, the computer's network program sends a ping to itself. If your PC seems to have trouble communicating over the Internet, type **ping 127.0.0.1** and *Enter*. If the ping fails, you know that the problem is internal and you need to focus on your network software's configuration. If the ping succeeds, then your computer is talking to the outside world at least.

6. Ping 127.0.0.1. Did it succeed?

TRACERT The Windows **tracert** program is like a super ping. It lists not only latency to a target host, but also each router along the way and the latency to that router. Actually, tracert shows three latencies for each router because it tests each router three times. To use tracert, type the tracert *hostname* and press *Enter*, or type the tracert *IP address* and press *Enter.* Again, hosts (and routers) behind firewalls will not respond.

7. Do a tracert on a host whose name you know and that you use frequently. You can stop the tracert process by hitting Control-C. a) What is the latency to the destination host? b) How many routers are there between you and the destination host? If this does not work because the host is behind a firewall, try reaching other hosts until you succeed.
8. Distinguish between the information that ping provides and the data that tracert provides.

Nslookup

The **nslookup** command allows you to send a DNS request message to your local name server. At the command prompt, type *nslookup hostname* [*Enter*], where *hostname* is a host name for which you wish to know the IP address. The information shown after you type your command is the IP address.

Your local DNS host may send you a non-authoritative IP address. Each DNS server has a group of IP addresses and host names for which it is the authoritative DNS server. For instance, the DNS servers for Hawaii.edu are authoritative for all host names ending in Hawaii.edu. Sometimes, a DNS server will happen to know the IP addresses for host names in other domains. This is a non-authoritative IP address.

9. Find the IP address for Microsoft.com and Apple.com.

RFCs

In networking, you frequently have to look up RFCs. Google, Bing, or other search tools let you do this.

10. a) Look up RFC 1149. b) In layperson's terms, what does this RFC specify? c) What are its sections? (This is a serious question. You should learn how RFCs are structured.) d) On what day was it created?

CHAPTER 2

Network Standards

LEARNING OBJECTIVES

By the end of this chapter, you should be able to:

- Provide the definitions of network; standards and protocols; message syntax, semantics, and order.
- Discuss message ordering in general and in HTTP and TCP.
- Discuss message syntax in general and in Ethernet frames, IP packets, TCP segments, UDP datagrams, and HTTP request and response messages.
- Explain encoding application messages into bits.
- Explain vertical communication on hosts.
- Discuss major standards architectures, especially TCP/IP, OSI, and the hybrid TCP/IP–OSI standards architecture.

INTRODUCTION

We looked at network standards briefly in Chapter 1. In this chapter, we will look at standards at a higher level, developing taxonomies of standards types. Much of the rest of this book focuses on specific standards; you will need to understand standards broadly to understand where those specific standards fit into the overall standards picture. This chapter also looks in some detail at the most important standards on the Internet and in corporate networks. These include Ethernet, IP, TCP, UDP, and HTTP.

Standard = Protocol

In this book, we use the terms *standard* and *protocol* to mean the same thing. In fact, if you look at the names of standards, you often see "protocol." Examples are the Hypertext Transfer Protocol, the Internet Protocol, the Transmission Control Protocol, and the User Datagram Protocol.

Network Standards

WHAT ARE NETWORK STANDARDS? As Figure 2-1 illustrates, **network standards** are rules of operation that govern the exchange of messages between two hardware or

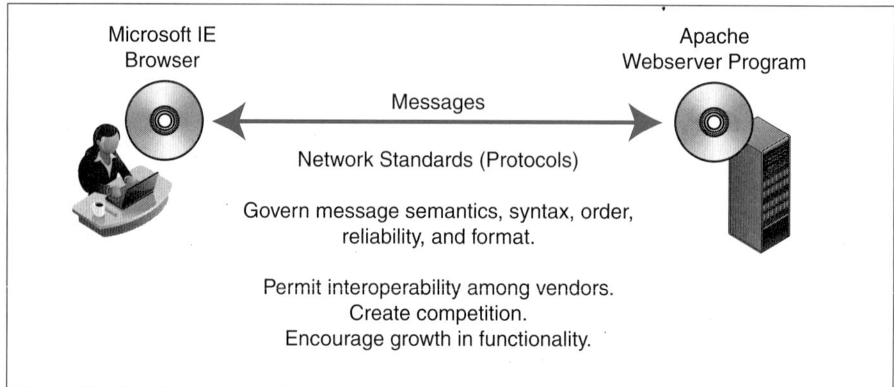

FIGURE 2-1 Network Standards

software processes. To give a human analogy, in my classes, the standard language is American English. Not all of my students are native English speakers, but we are able to communicate using a standard language. In this chapter, we will see that network standards govern a number of message characteristics, including semantics, syntax, message order, reliability, and connection orientation.

Network standards are rules of operation that govern the exchange of messages between two hardware or software processes. This includes message semantics, syntax, message order, reliability, and connection orientation.

NETWORK STANDARDS BRING COMPETITION Standards are important because they allow products from different vendors to **interoperate** (work together). In Figure 2-1, an Internet Explorer browser from Microsoft is working with an open source Apache webserver. Although these software sources are very different and may not even like each other, their products work together because they exchange messages using the Hypertext Transfer Protocol (HTTP) network standard.

With network standards, it is impossible for any company to maintain a monopoly by refusing to allow others to use its proprietary communication protocols. Competition drives down prices. It also spurs companies to add new features so that their products will not be pure commodities. These new features often appear in the next version of the standard.

Network standards are not only the key to competition. They are the key to networking in general. To work in networking, you need to understand individual standards so that you can design networks, set up network components, and troubleshoot problems. Learning networking is heavily about learning standards. In this chapter, we will look broadly at the general characteristics of standards and will also look at some key network standards.

Recap of Chapter 1 Standards Concepts

In Chapter 1, we saw that standards can be described in terms of their layer of operation.

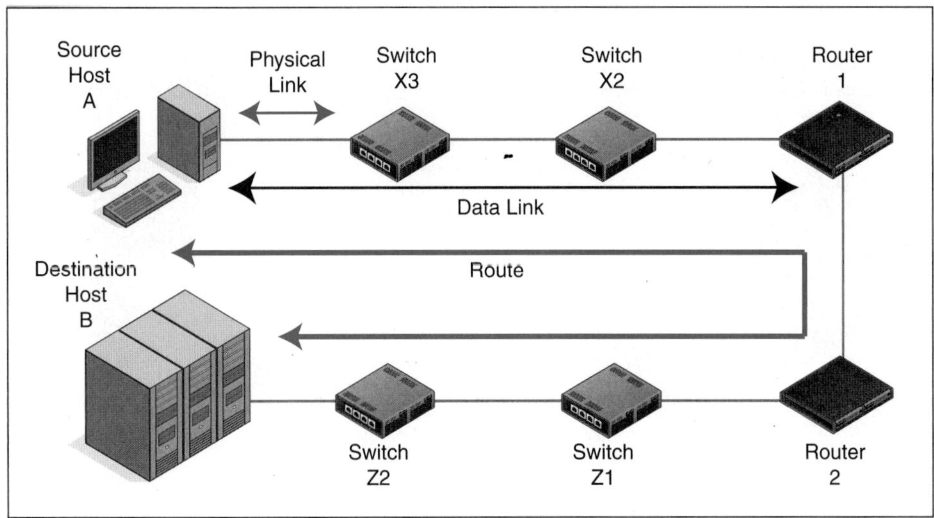

FIGURE 2-2 The Physical, Data Link, and Internet Layers

DELIVERY LAYERS As Figure 2-2 shows, three layers are involved in the transmission of packets between source hosts and destination hosts.

- Physical links are connections between adjacent devices, such as a host and a switch, two switches, two routers, a host and a switch, and so forth. Physical layer standards are not concerned with messages. Their job is to turn the bits of data link layer messages (frames) into signals.
- Data link layer standards govern the transmission of frames between two hosts, two routers, or a host and a router *across a single switched or wireless network*. The path that a frame takes is called a data link. This layer governs switch operation and frame organization.
- Internet layer standards govern the transmission of packets from the source host to the destination host, across multiple networks in an internet. The path that a packet takes between the two hosts is called its route. This layer governs router operation and packet organization.

A common source of confusion is that concepts are repeated at the data link and internet layer but with different terminology. This occurs because internetworking required the adding of a second layer of standards to those needed for transmission through single networks.

Also, recall that packets are carried inside frames. When a source host sends a packet to a destination host, the packet travels within a frame in each network along the way. If there are 19 single networks on the route between the source and destination hosts, a single packet will travel in 19 different frames.

THE TRANSPORT AND APPLICATION LAYERS The physical, data link, and internet layers are for standards that move packets along their way between the source host and the

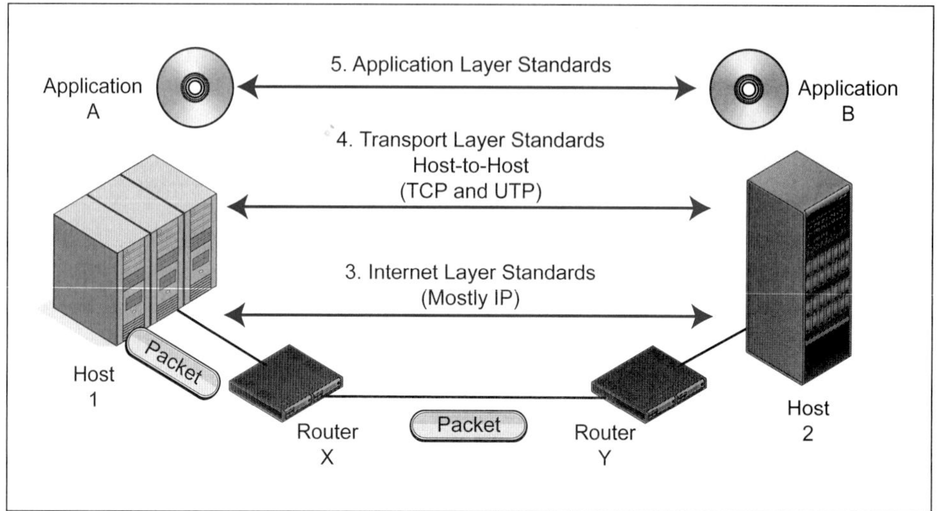

FIGURE 2-3 The Internet, Transport, and Application Layers

destination host. In contrast, Figure 2-3 shows that transport and application processes govern processes that only exist on the two communicating hosts.

- The transport layer supplements the internet layer. Internet layer operation typically is a best-effort service that does not guarantee that packets will be delivered. The transport layer is a "fix-up" layer that can add reliability and other desirable characteristics to transmission across an internet. In addition, the source host transport layer process fragments application messages so these fragments can be sent in individual packets. The destination host transport process reassembles the segments and passes the application message to the application.
- The application layer is for application standards. When two e-mail programs need to work together, they use an e-mail application standard. For webservice, HTTP is the application layer standard. There are more application layer standards than there are standards at all other layers combined because there are so many applications and because different applications usually need different application standards.

THE FIVE LAYERS Figure 2-4 recaps the five layers.

- Speaking broadly, the physical and data link layers govern transmission through single networks.
- Also speaking broadly, the internet and transport layer together govern transmission through an internet. The internet layer governs packet organization and raw packet delivery. The transport layer fixes up problems and does fragmentation and assembly.
- Finally, the application layer governs how two applications work together.

Test Your Understanding

1. a) Give the definition of network standards that this chapter introduced. b) In this book, do *standards* and *protocols* mean the same thing?

Broad Function	Layer	Name	Specific Function
Interoperability of application programs	5	Application	Application layer standards govern how two applications work with each other, even if they are from different vendors.
Transmission across an internet	4	Transport	Transport layer standards govern aspects of end-to-end communication between two end hosts that are not handled by the internet layer. These standards also allow hosts to work together even if the two computers are from different vendors or have different internal designs.
	3	Internet	Internet link layer standards govern the transmission of packets across an internet—typically by sending them through several routers along the route. Internet layer standards also govern packet organization, timing constraints, and reliability.
Transmission across a single switched or wireless network	2	Data Link	Data link layer standards govern the transmission of frames across a single switched or wireless network. Data link layer standards also govern frame organization, timing constraints, and reliability.
	1	Physical	Physical layer standards govern transmission between adjacent devices connected by a transmission medium.

FIGURE 2-4 Layers Recap

NETWORK STANDARD CHARACTERISTICS

Network standards govern communication. Figure 2-5 notes, more specifically, that standards govern five specific things about message exchanges: semantics, syntax, order, reliability, and format. In this section, we will focus on message order, semantics, and syntax, but we will also introduce the concept of whether a standard is reliable.

Message Ordering

In medicine and many other fields, a *protocol* is a prescribed series of actions to be performed in a particular order. In cooking, recipes work this way. If you do not put together the ingredients of a cake in the right amounts and prepare them in the right order, the cake is not likely to turn out very well.

Network Standards

Network standards are rules that govern the exchange of messages between hardware or software processes on different hosts, including messages (ordering, semantics, and syntax), reliability, and connection orientation.

Message Order

Turn taking, order of messages in a complex transaction, who must initiate communication, etc.

In the World Wide Web, the client program sends an HTTP request message.

The webserver program sends back an HTTP response message.

The client must initiate the interaction.

Other network standards have more complex turn taking; for instance TCP.

Human turn taking is loose and flexible.

Message order for network standards must be rigid because computers are not intelligent.

Message Semantics

Semantics = the meaning of a message

HTTP request message: "Please give me this file."

HTTP response message: Here is the file. (Or, I could not comply for the following reason.)

Network standards normally have a very limited set of possible message meanings.

For example, HTTP requests have only a few possible meanings

- GET: Please give me a file
- PUT: Store this file (not often used)
- A few more

Message Syntax (Organization)

Like human grammar, but more rigid

Header, data field, and trailer (Figure 2-8)

Not all messages have all three parts.

Field lengths are measured in bits or bytes.

Bytes are also called octets.

FIGURE 2-5 Network Standards Concepts

In this same way, network standards govern **message ordering**. For the Hypertext Transfer Protocol standard that we saw in the last chapter, message ordering is very simple. The client sends an HTTP request message, and the server sends back an HTTP response message. A server cannot send an HTTP response message unless that client has first sent an HTTP request message. Many protocols, including the Transmission Control Protocol (TCP) standard, which we will see in this chapter and in Chapter 9, involve many messages being sent in precise order.

Semantics

Computers are not intelligent (as you learned in your first programming course). Neither are switches, routers, hosts, or other network devices. To limit the complexity of software, protocols usually define only a few message types, and these types usually have only a few options. Put another way, network protocols greatly limit the **semantics** (meaning) of their messages. For example, the most common HTTP request message is a GET message, which requests a file. There is also a POST request message, which uploads a file to the webserver.

Semantics is what a message means.

Syntax

In addition, while human grammar is very flexible, network messages have very rigid **syntax**, that is, message organization. A little later in this chapter, we will look at the syntaxes of several important protocol messages.

Syntax is how a message is organized.

Test Your Understanding

2. a) What three things about message exchanges did we see in this section? b) Give an example not involving networking in which the order in which you do things can make a big difference. c) Distinguish between syntax and semantics.

EXAMPLES OF MESSAGE ORDERING

We will look at two examples of message ordering. We will look first at the very simple message ordering in HTTP. We will then look at the more complex message ordering in TCP.

Message Ordering in HTTP

Figure 2-6 illustrates an HTTP request–response cycle. The client sends a request, and the server sends a response. Note that the cycle is always initiated by the client, not by the server. As noted in the previous section, the server cannot transmit unless the client has sent it an HTTP request message. This is a very simple type of message ordering.

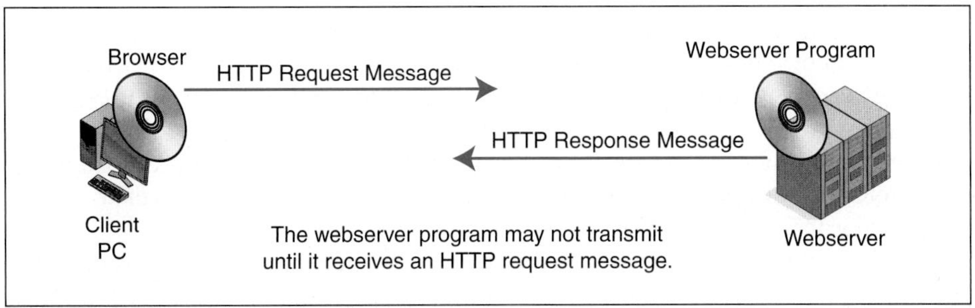

FIGURE 2-6 An HTTP Request–Response Cycle

Message Ordering and Reliability in TCP at the Transport Layer

Many protocols have much more complex rules for message ordering. We will look at the Transmission Control Protocol at the transport layer to see an example of this complexity.

THE SITUATION Figure 2-7 shows the transport layer processes on Host A and Host B. They are communicating via HTTP at the application layer. The Hypertext Transfer Protocol requires the use of TCP at the transport layer. The figure shows a sample communication session, which is called a connection.

SEGMENTS In TCP, messages are called **TCP segments** because each carries a segment of an application message (or is a control segment that does not carry application data).

THE THREE-STEP HANDSHAKE OPENING The communication begins with a **three-step handshake** to begin the communication.

- Host A, which is the client in the HTTP exchanges, initiates the communication. It transmits a TCP SYN segment to Host B. This indicates that Host A wishes to communicate.
- Host B sends back a TCP SYN/ACK segment. The SYN indicates that it also is willing to begin the communication. The ACK part is an acknowledgment of Host A's SYN message. In TCP, all segments are acknowledged, with the primary exception of pure ACKs. (If pure ACKs had to be acknowledged, there would be an endless series of ACKs.)
- Host A sends back a pure TCP ACK segment. This acknowledges Host B's SYN/ACK.

In TCP, all segments are acknowledged, with the primarily exception of pure ACKs.

CONNECTIONS TCP creates connections with distinct openings and closings. This is like a telephone call, in which you informally make sure that the other person can talk at the start of a call and mutually agree to end the call. In technical jargon, TCP is a **connection-oriented** protocol.

SEQUENCE NUMBERS In a connection-oriented protocol, each message is given a sequence number. This allows the receiver to ensure that no message is missing and allows the receiving process to deal with duplicate segments. (It simply discards duplicates.)

Sequence numbers are important because application messages are fragmented and delivered in separate packets. Sequence numbers allow the receiver to place the segments in order and reassemble them.

Note in Figure 2-7 that each side numbers its sequence numbers individually. For simplicity, we have called Host A's sequence numbers A1, A2, A3, and so forth. We have done the same with Host B's messages. So Host A's SYN segment is A1, while Host B's SYN/ACK is B1, and Host A's acknowledgment of the SYN/ACK is A2.

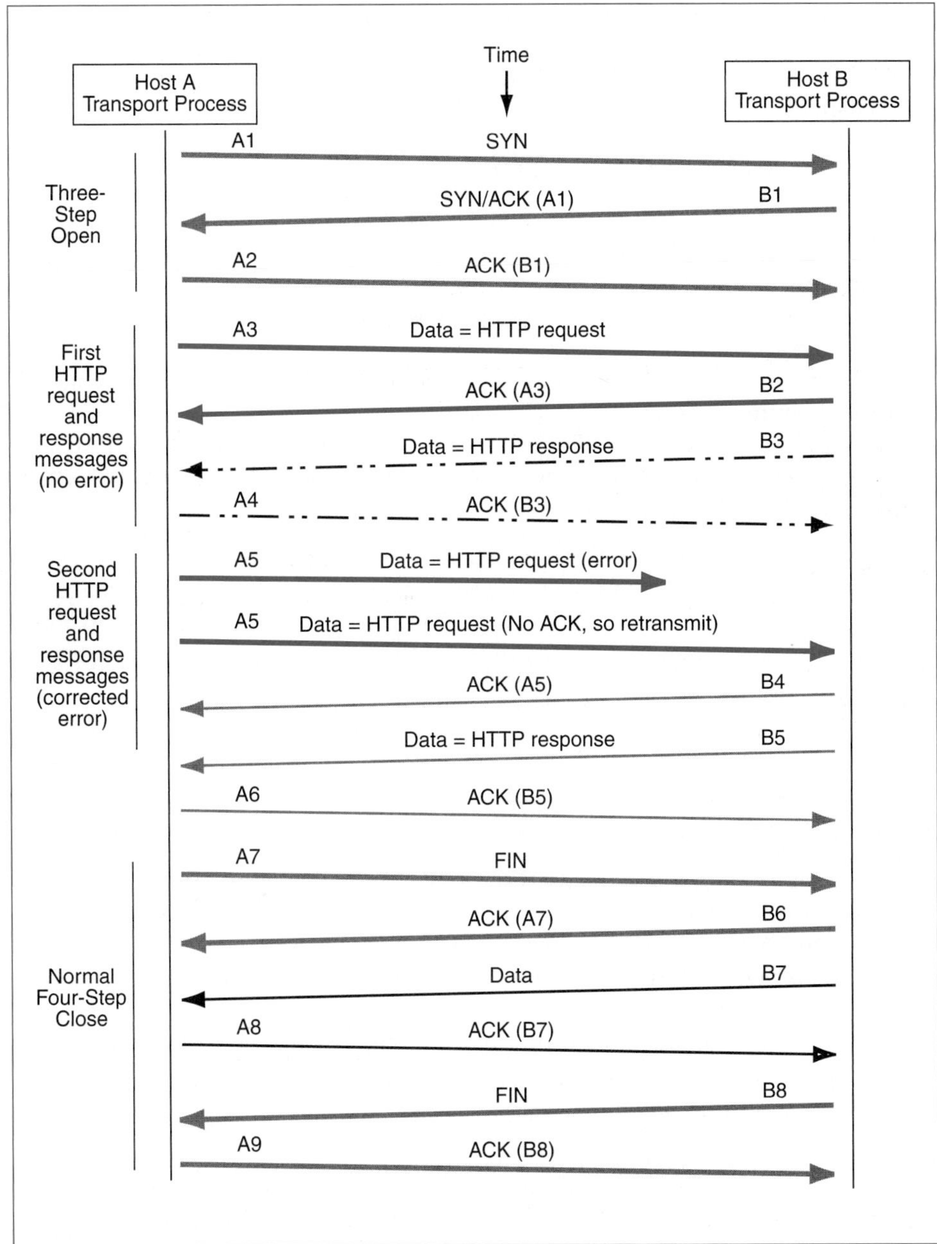

FIGURE 2-7 A TCP Connection

CARRYING APPLICATION DATA The next four segments (A3, B2, B3, and A4) constitute a request–response cycle.

- A3 carries an HTTP request.
- B2 is an ACK of A3.
- B3 carries the HTTP response message.
- A4 acknowledges the receipt of B3.

Usually, HTTP request messages are small enough to fit in a single TCP segment. However, most HTTP responses are long and must be sent in a number of TCP segments. This does not change the basic picture, however. There would simply be several more exchanges like B3 and A4.

RELIABILITY TCP is a **reliable** protocol. This means that it corrects errors.

- Segment A5 is sent but never reaches Host B.
- Host B does not send an acknowledgment, because ACKs are only sent when a segment is received correctly.
- Host A realizes that A5 has not been acknowledged. After a certain period of time, it retransmits A5.
- This time, the segment arrives correctly at Host B. Host B sends B4, which is an acknowledgment of A5.
- Finally, Host B sends an HTTP response message (B5) and receives an ACK (A6). Again, sending an HTTP response message tends to take several TCP data/acknowledgment cycles.

In this example, Segment A5 never reached the receiving transport process. What would have happened if A5 had reached the transport process but was damaged during transmission? In this case, the receiving transport process would discard the segment. It would not send an ACK. So there is a simple rule for ACKs. Unless a transport process receives a segment correctly, it does not send an acknowledgment.

> Unless a transport process receives a segment correctly, it does not send an acknowledgment.

THE FOUR-STEP HANDSHAKE CLOSING Host A has no more HTTP request messages to send, so it closes the connection. It does so by sending a FIN segment (A7), which Host B acknowledges (B6). This means that Host A will not send new data. However, it will continue to send ACKs to segments sent by Host B.

Host B has one more data segment to send, B7. When it sends this segment, Host A's transport process responds with an ACK (A8).

Now, Host B is finished sending data. It sends its own FIN segment (B8) and receives an acknowledgment (A9).

The connection is closed.

PERSPECTIVE TCP is a fairly complex protocol. It uses connections so that it can apply sequence numbers to segments. This allows it to fragment long application messages and deliver the segments with an indication of their order. It also uses connections so that it can provide reliable data to the application layer program above it.

We will see that almost all other protocols are unreliable. Many standards check for errors, but if they find an error, they simply discard the message. Discarded messages never get to the transport process on the other host, so they are never acknowledged. Receiving no acknowledgment, the sender resends them.

Why make only TCP reliable? There are two answers. First, TCP sits just below the application layer. This allows it to send clean data to the application program regardless of errors at lower levels, which are corrected by TCP resends.

Second, as Figure 2-3 shows, only the two hosts have transport layer processes, so error correction is only done once, on the two hosts. It is not done at each packet hop between routers or in each frame hop between switches. Error correction is a resource-consuming process, so it should be done as little as possible. Doing error correction at the transport layer processes on the two hosts accomplishes this.

Test Your Understanding

3. a) Describe the simple message ordering in HTTP. b) In HTTP, can the server transmit if it has not received a request message from the client? c) Describe the three-step handshake in TCP connection openings. d) What kind of message does the destination host send if it does not receive a segment during a TCP connection? e) What kind of message does the destination host send if it receives a segment that has an error during a TCP connection? f) Under what conditions will a source host TCP process retransmit a segment? g) Describe the four-step handshake in TCP connection closes. h) After a side initiates the close of a connection by sending a FIN segment, will it send any more segments? Explain.

EXAMPLES OF MESSAGE SYNTAX

We have just looked at message ordering. Now we will turn to message syntax. In this book, we will be looking at the syntax of many different types of messages. To give you a feeling for message syntax, we will look at the syntax of five important message types.

Syntax: General Message Organization

Figure 2-8 shows that message syntax in general has three parts—a header, a data field, and a trailer.

DATA FIELD The **data field** is the heart of the message. It contains the content being delivered by the message. In an HTTP response message, the data field contains the file that the response message is delivering.

The data field contains the content being delivered by a message.

HEADER The message **header**, quite simply, is everything that comes before the data field.

The message header is everything that comes before the data field.

TRAILER Some messages also have **trailers**, which consist of everything coming *after* the data field.

The message trailer is everything that comes after the data field.

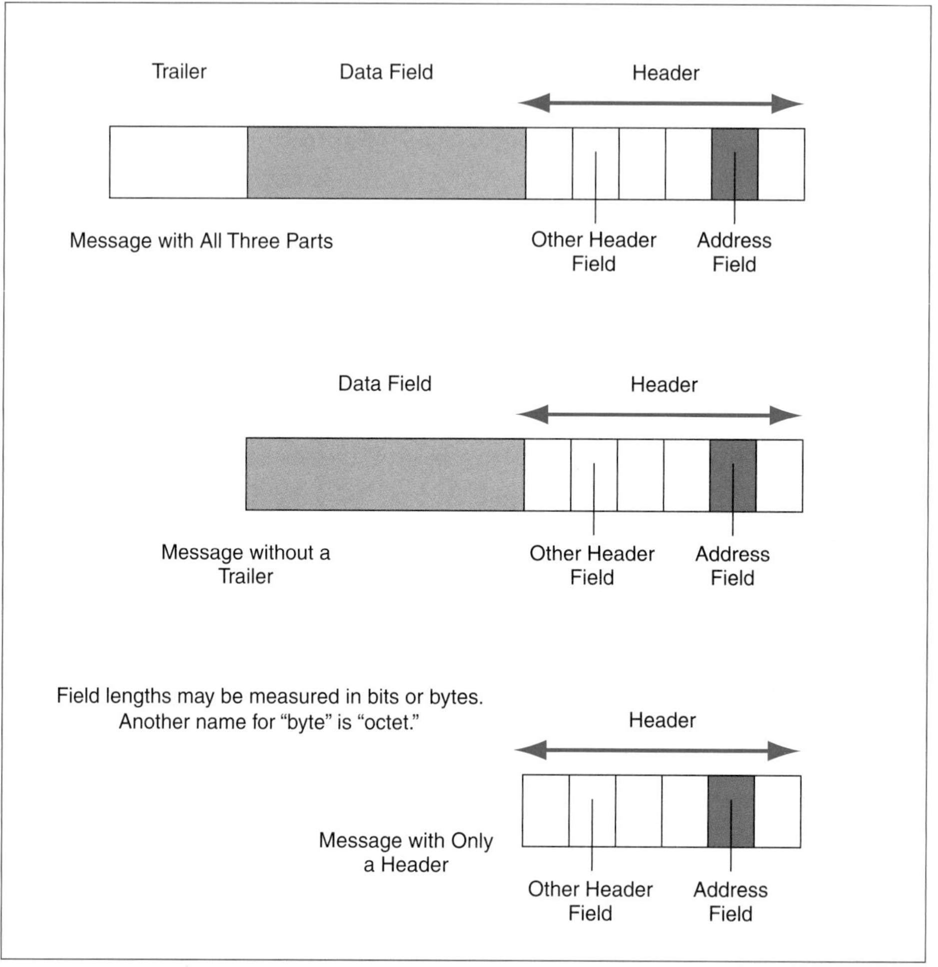

FIGURE 2-8 General Message Organization

NOT ALL MESSAGES HAVE ALL THREE PARTS HTTP messages demonstrate that only a header is present in all messages. Data fields are not always present but are very common. Trailers are not common.

FIELDS IN HEADERS AND TRAILERS The header and trailer usually contain smaller syntactic sections called **fields**. For example, a frame or packet has a destination address header field, which allows switches or routers along the way to pass on the frame or packet they receive.

> The header and trailer usually contain smaller syntactic sections called fields.

When we look at network standard messages in this chapter and in later chapters, we will be concerned primarily with header fields and trailer fields because these are the defining characteristics of standards.

OCTETS Field lengths can be measured in bits. Another common measure for field lengths in networking is the octet. An **octet** is a group of eight bits. Isn't that a byte? Yes, exactly. *Octet* is just another name for *byte*. The term is widely used in networking, however, so you need to become familiar with it. *Octet* actually makes more sense than *byte*, because *oct* means "eight." We have octopuses, octagons, and octogenarians.[1]

An octet is a group of eight bits.

Test Your Understanding

4. a) What are the three general parts of messages? b) What does the data field contain? c) What is the definition of a header? d) Is there always a data field in a message? e) What is the definition of a trailer? f) Are trailers common? g) Distinguish between headers and header fields. h) Distinguish between octets and bytes.

The Ethernet Frame Syntax

Messages at the data link layer are frames. In wired local area networks, the dominant network standard is Ethernet. Actually, Ethernet, like most "standards" is really a family of standards. Ethernet has many different physical layer protocols from which a company can choose. However, generally speaking, it has a single frame standard, which Figure 2-9 illustrates.

The fields in the frame are delimited by their lengths in octets or bits. The receiver or switch first receives the first bit of the preamble field. It then counts bits until it gets to the next field, the field after it, and so forth. Then it can process the frame.

This is called **synchronous communication**, because the senders' and receivers' clocks must be precisely synchronized for the receiver to read the message. For example, if the two clocks are not precisely synchronized, the receiver might believe that bit 1,000 is bit 999 or 1,001.

Ethernet actually has a rather complex frame. We will look at more of these in Chapter 6. There are only four fields that we need to emphasize at this time.

SOURCE AND DESTINATION ADDRESS FIELDS Ethernet has a destination MAC address field and a source MAC address field. These are like the address and return address on a postal envelope. Switches use the destination IP address to forward frames.

Note that Ethernet addresses are 48 bits long—not 32 bits long, like IP addresses. Different single network standards have different address lengths. Ethernet addresses are called MAC addresses for reasons we will see in Chapter 6. In that chapter, we will also see that an Ethernet MAC address has six pairs of symbols separated by dashes. An example would be B7-23-DD-6F-C8-AB.

---

[1] What is the eighth month? (Careful!)

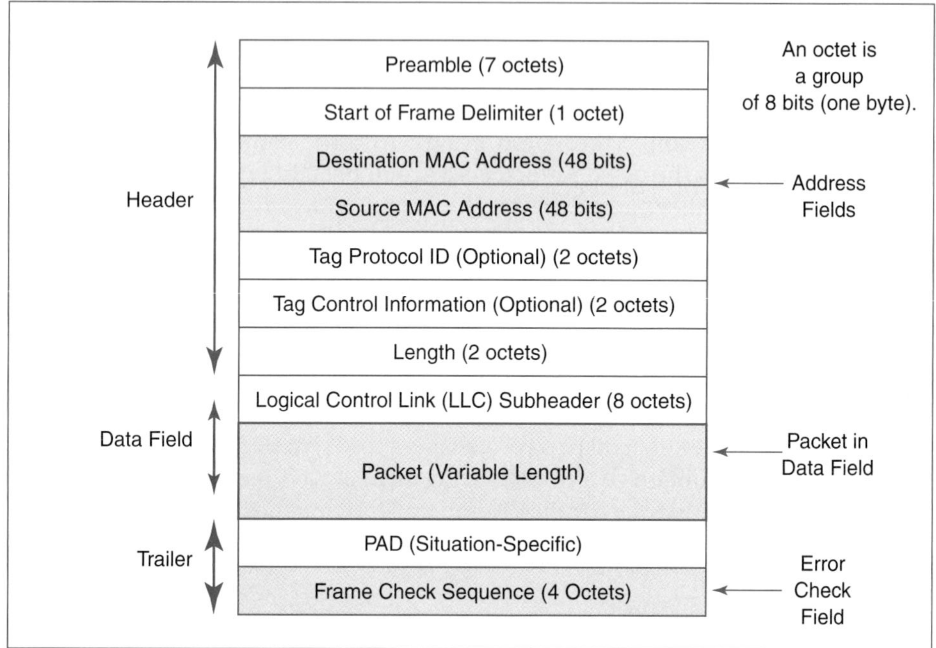

FIGURE 2-9 Ethernet Frame

PACKET IN THE DATA FIELD We saw in Chapter 1 that frames carry packets in their data fields. The figure illustrates how this occurs in Ethernet.

FRAME CHECK SEQUENCE FIELD The four-octet Frame Check Sequence field is used for error detection. The sending data link layer process computes a number based on other bits in the frame. The receiver recomputes this 32-bit number. If the recomputed number is the same as the number transmitted in the Frame Check Sequence field, then there have been no errors during transmission, and the receiver accepts the frame. If the numbers are different, there was a propagation error and the frame is corrupted. If so, the receiving data link layer process simply discards the frame. This is error detection but not full error correction. Ethernet is not a reliable protocol.

Test Your Understanding

5. a) Why is Ethernet transmission called synchronous transmission? b) How long are Ethernet MAC addresses? c) What devices read Ethernet destination MAC addresses? d) In what field is the IP address stored? e) If the receiver detects an error on the basis of the value in the Frame Check Sequence field, what does it do? f) Ethernet does error detection but not error correction. Is Ethernet a reliable protocol? Explain.

The Internet Protocol (IP) Packet Syntax

ILLUSTRATED WITH 32 BITS PER LINE An IP packet, like an Ethernet frame, is a long string of bits (1s and 0s). Unfortunately, drawing the packet this way would require a page several meters wide. Instead, Figure 2-10 shows that we usually depict an IP

<table>
<tr><td colspan="4">Bit 0</td><td colspan="3">Bit 31</td></tr>
<tr><td>Version Number (4 bits)</td><td>Header Length (4 bits)</td><td colspan="2">Diff-Serv (8 bits)</td><td colspan="3">Total Length (16 bits)</td></tr>
<tr><td colspan="4">Identification (16 bits)</td><td>Flags (3 bits)</td><td colspan="2">Fragment Offset (13 bits)</td></tr>
<tr><td colspan="3">Time to Live (8 bits)</td><td>Protocol (8 bits)</td><td colspan="3">Header Checksum (16 bits)</td></tr>
<tr><td colspan="7">Source IP Address (32 bits)</td></tr>
<tr><td colspan="7">Destination IP Address (32 bits)</td></tr>
<tr><td colspan="5">Options (if any)</td><td colspan="2">Padding (to 32-bit boundary)</td></tr>
<tr><td colspan="7">Data Field (dozens, hundreds, or thousands of bits)
Often contains a TCP segment</td></tr>
</table>

FIGURE 2-10 The Internet Protocol (IP) Packet

packet as a series of rows with 32 bits per row. In binary counting, the first bit is zero. Consequently, the first row shows bits 0 through 31. The next row shows bits 32 through 63. This is a different way of showing syntax than we saw with the Ethernet frame, but it is a common way of showing syntax in TCP/IP, so you need to be familiar with it.

32-BIT SOURCE AND DESTINATION ADDRESS FIELDS Each IP packet has source and destination IP addresses. These **IP addresses** are 32 bits long. An IP address gives a host's internet layer address on an internet consisting of multiple single networks. Routers use destination IP addresses to decide how to forward packets so that they will get closer to their destination.

For human comprehension, it is normal to express IP addresses in **dotted decimal notation**. As we saw in Chapter 1, a typical IP address in dotted decimal notation is 128.171.17.13—four numbers separated by dots. Each number must be between 0 and 255.

HEADER CHECKSUM FIELD The Header Checksum field is like the Frame Check Sequence field in the Ethernet frame. It also is used for error detection. As in the case of Ethernet frames, incorrect IP packets are simply discarded. There is no retransmission. So IP is not a reliable protocol.

Test Your Understanding

6. a) How many octets long is an IP header if there are no options? (Look at Figure 2-10.) b) List the first bit number on each header row in Figure 2-10, not including options. Remember that the first bit in Row 1 is Bit 0. c) What is the bit number of the first bit in the destination address field? (Remember that the first bit in binary counting is Bit 0.) d) How long are IP addresses? e) You have two addresses: B7-23-DD-6F-C8-AB and 217.42.18.248. Specify what kind of address each address is. f) What device in an internet besides the destination host reads the destination IP address? g) What is this device's purpose in doing so? h) Is IP reliable or unreliable?

Transmission Control Protocol (TCP) Segment Syntax

Earlier, we saw message ordering in the transmission of TCP segments. Now we will look at the syntax of TCP segments.

FIELDS IN TCP/IP SEGMENTS When IP was designed, it was made a very simple "best effort" protocol (although its routing tables are complex). The IETF left more complex internetwork transmission control tasks to TCP. Consequently, network professionals need to understand TCP very well. Figure 2-11 shows the organization of TCP messages, which are called TCP segments.

FLAG FIELDS TCP has six single-bit fields. Single-bit fields in general are called **flag fields**. If a flag field has the value 1, it is said to be **set**. (If it has the value 0, it is said to be **not set**.) In TCP, flag fields allow the receiving transport process to identify the kind of segment it is receiving. We will look at three of these flag bits:

- If the ACK bit is set, then the segment acknowledges another segment. If the ACK bit is set, the acknowledgment field also must be filled in to indicate which message is being acknowledged.
- If the SYN (synchronization) bit is set (has the value 1), then the segment requests a connection opening. Earlier in this chapter, we saw how SYN segments.
- If the FIN (finish) bit is set, then the segment requests a normal connection closing.

Single-bit fields are called flag fields. If a flag field has the value 1, it is said to be set. (If it has the value 0, it is said to be not set.)

Earlier, we talked about TCP SYN segments, ACK segments, and FIN segments. These are simply segments in which the SYN, ACK, or FIN bits are set, respectively.

SEQUENCE NUMBERS Each TCP segment has a unique 32-bit **sequence number** that increases with each segment. This allows the receiving transport process to put arriving TCP segments in order if IP delivers them out of order.

ACKNOWLEDGMENT NUMBERS Earlier in this chapter, we saw that TCP uses acknowledgments (ACKs) to achieve reliability. If a transport process receives a TCP segment correctly, it sends back a TCP segment acknowledging the reception. We saw earlier that if the sending transport process does not receive an acknowledgment, it transmits the TCP segment again.

The **acknowledgment number** field indicates which segment is being acknowledged. One might expect that if a segment has sequence number X, then the acknowledgment

TCP Segment

Bit 0 ... Bit 31

Source Port Number (16 bits)			Destination Port Number (16 bits)	
Sequence Number (32 bits)				
Acknowledgement Number (32 bits)				
Header Length (4 bits)	Reserved (6 bits)	Flag Fields (6 bits)	Window (16 bits)	
TCP Checksum (16 bits)			Urgent Pointer (16 bits)	
Options (if any)				Padding
Data Field				

Flag fields are 1-bit fields. They include SYN, ACK, FIN, and RST.

UDP Datagram

Bit 0 ... Bit 31

Source Port Number (16 bits)	Destination Port Number (16 bits)
UDP Length (16 bits)	UDP Checksum (16 bits)
Data Field	

FIGURE 2-11 TCP Segment and UDP Datagram

number in the segment that acknowledges it would have acknowledgment number X. Later in this book, we will see that the situation actually is more complex.

> The acknowledgment number indicates which segment is being acknowledged.

Test Your Understanding

7. a) Why is TCP complex? b) Why is it important for networking professionals to understand TCP? c) What are TCP messages called?

8. a) Why are sequence numbers good? b) What are 1-bit fields called? c) If someone says that a flag field is set, what does this mean? d) If the ACK bit is set, what other field must have a value? e) What is the purpose of the acknowledgement number field?

User Datagram Protocol (UDP) Datagram Syntax

Applications that cannot use the high functionality in TCP or that do not need this functionality can use the **User Datagram Protocol (UDP)** at the transport layer instead of TCP. UDP does not have openings, closings, or acknowledgments, and so it produces substantially less traffic than TCP. UDP messages are called datagrams. Because of UDP's simple operation, the syntax of the UDP datagram shown in Figure 2-11 is very simple. Besides two port number fields, which we will look at next, there are only two header fields.

- There is a **UDP length** field so that the receiving transport process can know how long the datagram is. The packet in the data field will have variable length, so the UDP datagram will have variable length.
- There also is a **UDP checksum** field that allows the receiver to check for errors in this UDP datagram.[2] If an error is found, however, the UDP datagram is discarded. There is no mechanism for retransmission.

Test Your Understanding

9. a) What are the four fields in a UDP header? b) Describe the third. c) Describe the fourth. d) Is UDP reliable? Explain.

Port Numbers

Both TCP and UDP use port numbers. Port numbers are needed because servers are multitasking computers. They can run multiple application programs at the same time. The server in Figure 2-12 is running three application programs—an SMTP (e-mail) application, an HTTP (webserver) application, and an FTP (file transfer) application.

The server assigns each application a **port number**. Well-known applications, such as HTTP, normally are associated with particular port numbers. For instance, SMTP normally is associated with Port 25, while HTTP normally is associated with Port 80, and FTP normally is associated with Ports 20 and 21. These are called **well-known port numbers**.

When a packet arrives containing a TCP segment (or UDP datagram) arrives at the server, the server's internet layer process removes the TCP segment and passes it to the transport layer program. The transport layer program looks at the destination port number, sees that it is 80, and passes it to the HTTP application. In Chapters 9 and 10, we will learn more about port numbers, including how clients use port numbers.

---

[2] If the UDP checksum field has the value zero, then error checking is not done at all.

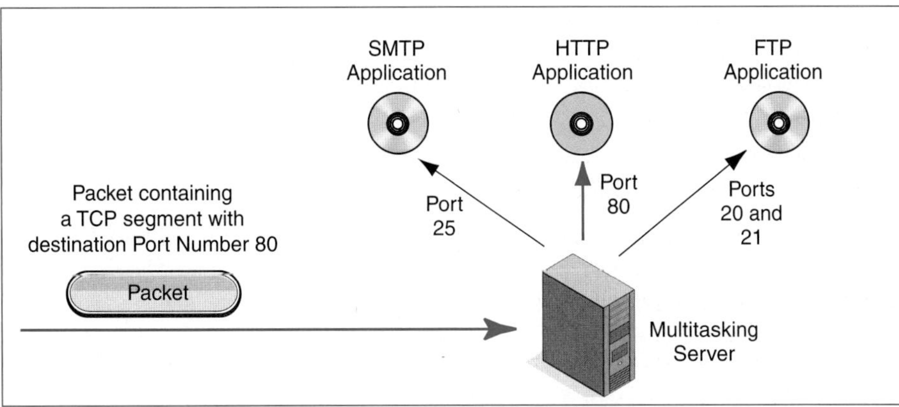

FIGURE 2-12 Port Numbers on Servers

Test Your Understanding

10. a) What message types have port numbers? b) What are a server's port numbers associated with? c) What kind of port numbers do well-known applications usually get? d) What is the well-known port number for HTTP? e) What is the well-known port number for SMTP? f) What are the well-known port numbers for FTP file transfer applications?

HTTP Request and Response Message Syntax

The highest layer is the application layer (Layer 5). Standards at this layer govern how application programs talk to one another. In our examples so far, we have used HTTP in examples most of the time. However, there are many application layer standards—more than there are standards at any other layer. There are application layer standards for e-mail, database queries, and every other application. After network professionals master the network and internetwork standards that this course presents, they spend much of the rest of their careers mastering application standards.

Unfortunately, some students become fixated on examples and lose sight of general principles. At the risk of feeding this fixation, we will look at the syntax for the Hypertext Transfer Protocol. Figure 2-13 illustrates the syntax of HTTP request and response messages.

HTTP REQUEST MESSAGE The HTTP request message is particularly simple. It consists of only two lines of text. Some HTTP request messages have additional lines, but it is rare to see an HTTP request message with more than a handful of lines.

- The first line specifies the GET method (this message requests a file retrieval), the location of the file to be retrieved (/Panko/home.htm), and the version of HTTP used by the sender (HTTP 1.1).
- The second line specifies the host to which this HTTP request message should be sent.

These two lines form the header of the message. The sender is transmitting no data, so there is no data field. Nor is there anything after a data field, so there is no trailer.

HTTP

The application layer is the highest layer.
It has more standards than at any other layer.
HTTP is *not* the only application layer standard; it is one of many.
Many application layer protocols, such as SMTP for e-mail, are much more complex than HTTP.

HTTP Request Message

GET/panko/home.htm HTTP/1.1[CRLF]
Host: voyager.shidler.hawaii.edu

HTTP Response Message

HTTP/1.1 200 OK[CRLF]
Date: Tuesday, 20-MAR-2011 18:32:15 GMT[CRLF]
Server: *name of server software*[CRLF]
MIME-version: 1.0[CRLF]
Content-type: text/plain[CRLF]
[CRLF]
File to be downloaded. A string of bytes that may be text, graphics, sound, video, or other content.

Notes

A relatively old feeling protocol
Fields ended by CRLF, which starts a new line
Based on e-mail (an old protocol) for rapid development

FIGURE 2-13 Hypertext Transfer Protocol Message Syntax

HTTP RESPONSE MESSAGE Figure 2-13 also shows the syntax of a relatively simple HTTP response message that responds to the HTTP request message we have just seen.

Header. First, there is a header. The header is everything that comes before the data field, which is the file being delivered.

- There is again a first line. It begins with HTTP 1.1 to show that it is willing to speak in this version of the standard. The 200 is a code that describes the response. The 200 code states that the message is delivering the requested file. The browser uses the code to know how to react. What about the final OK? The browser ignores it. HTTP is humanly readable, and the "OK" is designed to tell humans, who may not know what the 200 code means to know that everything is alright.
- This and other lines end with [CRLF]. This stands for carriage return/line feed. It means to start a new line.
- After the first line, other fields in the header have a very specific syntax. There is a keyword, a colon, and then a value for the keyword.
  - The date keyword, for instance, is followed by a colon and the date and time the HTTP response message was sent.

- The Server keyword, to give another example, describes the webserver software sending the response.
- The line containing the [CRLF] is a blank line. It indicates the end of the header.

Data Field. Following the header is the file being sent to the browser. This is a long byte stream that constitutes the text document, a photograph, a video clip, or other type of file being delivered.

Trailer. As usual, there is no trailer.

A TEXT PROTOCOL In contrast to the other protocols we have looked at, HTTP is a fairly primitive protocol. It delimits fields with carriage return/line feeds instead of having fields that end at a certain number of bits. Separating the header from the data field with a blank line also seems rather crude. Most application protocols are much more complex than HTTP.

Tim Berners-Lee, who created HTTP, based this new standard on e-mail standards. An e-mail header has a number of keywords (such as To and From) followed by a colon, the value for the keyword, and a new line. E-mail standards were very old, but they got the job done. HTTP also got the job done. In particular, new keywords can be added very easily given the robust way HTTP has of ending a field and starting a new field.

Test Your Understanding

11. a) Is the application layer standard always HTTP? b) Which layer has the most standards? c) At which layer would you find standards for voice over IP? (The answer is not explicitly in this section.) d) Are all application layer standards simple like HTTP? e) In HTTP response headers, what is the syntax of most lines (which are header fields)? f) In HTTP request and response message, how is the end of a field indicated? g) Do HTTP request messages have headers, data fields, and trailers? h) Do HTTP response messages that deliver files have headers, data fields, and trailers?

CONVERTING APPLICATION MESSAGES INTO BITS

Encoding

One function of application layer programs is to convert messages into bits. This conversion is called **encoding**. At the transport layer and lower layers, all messages consist of bits. Original application layer messages, in contrast, may have text, numbers, graphics images, video clips, and other types of information. It is the application layer's job to convert all of these into bits before putting them in the application layer message.

Test Your Understanding

12. a) What is encoding? b) At what layer is encoding done?

Encoding Text as ASCII

To convert text data to binary, applications use the **ASCII code**, whose individual symbols are each seven bits long. Seven bits gives 128 possibilities. This is enough for all keys on the keyboard plus some extra control codes.

Category	Meaning	7-Bit ASCII Code	8th bit in Transmitted Byte
Uppercase Letters	A	1000001	Unused
Lowercase Letters	a	1100001	Unused
Digits (0 through 9)	3	0110011	Unused
Punctuation	Period	0101110	Unused
Punctuation	Space	0100000	Unused
Control Codes	Carriage Return	0001101	Unused
Control Codes	Line Field	0001010	Unused

FIGURE 2-14 Encoding Text as ASCII

Figure 2-14 shows some ASCII codes. It shows that uppercase letters and lowercase letters have different ASCII codes. This is necessary because the receiver may need to know whether a character is an uppercase or lowercase letter. ASCII can also encode the digits from 0 through 9, as well as punctuation and other characters. There are even ASCII control codes that tell the receiver what to do. For example, when we looked at HTTP, we saw carriage returns and line feeds. A carriage return is 0101110, and a line feed is 0100000.

For transmission, the seven bits of each ASCII character are placed in a byte. The eighth bit in the byte is not used today.[3]

Test Your Understanding

13. a) Explain how many bytes it will take to transmit "Hello World!" without the quotation marks. (The correct total is 12.) b) If you go to a search engine, you can easily find converters to represent characters in ASCII. What are the 7-bit ASCII codes for "Hello" without the quotation marks? (Hint: H is 1001000.)

Whole Binary Numbers

Some application data consists of whole numbers. The sending application program represents whole numbers (integers) as whole **binary numbers**. Figure 2-15 shows how this is done.

With binary numbers, counting begins with 0. This is a source of frequent confusion. In counting, add 1 to each binary number to give the next binary number. There are four simple rules for addition in binary.

- If you add 0 and 0, you get 0.
- If you add 0 and 1, you get 1.
- If you add 1 and 1, you get 10 (carry the one).
- If you add 1, 1, and 1, you get 11 (carry the one).

[3] Early systems used the eighth bit in each byte as a "parity bit" to detect errors in transmission. This could detect a change in a single bit in the byte. At today's high transmission speeds, however, transmission errors normally generate multibit errors rather than single-bit errors. Consequently, parity is useless and is ignored.

Counting begins with 0, not 1
So the first three items are 0, 1, and 10 (10 is 2 in binary)

Basic Rules

				1
0	0	1	1	+1
+0	+1	+0	+1	+1
=0	=1	=1	=10	=11

Examples

1000	8
+1	+1
=1001	=9
+1	+1
=1010	=10
+1	+1
=1011	=11
+1	+1
=1100	=12

FIGURE 2-15 Binary Representations of Whole Numbers

These rules are simple to use.

- In binary numbers, 8 is 1000.
- The number 9 adds a 1 to the final 0, giving a 1, so 9 is 1001.
- Adding 1 to the final 1 gives us 10, so 10 is 1010.
- The number 11 adds a 1 to the final 0, giving 1011.
- The number 12 adds a 1 to the final 1. With carries, this gives 1100.[4]

Test Your Understanding

14. a) Does binary counting usually begin at 0 or 1? b) Give the binary representations for 13, 14, 15, 16, and 17 by adding one to successive numbers (12 is 1100).

Encoding Alternatives

Some application data can be expressed as alternatives, such as North, South, East, or West. The application layer process will create a field in the application layer message and represent each alternative as a group of bits. For instance, the four cardinal compass points can be represented by a two-bit field within the application message. North, South, East, and West can be represented as 00, 01, 10, and 11, respectively. (These are the binary numbers for 0, 1, 2, and 3). There is no order to the alternatives, so any choice can be represented by any pair of bits.

We just saw that having four alternatives requires a two-bit field. More generally, if a field has N bits, it can represent 2^N alternatives. We have just seen that a two-bit field can represent 2^2 alternatives, or 4. What if you need to represent 3 alternatives? One bit won't do it, because 2^1 is 2, and we need 3. We will have to use a 2-bit field, and one of the alternatives will go unused.

---

[4] In Chapter 1, we saw how to use the Microsoft Windows Calculator in scientific mode to convert between decimal, binary, and hexadecimal. You can also use it to check your binary calculations (obviously not on tests). First click on the Bin button to put the calculator in binary mode. Then use the calculator normally. For example, to add two binary numbers, click on Bin, enter the first binary number, hit the plus button, type the second binary number, and click on the equal sign to see the total.

If a field has N bits, it can represent 2^N alternatives.

Figure 2-16 illustrates how alternatives encoding is done for 1-, 2-, 4-, and 8-bit fields. It shows that with one bit, you can encode yes or no, male or female, or any other dichotomy. Two bits, as we just saw, are good for the four cardinal compass points. With four bits, you can have 16 alternatives. You need 4 bits to represent the top 10 security threats, because 3 bits will only encode eight alternatives. This will "waste" six alternatives. With 8 bits, you can represent 256 alternatives.

The 2^N rule is not only used at the application layer. In many layer messages, fields represent alternatives. A one-octet field has 8 bits, so it can represent 2^8 possible alternatives (256). You should memorize the number of alternatives that can be represented by 4, 8, and 16 bits, because these are common field sizes.

You should memorize the number of alternatives for four and eight bits. Each added bit doubles the number of alternatives, while each bit subtracted cuts the number of alternatives in half. So if 8 bits can represent 256 alternatives, 7 bits can represent 128 alternatives, while 9 bits can represent 512 alternatives.

Test Your Understanding

15. a) If a field is N bits long, how many alternatives can it represent? b) How many alternatives can you represent with a 4-bit field? c) For each bit you add to an alternatives field, how many additional alternatives can you represent? d) How many alternatives can you represent with a 10-bit field? (With 8 bits, you can represent 256 alternatives) e) If you need to represent 128 alternatives in a field, how many bits long must the field be? f) If you need to represent 18 alternatives in a field, how many bits long must the field be? g) Come up with three examples of things that can be encoded with 3 bits.

Bits in Field	Number of Alternatives that can be Encoded	Possible Bit Sequences	Examples
1	$2^1 = 2$	0, 1	Yes or No, Male or Female, etc.
2	$2^2 = 4$	00, 01,10, 11	North, South, East, West; Red, Green, Blue, Black
4	$2^4 = 16$	0000, 0001, 0010, . . .	Top 10 security threats. Three bits would only give 8 alternatives. (With 4 bits, 6 values go unused.)
8	$2^8 = 256$	00000000, 00000001, . . .	One byte per color gives 256 possible color levels.
16	$2^{16} = 65{,}536$	0000000000000000, 0000000000000001, . . .	Two bytes per color gives 65,536 color levels.
32	$2^{32} = 4{,}294{,}967{,}296$	0000000000000000 0000000000000000, etc.	Number of Internet Protocol Version 4 addresses
If a field has N bits, it can represent 2^N alternatives.			

FIGURE 2-16 Binary Encoding for a Number of Alternatives

VERTICAL COMMUNICATION ON HOSTS

So far, we have talked about what happens at individual layers. For instance, the transport process on the sending host sends TCP segments or UDP datagrams to the transport process on the receiving host.

Obviously, however, there is no direct connection between the two hosts at the transport layer. Barring software telepathy, all communication must somehow travel through the physical layer.

In Chapter 1, we saw that Layer 3 packets are carried in the data fields of Layer 2 frames in single networks. Networking people say that the packet is encapsulated (placed) in the data field of the frame. In general, **encapsulation** is placing a message in the data field of another message.

Encapsulation is placing a message in the data field of another message.

Figure 2-17 shows that encapsulation actually is a general process.

- In the figure, the source host's application layer process sends an HTTP message to the application layer process on the destination host. The source host's application

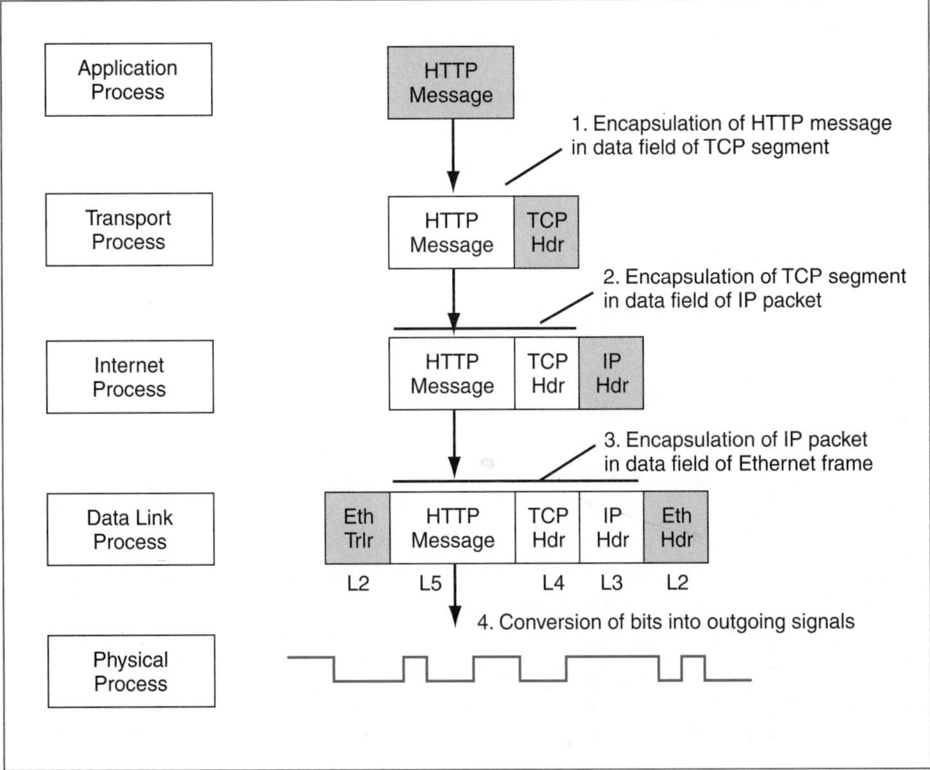

FIGURE 2-17 Layered Communication on the Source Host

process cannot deliver the HTTP message, so it passes the HTTP message down to the transport layer process on the source host.

- The transport layer process encapsulates the HTTP message in the data field of a TCP segment. The transport layer then passes the TCP segment down to the internet layer process.

You can see in the figure that encapsulation and passing down occur at the internet layer and then at the data link layer. The whole process is like a Russian nesting doll.

When the physical layer process receives the frame from the data link layer process, however, it does not do encapsulation. It merely converts the bits of the frame into signals and transmits its signal out the physical link connecting it to a switch, router, or host.

The purpose of this section has been to give you a light understanding of how layered communication works on the source host. However, there are many other aspects to vertical communication on source hosts, destination hosts, routers, and switches.

Test Your Understanding

16. a) What is encapsulation? b) Why is encapsulation necessary for there to be communication between processes operating at the same layer but on different hosts, routers, or switches? c) After the internet layer process in Figure 2-17 receives the TCP segment from the transport layer process, what two things does it do? d) After the data link layer process in Figure 2-17 receives the IP packet from the internet layer process, what two things does it do? e) After the physical layer process receives a frame from the data link layer process, what does the physical layer process do? f) If encapsulation occurs on the source host, what analogous process will occur on the destination host? (The answer is not in the text.)

MAJOR STANDARDS ARCHITECTURES

When you build a house, you do not design and build a bedroom, then move on to the next room. If you designed and built rooms one at a time, without having an overall plan for the house, you might end up having to go through the bathroom whenever you passed from the dining room to the kitchen. Instead, an architect begins with an overall plan for a house. This plan is called the architecture. The architecture specifies the functions of individual rooms (bedrooms, kitchens, etc.), how many rooms of each type are required, and how the rooms will fit together. Only after the architecture is finished does the architect begin to design individual rooms.

Network standards also need architectures. Network standards architectures break the standards functionality needed for communication into layers and define the functions of each layer. Only after they have done that do they develop individual standards. For instance, the TCP/IP standards architecture shown in Figure 2-18 has four layers. This architecture was created in the late 1970s. Since then, the Internet Engineering Task force has been adding individual standards, beginning with core standards such as IP, TCP, and UDP. It will continue to create individual standards for many years to come.

Broad Purpose	TCP/IP	OSI	Hybrid TCP/IP-OSI
Applications	Application	Application (Layer 7)	Application (Layer 5)
		Presentation (Layer 6)	
		Session (Layer 5)	
Internetworking	Transport	Transport (Layer 4)	TCP/IP Transport Layer (Layer 4)
	Internet	Network (Layer 3)	TCP/IP Internet Layer (Layer 3)
Communication within a single LAN or WAN	Use OSI	Data Link (Layer 2)	Data Link (OSI) Layer (Layer 2)
	Standards Here	Physical (Layer 1)	Physical OSI Layer (Layer 1)

Notes:

The Hybrid TCP/IP-OSI Architecture governs the Internet and dominates internal corporate networks.

OSI standards dominate the physical and data link layers (which govern communication within individual networks) almost exclusively.

TCP/IP dominates the internet and transport layer in internetworking and governs 80 to 90 percent of all corporate traffic above the data link layer.

FIGURE 2-18 The Hybrid TCP/IP–OSI Architecture

Network standards architectures break the standards functionality needed for communication into layers and define the functions of each layer. Only after they have done that do they develop individual standards.

Generally, all standards must be taken from the same architecture. For instance, electrical standards architectures in the different countries have different voltages, amperages, alternating current cycles per second, plug shapes, and socket shapes. If you took a device built for the U.S. architecture, you could not take it to another country and simply change the plug to fit that country's sockets. Actually you *could* do that physically, but if you plugged the modified device into the wall jack, you should not expect the device to work. You would need a converter that converts between all elements in the U.S. electrical standards architecture and all elements in the foreign electrical architecture.

Although there are several major network standards architectures, two of them dominate actual corporate use: OSI and TCP/IP. Figure 2-18 illustrates layering in both. As you can see, they are quite different. Although it generally is impossible to mix standards for different architectures, this kind of mixing is possible with TCP/IP and OSI. It is possible to use a hybrid TCP/IP–OSI architecture that uses TCP/IP standards at some layers and OSI standards at other layers. This hybrid standards architecture is the dominant standards architecture in real corporations.

TCP/IP and OSI are often portrayed as competitors. In fact, their standards agencies mostly specialize on different layers. When both are developing standards for a single layer, such as voice over IP standards at the application layer, their standards agencies often work together.

Test Your Understanding

17. a) What does a network standards architecture do? b) In what order are standards and standards architectures developed? c) What are the two dominant network standards architectures? d) What is the dominant network standards architecture in most real firms today? e) Are the two network standards architectures competitors?

The OSI Network Standards Architecture

OSI is the "Reference Model of Open Systems Interconnection." *Reference model* is another name for *architecture*. An open system is one that is open to communicating with the rest of the world. In any case, OSI is rarely spelled out, which is merciful.

STANDARDS AGENCIES: ISO AND ITU-T Standards architectures are managed by organizations called **standards agencies**. Figure 2-19 shows that OSI has two standards agencies.

- One is the **International Organization for Standardization (ISO)**, which generally is a strong standards organization for manufacturing, including computer manufacturing. By the way, do not confuse OSI and ISO. OSI is an architecture; ISO is a standards agency.

	OSI	TCP/IP
Standards Agency or Agencies	ISO (International Organization for Standardization) ITU-T (International Telecommunications Union–Telecommunications Standards Sector)	IETF (Internet Engineering Task Force)
Dominance	Nearly 100% at physical and data link layers	90% at the internet and transport layers
Documents Are Called	Various	Mostly RFCs (requests for comments)
Notes: Do not confuse OSI (the architecture) with ISO (the organization). The acronyms for ISO and ITU-T do not match their names, but these are the official names and acronyms.		

FIGURE 2-19 OSI and TCP/IP

Layer Number	OSI Name	Purpose	Use
7	Application	Governs remaining application-specific matters.	Some OSI applications are used
6	Presentation	Designed to handle data formatting differences and data compression and encryption. In practice, a category for general file format standards used in multiple applications.	Rarely used as a layer. However, many file format standards are assigned to this layer.
5	Session	Initiates and maintains a connection between application programs on different computers.	Rarely used
4	Transport	Generally equivalent to the TCP/IP transport layer. However, OSI transport layer standards are not compatible with TCP/IP transport layer standards.	Rarely used
3	Network	Generally equivalent to the TCP/IP internet layer. However, OSI network layer standards are not compatible with TCP/IP internet layer standards.	Rarely used
2	Data Link	End-to-end transmission in a single switched network. Frame organization. Switch operation.	Nearly 100% dominant
1	Physical	Physical connections between adjacent devices.	Nearly 100% dominant

FIGURE 2-20 OSI Layers

- The other is the **International Telecommunications Union–Telecommunications Standards Sector (ITU–T).**[5] Part of the United Nations, the ITU–T oversees international telecommunications standards.

Although ISO or the ITU–T must *ratify* all OSI standards, other organizations frequently *create* standards for inclusion in OSI. For instance, we will see in Chapter 6 that the IEEE creates Ethernet standards. These standards are not official until the ITU-T or ISO ratifies them, although this has always been a mere formality.

OSI's DOMINANCE AT LOWER LAYERS (PHYSICAL AND DATA LINK) Although OSI is a seven-layer standards architecture (see Figure 2-20), standards from Layers 3 through 6 are rarely used.

However, at the two lowest layers—the physical and data link layers—corporations use OSI standards almost universally in their networks. These two layers govern transmission within a switched or wireless network. Almost all single networks—both LANs and WANs—follow OSI standards at the physical and data link layers, regardless of what upper-layer standards they use. OSI standards are almost 100 percent dominant at the bottom two layers.

[5] No, the names and acronyms do not match for ISO and ITU–T, but these are the official names and acronyms for these two organizations.

Almost all switched and wireless networks—both LANs and WANs—follow OSI standards at the physical and data link layers, regardless of what upper-layer standards they use.

Other standards agencies, recognizing the dominance of OSI at the physical and data link layers, simply specify the use of OSI standards at these layers. They then create standards only for internetworking and applications.

OSI NETWORK AND TRANSPORT LAYERS The network layer functionality of OSI corresponds closely to the internet layer functionality of TCP/IP that we saw earlier in this chapter. The transport layer functionality of OSI, in turn, is very similar to the transport layer functionality of TCP/IP. However, while *functionality* may be similar between OSI and TCP at these layers, actual OSI and TCP/IP *standards* at these layers are completely incompatible. More importantly, OSI standards are rarely used at the network or transport layers by real organizations.[6]

OSI SESSION LAYER The **OSI session layer (OSI Layer 5)** initiates and maintains a connection between application programs on different computers. For instance, suppose that a single transaction requires a number of messages. If there is a connection break, the transmission can begin at the last session layer rollback point instead of restarting at the beginning. For example, if communication fails during a database transaction, the entire transaction does not have to be done over—only the work since the last rollback point. Unfortunately, while this is good for database applications and a few other applications, few applications benefit from the overhead added by the OSI session layer.

OSI PRESENTATION LAYER The **OSI presentation layer (OSI Layer 6)** is *designed* to handle data-formatting differences between the two computers. For example, most computers format character data (letters, digits, and punctuation signs) in the ASCII code. In contrast, IBM mainframes format them in the EBCDIC code. The presentation layer can handle this and other format translations.

The OSI presentation layer also is *designed* to be used for compression and data encryption for application data.

In practice, the presentation layer is rarely used for either data format conversion or compression and encryption. Rather, the presentation layer has become a category for general application file format standards used in multiple applications, including MP3, JPEG, and many other general OSI file format standards.

OSI APPLICATION LAYER The **OSI application layer (OSI Layer 7)** governs remaining application-specific matters that are now covered by the session and presentation layers. The OSI application layer, freed from session and presentation matters, focuses on concerns specific to the application in use.

---

[6] Although OSI physical and data link layer standards are dominant, and while many OSI application layer standards are used, almost no systems implement OSI standards at the network, transport, session, or presentation layers. Then why, you may ask, do you need to be able to describe these layers? The answer is that you need to get a job. In a very large percentage of all job interviews, an interviewer noting that you have taken a networking course, asks you to describe the OSI layers. I kid you not.

Test Your Understanding

18. a) What standards agencies are responsible for the OSI standards architecture? Just give the acronyms. b) At which layers do OSI standards dominate usage? c) Name and describe the functions of OSI Layer 5. d) Name and describe the intended use of OSI Layer 6. e) How is the OSI presentation layer actually used? f) Beginning with the physical layer (Layer 1), give the name and number of the OSI layers.

The TCP/IP Network Standards Architecture

The **TCP/IP** architecture is mandatory on the Internet at the internet and transport layers. TCP/IP is also widely used at these layers by companies for their internal corporate internets.

The TCP/IP architecture is named after two of its standards, TCP and IP, which we have already seen briefly. However, TCP/IP also has many other standards, including the UDP standard we have already seen. This makes the name TCP/IP rather misleading. Another confusing point about names is that TCP/IP is the *standards architecture,* while TCP and IP are individual *standards* within the architecture.

Note that TCP/IP is the standards architecture while TCP and IP are individual standards within the architecture.

THE INTERNET ENGINEERING TASK FORCE (IETF) TCP/IP's standards agency is the **Internet Engineering Task Force (IETF)**. Traditionally, the IETF has been viewed as being in competition with ISO and ITU–T for creating standards. However, as noted earlier in this chapter, the IETF and these other organizations have begun to cooperate in standards development. For instance, the IETF is working closely with ITU–T on VoIP transmission standards.

Having evolved from the Network Working Group described in the previous chapter, the IETF historically has been rather informal. Its committees traditionally focus on consensus rather than on voting, and technical expertise is the main source of most power within the organization. Although corporate participation has somewhat "tamed" the IETF, it remains a fascinating organization, although its members might argue that "organization" is too strong a word.

A great deal of the IETF's success is due to the fact that the IETF typically produces simple standards and then adds to their complexity over time. In fact, IETF standards often have the word *simple* in their name—for instance, the Simple Mail Transfer Protocol. Consequently, the IETF develops standards quickly. In addition, vendors can develop TCP/IP-based products quickly and inexpensively because of this speed of development and simplicity. "Inexpensive and fast to market" is almost always a good recipe for success. In contrast, OSI standards often take a very long time to be developed and often are so bloated with functionality that they are uneconomical to implement.

Note that the success of TCP/IP is *not* primarily due to its use on the Internet. Many corporations had already shifted many of their networks to TCP/IP before the Internet became a dominant force in the 1990s.

REQUESTS FOR COMMENTS (RFCS) As we saw in Chapter 1, most documents produced by the IETF have the rather misleading name **requests for comment (RFCs)**. Every few years, the IETF publishes a list of which RFCs are **Official Internet Protocol Standards**. Each list of standards adds some RFCs to the list and drops previously listed standards.

DOMINANCE AT THE INTERNETWORKING LAYERS (INTERNET AND TRANSPORT) As noted earlier, physical and data link layer standards govern the transmission of data within a single *switched or wireless network*. We saw earlier that OSI standards are completely dominant at these layers.

TCP/IP internet and transport layer standards, in turn, govern transmission across an entire internet, ensuring that any two host computers can communicate. TCP/IP application standards ensure that the two application programs on the two hosts can communicate as well.

TCP/IP is dominant in the internetworking layers (internet and transport). However, it is less dominant at these layers than OSI standards are dominant at the physical and data link layers. In most organizations, TCP/IP standards are used for about 90 percent of internet and transport layer traffic, and this dominance is growing. In a few years, the use of other architectures at the internet and transport layers in new products will be negligible.

TCP/IP is dominant in corporate networking at the internet and transport layers, although less dominant than OSI is at the physical and data link layers. In most organizations, TCP/IP standards are used for 90 percent of internet and transport layer traffic.

Test Your Understanding

19. a) Which of the following is an architecture: TCP/IP, TCP, or IP? b) Which of the following are standards: TCP/IP, TCP, or IP? c) What is the standards agency for TCP/IP? d) Why have this agency's standards been so successful? e) What are most of this agency's documents called? f) At which layers is TCP/IP dominant? g) How dominant is TCP/IP today at these layers compared with OSI's dominance at the physical and data link layers?

The Application Layer

OSI is completely dominant at the physical and data link layers, and TCP/IP is very dominant at the internet and transport layers. What about the application layer? The answer here is complex.[7] Overall, it seems best to say that no standards agency or

---

[7] Many application protocols come from the IETF. These include such popular standards as e-mail protocols (SMTP, POP, IMAP, etc.), the FTP standards, and the Simple Network Management Protocol (SNMP) in network management.

Some application standards come directly from OSI. This is particularly true of graphics file format standards. However, many OSI standards were too complex for widespread use. The IETF then produced simpler versions of these OSI standards. A good example is the Lightweight Directory Access Protocol (LDAP) for access to directory servers. LDAP evolved (devolved?) from the OSI directory access protocol standard.

Other standards agencies also produce application layer standards. HTTP and HTML standards come from the World Wide Web Consortium (W3C), although the IETF is producing some WWW standards. Most confusingly, incompatible Service Oriented Architecture Web services standards are being produced by several competing standards agencies.

architecture dominates at the application layer, although the IETF is particularly strong, especially for popular standards such as e-mail and file transfer.

Although many applications do not come from the IETF, they almost all run over TCP/IP standards at the internet and transport layers. This is true even for standards created by ISO and ITU–T.

At the application layer, as noted earlier, there is growing cooperation between ISO, the ITU–T, and the IETF. In voice over IP, the ITU–T and the IETF have harmonized several key standards. ISO and the IETF, in turn, have been cooperating in file format standards.

Test Your Understanding

20. a) Is any standards architecture dominant at the application layer? b) Do almost all applications, regardless of what standards architecture they come from, run over TCP/IP standards at the internet and transport layers?

TCP/IP *and* OSI: The Hybrid TCP/IP–OSI Standards Architecture

We have noted neither TCP/IP nor OSI dominates at all layers. Consequently, as discussed earlier, the most common standards pattern in organizations is to use OSI standards at the physical and data link layers, TCP/IP standards at the internet and transport layers, and application standards from several sources. This is very important for you to keep in mind because this **hybrid TCP/IP–OSI standards architecture** in Figure 2-18 will form the basis for most of this book.

> The most common standards pattern in organizations is to use OSI standards at the physical and data link layers, TCP/IP standards at the internet and transport layers, and application standards from several sources. This is the hybrid TCP/IP–OSI standards architecture.

Test Your Understanding

21. a) What layers of the hybrid TCP/IP–OSI standards architecture use OSI standards? b) What layers use TCP/IP standards? c) Do wireless LAN standards come from OSI or TCP/IP? Explain. (The answer is not explicitly in this section.) d) Do switched WAN standards come from OSI or TCP/IP? Explain. (Again, the answer is not explicitly in this section.)

A Multiprotocol World at Higher Layers

At the same time, quite a few networking products (especially legacy products) in organizations follow other architectures shown in Figure 2-21. Real corporations live and will continue to live for some time in a multiprotocol world in which network administrators have to deal with a complex mix of products following different architectures above the data link layer. In this book, we focus on OSI and TCP/IP because they are by far the most important and are becoming ever more so. However, in a typical organization, 10 percent to 20 percent of all Layer 3 and Layer 4 traffic still uses protocols from other standards architectures.

IPX/SPX The most widely used non-TCP/IP standards architecture found at upper layers in LANs is the **IPX/SPX architecture**. Older Novell NetWare file servers required

Architecture	Typical Usage at the Internet and Transport Layers
IPX/SPX	Older Novel Netware file servers.
Systems Network Architecture (SNA)	Older IBM mainframes.
AppleTalk	Older Apple products.

FIGURE 2-21 Other Major Standards Architectures

standards from this architecture at the internet and transport layers. Many NetWare users are switching to TCP/IP, which is the default installation in newer Novell Netware file servers. However, some firms that are well acquainted with Novell Netware IPX/SPX standards configure even newer Netware file servers with IPX/SPX at the internet and transport layers.

SYSTEMS NETWORK ARCHITECTURE (SNA) IBM mainframe computers traditionally used the Systems Network Architecture (SNA) standards architecture, which actually predates OSI and TCP/IP. Most firms, however, are transitioning or have already transitioned their mainframe communications from SNA to TCP/IP.

APPLETALK Apple desktop and notebook computers were originally designed to use Apple's proprietary **AppleTalk architecture** when they talk to other Apple products. AppleTalk protocols are rare in corporations today, and most recent Apple products speak TCP/IP.

Test Your Understanding

22. a) Under what circumstances might you encounter IPX/SPX standards? b) SNA standards? c) AppleTalk standards?

CONCLUSION

Synopsis

In this chapter, we looked broadly at standards. Most of this book (and the networking profession in general) will focus on standards, which are also called protocols. Standards govern message exchanges. More specifically, they place constraints on message semantics (meaning), message syntax (format), and message order.

Standards are connection-oriented or connectionless. In connection-oriented protocols, there is a distinct opening before content messages are sent and a distinct closing afterward. There also are sequence numbers, which allow fragmentation and are used in supervisory messages (such as acknowledgments) to refer to specific messages. In connectionless protocols, there are no such openings and closings. Connectionless protocols are simpler than connection-oriented protocols, but they lose the advantages of sequence numbers.

In turn, reliable protocols do error correction, while unreliable protocols do not. Although unreliable protocols may do error detection without error correction, this does not make them reliable. In general, standards below the transport layer are unreliable

Layer	Protocol	Connection-Oriented or Connectionless?	Reliable or Unreliable?
5 (Application)	HTTP	Connectionless	Unreliable
4 (Transport)	TCP	Connection-oriented	Reliable
4 (Transport)	UDP	Connectionless	Unreliable
3 (Internet)	IP	Connectionless	Unreliable
2 (Data Link)	Ethernet	Connectionless	Unreliable

FIGURE 2-22 Characteristics of Protocols Discussed in This Chapter

in order to reduce costs. The transport standard usually is reliable; this allows error correction processes on just the two hosts to correct errors at the transport layer and at lower layers, giving the application clean data. Figure 2-22 compares the main protocols we have seen in this chapter in terms of connection orientation and reliability.

To discuss message ordering in more detail, we looked at HTTP and TCP. Message ordering in HTTP is trivial. The browser must initiate the communication by sending an HTTP request message; afterward the webserver program may transmit. TCP, in contrast, has complex message ordering. A three-step handshake is needed to open a connection, and four messages are needed to close a connection. Correctly received TCP messages (called segments) are always acknowledged by the receiver. If the sender does not receive an acknowledgment promptly, it retransmits the unacknowledged segment. This gives reliability.

To discuss message syntax in more detail, we looked briefly at the syntax of Ethernet frames, IP packets, TCP segments, UDP datagrams, and HTTP request and response messages. We saw that they represent syntax in three different ways. We will be looking at the syntax of many messages in this course, so you should be familiar with all methods for representing syntax. In the discussion, we saw that octet is another name for byte. We also saw that application programs on multitasking servers are represented by well-known port numbers.

The application layer must convert text, graphics, video, and other application layer content into bits (1s and 0s). In this chapter, we looked at how application programs encode a number of alternatives, ASCII text, and whole numbers is done.

We looked at how layer processes work together on the source host. After each layer creates its message, it immediately passes the message down to the next-lower-layer process. The data link, internet, and transport processes take every message they are given and encapsulate it in a message suitable for that layer.

Individual standards are not created in isolation. Individual standards are created within broad plans called standards architectures. Only after the broad architecture is designed by specifying the functionality of each layer and ensuring that the architecture overall allows interoperability are individual standards developed.

The TCP/IP architecture has the Internet Engineering Task Force (IETF) as its standards agency. It has a four-layer architecture. The bottom layer basically says, "Use OSI standards for single switched and wireless networks."

The OSI architecture has ISO and ITU–T as its two standards agency. OSI has a seven-layer architecture. Two of its layers are unusual. The session layer manages a connection between application programs; if there is an interruption, only communication since the last "rollback point" needs to be repeated. The presentation layer was designed as a layer for converting between application formatting implementations in different operating systems and to provide encryption and compression. In fact, it has become a catch-all category for application standards, such as jpeg.

Most real corporations today use a hybrid TCP/IP–OSI standards architecture that combines layers from TCP/IP and OSI. This architecture, which Figure 2-18 illustrates, will be the focus of this book. At the physical and data link layers, which govern transmission through a single switched or wireless network, OSI standards are almost always used.

At the internet and transport layers, TCP/IP standards are used for about 90 percent of all communication. The remaining transmissions use internet and transport layer standards from IPX/SPX, SNA, AppleTalk, and other minor standards architectures. Application layer standards come from many standards architectures.

END-OF-CHAPTER QUESTIONS

Thought Questions

1. How do you think TCP would handle the problem if an acknowledgment were lost, so that the sender retransmitted the unacknowledged TCP segment, therefore causing the receiving transport process to receive the same segment twice?
2. a) In Figure 2-12, what will be the value in the destination port number field if a packet arrives for the e-mail application? b) When the HTTP program sends an HTTP response message to a client PC, in what field of what message will it place the value 80?
3. Binary for 50 is 10010. Give the binary for 51, 52, and 53.
4. You need to represent 1,026 different city names. How many bits will this take if you give each city a different binary number?

Brainteaser Questions

If you can get these, that's impressive; but it's not expected.

1. How can you make a connectionless protocol reliable? (Try to answer this one, but you may not be able to do so.)
2. Spacecraft exploring the outer planets need reliable data transmission. However, the acknowledgments would take hours to arrive. This makes an ACK-based reliability approach unattractive. Can you think of another way to provide reliable data transmission to spacecraft? (Try to answer this one, but you may not be able to do so.)

Perspective Questions

1. What was the most surprising thing you learned in this chapter?
2. What was the most difficult material for you in this chapter?

CHAPTER 2a

Hands-On: Wireshark Packet Capture

LEARNING OBJECTIVES

By the end of this chapter, you should be able to:

- Use the Wireshark packet capture program at a novice level.
- Capture packets in real time.
- Analyze the packets at a novice level.

INTRODUCTION

A good way to practice what you have learned in this chapter is to look at individual packets. Packet capture programs record packets going into and out of your computer. If you capture a brief webserver interaction, you can look at header fields, TCP three-step connection starts, and other information. There are several good packet capture programs. We will look at Wireshark, which is simple to use, popular, and free to download. (At least at the time of this writing.)

GETTING WIRESHARK

To get Wireshark, go to wireshark.org. Do *not* go to wireshark.com. Follow the instructions and download the program on your computer.

USING WIRESHARK

Getting Started

After installation, open the Wireshark program. You will see the opening screen. It will look like the screen in Figure 2a-1. There will be controls at the top with a blank area below them. You will soon fill this area with your packet capture.

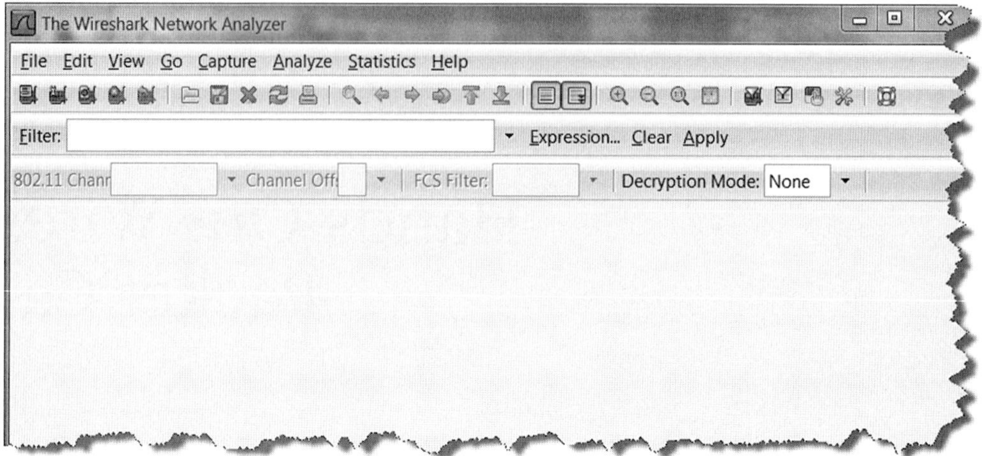

FIGURE 2a-1 Initial Wireshark Screen

Starting a Packet Capture

To start a packet capture, click on the Go menu item. Then, when the Wireshark: Capture Interfaces dialog appears, as Figure 2a-2 illustrates, select a network interface and click on Start.

Getting Data

Your browser should already be open. Switch to your browser and enter a URL. (In this example, the author went to Wikipedia.org.) This creates a flurry of packets between you and the host specified in the URL. These appear on the window below the controls, as shown in Figure 2a-3.

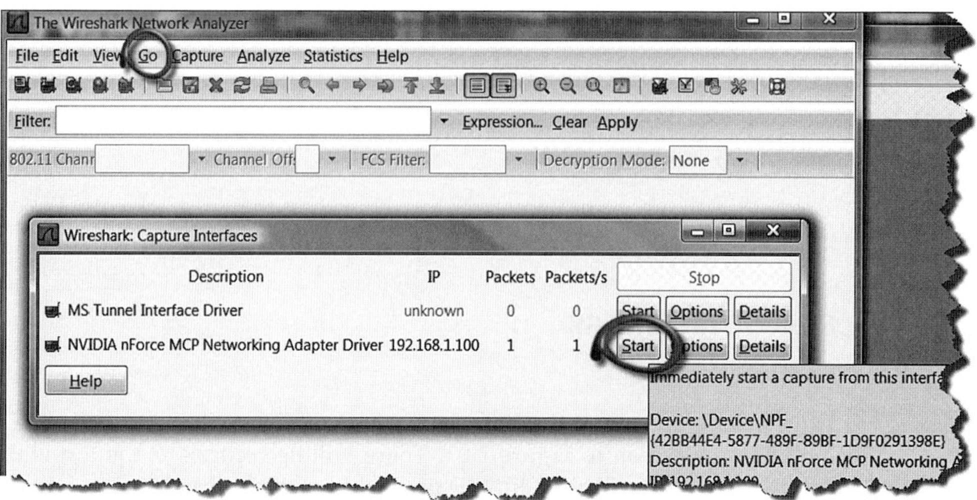

FIGURE 2a-2 Starting a Packet Capture in Wireshark

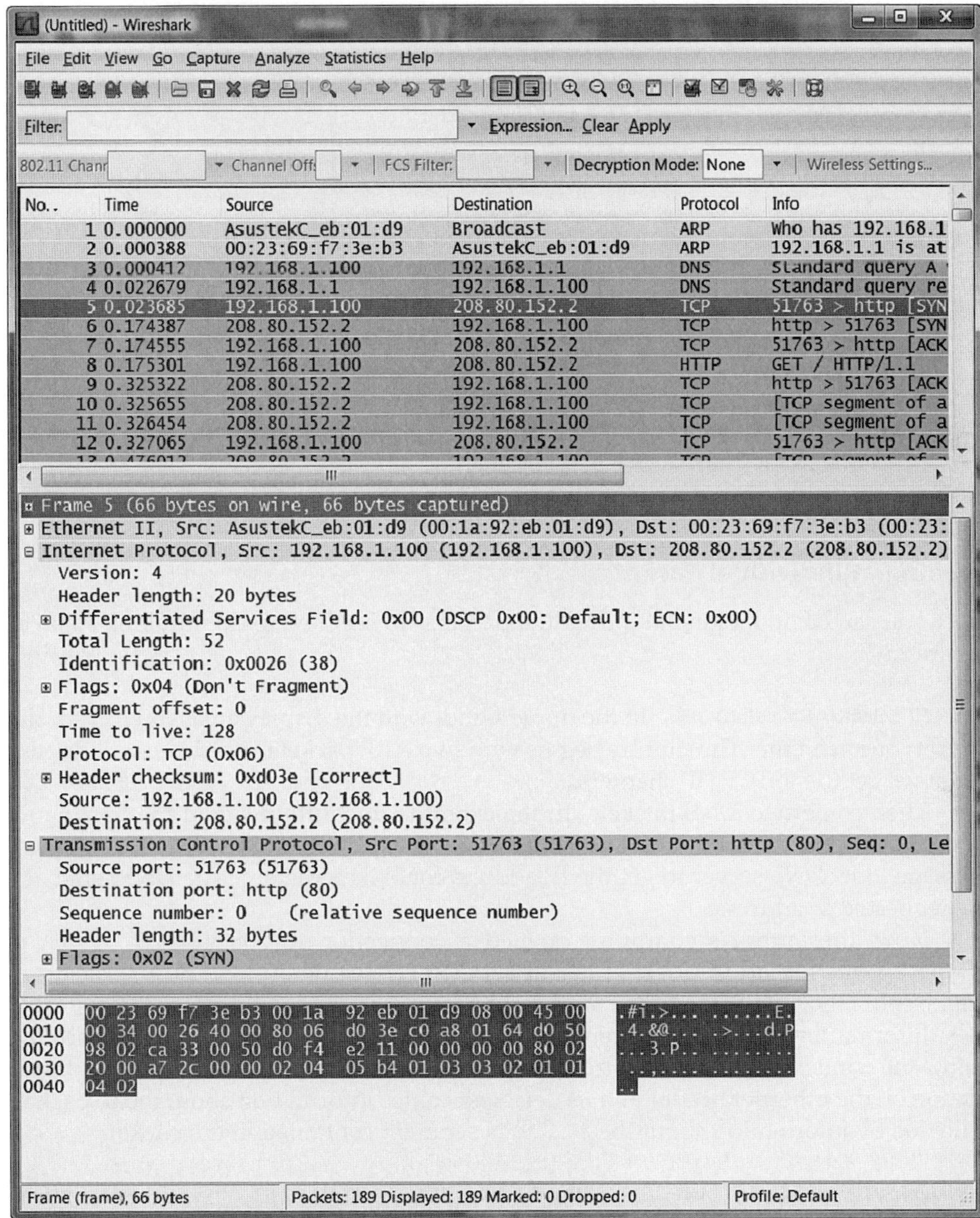

FIGURE 2a-3 Collecting Data

Stopping Data Collection

To stop the data collection, click on the Capture menu item, as Figure 2a-4 shows. When the drop-down menu appears, select *Stop*. You now have a packet stream to analyze.

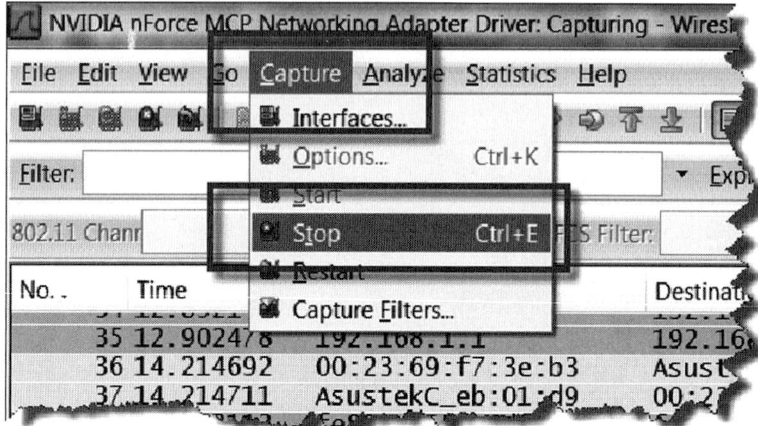

FIGURE 2a-4 Stopping the Data Collection

Looking at Individual Packets

Now you can begin looking at individual packets. To see how to do this, look again at Figure 2a-3.

PACKET SUMMARY WINDOW In the upper window in the display area, you can see the packets one at a time. The capture begins with two ARP packets, which we will discuss when we get to the TCP/IP chapters.

Then come two DNS packets. In the example, the author typed the host name Wikipedia.org in the URL. The author's computer (192.168.1.100) sent a DNS request message to its DNS server to get the IP address for Wikipedia.org. The DNS sent back the requested IP address.

Now, the author's computer opened a connection to 208.80.152.2, which is Wireshark.org's IP address.[1] It first sent a TCP SYN segment to 208.80.152.2. This is Frame 5. In Figure 2a-3, the frame has been selected.

Information about the contents of this particular frame is shown in a window below the window showing each frame on a single line. First, the window shows information on the Ethernet header and trailer. Next comes information about the IP packet, followed by information about the TCP SYN segment contained in the packet.

WINDOW WITH DETAILED INFORMATION ON THE SELECTED PACKET The Ethernet information has been minimized. Only the source and destination MAC addresses are shown. However, information about the IP packet has been maximized. You can see the values of the individual fields in the selected packet. For example, note that the Time to Live field in this packet had the value 128. In addition, the protocol field value indicates that the data field contains a TCP segment.

[1] If you try this, you may get a different IP address. Many firms have multiple physical webservers that they associate with a host name. A DNS response message returns the IP address of one of these physical servers.

The TCP segment information is also expanded, although only the first few fields are shown in the window. Note that the destination port is 80, indicating that the author was contacting the Wireshark.org webserver. Note also that the Flag Fields information says that the SYN bit is set, as one would expect.

To make life easier for you, Wireshark does as much translation as possible. For example, it interprets the information in the protocol field as indicating that there is a TCP segment in the packet's data field. It also indicates that Port 80 is HTTP.

The information on sequence number is highly simplified compared to the discussion in Chapter 2. This is the first TCP segment being sent. It is given the value 0 rather than its complex real value.

HEX WINDOW The lowest window shows the contents of the packet in hexadecimal (Base 16) format. Hex is difficult for new analysts to interpret, but it is very compact compared to the information in the middle window. Experienced packet analysts quickly learn the positions of important fields and learn to read the hex symbols for that field.

Options

Figure 2a-5 shows that Wireshark capture options allow you to control what packets are captured. If you are connected to multiple external servers simultaneously, this can allow you to capture only packets for a particular connection.

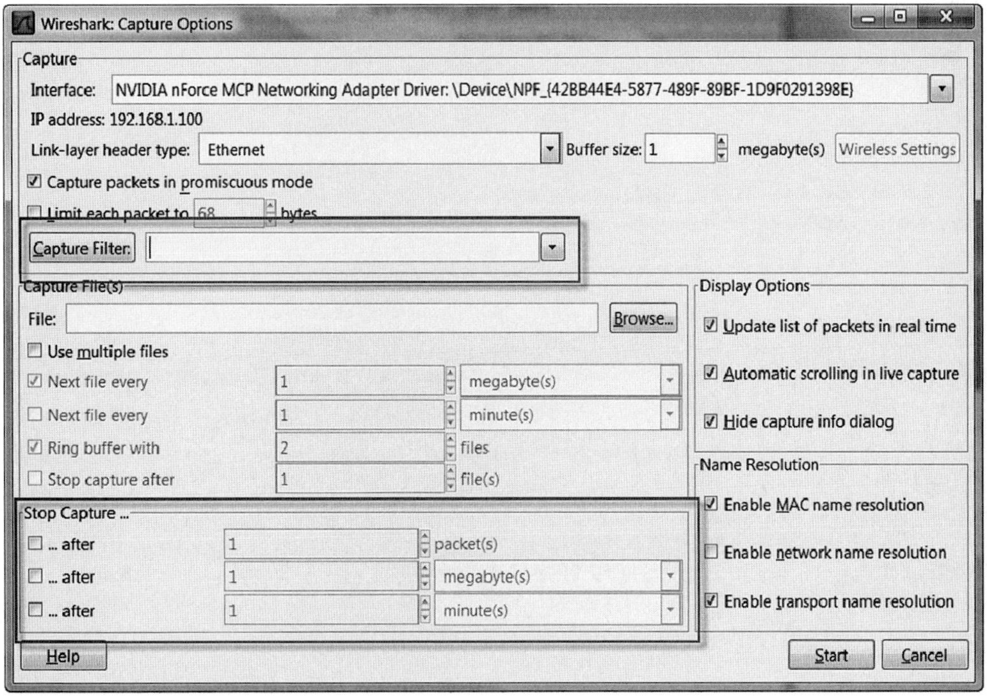

FIGURE 2a-5 Wireshark Options

EXERCISES

1. Do the following:
   - Download Wireshark.
   - Start Wireshark.
   - Turn on Wireshark capture.
   - Type a URL in your browser window (not Wikipedia.org).
   - After a few seconds, stop the capture.
   - Answer the following questions:

     1a. What URL did you use? What was the IP address of the webserver?

     1b. Find the frame in which your PC sent the SYN packet. List the source and destination IP address, the source and destination port numbers, and the header checksum.

     1c. Select the SYN/ACK packet. List the source and destination IP address, the source and destination port numbers, and the header checksum.

     1d. Select the packet that acknowledges the SYN/ACK segment. List the source and destination IP address, the source and destination port numbers, and the header checksum.

2. Change the options so that only packets you send are recorded. Do a capture. Click on the window containing Wireshark and hit Alt-Enter. This captures the window to your clipboard. Paste it into your homework.

CHAPTER 3

Network Security

LEARNING OBJECTIVES

By the end of this chapter, you should be able to:

- Describe the threat environment, including types of attackers and types of attacks.
- Explain the Plan-Protect-Respond cycle for security management.
- Explain in detail planning principles and policy-based security.
- Describe protection.
  - Evaluate authentication mechanisms, including passwords, smart cards, biometrics, digital certificate authentication, and two-factor authentication.
  - Describe firewall protection, including stateful inspection.
  - Explain in detail the protection of dialogues by cryptography, including symmetric key encryption for confidentiality, electronic signatures, and cryptographic system standards.
- Describe response: Reacting according to plan for successful compromises and disasters.

STEUBEN ARC

Steuben ARC is a nonprofit organization in Bath, New York, that provides care for developmentally disabled adults. In September 2009, cyberthieves stole nearly $200,000 from the company.[1]

The attack began when a cybercriminal sent a fake invoice in an e-mail message to one of the company's accountants. The message had an attachment, dhlinvoice.zip. When the accountant opened the attachment, it installed a very sophisticated keystroke logger on the accountant's computer. This program captured the accountant's username and password on the company's accounting server and sent it to the attacker.

Armed with this information, the thieves transferred the money out of the company's bank accounts in two batches. Instead of sending it to themselves, the thieves had

---

[1] Brian Krebs, "Cyber Gangs Hit Healthcare Providers," *Washington Post*, September 28, 2009. voices.washingtonpost.com/securityfix/2009/09/online_bank_robbers_target_hea.html?wprss=securityfix. Mary Pernham, "Alleged cyber-theft: Hackers take $50K from Arc," *The Corning Leader*, October 1, 2009. www.the-leader.com/news/x1699607673/Alleged-cyber-theft-Hackers-take-50K-from-Arc.

the banks send the money to 20 money mules around the country. These money mules forwarded the money to offshore accounts controlled by the attackers. For each transaction, the mules received a fee. Using money mules allowed the attackers to avoid shipping the money directly to offshore accounts, which could have raised the bank's suspicions.

The bank actually did become suspicious. It blocked some of the transfers to money mules and by money mules to offshore banks. However, only some of the money was recovered. Overall, it was a successful attack.

Test Your Understanding

1. a) How did the attacker get the credentials for the company's bank account? b) Why were money mules used? c) List indications that this was a sophisticated attack. d) How might the company have been able to avoid this compromise? e) What motivated the attacker? f) What would you say to executives in small companies who believe that they are too little to be attacked?

INTRODUCTION

This is the third of four introductory chapters. The fact that it deals entirely with security tells you how important security has become in networking. In the 1990s, the Internet blossomed, allowing billions of people to reach hundreds of millions of servers around the world. Unfortunately, the Internet also made all of these users potential victims. Security quickly became one of the most important IT management issues.

Figure 3-1 shows the outline of this book. In this chapter, we will begin looking at the threat environment. Sun Tzu's *The Art of War* admonishes defenders to "know your enemy." We will see that the **threat environment**—the attacks and attackers that companies face—is complex and rapidly changing.

The figure then shows the **plan-protect-respond cycle** that companies follow to deal with the threat environment. This cycle begins with the planning that companies must do to defend against these threats. This is followed by the Protect phase, in which

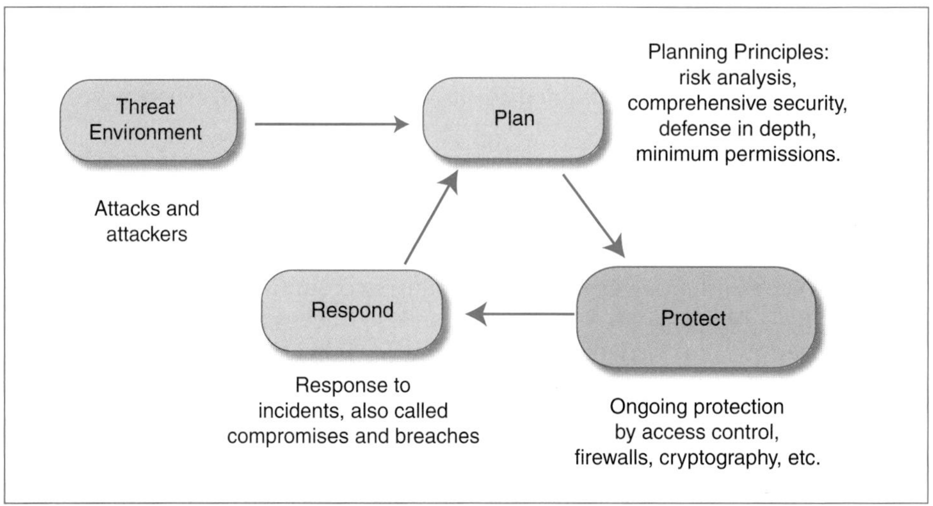

FIGURE 3-1 Threats and the Plan-Protect-Respond Cycle

companies implement the protections they have planned. The Protect box is larger than the others to emphasize that this phase accounts for most of the work in security. The Respond stage is needed when protections fail and attacks succeed. Successful attacks are called **compromises**, **incidents**, or **breaches**.

The main thing that sets security management apart from other aspects of IT management is that the company must battle against intelligent adversaries, not simply against errors and other forms of unreliability. Companies today are engaged in an escalating arms race with attackers, and security threats and defenses are mutating at a frightening rate.

In this chapter, we will look at security broadly. We will focus on network security, but network security is impossible to separate from general IT security.

Test Your Understanding

2. a) What are the two elements of the threat environment? b) Briefly explain each of the three stages in the plan-protect-respond cycle. c) Which of these three stages consumes the most corporate effort? d) Give three names for *successful attack*. e) What is the main thing that separates security from other aspects of IT?

TYPES OF ATTACKS

As just noted, we will begin by looking at the threat environment that corporations face. In this section, we will look at types of attacks. Later, we will look at the types of attackers.

Malware

A general name for evil software

Vulnerability-Specific versus Universal Malware

- Vulnerabilities are flaws in specific programs.
  - Vulnerabilities allow attacks against these specific programs.
- Vulnerability-specific malware requires a specific vulnerability to be effective.
  - Vendors release patches to close vulnerabilities.
  - However, users do not always install patches promptly or at all and so continue to be vulnerable.
  - Also, zero-day attacks occur before the patch is released for the vulnerability.
- Universal malware does not require a specific vulnerability to be effective.
  - Universal malware often requires risky human actions.

Viruses

- Pieces of code that attach themselves to other programs
  - Virus code executes when an infected program executes.
  - The virus then infects other programs on the computer.
- Propagation vectors
  - E-mail attachments
  - Visits to websites (even legitimate ones)

(continued)

FIGURE 3-2 Malware (Study Figure)

Social networking sites
Many others (USB RAM sticks, peer-to-peer file sharing, etc.)
Stopping viruses
Antivirus programs are needed to scan arriving files for viruses.
Antivirus programs also scan for other malware.
Patching vulnerabilities may help.

Worms

Stand-alone programs that do not need to attach to other programs
Can propagate like viruses through e-mail, etc.
This requires human gullibility, which is slow.
Directly propagating worms jump to victim hosts.
Worms can do this if target hosts have a specific vulnerability.
Directly propagating worms can spread with amazing speed.
Directly propagating worms can be thwarted by firewalls and by installing patches.
Not by antivirus programs

Mobile Code

HTML webpages can contain scripts.
Scripts are snippets of code that are executed when the webpage is displayed in a browser.
Scripts enhance the user experience and may be required to see the webpage.
Scripts are called mobile code because they are downloaded with the webpage.
Scripts are normally benign but may be damaging if the browser has a vulnerability.
The script may do damage or download a program to do damage.

Payloads

After propagation, viruses and worms execute their payloads.
Payloads erase hard disks or send users to pornography sites if they mistype URLs.
Often, the payload downloads another program.
An attack program with such a payload is called a downloader.
Many downloaded programs are Trojan horses.
Trojan horses are programs that disguise themselves as system files.
Spyware Trojans collect sensitive data and send the data to an attacker.
Website activity trackers
Keystroke loggers
Data mining software

FIGURE 3-2 Continued

Malware Attacks

We will begin with malware attacks. **Malware** is a name for any evil software. This includes viruses, worms, Trojan horses, and other dangerous attack software. Malware attacks are the most frequent attacks on companies. Nearly every firm has one or more significant malware compromises each year.

Malware is evil software.

Test Your Understanding

3. What is malware?

VULNERABILITY-SPECIFIC VERSUS UNIVERSAL MALWARE A **vulnerability** is a flaw in a program that permits a specific attack or set of attacks against this program. If the vulnerability is not present in the program, a **vulnerability-specific** attack aimed at that vulnerability will fail.

A vulnerability is a flaw in a program that permits a specific attack or set of attacks against this program.

When a vulnerability is discovered, the software vendor usually issues a **patch**, which is a small program designed to correct the vulnerability. After patch installation, the program is safe from attacks based on that particular vulnerability. Too often, however, users fail to install patches, and their programs continue to be vulnerable. Even if they do install patches, furthermore, they may delay doing so. This creates a long window of opportunity for attackers.

Of course, if attacks begin before the program vendor creates a patch (or even learns about the attack), the attacks will succeed. A vulnerability-specific attack that occurs before a patch is available is called a **zero-day attack**.

A vulnerability-specific attack that occurs before a patch is available is called a zero-day attack.

Not all malware is vulnerability-specific. **Universal malware** works whether or not the computer has a security vulnerability. In general, universal malware programs require the human victim to do something risky, such as downloading "free" software, pornography, or an electronic greeting card.

Test Your Understanding

4. a) What is a vulnerability? b) How can users eliminate a vulnerability in one of their programs? c) What name do we give to attacks that occur before a patch is available? d) What type of malware does not require a vulnerability?

VIRUSES Pieces of executable code that attach themselves to other programs are called **viruses**. Within a computer, whenever an infected program runs (executes), the virus attaches itself to other programs on that computer.

Viruses are pieces of executable code that attach themselves to other programs on that computer.

Propagation Vectors. The virus spreads between computers when an infected program is transferred to another computer via a USB RAM stick, an e-mail attachment, a webpage download, a peer-to-peer file-sharing transfer, a social networking site, or some other **propagation vector** (method for malware to move to a victim computer). Once on another machine, if the infected program is executed, the virus spreads to other programs on that machine.

More than 90 percent of viruses today spread via e-mail. Viruses find addresses in the infected computer's e-mail directories. They then send messages with infected attachments to all of these addresses. If a receiver opens the attachment, the infected program executes, and the receiver's programs become infected.

Another popular propagation vector is visiting a website and having the website download a virus (or other type of malware) to your computer. Obviously, the risk is greatest if you visit a high-risk website, such as a site for "free" software or pornography. However, even if you visit a known legitimate website daily, you may become infected one day if an attacker has planted malware on its webpages. (In 2009, this happened to subscribers who went to the *New York Times* website.) A substantial fraction of all infected websites are legitimate websites.

Social networking sites are already popular with virus writers. By their nature, social networking sites are designed for sharing, and if a malware writer can inject malware into the sharing process, spread can be very rapid. USB RAM sticks and peer-to-peer file transfers are two examples of these other propagation vectors.

Stopping Viruses. To stop viruses, a company must protect its computers with **antivirus programs** that scan each arriving e-mail message or file for patterns that identify viruses. Antivirus programs today also scan for other types of malware, but we still call them antivirus programs.

Antivirus programs also scan for other types of malware.

For some viruses (but not for all), it is also useful to patch security vulnerabilities. However, patching does not work for most viruses.

Firewalls are devices that examine each packet passing through a certain part of a network. Client PCs can have firewalls. These firewalls have major benefits, but they do not stop normally propagating worms and viruses.

Test Your Understanding

5. a) What is a propagation vector? b) How do viruses propagate within computers? c) How do viruses propagate between computers? d) In what two ways can viruses be stopped? e) Do firewalls usually stop viruses?

WORMS Another important type of malware is worms. We have just seen that viruses are pieces of code that must attach themselves to other programs. In contrast, **worms** are full programs that operate by themselves. Both viruses and worms can create mass epidemics that infect hundreds or even millions of computers.

Worms are full programs that operate by themselves.

Worms are capable of propagating, like viruses, through e-mail attachments, USB RAM sticks, and similar propagation vectors. These methods typically require human gullibility to succeed. Although human gullibility is sadly reliable, it is rather slow. Until someone opens an e-mail attachment, inserts an infected USB RAM stick, or takes some other action, nothing happens.

Unlike viruses, some (but not all) worms have another propagation vector. A **directly propagating** worm tries to jump from the infected computer to many other computers. Target computers that have a specific vulnerability will accept the directly propagating worm. They then become sites from which the worm spreads further.

A directly propagating worm tries to jump from the infected computer to many other computers.

Freed from the need for human intervention, directly propagating worms can spread with incredible speed. In 2003, the Blaster worm infested 90 percent of all vulnerable hosts on the entire Internet within 10 minutes. The nightmare scenario for security professionals is the prospect of a fast-spreading worm that exploits a vulnerability in a large percentage of all computers on the Internet.

Antivirus programs do nothing to stop directly propagating worms. However, firewalls and patching vulnerabilities can stop them.

Antivirus programs do nothing to stop directly propagating worms. However, firewalls and patching vulnerabilities can stop them.

Figure 3-3 summarizes the differences between how viruses and worms can be stopped. Note that directly propagating worms cannot be stopped using techniques used to stop traditionally propagating worms.

Test Your Understanding

6. a) How do viruses and worms differ? b) Distinguish how directly propagating worms and e-mail worms spread. c) Which can spread faster—viruses or directly propagating worms? Explain. d) How can directly propagating worms be stopped? e) Can antivirus programs usually stop directly propagating worms?

Propagation Vector	Antivirus Program	Firewall	Patching Vulnerabilities
Normally propagating virus or worm (e-mail, visiting website, etc.)	Yes	No	Sometimes
Directly-propagating worm	No	Yes	Yes

FIGURE 3-3 Stopping Viruses and Worms

MOBILE CODE An HTML webpage can contain a **script**, which is a group of commands written in a simplified programming language. Scripts are executed when the webpage is loaded. Scripts can enhance the user's experience, and many webpages will not work unless script execution is enabled, which it usually is by default.

Scripts are referred to as **mobile code** because they travel with the downloaded webpage from the webserver to the browser. Mobile code normally is safe and beneficial. However, if the user's browser has a vulnerability, a script may be able to do harm. A script may do damage itself or may download a more complex program to do damage.

Test Your Understanding

7. a) What is a script? b) Are scripts normally bad? c) Under what circumstances are scripts likely to be dangerous? d) Why are scripts on webpages called mobile code?

PAYLOADS In war, when a bomber aircraft reaches its target, it releases its payload of bombs. Similarly, after they spread, viruses, worms, and other types of malware may execute pieces of code called **payloads**. Malicious payloads can completely erase hard disks and do other significant damage. In some cases, they can take the victim to a pornography site whenever the victim mistypes a URL. In other cases, they can turn the user's computer into a spam generator or a pornography download site. Not all malware has malicious payloads or payloads at all.

TROJAN HORSES Often, the payload installs another program on the computer. A program that does this is called, as you might suspect, a **downloader**.

Often, the downloader retrieves and installs a **Trojan horse**, which is a program that disguises itself as a legitimate system file. This makes it difficult to detect. A Trojan horse cannot spread from one computer to another by itself. Rather, it relies on a virus, worm, hacker, or gullible user to install it on a computer. Once installed, the Trojan horse continues to exploit the user indefinitely.

A Trojan horse cannot spread from one computer to another by itself.

SPYWARE An especially problematic category of Trojan horses is **spyware**—a name given to Trojan horses that **surreptitiously** (without your knowledge) collect information about you and send this information to the attacker.

- Some spyware Trojans collect information about your Web surfing habits and send this information to advertisers.
- More dangerous are **keystroke loggers**, which record your keystrokes. Within these keystrokes, they look for passwords, social security numbers, and other information that can help the person who receives the keystroke logger's data.
- **Data mining** spyware, in contrast, searches through files on your hard drive for potentially useful information and sends this information to the attacker.

Test Your Understanding

8. a) What are payloads? b) What are Trojan horses? c) How do Trojan horses propagate to computers? d) What is spyware? e) What is a keystroke logger? f) What does data mining software do?

Attacks on Individuals

SOCIAL ENGINEERING As technical defenses have improved, malware writers have focused more heavily on **social engineering**, which is a fancy name for tricking the victim into doing something against his or her interests. Viruses and worms have long tried to do this with e-mail attachments—say, by telling the user that he or she has won a lottery and needs to open the attachment for the details. The range of social engineering attacks has expanded greatly in the last few years.

Social engineering is tricking the victim into doing something against his or her interests.

SPAM The most annoying type of malware on a day-in, day-out basis is **spam**,[2] which is unsolicited commercial e-mail. Spammers send the same solicitation e-mail message to millions of e-mail addresses in the hope that a few percent of all recipients will respond. Most users today have to delete several spam messages for every legitimate message they receive.

Spam is unsolicited commercial e-mail.

FRAUD Spam is not merely annoying. Attackers use spam to perpetrate damaging attacks. Few spam messages are really designed to sell legitimate products. Most are fraudulent attempts to get someone to send money for "investment opportunities" and goods that will not be delivered or that are effectively worthless. **Fraud** is lying to get victims to do something against their financial self-interest. Social engineering is the more general term; fraud is social engineering applied to financial interests.

Fraud is lying to get victims to do something against their financial self-interest.

E-MAIL ATTACHMENTS Some spam messages have damaging e-mail attachments. For instance, a spam message may say that it is an electronic greeting card. The user is told that a program must be downloaded to read the greeting card. The "reader" program, of course, is malware.

INCLUDING A LINK TO A WEBSITE THAT HAS MALWARE Spam can also take victims to dangerous websites. One way for spam to create problems is for messages to include a

[2] Except at the beginnings of sentences, e-mail *spam* is spelled in lowercase. This distinguishes unsolicited commercial e-mail from the Hormel Corporation's meat product, Spam, which should always be capitalized. In addition, Spam is *not* an acronym for "spongy pink animal matter."

Social Engineering

Tricking the victim into doing something against his or her interests

Spam

Unsolicited commercial e-mail

Fraud

Fraud is lying to the user to get the user to do something against his or her financial self-interest.

Spam often asks the victim to send money for products that will not be delivered or for false investment opportunities.

E-Mail Attachments

Including a Link to a Website That Has Malware

Phishing Attacks

A sophisticated social engineering attack in which an authentic-looking e-mail or website entices the user to enter his or her username, password, or other sensitive information

Credit Card Number Theft

Uses stolen credit card numbers in unauthorized transactions

Performed by carders

Identity Theft

- Involves collecting enough data to impersonate the victim in large financial transactions
  - Purchase a house or car
  - Obtain a loan
  - Commit fraud or other crime
- Can result in much greater financial harm to the victim than carding
- May take a long time to restore the victim's credit rating
- In corporate identity theft, the attacker impersonates an entire corporation
  - Accept credit cards in the company's name
  - Commit other crimes in the name of the firm
  - Can seriously harm a company's reputation

FIGURE 3-4 Attacks on Individuals (Study Figure)

link to a website. If the receiver clicks on the link, he or she will be taken to a website that will complete the fraud or download malware into the victim's computer.

PHISHING ATTACKS An especially effective form of spam does **phishing**,[3] which is the use of authentic looking e-mail or websites to entice the user to send his or her

[3] IT attackers often replace *f* with *ph*. For example, *phone freaking* (dialing long-distance numbers illegally) became *phone phreaking* and later just *phreaking*.

username, password, or other sensitive information to the attacker. One typical example of phishing is an e-mail message that appears to be from the person's bank. The message asks the person to "confirm" his or her username and password in a return message. Another typical example is an e-mail message with a link to what appears to be the victim's bank website but that is, in fact, an authentic-looking fake website.

Phishing is the use of authentic looking e-mail or websites to entice the user to send his or her username, password, or other sensitive information to the attacker.

CREDIT CARD NUMBER THEFT In fraudulent spam, the message may convince the user to type a credit card number to purchase goods. The attacker will not deliver the goods. Instead, the **carder** (credit card number thief) will use the credit card number to make unauthorized purchases. Most credit card firms will refund money spent by the carder, but this can be a painful process, and the victim must notify the credit card firm promptly to get a refund.

IDENTITY THEFT In other cases, thieves collect enough data about a victim (name, address, social security number, driver's license number, date of birth, etc.) to impersonate the victim during complex crimes. This impersonation is called **identity theft**. Thieves commit identity theft in order to purchase expensive goods, take out major loans using the victim's assets as collateral, commit crimes, obtain prescription medicines, get a job, enter the country illegally, and to do many other things. Identity theft is more damaging than credit card theft because it can involve large monetary losses and because restoring the victim's credit rating can take months. Some victims have even been arrested for crimes committed by the identity thief.

In identity theft, thieves collect enough data about a victim to impersonate the victim during complex crimes.

Test Your Understanding

9. a) What is social engineering? b) What is fraud? c) What is the definition of spam? d) How can spam be used to harm people who open spam messages? e) What is phishing? f) Distinguish between credit card number theft and identity theft. g) What are carders? h) Which tends to produce more damage—credit card theft or identity theft? Explain your answer.

Human Break-Ins (Hacking)

A virus or worm typically has a single method. If that method fails, the attack fails. However, human attackers often break into a specific company's computers manually. A human adversary can attack a target company with a variety of approaches until one succeeds. This flexibility makes human break-ins much more likely to succeed than malware break-ins.

Human Break-Ins

Viruses and worms only have a single attack method.

Humans can keep trying different approaches until they succeed.

Hacking

Informally, hacking is breaking into a computer.

Formally, hacking is intentionally using a computer resource without authorization or in excess of authorization.

Scanning Phase

Send attack probes to map the network and identify possible victim hosts (Figure 3-6).

Scan for IP addresses with active hosts.

Scan IP addresses that reply for programs for which the attacker has an attack method.

The Break-In

Uses an exploit—a tailored attack method that is often a program (Figure 3-6).

Normally exploits a vulnerability on the victim computer.

The act of breaking in is called an exploit.

The hacker tool is also called an exploit.

After the Break-In

The hacker downloads a hacker tool kit to automate hacking work.

The hacker becomes invisible by deleting log files.

The hacker creates a backdoor (way to get back into the computer).

- Backdoor account—account with a known password and full privileges
- Backdoor program—program to allow reentry; usually Trojanized

The hacker can then do damage at his or her leisure.

- Download a Trojan horse to continue exploiting the computer after the attacker leaves
- Manually give operating system commands to do damage

FIGURE 3-5 Human Break-Ins (Study Figure)

WHAT IS HACKING? **Hacking** is defined as intentionally using a computer resource without authorization or in excess of authorization. The key issue is authorization.[4] If you see a password written on a note attached to a computer screen, this does not mean that you have authorization to use it. Also, note that it is hacking even if a person is given an account but uses the computer for unauthorized purposes.

> Hacking is intentionally using a computer resource without authorization or in excess of authorization.

[4] Note also that the unauthorized access must be intentional. Proving intentionality is almost always necessary in criminal prosecution, and hacking is no exception.

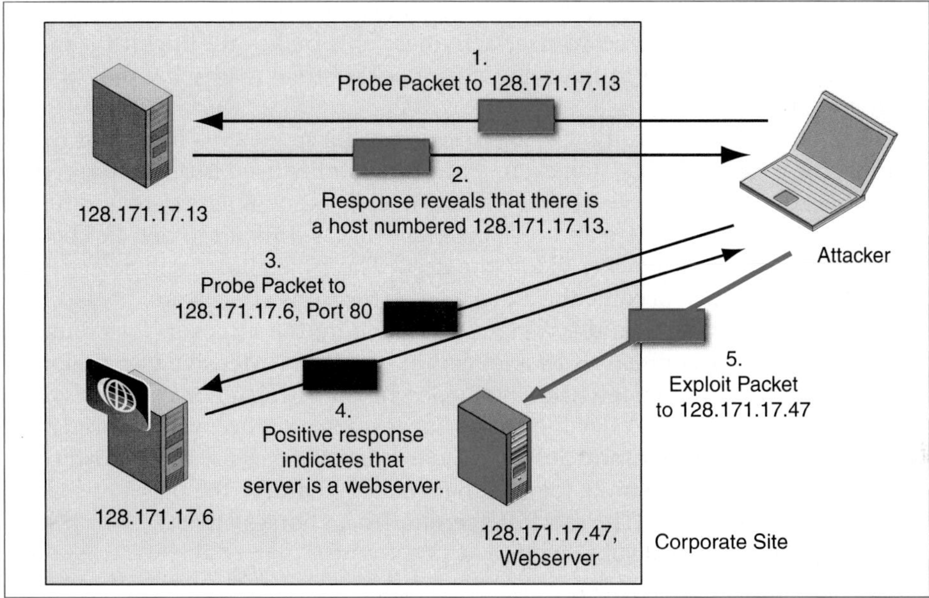

FIGURE 3-6 Scanning Probes and Exploit Packets

THE SCANNING PHASE When a hacker attacks a firm, he or she usually begins by **scanning** the network. Figure 3-6 shows that this involves sending **probe packets** into the firm's network. Responses to these probe packets tend to reveal information about the firm's general network design and about its individual hosts. Usually there are two phases to these probe attacks.

- The first probe packet in the figure is an IP address probe. It is sent to the IP address 128.171.17.13. If the host at that IP address responds, this means that there is a potential victim at that IP address. The attacker typically probes a large range of IP addresses to get a list of potential victims.
- The attacker then sends port number probes to previously identified IP addresses. This second round of probes is sent to particular ports on these hosts. In the figure, the probe packet is sent to Port 80. As we saw in Chapter 2, this is the well-known port number for webservers. If the server responds, the attacker knows that Host 128.171.17.6 is a webserver.

THE BREAK-IN The colored server is a webserver that the attacker has previously probed. The attacker has an **exploit** (attack method) for webservers. He or she uses this exploit to take over the host by sending exploit packets. Confusingly, the act of breaking into a computer is also called an exploit, as is the program the attacker uses during the break-in.

AFTER THE BREAK-IN After the break-in, the real work begins.

- Typically, the first thing a hacker does is to download[5] a **hacker toolkit** to the victim computer. The toolkit is a collection of tools that automate some tasks the hacker will have to perform after the break-in.

- Second, the hacker typically uses the hacker toolkit to erase the operating system's log files that record user activities. This makes it difficult for the computer's rightful owner to trace how the attacker broke in or what the hacker did after the break-in.
- Third, the hacker typically uses the hacker toolkit to create a **backdoor** that will allow the hacker back in later, even if the vulnerability used to break in is repaired. The backdoor may simply be a new account with a known password and full privileges. It can also be a Trojan horse program that is difficult to detect. The Trojan horse will allow the attacker to log into itself. The Trojan horse will have extensive permissions, which become the attacker's permissions after login.
- Fourth, once invisible and having a way back in, the attacker does damage at leisure by giving commands as a logged-in user with extensive permissions. For long-term exploitation, the hacker may download a Trojan horse, which will continue to cause damage after the hacker leaves. For instance, the Trojan may turn the host into a pornography download site or use the compromised host to attack other computers. Keystroke loggers that collect whatever the user types are also popular Trojan horses. The most dangerous Trojan horses are bots, which we will learn about in the next subsection.

Although hacker toolkits and Trojan horses automate a great deal of what the hacker wishes to do, hackers also work manually. With full access to the computer, the attacker can give ordinary operating system commands to read any file on the computer, change files, delete them, or do anything else that a legitimate user with extensive permissions can do.

Test Your Understanding

10. a) List the three main phases in human break-ins (hacks). b) What is hacking? c) What are the two purposes of probe packets? d) What is an exploit? e) What steps does a hacker usually take immediately after a break-in? f) What software does the hacker download to help him or her do work after compromising a system? g) After breaking in, what does a hacker do to avoid being caught? h) What is a backdoor? i) What are the two types of backdoors?

Denial-of-Service (DoS) Attacks Using Bots

Another type of attack, the denial-of-service attack, does not involve breaking into a computer, infecting it with a virus, or infesting it with a worm. Rather, the goal of **denial-of-service (DoS)** attacks is to make a computer or entire network unavailable to its legitimate users.

> The goal of denial-of-service (DoS) attacks is to make a computer or entire network unavailable to its legitimate users.

---

[5] Some students find the use of the term *download* to be confusing. Look at it this way. The hacker is now logged into the victim computer. So he or she downloads software from a toolkit server to the victim computer and installs the software on the victim computer.

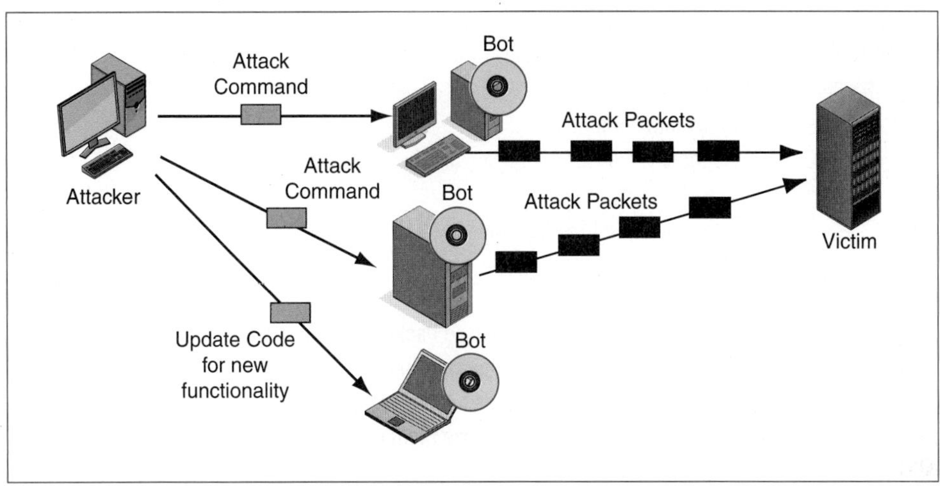

FIGURE 3-7 Distributed Denial-of-Service (DDoS) Attack Using Bots

As Figure 3-7 shows, most DoS attacks involve flooding the victim computer with attack packets. The victim computer becomes so busy processing this flood of attack packets that it cannot process legitimate packets. The overloaded host may even fail.

More specifically, the attack shown in the figure is a **distributed DoS (DDoS)** attack. In this type of DoS attack, the attacker first installs programs called bots on hundreds or thousands of PCs or servers. When the user sends these bots an attack command, they all begin to flood the victim with packets.

Bots are not limited to DDoS attacks. **Bots** are general-purpose exploitation programs that can be remotely controlled after installation. As Figure 3-7 shows, the attacker can send attack commands to the bots and can even upgrade them remotely with new capabilities.

Bots are general-purpose exploitation programs that can be remotely controlled after installation and can even be upgraded remotely with new capabilities.

Bots are extremely dangerous because they can engage in massive attacks that were previously possible only with relatively dumb and inflexible viruses and worms. Through upgrades, bots bring the flexibility of human thought into the attack, making them very dangerous.

Test Your Understanding

11. a) What is the purpose of a denial-of-service attack? b) What are bots? c) What gives bots flexibility? d) How do distributed DoS attacks work?

TYPES OF ATTACKERS

The threat environment consists of types of attacks and types of attackers. As Figure 3-8 shows, there are many different types of attackers facing organizations today.

Traditional Attackers

- Traditional hackers
  - Hackers break into computers.
  - Driven by curiosity, a desire for power, and peer reputation
- Malware writers
  - It usually is not a crime to write malware.
  - It is almost always a crime to release malware.
- Script kiddies
  - Use attack scripts written by experienced hackers and virus writers
  - Scripts usually are easy to use, with graphical user interfaces.
  - Script kiddies have limited knowledge and abilities.
  - But large numbers of script kiddies make them dangerous.
- Disgruntled employees and ex-employees
  - Have extensive access, knowledge of how systems work, and knowledge of how to avoid detection

Criminal Attackers

- *Most* attacks are now made by criminals.
- Crime generates funds that criminal attackers need to increase attack sophistication.
- Large and complex black markets for attack programs, attacks for hire services, bot rentals and sales, money laundering, and other activities

On the Horizon

- Cyberterror attacks by terrorists
- Cyberwar by nations
- Potential for massive attacks far larger than conventional attacks

FIGURE 3-8 Types of Attackers

Traditional Attackers

When most people think of attackers, they normally have three pictures in their minds: hackers driven by curiosity, virus writers, and disgruntled employees and ex-employees. Indeed, these used to be the three most important types of attackers.

HACKERS Traditionally, some **hackers** have been motivated primarily by curiosity and the sense of power they get from breaking into computers. In many cases, they are also motivated by a desire to increase their reputation among their hacker peers by boasting about their exploits. This typically is the image of hackers presented in Hollywood movies. However, these are not the typical hackers today.

MALWARE WRITERS **Malware writers**, as the name suggests, create malware. Malware writers appear to enjoy the excitement of seeing their programs spread rapidly. These malware writers tend to be blind to the harm that they do to people.

In most countries, including the United States, it generally is not illegal to *write* malware. These activities are protected under freedom of speech. However, *releasing* malware is illegal in nearly all countries.

SCRIPT KIDDIES Experienced hackers and virus writers often developed small programs, called **scripts,** to automate parts of their attacks. Over time, these programs grew more sophisticated. More importantly, they grew easier to use. Many now have graphical user interfaces and the look, feel, and reliability of commercial programs.

Some hackers and virus writers release or sell their scripts. This has led to the emergence of relatively nontechnical **script kiddie** attackers, who use these scripts developed by more experienced attackers. Although traditional attackers disparage script kiddies for their lack of skills, there are far more script kiddies than traditional hackers and virus writers, and script kiddies collectively represent a serious threat to corporations.

DISGRUNTLED EMPLOYEES AND EX-EMPLOYEES Other traditional types of attackers are **disgruntled employees** and **disgruntled ex-employees** who attack their own firms. Employee attackers tend to do extensive damage when they strike because they typically already have access to systems, have broad knowledge of how the systems work, and often know how to avoid detection.

The most dangerous employees of all are IT staff members and especially IT security staff members. They typically have far more access than other employees, have much better knowledge of corporate systems, and have extensive knowledge of how to avoid detection. In fact, they may even be in charge of identifying attackers. The ancient Roman question, "Quis custodiet ipsos custodes?" means "Who guards the guardians?" It is a serious question in security.

Criminal Attackers

Today, there are still many traditional attackers of the types we have just seen. However, even collectively they do not make up the majority of attackers today. Today, *most* attackers are **career criminals,** who steal credit card numbers to commit credit card fraud, who extort firms, and who steal trade secrets to sell to competitors.

Today, *most* attackers are career criminals.

Funded by their crimes, many criminals can afford to hire the best hackers and to enhance their own security-breaking skills. Consequently, criminal attacks are not just growing in numbers; they also are growing very rapidly in technical sophistication.

Cyberterrorists and National Governments

On the horizon is the danger of far more massive **cyberterror** attacks by terrorists and even worse **cyberwar** attacks by national governments. These could produce unprecedented damages in the hundreds of billions of dollars.

Cyberwar is not a theory. The United States has acknowledged that it has long had cyberwar capabilities, and it established a consolidated Cyberwar Command in 2009. It is clear that several other countries have these capabilities as well (especially China). Countries could use IT to do espionage to gather intelligence, conduct attacks on opponents' financial and power infrastructures, or destroy enemy command and control facilities during physical attacks.

A 2009 article in the *New York Times*[6] reported that before the 2003 invasion of Iraq, the United States considered an attack that would shut down Iraq's entire financial infrastructure. This attack was not approved, but this was not because it was unfeasible. It was not approved because its impact might have spread beyond Iraq and might even have damaged the U.S. financial system.

Cyberterror is also likely. During physical attacks, terrorists might disable communication systems to thwart first responders and to spread confusion and terror among the population. Cyberterrorists could also conduct purely IT-based attacks. While the United States was afraid of side-effects of cyberwar attacks on Iraq, terrorists would have no such qualms.

Test Your Understanding

12. a) Are most attackers today driven by curiosity and a sense of power? b) Is it generally illegal to write malware? c) For what three reasons are employees dangerous? d) What are the most dangerous types of employees? e) What type of attacker are most attackers today? f) What are cyberterror and cyberwar attacks? g) Why are cyberwar and cyberterror serious security concerns?

PLANNING

Security Is a Management Issue

People tend to think of security as a technological issue, but security professionals agree unanimously that security is primarily a management issue. Unless a firm does excellent planning, implementation, and day-to-day execution, the best security technology will be wasted. As Bruce Schneier, a noted security expert, has often said, "Security is a process, not a product."[7] Unless firms have good security processes in place, the most technologically advanced security products will do little good.

> Security is primarily a management issue, not a technology issue.

Test Your Understanding

13. Why is security primarily a management issue, not a technology issue?

Planning Principles

Perhaps more than any other aspect of IT, effective security depends on effective planning. Security planning is a complex process that we can only discuss briefly. We will note four key principles that must be used in planning.

RISK ANALYSIS In contrast to military security, which often makes massive investments to stop threats, corporate security planners have to ask whether applying a

---

[6] Markoff, John, and Shanker, Thom., "'03 Plan Displays Cyberwar Risk," *New York Times*, August 1, 2009. www.msnbc.msn.com/id/3032619/%2328368424.

[7] Schneier, Bruce, *Crypto-Gram Newsletter*, May 15, 2000. www.schneier.com/crypto-gram-0005.html.

Security Is a Management Issue, Not a Technical Issue.

- Without good management, technology cannot be effective.
- A company must have good security processes.

Security Planning Principles

- Risk analysis
  - Risk analysis is the process of balancing threats and protection costs for individual assets.
  - Annual cost of protection should not exceed the expected annual damage.
  - If the probable annual damage is $10,000 and the annual cost is $200,000, the protection is not worth the cost.
  - Goal is not to eliminate risk but rather to reduce it in an economically rational level.
- Comprehensive security
  - An attacker has to find only one weakness.
  - A firm needs comprehensive security to close all avenues of attack.
  - This requires very good planning.
- Defense in depth
  - Every protection breaks down sometimes.
  - An attacker should have to break through several lines of defense to succeed.
  - Providing several lines of defense is called defense in depth.
- Minimal permissions
  - Access control is limiting who can use resources and limiting their permission while using resources.
  - Permissions are things they can do with the resource.
  - People should be given minimum permissions—the least they need to do their jobs—so that they cannot do unauthorized things.

FIGURE 3-9 Security Planning (Study Figure)

countermeasure against a particular threat is economically justified. For example, if the probable annual loss due to the threat is $10,000 and the security measures needed to thwart the threat will cost $200,000 per year, the firm should not spend the money. Instead, it should accept the probable loss. **Risk analysis** is the process of balancing threats and protection costs for individual assets.

Risk analysis is the process of balancing threats and protection costs for individual assets.

Figure 3-10 shows a simple risk analysis. Without a countermeasure, the damage per successful attack is expected to be $1,000,000, and the annual probability of a successful attack is 20 percent. Therefore, the annual provable damage is $200,000 without a countermeasure. The net annual probable outlay therefore is $200,000.

Countermeasure A is designed to cut the seriousness of a successful attack in half. So the damage per successful attack is expected to be $500,000 instead of a million dollars. The countermeasure will not reduce the probability of a successful attack, so that continues to be 20 percent. With Countermeasure A, then, the annual probable damage

Countermeasure	None	A
Damage per successful attack	$1,000,000	$500,000
Annual probability of a successful attack	20%	20%
Annual probable damage	$200,000	$100,000
Annual cost of countermeasure	$0	$20,000
Net annual probable outlay	$200,000	$120,000
Annual value of countermeasure	$0	$80,000

FIGURE 3-10 Risk Analysis Example

is reduced to $100,000. However, the countermeasure is not free. It will cost $20,000 per year. So the net annual probable outlay is $120,000 with the countermeasure.

Countermeasure A, then, will reduce the net annual probable outlay from $200,000 to $120,000. The countermeasure has a value of $80,000 per year. This is positive, so Countermeasure A is justified economically.

Note that *the goal of security is not to eliminate risk*. That would be impossible. Despite efforts throughout human history, we still have theft, and we still have murder. Despite strong security efforts, in turn, there will still be some risk of a compromise. People with little to steal who live in gated communities with bars on their windows, expensive alarm systems, and a permanent armed security guard are not doing rational risk analysis. The goal of security is to reduce risk to a degree that is economically rational.

The goal of security is not to eliminate risk. The goal of security is to reduce risk to a degree that is economically rational.

COMPREHENSIVE SECURITY To be safe from attack, a company must close off *all* vectors of attack. In contrast, an attacker only needs to find one unprotected attack vector to succeed. Although it is difficult to achieve **comprehensive security**, in which all avenues of attack are closed off, it is essential to come as close as possible.

Comprehensive security is closing off all avenues of attack.

DEFENSE IN DEPTH Another critical planning principle is defense in depth. Every protection will break down occasionally. If attackers have to break through only one line of defense, they will succeed during these vulnerable periods. However, if an attacker has to break through two, three, or more lines of defense, the breakdown of a single defense technology will not be enough to allow the attacker to succeed. Having successive lines of defense that all must be breached for an attacker to succeed is called **defense in depth**.

Having several lines of defense that all must be breached for an attacker to succeed is called defense in depth.

ACCESS CONTROL WITH MINIMUM PERMISSIONS Security planners constantly worry about access to resources. People who get access to resources can do damage. Not surprisingly, companies work very hard to control access to their resources. **Access control** is limiting who may have access to each resource and limiting their permissions when using the resource.

Access control is limiting who may access each resource and limiting their permissions when using the resource.

One aspect of access control that we will see later is authentication—requiring users requesting access to prove their identity. However, just because you know who someone is does not mean that they should have unfettered access to your resources. (There undoubtedly are several people you know who you would not let drive your car.)

Permissions are the actions that a person given access to a resource is allowed to take. For example, although everyone is allowed to view the U.S. Declaration of Independence, they are not allowed to add their own signatures at the bottom.

Permissions are the actions that a person given access to a resource is allowed to take.

An important principle in assigning permissions is to give each person **minimum permissions**—the least permissions that the user needs to accomplish his or her job. In the case of access to team documents, for example, most team members may only be given read-only access, in which they can read team documents but not change them. It is far less work to give the user extensive or full permissions so that they do not have to be given additional permissions later. However, it is not safe to do so.

Minimum permissions are the least permissions that the user needs to accomplish his or her job.

Test Your Understanding

14. a) List the four major planning principles. b) What is risk analysis? c) Repeat the risk analysis described in this section, this time with Countermeasure B that does not affect damage severity but that reduces the likelihood of an attack by 75 percent. The annual cost of Countermeasure B is $175,000. Show the full table. d) Comment on the statement, "The goal of security is to eliminate risk." e) What is comprehensive security? f) Why is comprehensive security important? g) What is defense in depth? h) Why is defense in depth necessary? i) What is access control? j) What are permissions? k) Why should people get minimum permissions?

Policy-Based Security

POLICIES The heart of security management is the creation and implementation of security policies. Figure 3-11 illustrates how policies should be used. **Policies** are broad statements that specify what should be accomplished. For example, a policy might be, "All information on USB RAM sticks should be encrypted."

Policies are broad statements of what should be accomplished.

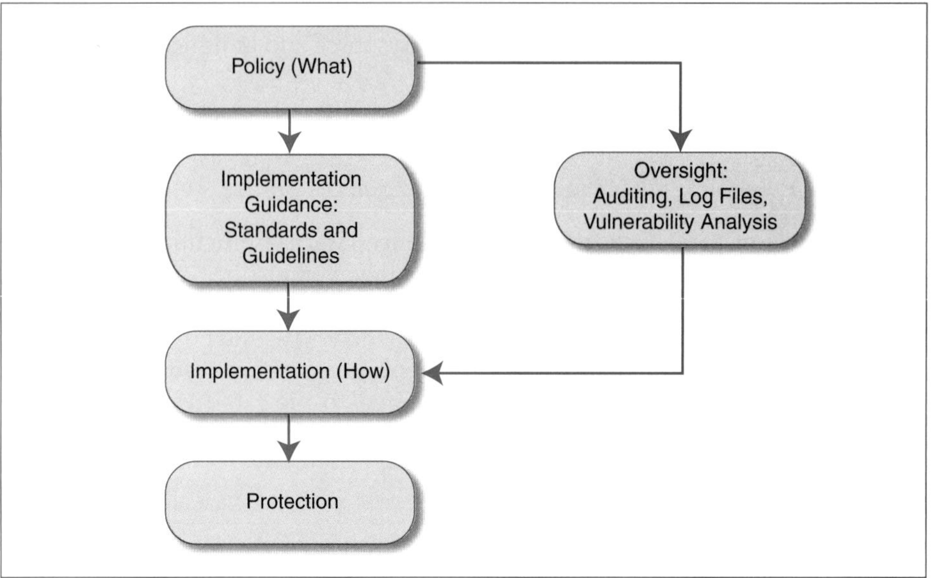

FIGURE 3-11 Policy-Based Security

POLICY VERSUS IMPLEMENTATION Note that the policy does not specify what encryption technology should be used or other implementation details. Put another way, policies describe *what* (should be done), not *how* (to do it).

Policies describe what (should be done), not how (to do it).

This separation of policy from implementation permits the implementers to implement the policy in the best way possible. Policymakers have the overview knowledge that operational people may not have. For instance, policymakers may know that new laws create serious liability unless USB RAM sticks are encrypted. However, people who do implementation are likely to know more about the specific technologies and the local situation than do policymakers. They have the specific knowledge that policymakers do not, including technical knowledge.

The separation of policy from implementation does not mean that policy is irrelevant to implementation. It is easy to get lost in implementation details. Having a clear policy permits everybody involved in implementation to stay synchronized by checking frequently whether what they are doing will lead to the successful implementation of the policy.

IMPLEMENTATION GUIDANCE In many cases, the policymaker will only specify the policy. However, in some cases, the policymaker will also create some implementation guidance. **Implementation guidance** consists of instructions that are more specific than policies but less specific than implementation.

Implementation guidance consists of instructions that are more specific than policies but less specific than implementation.

For example, after establishing a policy that USB RAM sticks must be encrypted, implementation guidance might be added in the form of a directive that the encryption must be strong encryption. This ensures that implementers will not have the latitude to choose weak encryption that can be defeated by an attacker.

There are two general forms of implementation guidance: standards and guidelines. **Standards** are mandatory directives that *must* be followed. Requiring strong encryption is a standard. It is mandatory for implementers to follow the directive.

Standards are mandatory directives that *must* be followed.

In turn, **guidelines** are directives that *should* be followed but that need not be followed, depending on the context.[8] For example, a directive that security staff members should have three years of security work experience indicates that someone hiring a security staff member must consider that three years of experience is a good indicator of competence. If the person doing the hiring selects someone with only two years of work experience, it would be legitimate to ask the person doing the hiring if he or she felt that less than three years of work experience was acceptable. Following guidelines is optional, but seriously considering guidelines is mandatory.

Guidelines are directives that *should* be followed but that need not be followed, depending on the context.

OVERSIGHT Figure 3-11 also shows that policymakers cannot merely toss policies and implementation guidance out and ignore how implementation is done. It is essential for management to exercise **oversight,** which is a collection of methods to ensure that policies have been implemented properly.

Oversight is a collection of methods to ensure that policies have been implemented properly.

One form of oversight is an audit. An **audit** samples actions taken within the firm to ensure that policies are being implemented properly. Note that an audit only *samples* actions. It does not look at everything, which would be impossible to do. However, if the sampling is done well, the auditor can issue an opinion on whether a policy is being carried out appropriately based on well-considered data.

An audit samples actions taken within the firm to ensure that policies are being implemented properly.

[8] In the *Pirates of the Caribbean* movies, there was a running joke that the Pirate's Code is "more like a guideline, really."

Another form of oversight is **reading log files.** Whenever users take actions, their actions should be recorded in log files. Reading log files allows improper behavior to be discovered. Of course, if these log files are not read, they are useless. Consequently, it is critical to read log files frequently. Important log files should be read daily or even several times each day.

Reading log files allows improper behavior to be discovered.

Another important oversight mechanism is vulnerability testing. Simply put, **vulnerability testing** is attacking your own systems before attackers do, so that you can identify weaknesses and fix them before they are exploited by attackers.

Vulnerability testing is attacking your own systems before attackers do, so that you can identify weaknesses and fix them before they are exploited by attackers.

Note that the policy drives both implementation and oversight. Implementers who attempt to implement the policy must interpret the policy. Auditors and other oversight professionals must also interpret the policy. If the implementers are lax, the auditors should be able to identify this. However, if oversight practitioners and implementers disagree, this may simply mean that they are interpreting the policy differently. Policy makers may find that one or the other has made a poor choice in interpreting the policy. They may also find that the policy itself is ambiguous or simply wrong. The important thing is to identify problems and then resolve them.

Policies drive both implementation and oversight.

EFFECTIVE PROTECTION Policies certainly do not give protection by themselves. Neither may unexamined implementations. Protection is most likely to be effective when excellent implementation is subject to strong oversight.

Test Your Understanding

15. a) What is a policy? b) Distinguish between policy and implementation. c) Why is it important to separate policies from implementation? d) Why is oversight important? e) Compare the specificity of policies, implementation guidance, and implementation. f) Distinguish between standards and guidelines. g) Must guidelines be considered? h) List the three types of oversight listed in the text. i) What is vulnerability testing, and why is it done? j) Why is it important for policy to drive both implementation and oversight?

AUTHENTICATION

The most complex element of access control is authentication. Figure 3-12 illustrates the main terminology and concepts in authentication. The user trying to prove his or her identity is the **supplicant**. The party requiring the supplicant to prove his or her identity is the **verifier**. The supplicant tries to prove his or her identity by providing **credentials** (proofs of identity) to the verifier.

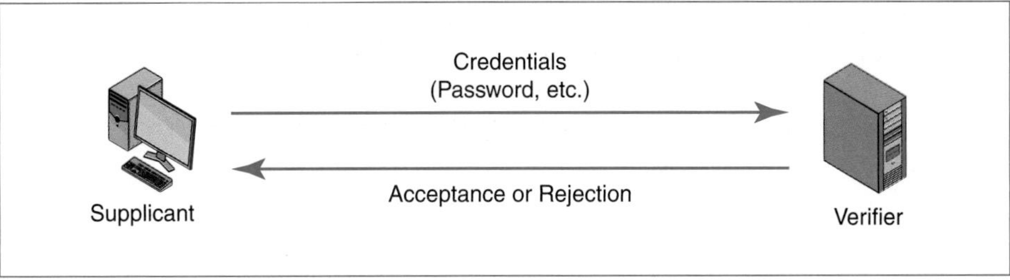

FIGURE 3-12 Authentication

The type of authentication tool that is used with each resource must be appropriate for the risks to that particular resource. Sensitive personnel information should be protected by very strong authentication methods. However, strong authentication is expensive and inconvenient. For relatively nonsensitive data, weaker but less expensive authentication methods may be sufficient. Strength of authentication, like everything else in security, is a matter of risk management.

Test Your Understanding

16. a) What is authentication? b) Distinguish between the supplicant and the verifier. c) What are credentials? d) Why must authentication be appropriate for risks to an asset?

Reusable Passwords

The most common authentication credential is the **reusable password**, which is a string of characters that a user types to gain access to the resources associated with a certain **username** (account) on a computer. These are called *reusable* passwords because the user will type the password each time he or she needs access to the resource. The reusable password is the weakest form of authentication, and it is appropriate only for the least sensitive assets.

The reusable password is the weakest form of authentication, and it is appropriate only for the least sensitive assets.

EASE OF USE AND LOW COST The popularity of password authentication is hardly surprising. For users, passwords are familiar and relatively easy to use. For corporate IT departments, passwords add no additional cost because operating systems and many applications have built-in password authentication.

WORD/NAME PASSWORDS AND DICTIONARY ATTACKS The main problem with passwords is that most users pick very weak passwords. They often pick ordinary **dictionary words** or the **names** of family members, pets, sports teams, or celebrities. Dictionary-word and name passwords often can be **cracked** (guessed) in a few seconds if the attacker can get a copy of the password file (which contains an encrypted list of account names and passwords). The attacker uses a **dictionary attack**, trying all words or names in a standard or customized dictionary. There are only a few thousand dictionary words and names in

Reusable Passwords

Passwords are strings of characters.
They are typed to authenticate the use of a username (account) on a computer.
They are used repeatedly and so are called reusable passwords.

Benefits

Ease of use for users (familiar)
Inexpensive because they are built into operating systems

Often Weak (Easy to Crack)

Word and name passwords are common.
They can be cracked quickly with dictionary attacks.

Passwords Should Be Complex

Should mix case, digits, and other keyboard characters ($, #, etc.)
Complex passwords can be cracked only with brute force attacks (trying all possibilities).

Passwords Also Should Be Long

Should have a minimum of eight characters
Each added character increases the brute force search time by a factor of about 70.

Other Concerns

If people are forced to use long and complex passwords, they tend to write them down.
People should use different passwords for different sites.
Otherwise, a compromised password will give access to multiple sites.

FIGURE 3-13 Password Authentication (Study Figure)

any language, so dictionary attacks can crack dictionary-word and name passwords almost instantly.

The main problem with passwords is that most users pick very weak passwords.

Dictionary attacks also have **hybrid modes**, in which they look for simple variations on words, such as a word with the first letter capitalized, followed by a single digit (for instance, Dog1). Hybrid word or name passwords are cracked almost as quickly as passwords made of simple words and names.

Names, words, and simple variants of words and names that can be cracked by hybrid mode dictionary attacks are never adequate as passwords, regardless of how long they are. They can always be cracked too quickly for safety.

Names, words, and simple variants of words and names that can be cracked by hybrid mode dictionary attacks are never adequate as passwords, regardless of how long they are.

COMPLEX PASSWORDS AND BRUTE FORCE ATTACKS Dictionary attacks fail if passwords are more complex than dictionary words, names, and simple variations. Good complex passwords have all of the following:

- Lowercase letters.
- Uppercase letters, not simply at the start of the password.
- The digits from 0 to 9, not simply at the end of the password.
- Other keyboard symbols, such as & and #, which serve as swear words in cartoons—not simply at the end of the password.

Complex passwords can only be cracked by **brute force attacks** that try all possible combinations of characters. First, all combinations of a single character are tried, all combinations of two characters, all combinations of three characters, and so forth. Brute force attacks take far longer than dictionary attacks.

Unfortunately, complex passwords are difficult for users to remember, so they tend to write them on a sheet of paper that they keep next to their computers. This makes passwords easy to steal so that there is no need to crack them by dictionary or brute force attacks.

COMPLEX PASSWORD LENGTH Increasing **password length** (the number of characters in the password) makes a complex password stronger. If the password has a combination of uppercase and lowercase letters, digits, and other keyboard characters, then each additional character increases cracking time by a factor of about 70. Given the speed of brute force cracking today, passwords should be complex and at least eight characters long to be considered adequate. Even longer passwords are highly desirable.

Complex passwords should be at least eight characters long to be considered adequate.

Test Your Understanding

17. a) Distinguish between usernames and reusable passwords. b) Why are passwords widely used? c) What types of passwords are susceptible to dictionary attacks? d) What types of passwords are susceptible to dictionary attacks in hybrid mode? e) Can a password that can be broken by a dictionary attack or a dictionary attack in hybrid mode be adequately strong if it is very long? f) What is a brute force attack? g) What types of passwords can be broken only by brute force attacks? h) Why is password length important? i) How long should passwords be?

18. Critique each of the following passwords. First, describe the type of attack that would be used to crack it, justifying your answer. Second, say whether or not it is of adequate strength, justifying your answer. a) Viper1 b) R7%t& c) NeVeR

Other Forms of Authentication

Companies are beginning to look for stronger types of authentication for most of their resources. This will allow them to replace most or all of their reusable password access systems. We only have space to mention the few types of authentication shown in Figure 3-14.

Perspective

Goal is to replace reusable passwords.

Access Cards

Permit door access

Proximity access cards do not require scanning.

Need to control distribution and disable lost or stolen access cards

Biometrics

Biometrics uses body measurements to authenticate you.

Vary in cost, precision, and susceptibility to deception

Fingerprint scanning

- Inexpensive but poor precision, deceivable
- Sufficient for low-risk uses
- On a notebook, may be better than requiring a reusable password

Iris scanning

- Based on patterns in the colored part of your eye
- Expensive but precise and difficult to deceive

Facial scanning

- Based on facial features
- Controversial because can be done surreptitiously—without the supplicant's knowledge

Digital Certificate Authentication

Components

- Everyone has a private key that only he or she knows.
- Everyone also has a public key that is not secret.
- Public keys are available in unalterable digital certificates.
- Digital certificates are provided by trusted certificate authorities.

Operation

- Supplicant does a calculation with his or her private key.
- Verifier checks this calculation with a public key in a digital certificate.
- Verifier uses the digital certificate of the true party—the person the supplicant claims to be.
- If the calculation check works, the supplicant must have the true party's private key, which only the true party should know. The supplicant must be the true party.

Two-Factor Authentication

Supplicant needs two forms of credentials.

Example: debit card and pin

Strengthens authentication

Fails if attacker controls user's computer or intercepts authentication communication

FIGURE 3-14 Other Forms of Authentication

ACCESS CARDS To get into your hotel room, you may have to swipe your **access card** through a card reader before being allowed through. For door and computer access, many companies also use these handy access cards, including **proximity access cards** that use radio signals and can be read with a simple tap against a reader. Companies need to control the distribution of access cards, and they need to rapidly disable any access card that has been lost or stolen.

BIOMETRICS Access cards are easy to use, but if you lose your access card, you cannot get entry. In hotels, of course, you simply walk down to the front desk. They disable the code on your room card reader and give you a new card that will open your room. In corporate environments, the process takes a good deal longer.

In biometrics, in contrast, access control is granted based on something you always have with you—your body. **Biometrics** is the use of body measurements to authenticate you.

Biometrics is the use of body measurements to authenticate you.

There are several types of biometrics that differ in cost, precision, and susceptibility to deception by someone wishing to impersonate a legitimate user.

- At the low end on price, precision, and the ability to reject deception is **fingerprint scanning**, which looks at the loops, whorls, and ridges in your fingerprint. Although fingerprint scanning is not the strongest form of authentication, its low price makes it ideal for low-risk applications. Even for protecting laptop computers and smart phones, fingerprint scanning may be preferred to reusable passwords given the tendency of people to pick poor passwords and forget them.
- At the high end of the scale on price, precision, and the ability to reject deception is **iris scanning**,[9] which looks at the pattern in the colored part of your eye. Although extremely precise, iris scanners are too expensive to use for computer access. They are normally used for access to sensitive rooms.
- One controversial form of biometrics is **facial scanning**, in which each individual is identified by his or her facial features. This is controversial because facial scanning can be done **surreptitiously**—without the knowledge of the person being scanned. This raises privacy issues.

DIGITAL CERTIFICATE AUTHENTICATION The strongest form of authentication is digital certificate authentication. Figure 3-15 illustrates this form of authentication.

- In this form of authentication, each person has a secret **private key** that only he or she knows.
- Each person also has a **public key**, which anyone can know.
- A trusted organization called a **certificate authority** distributes the public key of a person in a document called a **digital certificate**. A digital certificate cannot be changed without this change being obvious.

---

[9] In science fiction movies, eye scanners are depicted as shining light into the supplicant's eye. This does not really happen. Iris scanners merely require the supplicant to look into a camera. In addition, science fiction movies use the term *retinal scanning*. The retina is the back part of the eye and has distinctive vein patterns. Retinal scanning is not used frequently because the supplicant must press his or her face against the scanner.

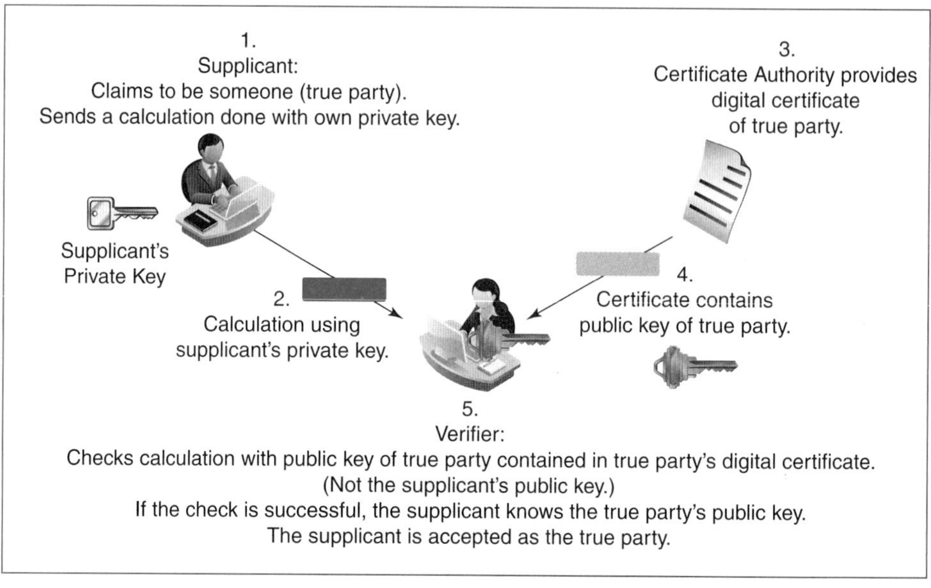

FIGURE 3-15 Digital Certificate Authentication

First, the supplicant claims to be someone we will call the true party. To prove this claim, the supplicant does a calculation[10] with his or her private key and sends this calculation to the verifier.

Second, the verifier gets the true party's digital certificate, which contains the true party's public key. The verifier tests the calculation with the public key of the true party—the person the supplicant claims to be.[11] If the test works, then the supplicant must know the true party's private key, which only the true party should know. The supplicant must be the true party.

Note that the verifier uses the public key of the true party—not the supplicant's public key. If the verifier used the supplicant's public key, the test would always succeed—even if the supplicant is an impostor.

Note that the verifier uses the public key of the true party—not the supplicant's public key.

TWO-FACTOR AUTHENTICATION Debit cards are potentially dangerous because if someone finds a lost debit card, the finder might be able to use it to make purchases. So

[10] To be more specific, the verifier sends the supplicant a challenge message that is a random stream of bits. The calculation that the supplicant does is to encrypt the challenge message with the supplicant's private key. The result is the response message, which the supplicant sends to the verifier.

[11] To be more specific, the verifier decrypts the response message with the true party's public key. If someone encrypts something with his or her private key, this can be decrypted with the true party's public key. If the supplicant is the true party, the true party's public key will decrypt the response message back to the challenge message. The supplicant will be authenticated as the true party. If the supplicant is an impostor, when the verifier decrypts the response message, the result will not be the challenge message that the verifier originally sent to the supplicant. The supplicant will then be rejected as an impostor.

possession of the debit card is not enough to use it. To use a debit card, the user must type a **personal identification number (PIN)**, which usually is four or six digits long. Requiring two credentials for authentication is called **two-factor authentication**. Two-factor authentication increases the strength of authentication.

However, if a user's computer is compromised, the attacker typically controls both aspects of communication. Two-factor authentication may also break down if an eavesdropper can intercept authentication communication between the two parties.

Two-factor authentication requires two forms of authentication.

Test Your Understanding

19. a) What security problem do access cards have? b) What is biometrics? c) By what three criteria should biometric methods be judged? d) Why may fingerprint scanning be used to authenticate access to a laptop? e) Why is iris scanning desirable? f) Why is face recognition controversial?
20. a) In digital certificate authentication, what does the supplicant do? b) What does the verifier do? c) Does the verifier use the true party's public key or the supplicant's public key? d) How does the verifier get the public key? e) From what type of organization does the verifier get the digital certificate?
21. a) Why is two-factor authentication desirable? b) Will two-factor authentication still be strong if the attacker controls the supplicant's computer? c) Will two-factor authentication still be strong if the attacker can intercept all authentication communication?

FIREWALLS

In hostile military environments, travelers must pass through one or more checkpoints. At each checkpoint, their credentials will be examined. If the guard finds the credentials insufficient, the guard will stop the arriving person from proceeding and note the violation in a checkpoint log.

DROPPING AND LOGGING PROVABLE ATTACK PACKETS Figure 3-16 shows that firewalls operate in similar ways. Whenever a packet arrives, the **firewall** examines the packet. If the firewall identifies a packet as a **provable attack packet**, the firewall discards it. On the other hand, if the packet is not a provable attack packet, the firewall allows it to pass.

If a firewall identifies a packet as a provable attack packet, the firewall discards it.

The firewall copies information about the discarded packet into a **firewall log file**. Firewall managers should read their firewall log files every day to understand the types of attacks coming against the resources that the firewall is protecting.

Note that firewalls pass *all* packets that are not provable attack packets. Some attack packets will not be provable attack packets. Consequently, some attack packets inevitably get through the firewall to reach internal hosts. It is important to harden all

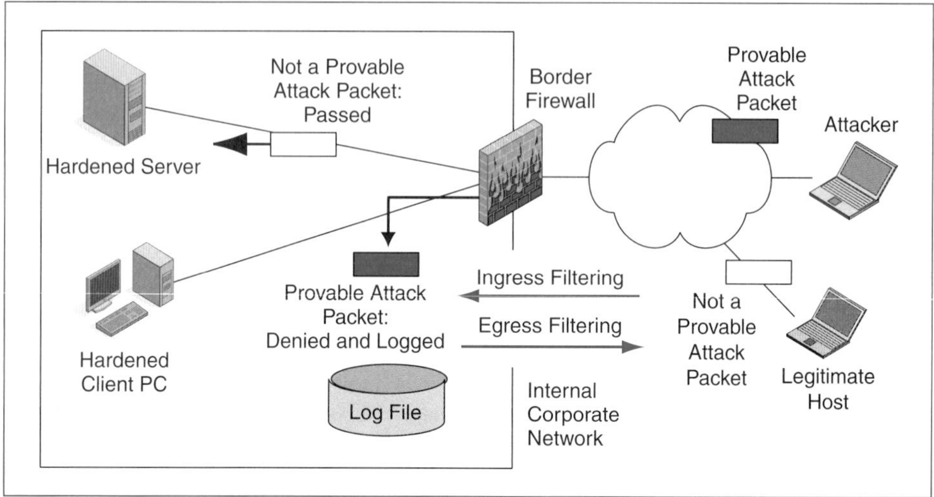

FIGURE 3-16 Firewall Operation

internal hosts against attacks by adding firewalls, adding antivirus programs, installing all patches promptly, and taking other precautions. This chapter focuses on network security, rather than IT host security, so we will not consider host hardening.

INGRESS AND EGRESS FILTERING When most people think of firewalls, they think of filtering packets arriving at a network *from the outside*. Figure 3-16 illustrates this **ingress filtering**.

Most firms also do **egress filtering**—that is, they filter packets going from the network *to the outside*. By doing egress filtering, the corporation is acting as a good citizen, ensuring that its computers are not used in attacks against outside firms. Egress filtering also attempts to prevent sensitive corporate information from being sent outside the firm.

Test Your Understanding

22. a) What does a firewall do when a packet arrives? b) Does a firewall drop a packet if it probably is an attack packet? c) Why is it important to read firewall logs daily? d) Distinguish between ingress and egress filtering.

Stateful Firewall Filtering

We have used the term *firewall filtering* up until now without explaining it. We did this because different firewalls use several different **filtering methods**. Most firewalls today, however, use stateful firewall inspection, which was invented in the early 1990s and has now become the dominant firewall filtering method.

STATES AND FILTERING INTENSITY When you talk with someone on the telephone, there are two basic stages to your conversation.

- At the beginning of a call, you need to identify the other party and decide whether you are both willing to have a conversation.

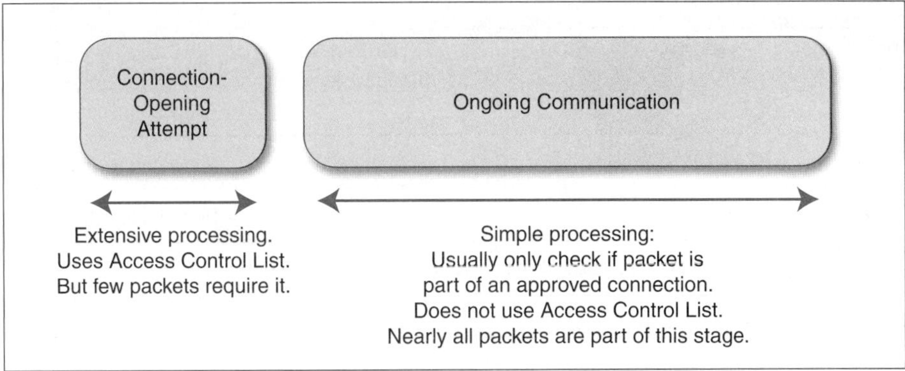

FIGURE 3-17 States in Stateful Packet Inspection (SPI)

- Afterward, if you do decide to talk, you usually don't have to constantly worry about whether the conversation should go on with this person.

The key point here is that you do different things in different stages of a conversation. In the first stage, you have to pay careful attention to identifying the caller and making a decision about whether it is wise to talk. After that, you simply talk and normally do not have to spend much time thinking about whether to talk to the person.

Most firewalls today use **stateful packet inspection (SPI)** filtering, which uses the insight that there are also stages in network conversations and that not all stages require the same amount of firewall attention. At the simplest level, Figure 3-17 shows that there are two stages, which SPI firewalls call **states**: opening a connection (conversation) and ongoing communication.

SPI FILTERING IN THE CONNECTION OPENING STATE SPI firewalls focus heavily on the opening state. They have complex rules to tell them whether or not to allow conversation. If they decide to allow a connection, however, they give minimal attention to packets in the ongoing communication state. This makes sense because the decision to allow a connection is the most complex and dangerous stage in the connection.

For example, suppose that a packet arriving at a firewall contains a TCP SYN segment. This clearly is a connection-opening request to the destination host. So the firewall compares the features of the packet to the rules in its **access control list (ACL)**. Figure 3-18 shows a very simplified access control list. This ACL only has three rules.

- Rule 1 allows connections to all hosts (all IP addresses) on Port 25. We saw in Chapter 2 that Port 25 is the well-known port number for SMTP. This rule permits connections to all internal mail servers.
- Rule 2 permits connections to a single internal host, 10.47.122.79, on Port 80. This rule allows access to a single internal webserver—the webserver at IP address 10.47.122.79. This is safer than Rule 1 because Rule 1 opens the firewall to *every* internal mail server, while Rule 2 only opens the firewall to connections to *a single* server.

Rule	Destination IP Address or Range	Service	Action
1	ALL	25	Allow connection
2	10.47.122.79	80	Allow connection
3	ALL	ALL	Do not allow connection
Note: ACLs are only applied to packets that attempt to open a connection.			

FIGURE 3-18 Stateful Inspection Firewall Access Control List (ACL) for Connection-Opening Attempts

- The last rule is the default rule for incoming packets that try to open a connection. (The default is what you get if you do not explicitly specify something else.) This last rule ensures that unless a packet is explicitly allowed by an earlier rule, it is dropped and logged.

Although ACL rules generally are not very complex, there tend to be many of them in real ACLs. Running each connection-opening attempt through the access control list can be fairly time consuming. Fortunately, only a very small percentage of all packets arriving at a firewall are connection opening attempts.

HANDLING PACKETS DURING ONGOING COMMUNICATION If a packet does not attempt to open a connection or is not part of a connection-opening attempt, then either the packet must be part of the ongoing communication state of an approved connection or the packet is spurious. When a packet arrives that does not attempt to open a connection, then the stateful firewall does the following (see Figure 3-19).

- If the packet is part of an established connection, it is passed without further inspection. (However, these packets can be further filtered if desired.)
- If the packet is not part of an established connection, then it must be spurious. It is dropped and logged.

These rules for ongoing communication are very simple to implement. Consequently, most packets are handled with very little processing power. This makes stateful firewalls very inexpensive.

If the packet does not attempt to open a connection,

Then if the packet is part of an established connection,

It is passed without further inspection.

(However, these packets can be given additional filtering if desired.)

Otherwise, it is dropped and logged.

Nearly all packets are not part of connection-opening attempts

This simplicity makes the cost of processing most packets minimal.

FIGURE 3-19 Stateful Inspection for Packets that Do Not Attempt to Open a Connection

PERSPECTIVE Although the simple operation of stateful inspection makes it inexpensive, stateful filtering provides a great deal of protection against attacks coming from the outside. This combination of low cost and strong security is responsible for the dominance of stateful inspection today.

Test Your Understanding

23. a) Why are states important? b) Why are ACLs needed for stateful firewalls? c) When a packet that is part of an ongoing connection arrives at a stateful inspection firewall, what does the firewall usually do? d) When a packet that is not part of an ongoing connection and that does not attempt to open a connection arrives at a stateful inspection firewall, what does the firewall do? e) Why are stateful firewalls attractive? f) What type of firewalls do most corporations use for their main border firewalls?

THE PROTECTION OF DIALOGUES WITH CRYPTOGRAPHIC SYSTEMS

We now continue our discussion of the protection phase in the plan-protect-respond cycle by looking at cryptographic protections for dialogues involving the exchange of many messages. Cryptography is the use of mathematics to protect dialogues.

Cryptography is the use of mathematics to protect dialogues.

Symmetric Key Encryption for Confidentiality

ENCRYPTION FOR CONFIDENTIALITY When most people think of cryptographic protection, they think of encryption for confidentiality. Encryption for confidentiality is the scrambling of messages so that communication is **confidential** (cannot be read by eavesdroppers). Encryption methods are called **ciphers.** The receiver **decrypts** (unscrambles) the message in order to read it. Figure 3-20 illustrates encryption for confidentiality.

SYMMETRIC KEY ENCRYPTION Most encryption for confidentiality uses **symmetric key encryption** ciphers, in which the two sides use the same key to encrypt messages to each other and to decrypt incoming messages. As Figure 3-20 shows, symmetric key encryption ciphers use only a single key for encryption by Party A and decryption by Party B. When Party B sends to Party A, in turn, Party B uses the single key to decrypt, while Party A uses the single key to decrypt. The dominant symmetric key encryption cipher today is the Advanced Encryption Cipher (AES).

KEY LENGTH Earlier, we looked at brute force password guessing. Symmetric and keys also can be guessed by the attacker's trying all possible keys. This is called **exhaustive search.** The way to defeat exhaustive key searches is to use long keys, which are merely binary strings. For symmetric key ciphers, lengths of 100 bits or greater are considered to be strong keys. AES supports multiple strong key lengths up to 256 bits.

Keys are long strings of bits.

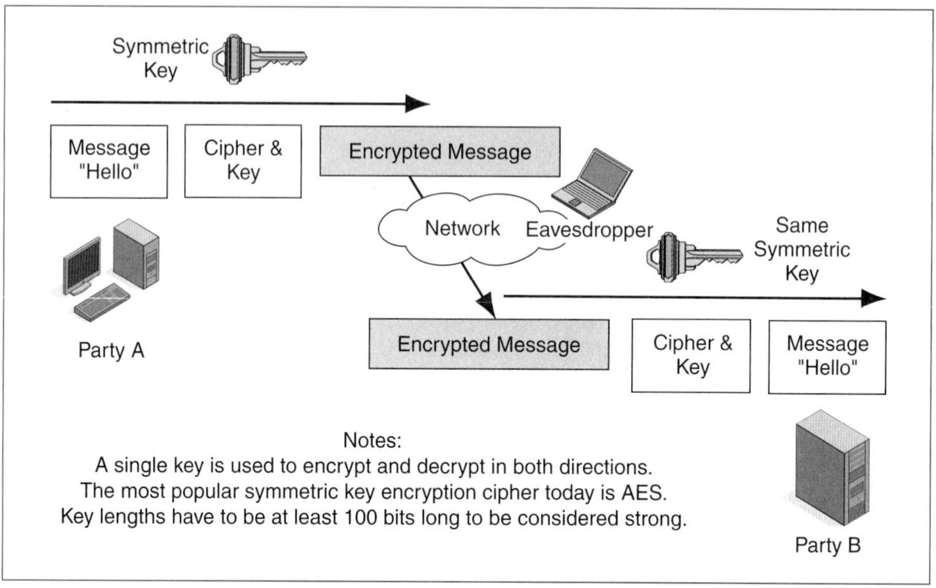

FIGURE 3-20 Symmetric Key Encryption for Confidentiality

Test Your Understanding

24. a) What is a cipher? b) What protection does confidentiality provide? c) In two-way dialogues, how many keys are used in symmetric key encryption? d) What is the minimum size for symmetric keys to be considered strong?

Electronic Signatures

AUTHENTICATION AND MESSAGE INTEGRITY In addition to encrypting each packet for confidentiality, cryptographic systems normally add **electronic signatures** to each packet. This is illustrated in Figure 3-21. Electronic signatures are small bit strings that

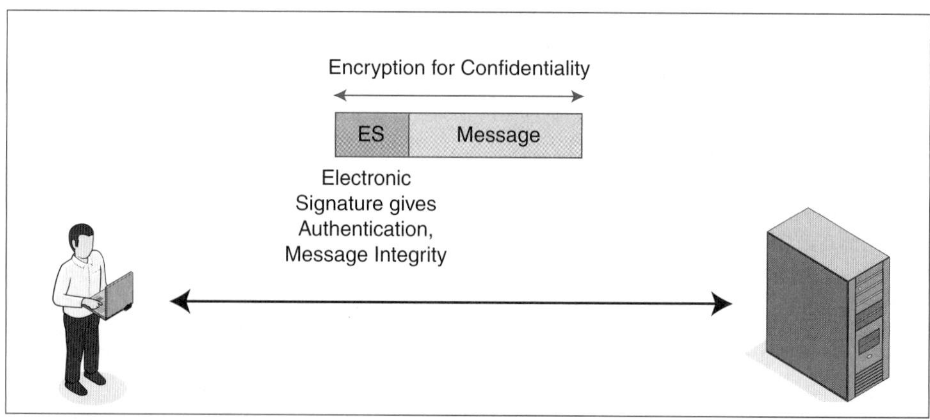

FIGURE 3-21 Electronic Signature

provide message-by-message authentication, much as people use signatures to authenticate individual written letters. An electronic signature allows the receiver to detect a message added to the dialogue by an impostor.

Electronic signatures also provide **message integrity,** meaning that the receiver will be able to detect it if the packet is changed in transit. Consequently, cryptographic systems provide three protections to every packet: message-by-message confidentiality, authentication, and message integrity.[12]

Test Your Understanding

25. What two protections do electronic signatures provide?

Cryptographic System Standards

Protecting dialogues requires the use of multiple cryptographic protections—even more than we have seen in this section. Users cannot be expected to master all of these protections. Fortunately, these protections come bundled in **cryptographic systems**, which provide all of these protections in an integrated way and in a way that is transparent to users. They provide initial authentication at the start of a dialogue. They also provide confidentiality, authentication, and message integrity for individual packets, plus a number of other protections.

In this book, we will see two major cryptographic system standards. One is SSL/TLS.[13] You may never have heard of it, but you have used it many times. **SSL/TLS**

Cryptographic Systems

- Packages of Cryptographic Protections
- Users do not have to know the details.
- Defined by cryptographic system standards

SSL/TLS

- Cryptographic system standard widely used in sensitive browser–webserver communication
- Medium-strength security
- Easy to implement

IPsec

- Protects IP packets and all of their embedded contents
- Very strong security
- Expensive to implement

FIGURE 3-22 Cryptographic Systems (Study Figure)

[12] There are two common types of electronic signatures—digital signatures and key-hashed message authentication codes. Many writers focus on digital signatures, but digital signatures are not very common because they are expensive to implement.

[13] Originally, this standard was called SSL (Secure Sockets Layer). Later, this was changed to TLS (Transport Layer Security). This book calls it SSL/TLS to reflect both names.

is the nearly universal cryptographic system standard for communication between browsers and webservers. Nearly all e-commerce sites use SSL/TLS to secure financial transactions. In addition, many web-based e-mail systems use SSL/TLS to secure communication between the browser and the webserver. Although SSL/TLS only offers medium security, it is sufficient for most consumer needs and many corporate needs, and it is essentially free, being built into every browser and webserver.

A more secure (and more expensive to implement) cryptographic system standard is **IPsec**, which is short for IP security. This IETF standard protects individual IP packets, including all of the transport and application layer messages encapsulated in the packet. Although IPsec has very high security, it is expensive to implement.

Test Your Understanding

26. a) Users are not experts in cryptography. How can users be given cryptographic protection without the need to know cryptography? b) What three protections do cryptographic systems provide for individual messages? c) In what way is SSL/TLS more attractive than IPsec? d) In what way is IPsec more attractive than SSL/TLS?

RESPONDING

The last stage in the plan-protect-respond cycle is *responding*. Inevitably, some attacks will succeed and will require corrective action. **Response** is reacting to a security incident according to plan. The "according to plan" part is crucial. The amount of damage done in these compromises depends heavily on how quickly and how well the organization responds. Without strong and well-rehearsed response plans, response will take far too long and is likely to be less effective than it should be. For example, the case study at the end of this chapter shows that Walmart has a well-rehearsed response plan for natural disasters. In Hurricane Katrina, while the Federal Emergency Response Agency stumbled badly, Walmart responded quickly and effectively.

Response is reacting to a security incident according to plan.

Stages

There are four general stages in responding to an attack.

DETECTING THE ATTACK The first stage is detecting the attack. Detection can be done by technology or simply by users reporting apparent problems. Obviously, until an attack is detected, the attacker will be able to continue doing damage. Companies need to develop strong procedures for identifying attacks quickly.

STOPPING THE ATTACK The second stage is stopping the attack. The longer an attack has to get into the system, the more damage the hacker can do. Reconfiguring corporate firewall ACLs may be able to end the attack. In other cases, attack-specific actions will have to be taken.

REPAIRING THE DAMAGE The third stage is repairing the damage. In some cases, this is as simple as running a cleanup program or restoring files from backup tapes. In other

Stages

- Detecting the attack
- Stopping the attack
- Repairing the damage
- Punishing the attacker?

Major Attacks and CSIRTs

- Major incidents are those the on-duty staff cannot handle.
- Computer security incident response team (CSIRT)
  - Must include members of senior management, the firm's security staff, members of the IT staff, members of functional departments, and the firm's public relations and legal departments

Disasters and Disaster Recovery

- Natural and humanly-made disasters
- IT disaster recovery for IT
  - Dedicated backup sites and transferring personnel
  - Having two sites that mutually back up each other
- Business continuity recovery
  - Getting the whole firm back in operation
  - IT is only one player.

Rehearsals

- Rehearsals are necessary for speed and accuracy in response.
- Time literally is money.

FIGURE 3-23 Incident Response (Study Figure)

cases, it may involve the reformatting of hard disk drives and the complete reinstallation of software and data.

PUNISHING THE ATTACKER? The fourth general stage is punishing the attacker, but this stage often is skipped. Punishing the attacker is relatively easy if the attacker is an employee. In general, however, attackers are extremely difficult to track down. Even if they are found, prosecution may be difficult or impossible.

If legal prosecution is to be pursued, it is critical for the company to use proper forensic procedures to capture and retain data. **Forensic** procedures are ways to capture and safeguard data in ways that fit rules of evidence in court proceedings. These rules are very complex, and it is important for the firm to use certified forensics professionals. Even if an employee is fired, it is desirable for the company to use good forensic procedures to avoid a potential lawsuit.

Major Incidents and CSIRTs

Minor attacks can be handled by the on-duty IT and security staff. However, during **major incidents**, such as the theft of thousands of credit card numbers from a corporate host, the company must convene the firm's **computer security incident response team (CSIRT)**, which is trained to handle major incidents.

The key to creating CSIRTs is to have the right mix of talents and viewpoints. Major attacks affect large parts of the firm, so the CSIRT must include members of senior management, the firm's security staff, members of the IT staff, members of functional departments, and the firm's public relations and legal departments.

Disasters and Business Continuity

When natural disasters, terrorist attacks, or other catastrophes occur, the company's basic operations may be halted. This can be extremely expensive. Companies must have active disaster recovery plans to get their systems working quickly.

IT disaster recovery is the reestablishment of information technology operations. Many large firms have dedicated backup sites that can be put into operation very quickly, after data and employees have been moved to the backup site. Another option, if a firm has multiple server sites, is to do continuous data backup across sites. If one site fails, the other site can take over immediately or at least very rapidly.

More broadly, **business continuity recovery** goes beyond IT disasters to deal with events that affect enough of a firm to pause or stop the functioning of the business. IT security is only one player in business continuity recovery teams.

Rehearsals

"Practice makes perfect" is time-honored advice. It certainly is true for major attacks that must be handled by CSIRTs, and it is doubly true for disaster recovery. It is important for the company to establish CSIRT and disaster teams ahead of time and to have them rehearse how they will handle major attacks and disasters. Although practice does not really make perfect, it certainly improves response speed and quality. During the first two or three rehearsals, team members will work together awkwardly, and there will be many mistakes. Rehearsals will also reveal flaws in the company's major incident and disaster response plans. It is important to work through these problems before the firm is in a real crisis.

Test Your Understanding

27. a) What is the definition of response? b) What are the two benefits of a well-rehearsed response plan? c) What are the four response phases when attacks occur? d) What is the purpose of forensic tools? e) Why are CSIRTs necessary? f) Should the CSIRT be limited to security staff personnel? g) Distinguish between disaster recovery and business continuity recovery. h) Explain how firms use backup sites in disaster recovery.

CONCLUSION

Synopsis

ATTACKS Companies today suffer compromises from many different types of attacks.

- Viruses attach themselves to other programs and need human actions to propagate—most commonly by users opening e-mail attachments that are infected programs. Worms are full programs; they can spread by e-mail, but directly propagating worms can propagate on their own, taking advantage of unpatched vulnerabilities in victim hosts. Some vulnerability-enabled worms can spread through the Internet host population with amazing speed. Many worms and viruses carry damaging payloads. Often, payloads place Trojan horse programs or other types of exploitation software on the victim computer. Malware is the general name for evil software.

- Viruses, worms, and Trojan horses are not the only attacks that are aimed at individuals. Spam deluges the victim with unsolicited commercial e-mail, and messages often are fraudulent. Spyware collects information about users and sends this information to an attacker. Phishing attacks use an official-looking e-mail message or website to trick users into divulging passwords and other special information. Attacks on individuals, including e-mail virus and worm attacks, often depend on social engineering—tricking the victim into doing something against his or her best interests. Two common goals of attacks on individuals are credit card number theft, in which a credit card number is stolen, and identity theft, in which enough private information is stolen to enable the attacker to impersonate the victim in large financial transactions.
- Hacking is the intentional use of a computer resource without authorization or in excess of authorization. Hacking break-ins typically require a prolonged series of probing actions on the part of the attacker.
- Denial-of-service (DoS) attacks overload victim servers so that they cannot serve users. Distributed DOS (DDoS) attacks use bots to carry out the attack. Bots can be updated to take on new functionality.

ATTACKERS Traditionally, most attackers were curiosity-driven hackers and disgruntled employees and ex-employees. Now, criminals dominate the attack world, and the money their crimes generate enables them to invest in new technology and hire top hackers. On the horizon, cyberterror attacks by terrorists and cyberwar attacks by national governments could do unprecedented levels of damage.

SECURITY MANAGEMENT Security is primarily a management issue, not a technical issue. Planning involves risk analysis (balancing the costs and benefits of protections), creating comprehensive security (closing all avenues of attack), using defense in depth (establishing successive lines of defense in case one line of defense fails), and access control using minimum permissions.

We looked at policy-based security in which a high-level policy group creates security policies and lower-level staff members implement the policy. Policies specify what (is to be done). Implementation focuses on how (to do it). This division of labor works because high-level policy people have a broad understanding of security risks and must create policies that will give comprehensive security. Implementation is done by lower-level staff members who know the technology and local situation in detail. Sometimes, the policy group creates intermediate implementation guidance consisting of standards (which must be followed) and guidelines that must be considered, although they do not have to be followed if there is good reason not to.

ACCESS CONTROL AND AUTHENTICATION Firms need to control access to their assets. Access control normally requires authentication—proving the identity of the person wishing access. The person requesting access is the supplicant, and the device requiring proof of identity is the verifier. The supplicant sends credentials to the verifier to prove the supplicant's identity. For consistency, a central authentication server is used to do the credential checking. There are several common authentication technologies.

- Passwords are inexpensive and easy to use, but users typically choose poor passwords that are easy to crack. Passwords should be used only for low-sensitivity resources.
- Access cards are often used for door entry.
- Biometrics promises to use bodily measurements to authenticate supplicants, replacing other forms of authentication. Concerns with biometrics include cost, error rates, and the effectiveness of deliberate deception by supplicants.
- Digital certificate authentication at the other extreme gives the strongest authentication, but it is complex and expensive to implement.

FIREWALLS Firewalls examine packets passing through the firewall. If a firewall finds provable attack packets, it drops them and records information about them in a log file. If a packet is not a provable attack packet—even if it really is an attack packet—the firewall will not drop it. Ingress filtering examines packets coming into the firm; egress filtering examines packets going out of the firm.

Most firewalls use stateful inspection, which divides communication into stages called states. During the risky connection-opening state, the firewall does extensive work to decide whether to allow connections by passing packets attempting to open a connection. This requires examining every connection-opening packet against rules in an access control list (ACL).

If a packet arrives that does not attempt to open a connection, then it is checked against allowed connections. If it is part of an approved connection, it is passed—usually with little or no additional filtering. It the packet that does not try to open a connection is not part of an approved connection, it is dropped.

Stateful packet inspection firewalls give strong security during the connection-opening state. This is processing intensive, but few packets are parts of connection-opening attempts. For most packets, SPI firewalls do simple inspection, which requires little processing power.

CRYPTOGRAPHIC SYSTEMS Cryptography is the use of mathematics to protect message dialogues. One key protection is encryption for confidentiality, which encrypts messages to prevent attackers from reading any messages that they intercept. Encryption methods are called ciphers. In symmetric key encryption, both sides encrypt and decrypt with a single key. To be strong, a symmetric encryption key needs to be more than 100 bits long.

In addition to providing message-by-message encryption, cryptographic systems also provide message-by-message authentication by adding an electronic signature to each message. Electronic signatures also give message integrity.

Cryptographic protections are organized into cryptographic system standards. SSL/TLS provides medium-strength protection; it is built into all browsers and webservers. IPsec is a very strong cryptographic system standard. It protects IP packets and encapsulated transport layer and application layer messages.

RESPONSE Protections occasionally break down. Response is reacting to compromises according to plan. The stages in response to attack typically include identifying the attack, stopping the attack, recovering from the attack, and (sometimes) punishing the attacker. Major incidents require the convening of a computer security incident response team (CSIRT). IT disaster recovery requires getting IT back in operation at another site, while business continuity recovery involves getting the entire firm back in operation. It is important for recovery teams to conduct rehearsals before problems occur so that they can respond quickly and correctly.

END-OF-CHAPTER QUESTIONS

Thought Questions

1. a) Suppose that an attack would do $100,000 in damage and has a 15 percent annual probability of success. Spending $9,000 per year on "Measure A" would cut the annual probability of success by 75 percent. Do a risk analysis comparing benefits and costs. Show your work clearly. b) Should the company spend the money? Explain. c) Do another risk analysis if Measure A costs $20,000 per year. Again, show your work. d) Should the company spend the money? Explain.
2. a) What form of authentication would you recommend for relatively unimportant resources? Justify your answer. b) What form of authentication would you recommend for your most sensitive resources?
3. For each of the following passwords, first state the kind of attack that would be necessary to crack it. Justify your answer. Then say whether or not it is an adequate password, again giving specific reasons. a) angelfish b) Surpassing1 c) BuFfALo d) 2T%v e) 5g*7sN8$^1
4. Keys and passwords must be long. Yet most personal identification numbers (PINs) that you type when you use a debit card are only four or six characters long. Yet this is safe. Why?
5. Revise the ACL in Figure 3-18 to permit access to an FTP server with IP address 10.32.67.112.
6. In digital certificate authentication, the supplicant could impersonate the true party by doing the calculation with the true party's private key. What prevents impostors from doing this?

Online Exercise

1. Go to http://www.cybercrime.gov. Go to the section on computer crimes. Select one of the cases randomly. Describe the type of attacker and the type of attack(s).

Case Study: Patco

In 2009, the Patco Construction Company had $588,000 drained from its bank accounts at Ocean Bank. The theft involved six withdrawals on May 8, May 11, May 12, May 13, May 14, and May 15. The money in each withdrawal was sent to a group of money mules.

After thieves stole all of the company's cash, they continued to make withdrawals. Patco's bank continued to allow withdrawals, covering them with over $200,000 from Patco's line of credit. Although the bank was able to recover or block $243,406 in transfers, Patco was still out $345,400. In addition, the bank began charging Patco for interest on the money that had been withdrawn using Patco's line of credit.

Although the transactions were far larger than Patco normally made, Ocean Bank did not inform Patco of any problems until one of the account numbers entered by the thieves was invalid. It sent a notification by mail, and it did not arrive at Patco until several days later. Patco notified the bank of problems the next morning. However, the bank had already sent out $111,963 that day, some of which was recovered.

The bank used account numbers and passwords. For transactions over $1,000, Patco employees had to answer two challenge questions. Most withdrawals were over $1,000, so employees had to answer these same challenge questions many times. Patco believes that these challenge messages were too easy.

The State of Maine has stringent banking laws. The Federal Financial Institutions Examination Council in 2005 required banks to use at least two-factor authentication and specifically noted that usernames and passwords were not enough. Patco sued People's United Bank for its losses, claiming that the challenge questions were nothing more than a second set of passwords and that the bank should have required much stronger credentials.

Patco also claimed that Ocean Bank should have been suspicious when such large unprecedented withdrawals were made and when they were sent to 30 different accounts. Normally, Patco only withdrew money for payrolls on Fridays. Its previous largest single-day withdrawal had been under $37,000. Patco's complaint stated that based on belief and information from the bank, Patco assumed that antifraud monitoring was being done by the bank.

Ocean Bank did not comment on the case, but most banks in a similar situation use the defense that they were not negligent. A bank can only be found negligent if it has lower protections than is the norm in the industry.

Caution. The information in this case is based on Patco's complaint.[14] Consequently, the statements made in the case are unvalidated and may be disputed by Ocean Bank as being nonfactual. Analyze the case based on Patco's allegations, but do not draw firm conclusions against the bank.

---

1. a) According to the information in the case, do you think the bank satisfied the requirement to use two-factor authentication? b) According to the information in the case, do you think the bank was doing antifraud monitoring? c) According to the information in the case, do you think Ocean Bank was negligent? d) According to the information in the case, if you were the head of Ocean Bank, what would you do to prevent the reoccurrence of this problem?

Case Study: Walmart

In 2005, Hurricane Katrina slammed into Louisiana and Mississippi, devastating New Orleans and many other cities along the U.S. Gulf Coast. Shortly afterward, the fourth most intense Atlantic hurricane in history, Rita, added enormously to the destruction. The Federal Emergency Management Agency (FEMA) became notorious for its handling of the crisis, responding belatedly and acting ineptly when it did respond.

Many businesses collapsed because they were poorly prepared for the hurricanes. One company that *did* respond effectively was Walmart.[15] In its Brookhaven, Mississippi, distribution center, the company had 45 trucks loaded and ready for delivery even before Katrina made landfall. The company soon supplied $20 million in cash donations, 100,000 free meals, and 1,900 truckloads full of diapers, toothbrushes, and other emergency supplies to relief centers. The company also supplied

---

[14] Patco Construction Company, Inc., plaintiff, v. People's United Bank, d/b/a Ocean Bank, defendant. State of Maine, York SS Superior Court Civil Action, Docket No. 09-CV.

[15] Liza Featherstone, "Wal-Mart to the Rescue!" *The Nation*, September 26, 2005. www.thenation.com/doc/20050926/featherstone. Michael Barbaro and Justin Gillis, "Wal-Mart at Forefront of Hurricane Relief," *Washington Post.com*, September 6, 2005. www.washingtonpost.com/wp-dyn/content/article/2005/09/05/AR2005090501598.html. *NewsMax.com*, "Wal-Mart Praised for Hurricane Katrina Response Efforts," www.newsmax.com/archives/ic/2005/9/6/164525.shtml. Ann Zimmerman and Valerie Bauerlein, "At Wal-Mart, Emergency Plan has Big Payoff," *Wall Street Journal*, September 12, 2005, B1.

flashlights, batteries, ammunition, protective gear, and meals to police and relief workers.

Although the relief effort was impressive, it was merely the visible tip of Walmart's disaster recovery program. Two days before Katrina hit, Walmart activated its business continuity center. Soon, 50 managers and experts in specific areas such as trucking were hard at work. Just before the storm knocked out the company's computer network, the center ordered the Mississippi distribution center to send out recovery merchandise such as bleach and mops to its stores. The company also sent 40 generators to its stores so that stores that lost power could open to serve their customers. It also sent out many security employees to protect stores.

After computer networks failed, the company relied on the telephone to contact its stores and other key constituencies. Most stores came back immediately, and almost all stores were able to serve their customers within a few days. Lines of customers were long, and Walmart engaged local law enforcement to help maintain order.

Walmart was successful because of intensive preparation. The company has a full-time director of business continuity. It also has detailed business continuity plans and clear lines of responsibility. In fact, while the company was still responding to Katrina and Rita, it was monitoring a hurricane off Japan, preparing to take action there if necessary.

1. a) Why was Walmart able to respond quickly? b) List at least three actions that Walmart took that you might not have thought of.

Perspective Questions

1. What was the most surprising thing you learned in this chapter?
2. What was the most difficult part of this chapter for you?

CHAPTER 4

Network Management

LEARNING OBJECTIVES

By the end of this chapter, you should be able to:

- Explain general concepts in network management, including a focus on the system life cycle, the importance of cost, and strategic network planning.
- Use quality-of-service (QoS) criteria in product selection.
- Do basic design work, including doing traffic analysis with and without redundancy, knowing common topologies, selecting topologies, understanding how to handle momentary traffic peaks, understanding how to reduce capacity needs through traffic shaping and compression, searching for natural designs, and understanding why network simulation software is used.
- Evaluate alternatives using multicriteria decision making and specifying costs.
- Describe operational management, including OAM&P, the Simple Network Management Protocol (SNMP), and network management software.

INTRODUCTION

Today, we can build much larger networks than we can manage easily. For example, even a mid-size bank is likely to have 500 Ethernet switches and a similar number of routers. Furthermore, network devices and their users are spread out over large areas—sometimes international areas. While network technology is exciting to talk about and concrete conceptually, network management is where the rubber reaches to road.

SDLC versus SLC

In programming and database courses, you focus on the systems development life cycle (SDLC), which looks at your information system from its moment of conception to its implementation. When you think about this, it is rather strange. Teaching the SDLC is like training new doctors in obstetrics and ignoring all training for illnesses after birth. Although the SDLC discusses episodic software maintenance activities, Figure 4-1 shows that the SDLC is far more limited than the **systems life cycle (SLC)**, which lasts from conception until death.

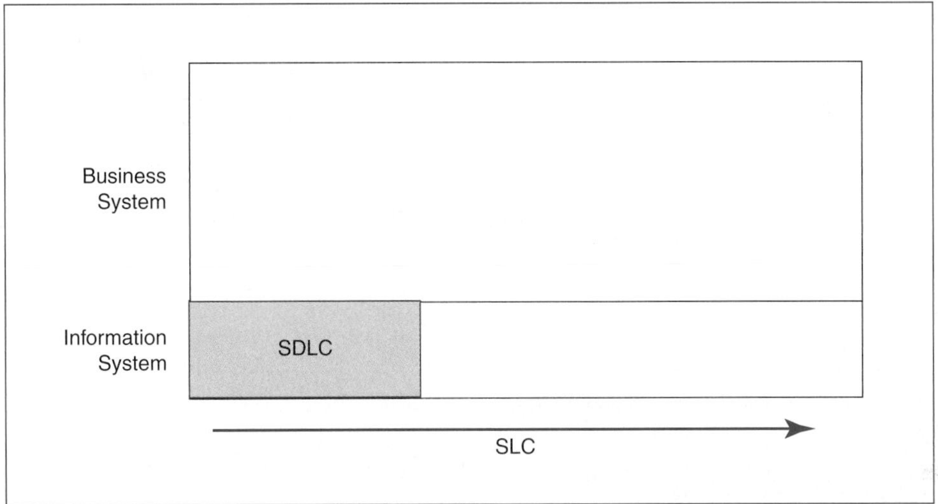

FIGURE 4-1 The Systems Development Life Cycle versus Business Systems and the Systems Life Cycle

In networking, the creation of new networks and the modification of old networks are important, but the real work of networking professionals is the administration of ongoing networks. While systems administrators who manage servers have a difficult job, especially in today's world of virtualization, network administrators are faced with very complex tasks that require a high level of understanding of how networks work. The operational phase of a network after its creation is an enormous part of the network professional's job.

Figure 4-1 also shows that while database and programming professionals have to focus on the information system, the network professional often has to focus on the broader business system in which the information system is embedded. This is particularly true in network security.

Test Your Understanding

1. a) Why must networking professionals be concerned with the SLC rather than the SDLC? b) Why is a focus on information systems insufficient in networking?

Cost

In networking, you can never say, "Cost doesn't matter." Figure 4-2 illustrates that network demand is likely to grow rapidly in the future, just as it has always done in the past. The figure also illustrates that network budgets are growing very slowly (if they are growing at all).

Taken together, these curves mean that network budgets are always stretched thin. If the network staff spends too much money on one project, it will not have enough money left over to do another important project. Although there are many concerns beyond costs, cost is always an important consideration in network management.

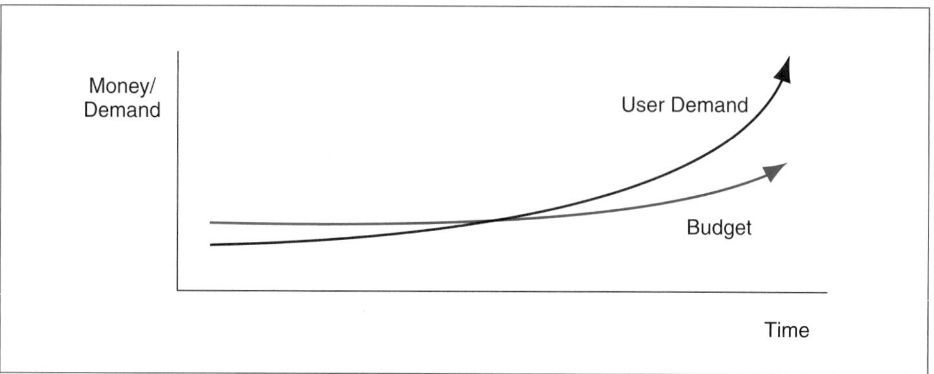

FIGURE 4-2 Network Demand and Budget

Test Your Understanding

2. a) Compare trends in network demand and network budgets. b) What are the implications of these trends?

Strategic Network Planning

One of the most important things to plan in networking is the **technological infrastructure**—the firm's arrangement of hardware, software, and transmission lines that allows the network to carry information. Figure 4-3 shows how organizations do strategic network planning.

WHAT-IS ANALYSIS Planning for changes in the technological infrastructure must begin with the "what is," that is, the current state of the company's network. This may sound like an easy task, but most firms do not have a thorough understanding of their

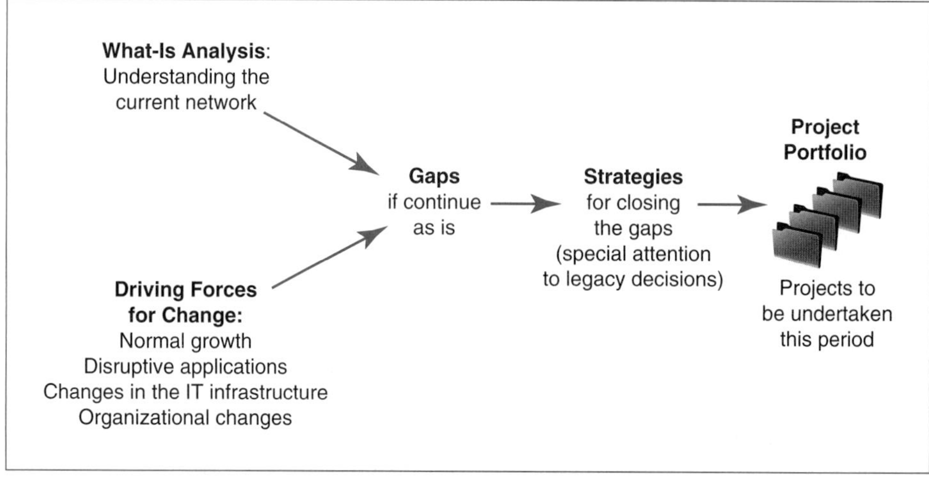

FIGURE 4-3 Strategic Network Planning

network components and interactions, much less of their trouble spots. What-is analysis begins with an exhaustive inventory of the network's components and their interrelationships. This sounds simple. It is not.

DRIVING FORCES FOR CHANGE While it is important to understand the current state of the network, companies must remember that today's technological infrastructure will not be sufficient for the future because many things will change. Companies need to consider the major driving forces that will require changes in the network. Some of these driving forces are:

- The normal continuing growth of application traffic demand. In most firms, traffic has been growing at an increasing rate. This will certainly continue in the future.
- The introduction of disruptive applications, which may create major surges in demand far beyond traditional patterns. Voice over IP is an obvious example. However, if video applications begin to grow, capacity planning will become extremely difficult.
- Changes in other elements of the IT technological infrastructure can also require extensive network changes. One long-term trend has been the consolidation of data centers from many to few. This can radically change traffic flows within the corporate network.
- Organizational change can be a major driving force. If a company is adding a site, not only will the site have to be served but communication between different parts of the firm will change, depending on what units are moved there. In fact, all corporate reorganizations are likely to impact network planning. At the extreme are nightmare scenarios that exist if the company is bought out or buys out another firm.

GAPS ANALYSIS Comparing driving forces to the what-is network will create inevitable gaps between what the firm will need and what the current network can provide. These gaps must be identified, characterized, and documented.

STRATEGIES FOR CLOSING THE GAPS The firm then needs to develop strategies for closing the gaps. It must consider multiple technologies and multiple topologies (physical connections) for each gap. As we will see later in this chapter, developing a strategy for closing a gap may benefit from network simulation programs.

SELECTING A PROJECT PORTFOLIO The network staff will not have the budget to close all gaps satisfactorily. So it is critical for the networking staff to be very selective in how it spends its money. The final stage in strategic planning is to create a **project portfolio**—a selection of projects that the firm will implement during the plan's initial period.

A strong consideration for any project is whether it involves **legacy decisions** that will lock the company into a specific vendor or technology option for several years. Making legacy decisions is not wrong. In fact, making a legacy decision is often necessary. However, because of its implications for the firm, companies need to give special scrutiny to potential projects that require legacy decisions.

Test Your Understanding

3. a) What is what-is analysis? b) List the four driving forces for change. c) For each, give an example not listed in the text. d) What is gaps analysis? e) Why is it necessary to create a project portfolio? f) What are legacy decisions, and why must projects that involve legacy decisions be judged with great care?

NETWORK QUALITY OF SERVICE (QoS)

In the early days of the ARPANET and the Internet, networked applications amazed new users. However, these users soon said, "Too bad this thing doesn't work better." Today, networking is a mission-critical service for corporations. If the network breaks down, much of the organization comes to a halt. Today, networks must work, and they must work *well*. Companies are concerned with network **quality-of-service (QoS) metrics**, that is, quantitative measures of network performance. Figure 4-4 shows that companies typically use a number of QoS metrics to quantify their quality of service so that they can set targets and determine if they have met those targets.

Test Your Understanding

4. a) What are QoS metrics? (Do not just spell out the acronym.) b) How are QoS metrics used?

Transmission Speed

There are many ways to measure how well a network is working. The most fundamental metric, as we saw in Chapter 1, is speed. While low speeds are fine for text messages, the need for speed becomes very high as large volumes of data must be sent, and video transmission requires extremely high transmission speed.

BITS PER SECOND (BPS) As we saw in Chapter 1, transmission speed[1] normally is measured in bits per second (bps). A bit is either a one or a zero. Obviously, a single bit

Quality-of-Service (QoS) Metrics

Quantifiable measures of network performance

Examples

Speed
Availability
Error rates
. . .

FIGURE 4-4 Network Quality of Service (QoS)

[1] Purists correctly point out that *speed* is the wrong word to use to describe transmission rates. At faster transmission rates, bits do not physically travel faster. The sender merely transmits more bits in each second. Transmission rates are like talking faster, not running faster. However, transmission rates are called transmission speeds almost universally, so we will follow that practice in this book.

cannot convey much information. Speeds today range from thousands of bits per second to billions of bits per second. To simplify the writing of transmission speeds, professionals add metric prefixes to the base unit, bps. For example, Figure 4-5 shows that in increasing factors of 1,000 (not 1,024 as with computer memory), we have kilobits per second (kbps), megabits per second (Mbps), gigabits per second (Gbps), and terabits per second (Tbps).

Speeds are measured in factors of 1,000, not 1,024.

Consistent with metric notation, kilo is abbreviated as lowercase k instead of uppercase K. However, megabits per second is Mbps, gigabits per second is Gbps, and terabits per second is Tbps.

Speed

Normally measured in bits per second (bps)

Not bytes per second

Metric prefixes increase in factors of 1,000 (not 1,024)

The metric abbreviation for kilo is lowercase k

Abbreviation	Meaning	Name	Example
1 kbps	1,000 bps	kilobits per second	3 kbps is 3,000 bps
1 Mbps	1,000 kbps	megabits per second	3.4 Mbps is 3,400,000 bps
1 Gbps	1,000 Mbps	gigabits per second	62 Gbps is 62,000,000,000 bps
1 Tbps	1,000 Gbps	terabits per second	

Sometimes speed is measured in bytes per second, Bps, compared to bps.

Bps usually only for file transfers

Expressing Speed in Proper Notation

As Written	Places Before Decimal Point	Space Between Number and Prefix?	Properly Written
23.72 Mbps	2	Yes	23.72 Mbps
2,300 Mbps	4	No	2.3 Gbps
0.5 Mbps	0 (leading zeros do not count)	No	500 kbps

There must be one to three spaces before the decimal point.

Leading zeros do not count.

(continued)

FIGURE 4-5 Transmission Speed (Study Figure)

There must be a space between the number and the units.

12 Mbps is proper; 12Mbps is improper.

If the number is decreased by 1,000 (4,523 becomes 4.523), then the prefix must be increased by a thousand (kbps to Mbps).

4,523 kbps becomes 4.523 Mbps.

(4,523 / 1000 * kbps * 1000)

If the number is increased by 1,000 (0.45 becomes 450), then the prefix must be decreased by 1,000 (Mbps to kbps).

0.45 Mbps becomes 450 kbps.

Rated Speed and Throughput

Rated Speed

The speed a system *should* achieve.

According to vendor claims or to the standard that defines the technology

Throughput

The data transmission speed a system *actually* provides to users.

Aggregate versus Individual Throughput on Shared Lines

The aggregate throughput is the total throughput available to all users.

The individual throughput is an individual's share of the aggregate throughput.

FIGURE 4-5 Continued

WRITING NUMBERS IN PROPER NOTATION Networking professionals write speeds in a very specific way. The basic rule for writing speeds (and metric numbers in general) in proper notation is that there should be one to three places before the decimal point and that there should be a space between the number and the units. Figure 4-5 illustrates how to write speeds properly.

> To write a speed in proper notation, there should be one to three places before the decimal point, and there should be a space between the number and the units.

- Given this rule, 23.72 Mbps is fine (two places before the decimal point and a space between the number and the metric prefix).
- However, 2,300Mbps has four places before the decimal point (2,300.00), so it should be rewritten as 2.3 Gbps (one place). Note that a space has been added between the number and its metric prefix. In turn, 0.5 Mbps has zero places to the left of the decimal point. (Leading zeros do not count.) It should be written as 500 kbps (three places).

When you look at a number in metric notation, remember that if $a = b*c$, then a also equals b*1000 * c/1000 or b/1000 *b*1000. Think of the metric prefix k as 1,000 and the metric prefix M as 1,000,000.

Suppose you have the speed 4,523 kbps. To write the number properly, you divide it by 1,000 to get 4.523. If you divide the number by 1,000, then you must multiply the prefix by 1,000. Multiplying kbps by 1,000 gives Mbps, so the number in proper notation becomes 4.523 Mbps.

To give another example, suppose you have 0.45 Mbps. You need to multiply the number by 1,000, getting 450. You then have to divide the prefix (Mbps) by 1,000 to give you kbps. The number in proper notation, then, is 450 kbps.

RATED SPEED VERSUS THROUGHPUT

Note: Some students find the distinction between rated speed and throughput difficult to learn. However, we must use this distinction throughout this book, so be sure to take the time to understand it.

Talking about transmission speed can be tricky. A network's **rated speed** is the speed it *should* achieve based on vendor claims or on the standard that defines the technology. For a number of reasons, networks often fail to deliver data at their rated speeds. In contrast to rated speed, a network's **throughput** is the data transmission speed the network *actually* provides to users.

Throughput is the data transmission speed a network *actually* provides to users.

AGGREGATE VERSUS INDIVIDUAL THROUGHPUT When a transmission line on a network is multiplexed, this means that several conversations between users will share the line's throughput. Consequently, it is important to distinguish between a line's **aggregate throughput**, which is the total it provides to all users who share it, and the **individual throughput** that single users receive as their shares of the aggregate throughput. As you learned as a child, sharing is theoretically good, but it means that you get less.

Test Your Understanding

5. a) In what units is transmission speed normally measured? b) Is speed normally measured in bits per second or bytes per second? c) Give the names and abbreviations for speeds in increasing factors of 1,000. d) What is 55,000,000,000 bps with a metric prefix? e) Write out 100 kbps in bits per second (without a metric prefix). f) Write the following speeds properly: 0.067 Mbps, 23,000 kbps, and 48.62Gbps.

6. a) Distinguish between rated speed and throughput. b) Distinguish between individual and aggregate throughput.

Other Quality-of-Service Metrics

Although network speed is important, it is not enough to provide good quality of service. We will look briefly at other important QoS metric categories.

AVAILABILITY One of these other metrics is **availability**, which is the percentage of time that the network is available for use. In contrast, **downtime** is the percentage of time that the network is not available.

Availability

The percentage of time a network is available for use

Downtime is the amount of time a network is unavailable (minutes, hours, days, etc.).

Error Rates

Require retransmissions

When an error occurs, TCP assumes there is congestion and slows its rate of transmission.

Packet error rate: The percentage of packets that have errors

Bit error rate: The percentage of bits that have errors

Latency and Jitter

- Latency
  - Delay, measured in milliseconds
- Jitter (Figure 4-7)
  - Variation in latency between successive packets
  - Makes voice sound jittery

Application Response Time

The time from when the user hits a key to when the user receives a response. (Figure 4-8)

Includes the two-way network latency

Includes contributions from the host or application program

Often, configuration problems produce application response time problems.

Improvement requires cooperation between networking and host administration.

Service Level Agreements

- Guarantees for performance
- Penalties if the network does not meet its service metrics guarantees
- Guarantees specify worst cases (no worse than)
  - Lowest speed (e.g., no worse than 1 Mbps)
  - Maximum latency (e.g., no more than 125 ms)
- Often written on a percentage basis
  - No worse than 100 Mbps 99.5% of the time

FIGURE 4-6 Quality of Service II (Study Figure)

Ideally, systems would be available 100 percent of the time, but that is impossible in reality. On the Public Switched Telephone Network, the availability target usually is 99.999 percent. This is known as the "five nines." Data networks generally have lower availability but are under pressure to improve their availability given the cost of network downtime to firms today.[2]

---

[2] On a more detailed basis, availability can be discussed in terms of the mean time to failure (MTTF) and the mean time to repair (MTTR). The former asks how frequently downtime occurs. The latter asks how long service is down after a failure begins. More short failures may be preferable to infrequent but very long outages.

ERROR RATES We will see later in this chapter that hosts send data in small messages called packets. Ideally, all packets would arrive intact, but this does not always happen. The **packet error rate** is the percentage of packets that are lost or damaged during delivery. The **bit error rate**, in turn, is the percentage of bits that are lost or damaged.

Most networks today have very low average error rates. However, when the network is overloaded, error rates can soar because the network has to drop the packets it cannot handle. Companies must measure error rates when traffic levels are high in order to have a good understanding of error rate risks.

The impact of even small error rates can be surprising. TCP is designed to avoid network congestion by generating TCP segments slowly at the beginning of a connection. If the segments arrive correctly, TCP generates segments more quickly. However, if there is an error or if an acknowledgment is lost, the TCP process assumes that the network is overloaded. It falls back to its initial slow start rate for creating TCP segments. Consequently, even a small error rate can produce a major drop in throughput for applications.

LATENCY AND JITTER When packets move through the network, they will encounter some delays. The amount of delay is called **latency**. Latency is measured in **milliseconds (ms)**. A millisecond is a thousandth of a second. When latency reaches about 125 milliseconds, turn-taking in telephone conversations becomes difficult.

A related concept is **jitter**, which Figure 4-7 illustrates. Jitter occurs when the latency between successive packets varies. Some packets will come too far apart in time, others too close in time. While jitter does not bother most applications, VoIP and streaming media are highly sensitive to jitter. If the sound is played back without adjustment, it will speed up and slow down. These variations often occur over millisecond time periods, and, as the name suggests, variable latency tends to make voice sound jittery.

Most networks were engineered to carry traditional data such as e-mail and database transmissions. In traditional applications, latency was only slightly important, and jitter was not important at all. However, as voice over IP and video over IP, which are sensitive to jitter and to some extent latency, have grown in importance, companies have begun to worry more about latency and jitter. They are finding that extensive network

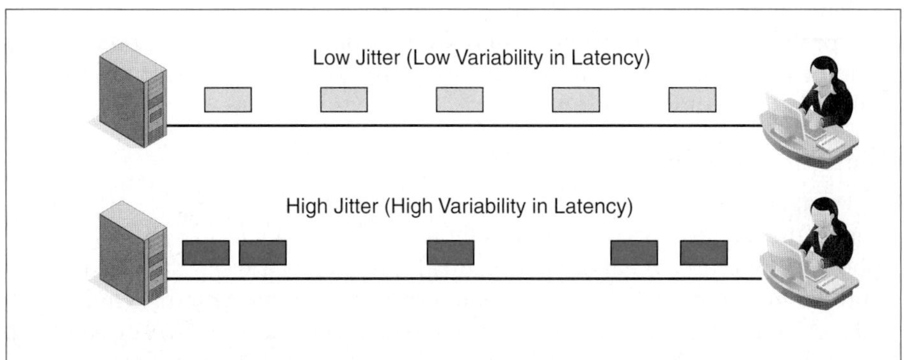

FIGURE 4-7 Jitter

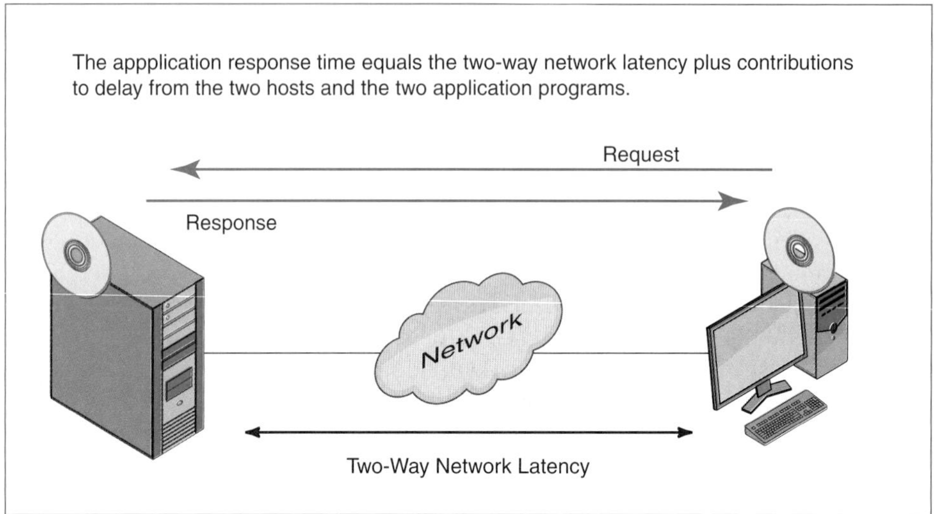

FIGURE 4-8 Application Response Time

redesign may be needed to give good control over latency and jitter. This may include fork-lift upgrades for many of its switches and routers.

APPLICATION RESPONSE TIME The most challenging QoS metric is application **response time**. This is the duration between the time the user presses a key (or clicks on the page) and the time he or she sees a response. Figure 4-8 shows how application response time is different from network latency.

The figure shows that network latency is only one factor in user response time. Most obviously, the delay at the client and server ends to do processing in the application software is important. However, there can be many other factors in response time. For example, poorly configured networking software in the client or server operating system may add delay. So may firewall filtering, the need for encryption (which is a heavy process), and other security matters.

Application response time planning and execution are complicated by the fact that systems administrators and application professionals often work separately from network professionals. In fact, they often know little about what the other side does. Application response time management requires strong and effective cooperation across these organizational boundaries.

Test Your Understanding

7. a) What is availability? b) What is downtime? c) What are the "five nines"? d) Does corporate network availability usually meet the five-nines expectation of the telephone network? e) What are packets? f) Distinguish between the packet error rate and the bit error rate. g) When should error rates be measured? Why? h) What is latency? i) In what units is latency measured? j) What is jitter? k) For what applications is jitter a problem? l) How does application response time differ from latency? m) Why is application response time difficult to improve?

Service Level Agreements (SLAs)

When you buy some products, you receive a guarantee promising that they will work and specifying penalties if they do not work. In networks, service providers often provide **service level agreements (SLAs)**, which are contracts that guarantee levels of performance for various metrics such as speed and availability. If a service does not meet its SLA guarantees, the service provider must pay a penalty to its customers.

WORST-CASE SPECIFICATION SLA guarantees are expressed as *worst cases*. For example, an SLA for speed might guarantee that speed will be *no lower* than a certain amount. If you are downloading webpages, you want at least a certain level of speed. If speed falls below your needs, that is a service violation. You certainly would not want a speed SLA to specify a maximum speed. More speed is good. Why would you want to limit the maximum speed? Many students have a difficult time with worst-case thinking. This is because speed is good, so they want the SLA to guarantee them high speed. That is not the SLA's job. SLA is like an insurance policy. It guarantees that speed will not fall below a certain level.

SLA guarantees are expressed as *worst cases*. Service will be no worse than a specific number.

For latency, then, the SLA will require that latency will be no *higher* than a certain value. You might specify an SLA guarantee of 125 ms (milliseconds). This means that you will not get worse latency.

PERCENTAGE-OF-TIME ELEMENTS In addition, SLAs have percentage-of-time elements. For instance, an SLA on speed might guarantee a speed of at least 480 Mbps 99.9 percent of the time. This means that the speed will nearly always be at least 480 Mbps but may fall below that 0.1 percent of the time without incurring penalties. A smaller exception percentage might be attractive to users, but it would require a more expensive network design. Nothing can be guaranteed to work properly 100 percent of the time.

Test Your Understanding

8. a) What are service level agreements? b) Does an SLA measure the best case or the worst case? c) Would an SLA specify highest latency or lowest latency? d) Would an SLA specify a lowest availability or a highest availability? e) What happens if a carrier does not meet its SLA guarantee? f) If carrier speed falls below its guaranteed speed in an SLA, under what circumstances will the carrier not have to pay a penalty to the customers?

DESIGN

Implementing a network project requires a company to go through all phases of the systems development life cycle. In most cases, these stages are similar to those for other IT projects. One special area in the SDLC is the design of a new network or of a modified network.

Network Drawing Tools

Network managers often draw pictures of their networks to help them comprehend how these networks function. Many draw these pictures by hand or use a general-purpose graphics tool such as PowerPoint. However, most turn to **network drawing tools** that are specifically created to build diagrams. The best are optimized for building network diagrams.

Figure 4-9 shows that most network drawing programs give you a canvas on which to draw. They also give you a palate of icons to drag onto the canvas. In the figure, the user is dragging a node icon onto the canvas. Nodes include switches, routers, firewalls, clients, servers, and other network devices. After the user has placed node icons on the canvas, he or she can still move them around.

Once the nodes are on the canvas in roughly their proper positions, the designer can connect them with transmission lines (see Figure 4-10). Again, transmission lines are simply icons that are dragged onto the canvas to connect two nodes. Even after the transmission lines are connected, the node icons can still be shifted around on the canvas. Their transmission connections can follow them.

The most popular network drawing program is Microsoft Office Visio. At its core, Visio is a general-purpose diagramming tool. The professional version, however, comes with a large collection of network-specific icons. Many of these icons are small pictures of individual products rather than pictures of generic routers, switches, and other devices. If you can obtain the professional version of Visio, it will be easier to do your homework and explore various options in network design.[3]

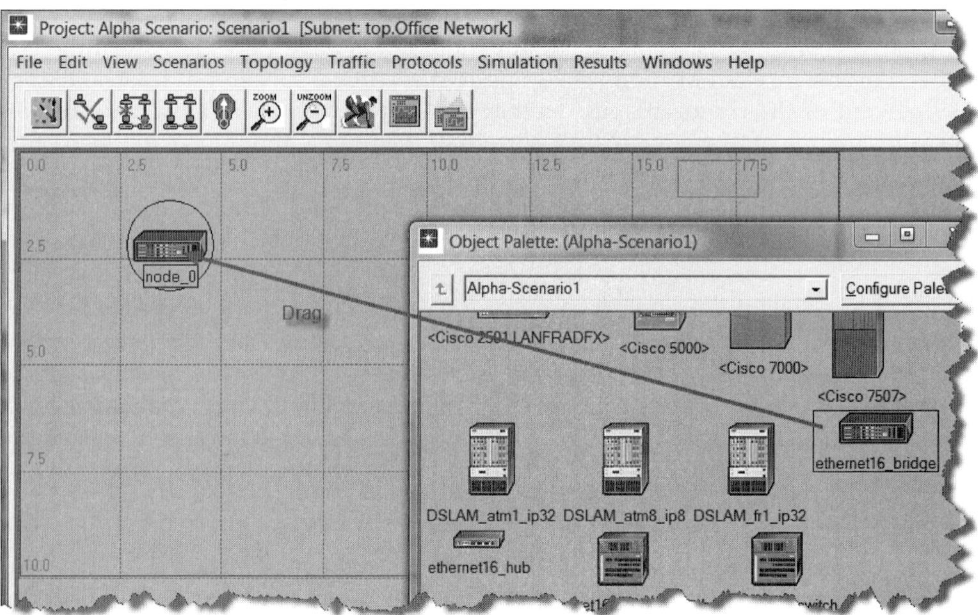

FIGURE 4-9 Adding Node Icons in a Network Drawing Program

[3] Unfortunately, Prentice-Hall cannot use Visio diagrams in its books. Rats.

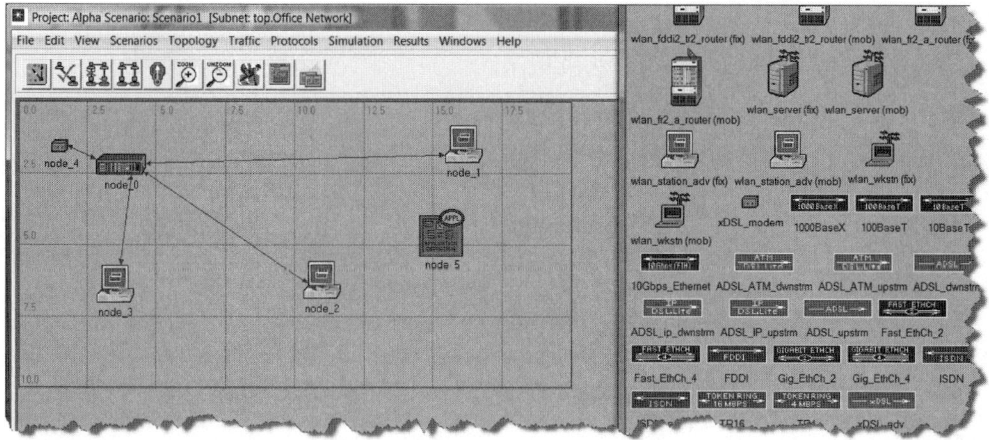

FIGURE 4-10 Connecting Node Icons with Link Icons

The two figures were created in a different tool, OPNET's IT Guru. This is a tool specifically for drawing networks. We will see later that it has some features that are far more advanced than those in Visio.

Test Your Understanding

9. How do you draw a network diagram in a network drawing program?

Traffic Analysis

In network design, traffic analysis asks how much traffic must flow over the network's many individual transmission lines. Figure 4-11 shows a trivial traffic analysis. A company only has two sites, A and B. A needs to be able to transmit to B at 100 Mbps. B needs to transmit to A at 1 Gbps. Transmission lines usually are symmetric, meaning that they have the same speed in both directions. Obviously, the company must install a transmission line that can handle 1 Gbps.

As soon as the number of sites becomes larger than two, traffic analysis becomes difficult. Figure 4-12 shows a three-site traffic analysis. For simplicity, we will assume that transmission is symmetric between each pair of sites.

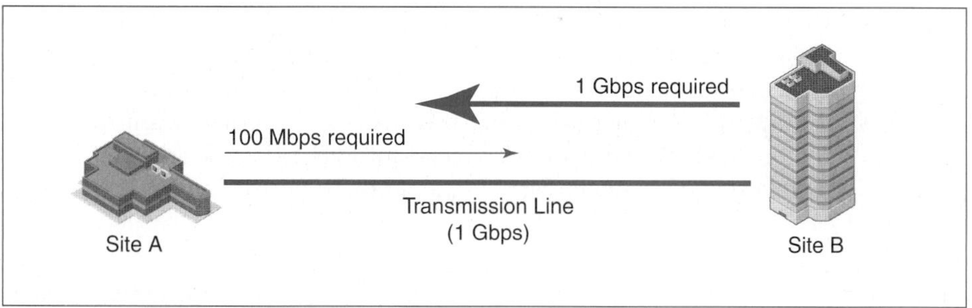

FIGURE 4-11 Two-Site Traffic Analysis

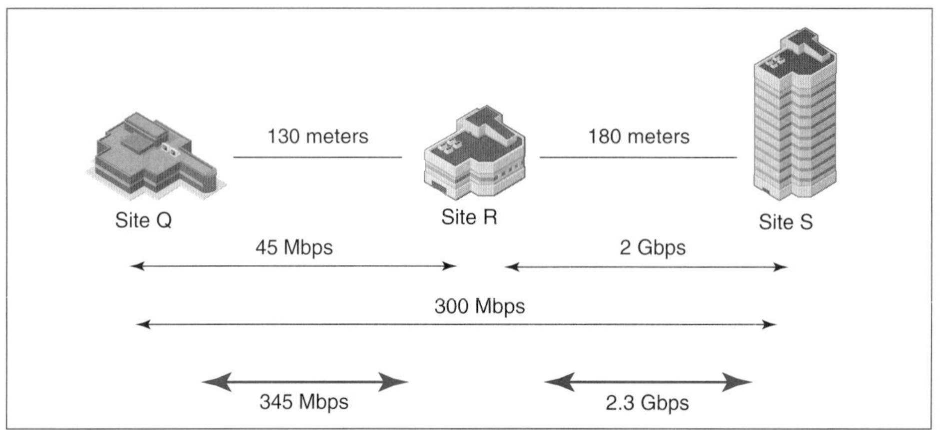

FIGURE 4-12 Three-Site Traffic Analysis

The figure shows that Site Q attaches to Site R, which attaches to Site S. Site Q is 130 meters west of site R. Site S is 180 meters east of Site R. Site Q needs to be able to communicate with Site R at 45 Mbps. Site R needs to be able to communicate with Site S at 2 Gbps. Site Q needs to be able to communicate with Site S at 300 Mbps.

Are you confused by the last paragraph? Anyone would be. In traffic analysis, it is critical to DTP—draw the picture! Figure 4-12 shows how the three sites are laid out. After laying out the sites, you draw the three required traffic flows.

Note that the line between Q and R must handle both Q–R traffic (45 Mbps) and the Q–S traffic (300 Mbps). It does not handle any of the traffic between R and S, however. Consequently, the line between Q and R must be able to handle 345 Mbps.

Similarly, the line between R and S must be able to handle R–S traffic (2 Gbps) and Q–S traffic (300 Mbps.) This means that the transmission line between R and S must be able to handle 2.3 Gbps.

If a company has many sites rather than just two or three, then doing traffic analysis manually becomes impossible. Companies use simulation programs that determine what site pair traffic will flow over what transmission lines. However, you need to understand what the program is doing, and the way to do that is to work through a few examples with only a few sites.

Test Your Understanding

10. Do a three-site traffic analysis. Site X attaches to Site Y, which attaches to Site Z. Site X is 130 meters east of Site Y. Site Z is 180 meters west of Site Y. Site X needs to be able to communicate with Site Y at 3 Gbps. Site Y needs to be able to communicate with Site Z at 1 Gbps. Site X needs to be able to communicate with Site Z at 700 Mbps. Supply your picture giving the analysis. You may want to do this in Office Visio or Windows Draw, then paste it into your homework. a) What traffic capacity will you need between Sites X and Y? b) Between Y and Z?

Redundancy

Transmission lines sometimes fail. Suppose that the transmission line between R and S in Figure 4-12 failed. Then Q would still be able to talk to R, but Q and R would not be able to talk to S. Obviously, this is highly undesirable.

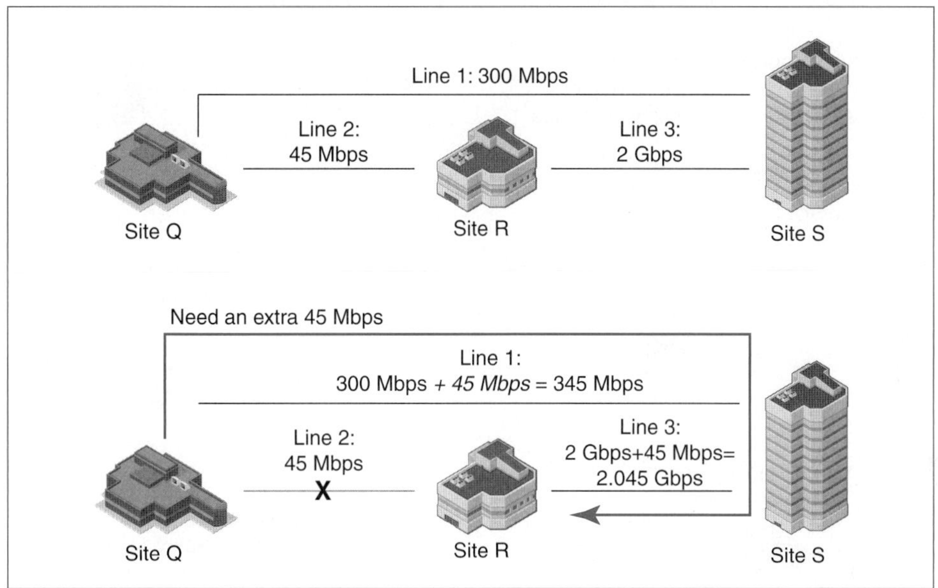

FIGURE 4-13 Three Sites with Redundancy

The solution is to install redundant transmission lines. Redundant transmission lines are extra transmission lines that are not necessary for the system to function but that provide backup paths in case another line fails. For example, Figure 4-13 again shows Sites Q, R, and S. This time, a redundant line has been added between Q and R.

What happens if the line between Q and R fails? The answer is that Site Q can still talk to Site S through the direct line. Also, Q can still talk to R by sending its transmissions to S, which will send them on to R.

When redundancy is used, lines must be given extra capacity in case of failures. For instance, if the line between Q and R is only 300 Mbps, this will be enough if there are no failures. However, if the line Q–R fails, the line will need another 45 Mbps. So it will need to have 345 Mbps of capacity to handle a Q–R failure. The R–S line will also need 45 Mbps more capacity. It will need 2.045 Gbps of capacity to handle both R–S traffic and Q–R traffic.

Test Your Understanding

11. a) What is the purpose of redundancy in transmission lines? b) If the line between R and S fails in Figure 4-13, how much capacity will the line between Q and S need? c) What about the line between Q and R?

Topology

Network design focuses heavily on network topology. The term **network topology** refers to the physical arrangement of a network's computers, switches, routers, and

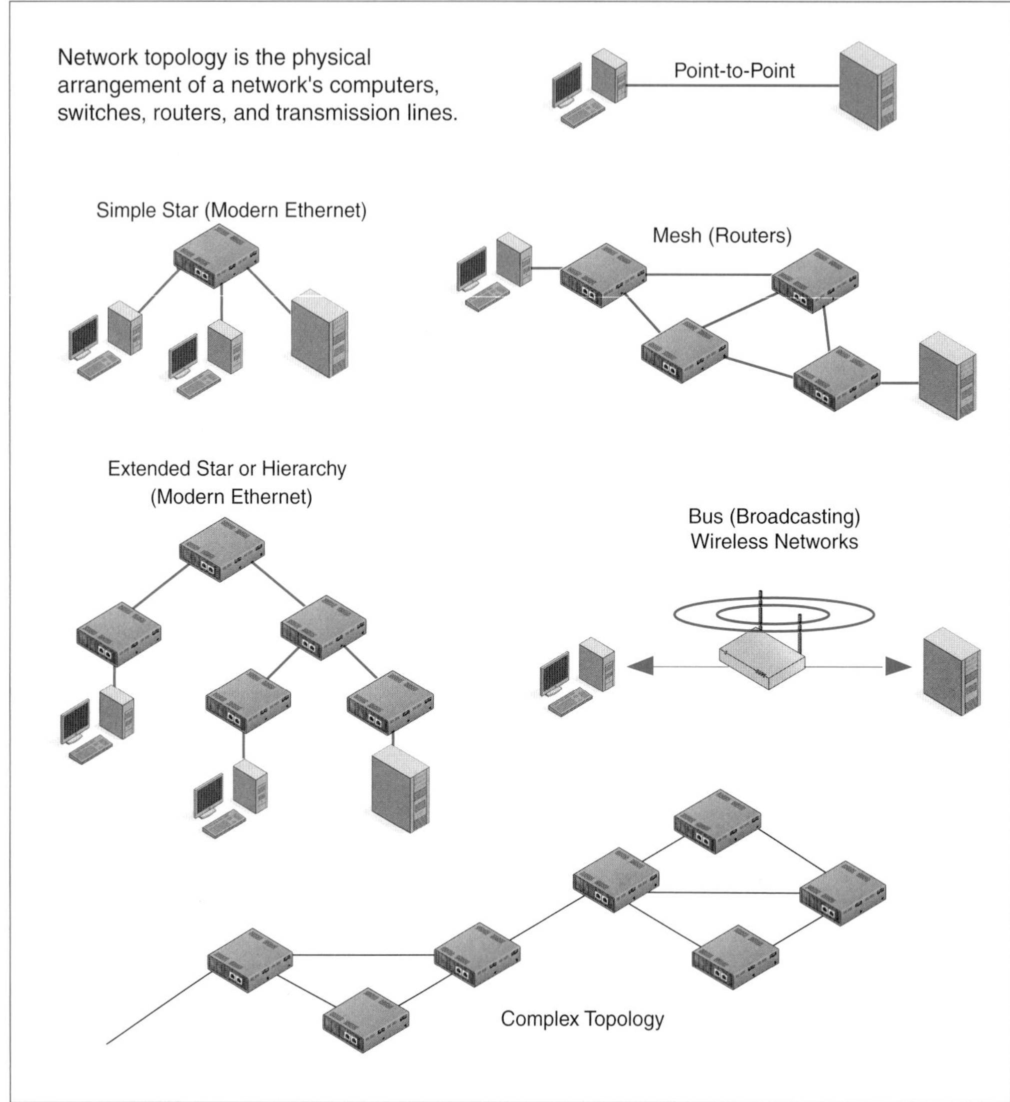

FIGURE 4-14 Major Topologies

transmission lines. Topology, then, is a physical layer concept. Different network (and internet) standards specify different physical topologies. Figure 4-14 shows the major "basic" topologies specified by network standards. Real networks often have complex topologies that involve a mixture of these basic topologies.

> Network topology is the physical arrangement of a network's computers, switches, routers, and transmission lines. It is a physical layer concept.

POINT-TO-POINT TOPOLOGY The simplest network topology is the **point-to-point topology**, in which two nodes are connected directly. Although some might say that a point-to-point connection is not a network, companies often connect a pair of sites with a point-to-point private leased line provided by a telephone carrier.

STAR TOPOLOGY AND EXTENDED STAR (HIERARCHY) TOPOLOGY Modern versions of Ethernet, which is the dominant LAN standard, use the star and extended star topologies. In a **simple star topology**, all wires connect to a single switch. In an **extended star (or hierarchy) topology**, there are multiple layers of switches organized in a hierarchy. We will see Ethernet hierarchies in Chapter 6. An important characteristic of hierarchical standards is that there is only a single possible path between any two end nodes.

MESH TOPOLOGY In a **mesh topology**, there are many connections among switches or routers, so there are many alternative paths to get from one end of the network to the other. The TCP/IP standards are designed for a mesh router topology.

BUS (BROADCAST) TOPOLOGIES In a **bus topology**, when a computer transmits, it broadcasts to all other computers. Wireless LANs and WANs, which we will see in Chapters 7 and 8, broadcast their signals and so have bus topologies.

REAL NETWORK TOPOLOGIES Some network technologies require a pure basic taxonomy. For example, the Ethernet technology we will see in Chapter 6 requires a strict hierarchy. Most networks, however, have **complex topologies** that use different basic topologies in different parts of the network.

Test Your Understanding

12. a) What is a network topology? b) At what layer do we find topologies? c) In what topology are there only two nodes? d) In what topologies is there only a single path between any two end nodes? e) In what topology are there usually many paths between any two end nodes? f) In what topology is broadcasting used? g) What topologies can be used in complex networks?

Leased Line Network Topologies

Figure 4-15 shows that companies have traditionally used leased lines to interconnect their sites. As we saw in Chapter 1, leased lines are high-speed, point-to-point, always-on carrier circuits. Because the telephone system only provides raw bandwidth between points, the company must design the overall data network.

FULL-MESH TOPOLOGY Should many or all pairs of sites be connected to each other, or should there be as few connections as possible? Figure 4-15 shows two topological extremes for building leased line networks.

The first is a **full-mesh topology**, which provides direct connections between every pair of sites. This provides many redundant paths so that if one site or leased line fails, communication can continue unimpeded.

Unfortunately, as the number of sites increases, the cost of a full mesh grows exponentially. For example, if there are N sites, a pure mesh will require $N*(N-1)/2$ leased

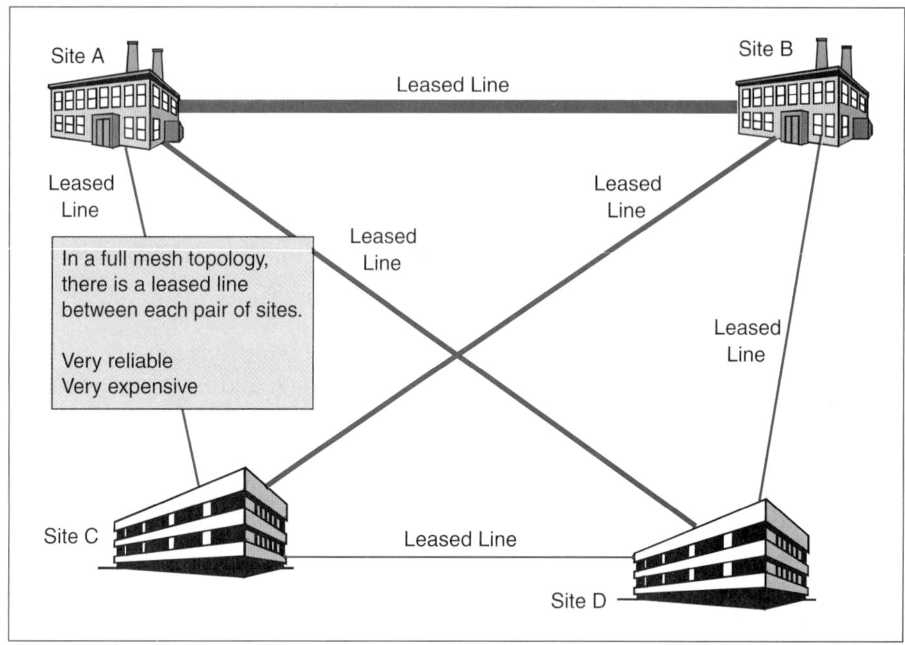

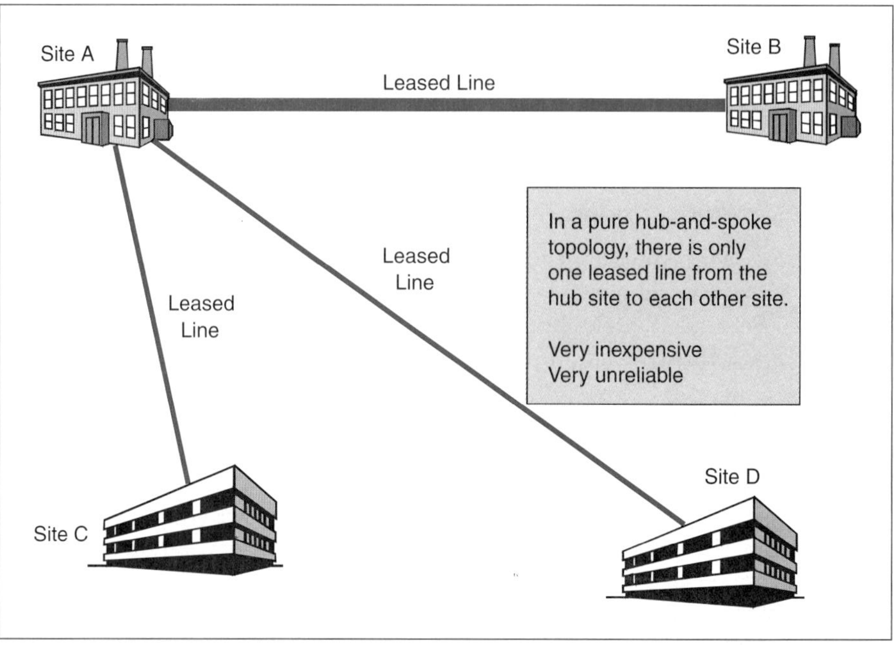

FIGURE 4-15 Full-Mesh and Pure Hub-and-Spoke Topologies for Leased Line Data Networks

lines. So a 5-site pure mesh will require 5*(5 – 1)/2 (10) leased lines, a 10-site pure mesh will require 45 leased lines, and a 20-site pure mesh will require 190 leased lines. Full meshes, while reliable, are prohibitively expensive if a company has many sites.

HUB-AND-SPOKE LEASED LINE NETWORKS The second extreme topology for building leased line networks is the **pure hub-and-spoke topology**. Figure 4-15 also shows this topology. In a pure hub-and-spoke topology, all communication goes through one site. This dramatically reduces the number of leased lines required to connect all sites compared to a full mesh, and so this kind of topology minimizes cost. However, it also reduces reliability. If a line fails, there are no alternative paths for reaching an affected site. More disastrously, if the hub site fails, the entire network goes down.

MIXED DESIGNS As you might suspect, full meshes and pure hub-and-spoke topologies represent the extremes of cost and reliability. Most real networks use a mix of these two pure topologies. Real networks must trade off reliability against cost.

Test Your Understanding

13. a) What is the advantage of a full-mesh leased line network? b) What is the disadvantage? c) What is the advantage of a pure hub-and-spoke leased line network? d) What is the disadvantage? e) Do most leased line networks use a full-mesh or a pure hub-and-spoke topology? Explain.

14. A company has three sites: Micah, Mallory, and William. Micah and Mallory need 100 Mbps of transmission capacity between them. Mallory and William need 200 Mbps of transmission capacity between them. Micah and William need 300 Mbps of transmission capacity between them. a) Create a hub-and-spoke network with Micah as the hub. What links will there be, and how fast will they need to be? Explain your reasoning. b) For the same situation, create a full-mesh network. What speeds will the links need to have if you are not concerned with redundancy in case of line failure? Explain your reasoning. c) Building on the last part of this question, add redundancy so that a failure of the line between Mallory and William will not bring down the network. (Hint: DTP.)

Handling Momentary Traffic Peaks

One fact of networking life that can never be ignored in designs is that traffic volume varies widely. Peak periods of traffic can overwhelm the network's switches and transmission lines. Given the statistical nature and high variability of traffic, **momentary traffic peaks** lasting a fraction of a second to a few seconds are bound to occur, and a firm must have a plan for managing momentary traffic peaks. Corporations can use several traditional traffic management methods to respond to momentary traffic peaks, as Figure 4-16 shows. Network planners must select which approach to use in the corporate network or in different parts of the corporate network.

OVERPROVISIONING ETHERNET LANS One approach is overprovisioning—adding much more switching and transmission line capacity than will be needed most of the time. With overprovisioning, it will be very rare for momentary traffic peaks to exceed capacity. This means that no regular ongoing management is required. The downside of

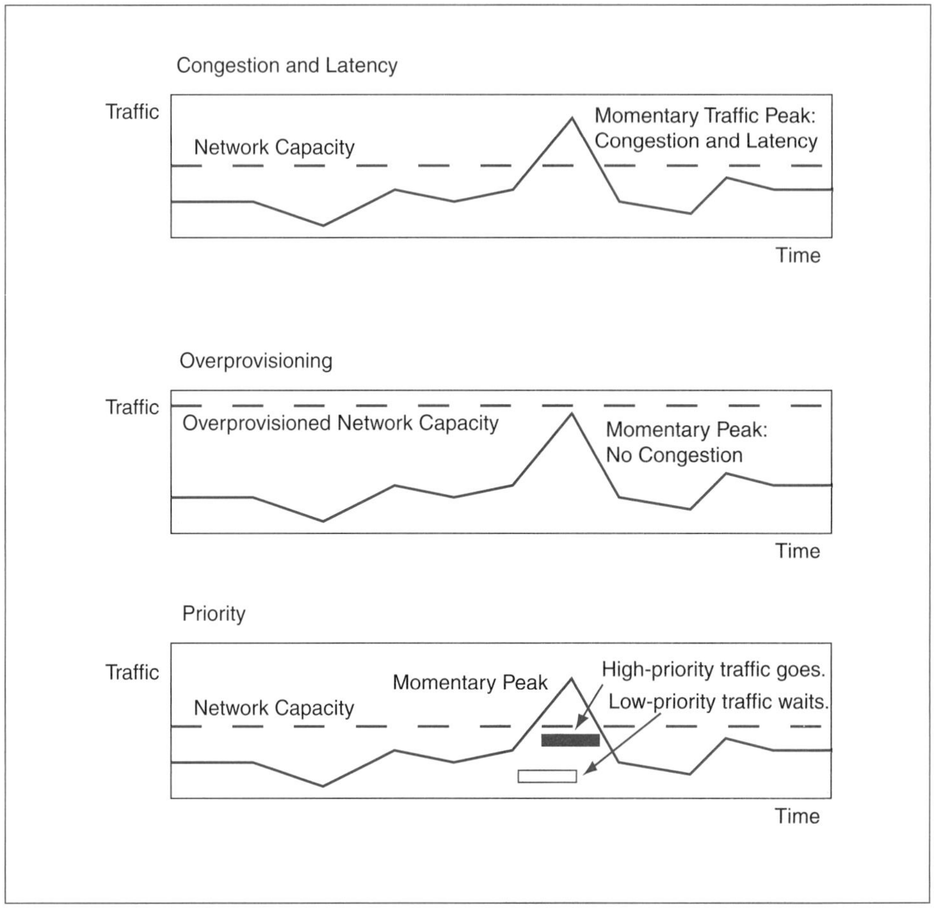

FIGURE 4-16 Handling Momentary Traffic Peaks

overprovisioning, of course, is that it is wasteful of capacity. Today, the simplicity of overprovisioning and the relatively low cost of overprovisioning on LANs make overprovisioning attractive on LANs. On WANs, however, where the cost per bit transmitted is very high, overprovisioning is too expensive to consider.

PRIORITY Priority, in turn, assigns high priority to latency-intolerant applications, such as voice, while giving low priority to latency-tolerant applications, such as e-mail. Whenever congestion occurs, high-priority traffic is sent through without delay. Low-priority traffic must wait until the momentary congestion clears. Priority allows the company to work with lower capacity than overprovisioning but requires more management labor.

QoS GUARANTEES Quality-of-service guarantees take a step beyond priority, reserving capacity on each switch and transmission line for certain types of traffic. This allows the firm to satisfy QoS service-level agreements for selected traffic by providing guarantees for minimum throughput, maximum latency, and even maximum jitter.

QoS guarantees require extremely active management. In addition, traffic with no QoS guarantees only gets whatever capacity is left over after reservations. This may be too little, even for latency-tolerant traffic.

TRAFFIC SHAPING Even with priority and overprovisioning, sufficient capacity must be provided for the total of all applications apart from momentary traffic peaks. Even more active management is needed to control the amount of traffic entering the network *in the first place*. Restricting traffic entering the network at access points is called **traffic shaping**, which is shown in Figure 4-17.

Filtering. Traffic shaping has two options. The first is **filtering** out unwanted traffic at access switches. Some traffic has no business on the corporate network, such as downloading MP3 and video files, game playing, and illegal file sharing.

Capacity Percentages. The second option in traffic shaping is to assign certain **percentages of capacity** to certain applications arriving at access switches. Even if file sharing has legitimate uses within a firm, for instance, the firm may wish to restrict the amount of capacity that file sharing can use. Typically, each application or application category is given a maximum percentage of the network's capacity. If that application attempts to use more than its share of capacity, incoming frames containing the application messages will be rejected.

Perspective on Traffic Shaping. Overprovisioning, priority, and QoS guarantees merely attempt to deal with incoming traffic. Traffic shaping actually reduces the amount of incoming traffic. Only traffic shaping can dramatically reduce network cost.

Although traffic shaping is very economical in terms of transmission capacity, it is highly labor intensive. It is used today primarily on high-cost WAN links. However, as management software costs fall in price and require less labor to operate, traffic shaping should see increasing use.

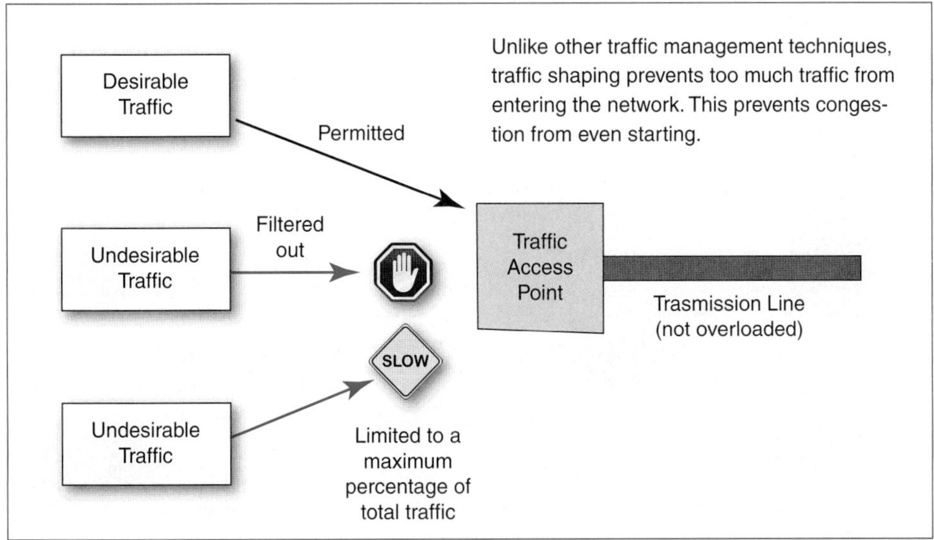

FIGURE 4-17 Traffic Shaping

Another issue that arises when traffic shaping is used is politics. Telling a department that its traffic will be filtered out or limited in volume is not a good way to make friends. Priority and QoS reservations also raise political problems, but in traffic shaping, these problems are particularly bad.

Test Your Understanding

15. a) How long are momentary traffic peaks? b) Distinguish between overprovisioning and priority. c) Distinguish between priority and QoS guarantees. d) What problem can QoS create? e) How is traffic shaping different from traditional approaches to handling momentary traffic overloads? f) In what two ways can traffic shaping reduce traffic?

Reducing Capacity Needs

We have just looked at the design implications of momentary traffic peaks, which usually only last a fraction of a second. More generally, we would like to reduce the overall traffic the network must carry. This would reduce the cost of network services directly.

TRAFFIC SHAPING Traffic shaping is not just a tool for dealing with momentary traffic peaks. By eliminating some types of traffic entirely and by limiting other types of traffic to small percentages of total network traffic, traffic shaping can substantially reduce the overall traffic a network must handle at all times.

COMPRESSION Another way to reduce traffic is to compress traffic before it enters a network. **Compression** exploits redundancy in data to recode the data into fewer bits. This means fewer bits to transmit. At the other end, the traffic must be decompressed to put it back in its original form. To give a rough analogy, dehydrated food with water removed is much lighter than normal food. Adding water later reconstitutes the dehydrated food.

Figure 4-18 shows there are two incoming data streams. The first is 3 Gbps. The second is 5 Gbps. This is a total of 8 Gbps. The capacity of the transmission line is only 1 Gbps. We have a problem.

Before the incoming data enters the transmission line, however, the data will be compressed by 10:1, which is often possible for typical data streams. This reduces the

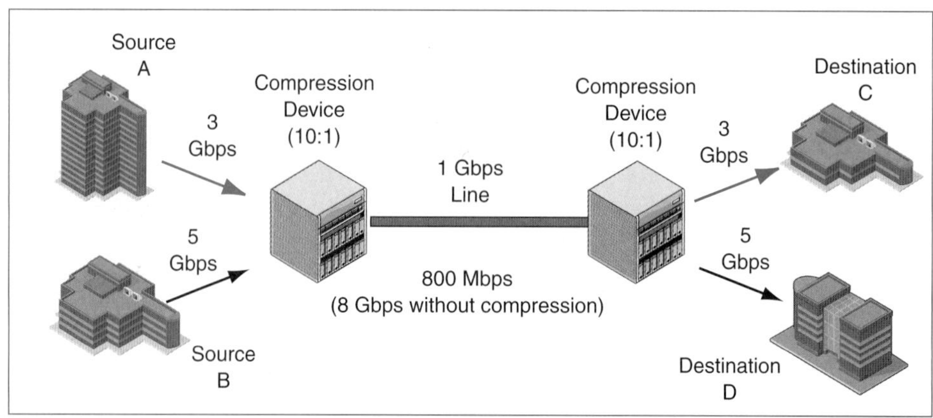

FIGURE 4-18 Compression

compressed data streams to 0.3 Gbps (300 Mbps) and 0.5 Gbps (500 Mbps), for a total of only 800 Mbps. This traffic can easily fit on a 1 Gbps line with ample room for other traffic.

At the other end, another device decompresses the data streams. It sends the 3 Gbps data stream to one destination and the 5 Gbps data stream to another destination.

One requirement for compression is that you must have compatible equipment at the two ends of the network. Given a frequent lack of vendor compatibility, compression tends to lock the company into a single vendor's products.

Test Your Understanding

16. a) Why is traffic shaping a more general tool than just being a way to handle momentary traffic peaks? b) Why can compression help in traffic management? c) What makes compression possible?

Natural Designs

We have been discussing general design principles. In many cases, however, designers must choose designs that are natural for their environments. For example, Figure 4-19

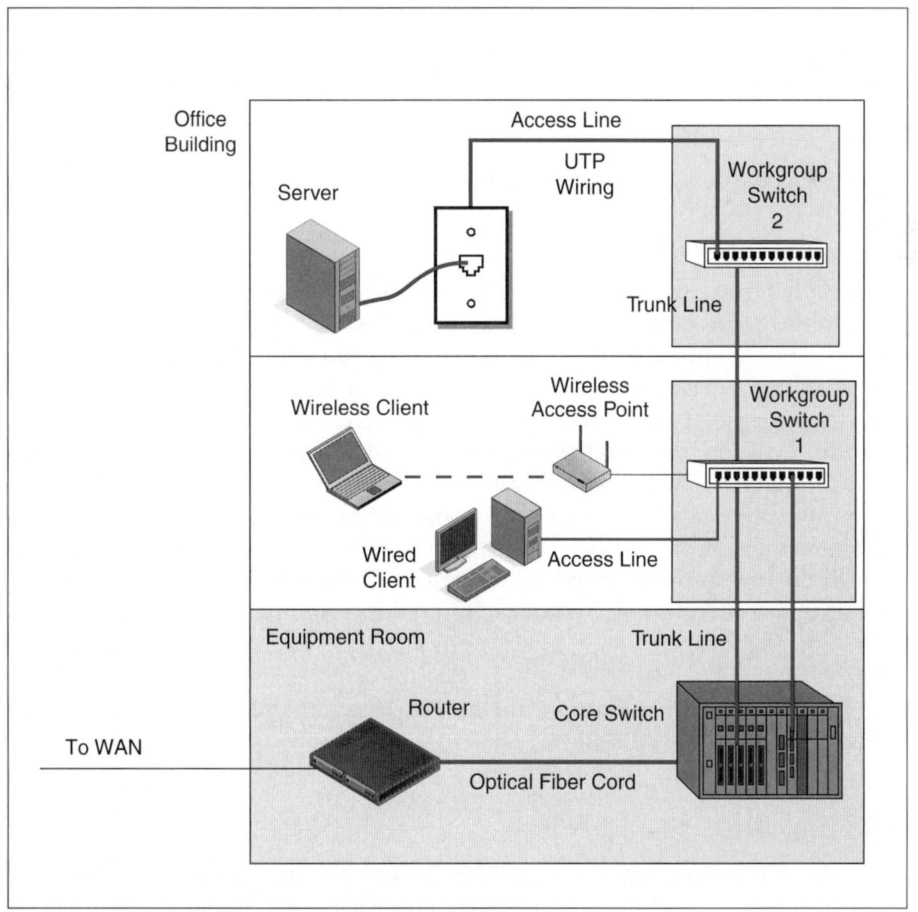

FIGURE 4-19 Natural Design for a Building LAN

Drawing Tools

Show the elements in the network and how they are interconnected

Network Simulation Software

Creates a computer model of the network, not just a picture
Contains the configuration of each device

What-If Analysis

Try alternatives to close gaps
Select the optimum design

FIGURE 4-20 Network Simulation Software

shows a natural design for a building LAN. This building has multiple floors. It will simply make everyone's life easier if each floor is given an Ethernet workgroup switch that serves the hosts and wireless access points on that floor. It is also natural to place a core switch in the basement and have all communication between switches go through the core switch. The core switch can then connect to a router that acts as a gateway to the outside world.

Test Your Understanding

17. Why was the design in Figure 4-19 selected?

Network Simulation Software

While network drawing tools such as Visio can help you create diagrams, network managers need more powerful **network simulation** tools that can allow them to envision the impacts of alternative designs so that they can select the best design. With network simulation software, designers can conduct what-if analyses without going through the expense of implementing alternatives to assess how each one works.

IT Guru, which was used to draw Figure 4-9 and Figure 4-10, actually is a full-network simulation tool. IT Guru creates a model of the network in which the specific configurations and capabilities of each node and transmission line are entered. The network manager selects a design or design modification, injects simulated application data traffic into the network, and sees the results. For each design or redesign case, the simulation runs several times in order to model uncertainties in the data.

Test Your Understanding

18. a) Why is network simulation valuable? b) What capability does IT Guru have that Visio does not?

EVALUATING ALTERNATIVES

When a design is completed, it is usually necessary to select between products offered by different vendors and perhaps between competing technologies. In much of this book, we will see descriptions of multiple technologies with a special focus on relative

Comparing Alternatives

In designs, must select among competing designs and even competing technologies
When learning about technologies, you need to understand pros and cons.

Minimum Requirements

Specifications that must be met
Noncompliant products that do not need a minimum requirement cannot be considered.
Scalability is a concern. (Figure 4-22)

Multicriteria Decision Making

Must look at all aspects of each alternative and evaluate each aspect (Figure 4-23)

Cost

Cost is difficult to measure. (Figure 4-24)

FIGURE 4-21 Product Selection

strengths and weaknesses. It is not enough to know the individual technologies. You must be able to discuss the pros and cons of competing technologies.

Minimum Requirements

Sometimes there are **minimum requirements** that will exclude a certain product or technology from final consideration. For example, if you need an e-mail server that will support at least 10,000 users in your company, you cannot consider an e-mail server product that will only support 2,000 users—even if it is very inexpensive. If security is a high concern, furthermore, you may only want to use a wired technology or routers that require the encryption of supervisory communication.

A special concern is scalability. Some choices simply do not **scale**, meaning that they are not useful beyond a certain traffic volume. As Figure 4-22 shows, a technology may be cost effective when its use is small but may grow too expensive at higher traffic volumes. **Scalable** solutions retain their cost advantage as volume grows.

Scalable solutions grow slowly in cost as traffic volume increases.

Another scalability problem is a complete inability to grow enough to meet a company's traffic volume regardless of how much a company spends. The example of the e-mail server that does not meet a minimum requirement is an example of a failure to scale enough.

Test Your Understanding

19. a) Should products that fail to meet minimum requirements be dropped from consideration? b) In what two ways can solutions fail to be scalable?

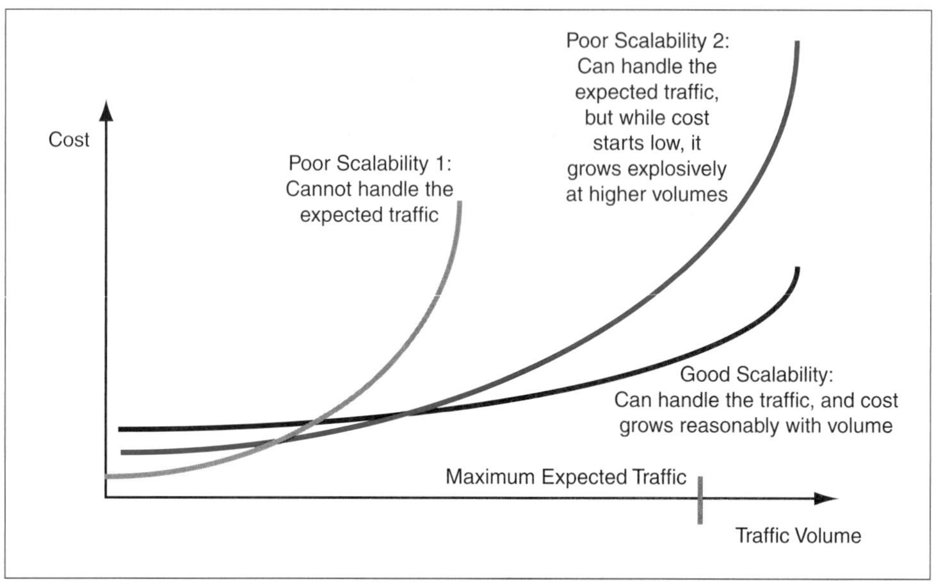

FIGURE 4-22 Scalability

Product Selection with Multicriteria Decision Making

Once a project is selected and initiated, the network staff must go through the traditional systems development life cycle to implement the project. Given that almost all readers know about the systems development life cycle, we will not discuss it in detail.

In software development projects, there usually is a **make-versus-buy decision**. Should the programming staff create the software itself, or should the company purchase the software? In networking projects, this first option rarely makes sense. User companies like banks and retail stores do not have the technical expertise to make their own switches and routers. Instead, they must *select* and *buy* these technologies. Consequently, in this book, we will look at the factors you need to understand in purchasing decisions that involve different technologies.

When making purchasing decisions, companies tend to use **multicriteria decision making**, which Figure 4-23 illustrates. In this approach, the company decides what product characteristics will be important in making the purchase. Things that are important in the purchasing decision are called **criteria**.

Of course, costs are important—both purchase costs and ongoing costs. However, other decision criteria are also important. In Figure 4-23, the criteria for the product are functionality, availability, cost, ease of management, and electrical efficiency.

Next to each criterion is the **criterion weight**. This weight gives the relative importance of each criterion compared to other criteria. Here, weights range from 1 to 5. Note that cost and functionality have the largest weights (5), emphasizing their importance.

For each product (there are only two in the figure), the evaluation team gives a **rating** for each decision criterion. In this example, the ratings range from 1 to 10, with higher values indicating higher value. More functionality is better, so higher numbers in ratings reflect greater functionality. In contrast, for cost, lower cost is better, so higher rated values must indicate lower cost.

		Product A		Product B	
Criterion	Criterion Weight (Max: 5)	Product Rating (Max: 10)	Criterion Score	Product Rating (Max: 10)	Criterion Score
Functionality	5	9	45	7	35
Availability	2	7	14	7	14
Cost	5	4	20	9	45
Ease of Management	4	8	32	6	24
Electrical Efficiency	1	9	9	8	8
Total Score			120		126

FIGURE 4-23 Multicriteria Decision Making in Purchase Decisions

After filling in the ratings on all criteria for all products, the network staff computes the **criterion score** for each product. To do this, the staff multiplies the criterion weight by the rating for that product in that criterion. It then totals the criterion scores into a **total score**.

In Figure 4-23, Product A has a total score of 120, while Product B has a total score of 126. Speaking simplistically, Product B appears to be a better choice. However, the two total scores are very close. Numbers should never drive out thinking. A closer look shows that Product A has very good functionality and ease of management, although its cost is high. Product B has poorer scores on functionality and ease of management. It may be possible to negotiate a lower price on Product A and redo the analysis.

Test Your Understanding

20. a) What is the make-versus-buy decision? b) For routers and switches, do firms usually make or buy? c) We are considering products A, B, and C. Our criteria are price, performance, and reliability with weights of 20 percent, 40 percent, and 40 percent, respectively. Product A's evaluation scores on these three criteria are 8, 6, and 6, respectively. For B, the values are 6, 8, and 8. For C, they are 7, 7, and 7. Present a multicriteria analysis of the decision problem in tabular form and showing all work. Interpret the table.

Cost

Although cost is only one factor in product selection, it is often a critical factor. If you are a mobile phone user, you know how difficult it is to figure your total cost for a month. Figuring the cost of a network alternative is equally complex.

SYSTEMS DEVELOPMENT LIFE CYCLE (SDLC) COSTS When you begin a project, you need to consider the cost of the system during its development life cycle.

Hardware Costs. Consider what happens when you buy a personal computer. You first have to take hardware into account. When you look at the price of a computer, this

Cost is Difficult to Measure

Systems Development Life Cycle Costs

- Hardware: Full price: base price and necessary components
- Software: Full price: base price and necessary components
- Labor costs: Networking staff and user costs
- Outsourcing development costs
- Total development investment

Systems Life Cycle (SLC) Costs

- System development life cycle versus system life cycle
- Total cost of ownership (TCO)
    - Total cost over entire life cycle
    - SDLC internal costs plus carrier costs
        - Carrier pricing is complex and difficult to analyze.
        - Must deal with leases

FIGURE 4-24 Cost (Study Figure)

may not include a display, and it usually does not include a printer. The **base price** is the price before adding components that will be needed in actual practice. In contrast, the **full price** of the hardware is the price of a complete working system. The distinction between base price and full price is also applicable to network hardware, including the switches and routers we will see later in this term.

Software Costs. After your first computer purchase, you realized that the software can be almost as expensive as the hardware. You have to consider the software you will need very carefully and understand the cost of that software. Individual software products, furthermore, often have misleading base prices that do not include all necessary components. Network product software decisions are similarly complex.

Labor Costs in Development. Although hardware and software costs are complex and difficult to measure, these problems pale beside the problems involved in estimating labor costs in development. Planning, procurement, installation, configuration, testing, programming, and other labor costs can easily exceed hardware and software costs.

User costs should also be considered as an aspect of labor cost because the time that users spend on the system's development during requirements definition and later development states is substantial. This time is not free to the company, any more than network staff time is free.

Outsourcing Development Costs. If the company outsources some or all of the development costs, then outsourcing costs need to be considered in the overall picture.

Total Development Investment. To evaluate potential projects, the networking staff must forecast the total development investment—the total of hardware, software, labor, and outsourcing costs during development. These expenditures truly are investments that should pay off over the life of the project.

SYSTEMS LIFE CYCLE (SLC) COSTS As noted earlier in this chapter, the systems development life cycle is only part of the overall systems life cycle, which lasts from conception to termination. It is important to consider **systems life cycle costs**, which are costs over a system's entire life—not just during the systems development period. The cost of a system over its entire life cycle is called the **total cost of ownership (TCO).**

Operating and management costs usually are very important over the system's life cycle. When making equipment and software purchases, it is important to consider how much labor is involved in operating and managing the equipment and software. These costs must be considered very carefully in product selection.

In particular, a new leading-edge technology may give fantastic performance, but leading-edge technologies tend to be immature and tend to create far higher support costs than established technologies.

One new factor in systems life cycle analysis is carrier costs. If you must deal with a communications carrier to carry your signals from one corporate site to another, then you also have to consider carrier pricing. This is rarely easy to do, and it is even harder to compare the prices of alternative carriers offering roughly the same service because of the wording in their contracts. In addition, you usually have to sign equipment leases or service agreements that lock you in for various periods of time, sometimes up to several years.

Test Your Understanding

21. a) What period of a network's life does the SDLC cover? b) Why are hardware and software base prices often misleading? c) List the four categories of SDLC costs. d) Why must user costs in development be considered?

22. a) Distinguish between the systems development life cycle and the systems life cycle. b) What is the total cost of ownership (TCO)? c) Why should operating and management costs be considered in addition to hardware, software, and transmission costs in purchasing decisions? d) What additional cost factor comes into SLC costs, compared to SDLC costs?

OPERATIONAL MANAGEMENT

In the networking systems life cycle, a great deal of the work takes place after the development finishes. This requires networking professionals to be able to do operational management as well as development.

OAM&P

After a network component is in place, it probably will be used for many years. During its **operational life**, there will be substantial labor costs. We will classify these costs in a way that telecommunications carriers have traditionally done—in terms of **operations, administration, maintenance, and provisioning (OAM&P).**

Managing the Network as it Provides Service

The most important (and expensive) part of the systems life cycle
Tasks described as OAM&P

Operations

Moment-by-moment traffic management
Network operations center (NOC)

Maintenance

Fixing things that go wrong
Preventative maintenance
Should be separate from the operations staff

Provisioning (Providing Service)

Includes physical installation
Includes setting up user accounts and services
Reprovisioning when things change
Deprovisioning when accounts and services are no longer permitted
Collectively extremely expensive

Administration

Paying bills, managing contracts, etc.

FIGURE 4-25 Ongoing Management (OAM&P)

OPERATIONS You probably have seen pictures of **network operations centers (NOCs)** for major telecommunications carriers. These are large rooms with dozens of monitors showing the conditions of various parts of the network. Most corporations also have network operations centers. These corporate NOCs are smaller, usually having only about a half dozen monitors. NOCs manage the network on a moment-by-moment basis.

MAINTENANCE You have undoubtedly seen telephone company maintenance trucks driving on their way to downed transmission lines, broken transformers, or other trouble spots. In addition to fixing equipment failures, telephone companies do preventative maintenance to prevent future failures.

In the same way, companies often have to fix their internal corporate network switches and other physical components. They also have to handle software problems. Although the network operations center can fix some problems remotely, most firms have separate NOC and maintenance staffs. The NOC staff usually is heavily occupied with the moment-by-moment operation of the network, so it makes sense to have other networking professionals focus on maintenance.

PROVISIONING If you get cable television service, the cable company has to **provision** your residence, that is, set up service. This includes physical setup (running the coaxial cable into your home). It also involves setting up your account on the company's computers. The cable company also has to reprovision customers when they change their service by adding channels, dropping optional services, or switching pricing plans.

Within a corporate network, provisioning may involve the installation of additional switches, routers, and transmission lines to serve new users. In networks, every time a new user joins the firm, the company has to provision service for that user. In fact, provisioning has to be done for every user account on every server and access point on the network.

Furthermore, once a user is provisioned for a particular resource, he or she may have to be **reprovisioned** if his or her authorizations change—say, if he or she is upgraded from read-only data access to full read/write access. The user also has to be reprovisioned if he or she changes jobs within a firm, joins project teams, or does many other things. Users also have to be **deprovisioned** when they leave project teams or leave the company entirely. Contractors and other outside organizations also have to be provisioned, reprovisioned, and deprovisioned when they start to work, change the way they work, or stop working with a company. Collectively, provisioning is extremely expensive.

ADMINISTRATION Operations, maintenance, and provisioning involve real-time work to keep the network running. In contrast, administrative work is dominated by such mundane tasks as paying bills to vendors and telephone companies, managing proposals and contracts, doing network budgeting, comparing network budgets to actual costs, and doing other dull but necessary tasks.

Test Your Understanding

23. a) For what is OAM&P an abbreviation in ongoing management? b) Distinguish between operations and maintenance. c) What is provisioning? d) When may reprovisioning be necessary? e) When may deprovisioning be necessary? f) Into which of the four categories would you classify the task of comparing the inventory of parts with the inventory list on the computer?

Network Management Software

Given the complexity of networks, network managers need to turn to **network management software** to support much of their work. Many of these are **network visibility** tools, which help managers comprehend what is going on in their networks.

PING The oldest network visibility tool is the basic ping command available in all operating systems. If a network is having problems, a network administrator can simply ping a wide range of IP addresses in the company. By analyzing which hosts and routers respond or do not respond, then drawing the unreachable devices on a map, the administrator is likely to be able to see a pattern that indicates the root cause of the problem. Of course, manually pinging a wide range of IP addresses could take a prohibitive amount of time. Fortunately, there are many programs that ping a range of IP addresses and portray the results.

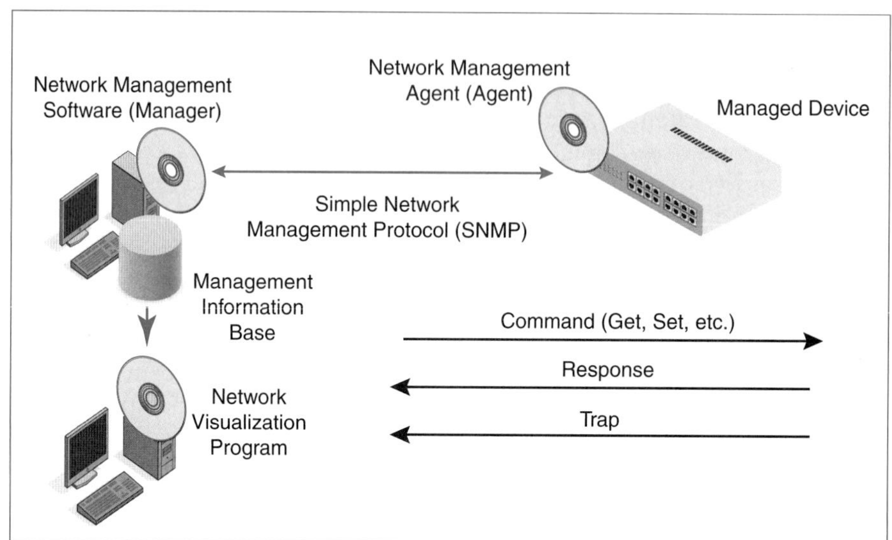

FIGURE 4-26 Network Management Software

THE SIMPLE NETWORK MANAGEMENT PROTOCOL (SNMP) Ping can tell you if a host is available. It can also tell you the latency in reaching that host. For remote device management, most network operation centers use more powerful network visualization products based on the **simple network management protocol (SNMP)**, which is illustrated in Figure 4-26. In the NOC, there is a computer that runs a program called the **manager**. This manager manages a large number of **managed devices**, such as switches, routers, servers, and PCs.

Actually, the manager does not talk directly with the managed devices. Rather, each managed device has an **agent**, which is hardware, software, or both. The manager talks to the agent, which in response talks to the managed device. To give an analogy, recording stars have agents who negotiate contracts with studios and performance events. Agents provide a similar service for devices.

The network operations center constantly collects data from the managed devices using SNMP **Get** commands. It places this data in a **management information base (MIB)**. Data in the MIB allows the NOC managers to understand the traffic flowing through the network. This can include failure points, links that are approaching their capacity, or unusual traffic patterns that may indicate attacks on the network.

In addition, the manager can send **Set** commands to the switches and other devices within the network. Set commands can reroute traffic around failed equipment or transmission links, reroute traffic around points of congestion, or turn off expensive transmission links during periods when less expensive links can carry the traffic adequately.

Normally, the manager sends a command and the agent responds. However, if the agent senses a problem, it can send a **trap** command on its own initiative. The trap command gives details of the problem.

There is one more program in the figure—a **network visualization program**. This program takes results from the MIB and interprets the data to display results in maps, find root causes for problems, and do other tasks. Note that this functionality is *not* included in the simple network management protocol. SNMP simply collects the data in a way that network visualization programs can use. This lack of specification allows network visualization program vendors to innovate without being constrained by standards. What do network visualization programs do?

AUTOMATION Many other network management chores can be automated to reduce the amount of work that network managers need to spend on minutia. For example, many routers are given a standard corporate configuration when they are installed. Doing this manually can take an hour or more per router. However, it may be possible to create a standard configuration, store it, and simply download it onto new routers. In addition, if corporate standard configurations change or a patch must be installed on all routers, it may be possible simply to "push out" these changes to all routers.

Test Your Understanding

24. a) List the main elements in SNMP. b) Does the manager communicate directly with the managed device? Explain. c) Distinguish between Get and Set commands. d) Where does the manager store the information it receives from Get commands? e) What kinds of messages can agents initiate?

CONCLUSION

Synopsis

This is the last of four introductory chapters. This chapter looked at network management. It began with a discussion of basic concepts, including the need to focus on the systems life cycle, the need to be efficient, and the strategic network planning process.

Strategic network planning includes doing what-is analysis (understanding the current situation), understanding driving forces for change, identifying gaps that will appear if the current system is not changed, creating strategies for closing gaps, and selecting a portfolio of projects to close as many gaps as possible.

Networks must not just work. They must work well. Networks must meet goals for quality-of-service (QoS) metrics. We focused heavily on speed, including how to write speeds properly. We also looked at availability, error rates, latency, jitter, and application response time. After discussing individual QoS metrics, we looked at service level agreements (SLAs), which guarantee levels for certain QoS metrics for a certain percentage of time. Many find it confusing that QoS metrics specify that service will be no worse than certain values. For example, SLAs will specify a minimum speed, not a maximum speed.

There was a long section on design. We looked at traffic analysis, which assesses the traffic that various transmission links must sustain, including redundancy in case of link failures. We looked in detail at a number of alternative topologies, including hierarchical topologies, mesh topologies, and bus (broadcast) topologies. We also looked at the handling of momentary traffic peaks through overprovisioning, priority, and QoS guarantees. Finally, we looked at reducing capacity needs through traffic shaping and compression. We discussed how network simulation can reduce the work of doing network design, including what-if analysis to compare alternatives.

When considering alternatives, it is important to evaluate them systematically. We discussed how to do this with multicriteria decision making. We looked in some detail at estimating costs realistically.

The chapter ended with a discussion of operational management. We looked at the four main traditional elements of network management—operations, administration, maintenance, and provisioning. We looked at the simple network management protocol (SNMP) and saw how important it is in collecting information needed for management and even for doing remote changes on network devices.

END-OF-CHAPTER QUESTIONS

Thought Question

1. Assume that an average SNMP response message is 100 bytes long. Assume that a manager sends 50 SNMP *Get* commands each second. a) What percentage of a 100 Mbps LAN link's capacity would the resulting response traffic represent? b) What percentage of a 128 kbps WAN link would the response messages represent? c) What can you conclude from your answers to this question?
2. The telephone network has long boasted that it has the "five nines" (99.999% availability). a) How much downtime is this per year? Express downtime in days, hours, minutes, etc. as appropriate. b) How much downtime is there per year with 99% availability?

Perspective Questions

1. What was the most surprising thing you learned in this chapter?
2. What was the most difficult part of this chapter for you?

CHAPTER 4a

Hands-On: Microsoft Office Visio

LEARNING OBJECTIVES

By the end of this chapter, you should be able to:

- Create a simple Visio diagram.

WHAT IS VISIO?

Microsoft Office Visio is a drawing program. The professional version has special symbols for drawing network diagrams. Visio is widely used by network professionals to visualize networks they are designing.

USING VISIO

Visio is part of the Microsoft Office family. Installing Visio is like installing any other Office product.

Figure 4a-1 shows how to start a Visio drawing. Of course, this begins by selecting File and then New. In the figure, Network has been selected for the type of drawing. Basic Network Diagram has been selected.

As Figure 4a-2 shows, this brings up a window with a canvas on which you can drag shapes. In the figure, the shape of a generic server has been dragged on to the screen. As you can see, many other network diagramming shapes can be dragged onto the screen.

After you have added the devices you need, it is time to begin showing how they are connected. As Figure 4a-3 shows, there is a connector icon at the top of the screen. Select the connector tool. Then drag between the two icons to connect them. After you have connected them, try dragging one of the connected devices. You will see that the connectors move with them.

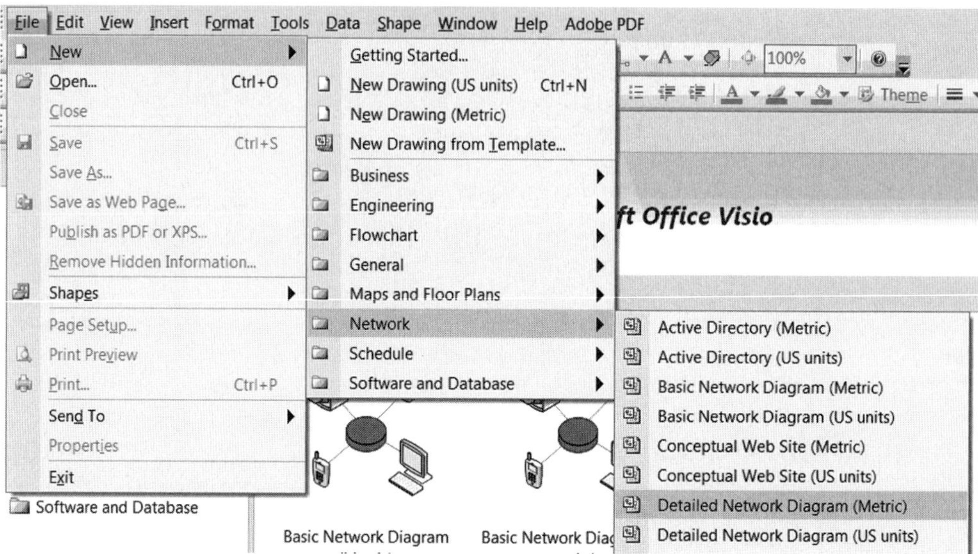

FIGURE 4a-1 Starting a Visio Drawing

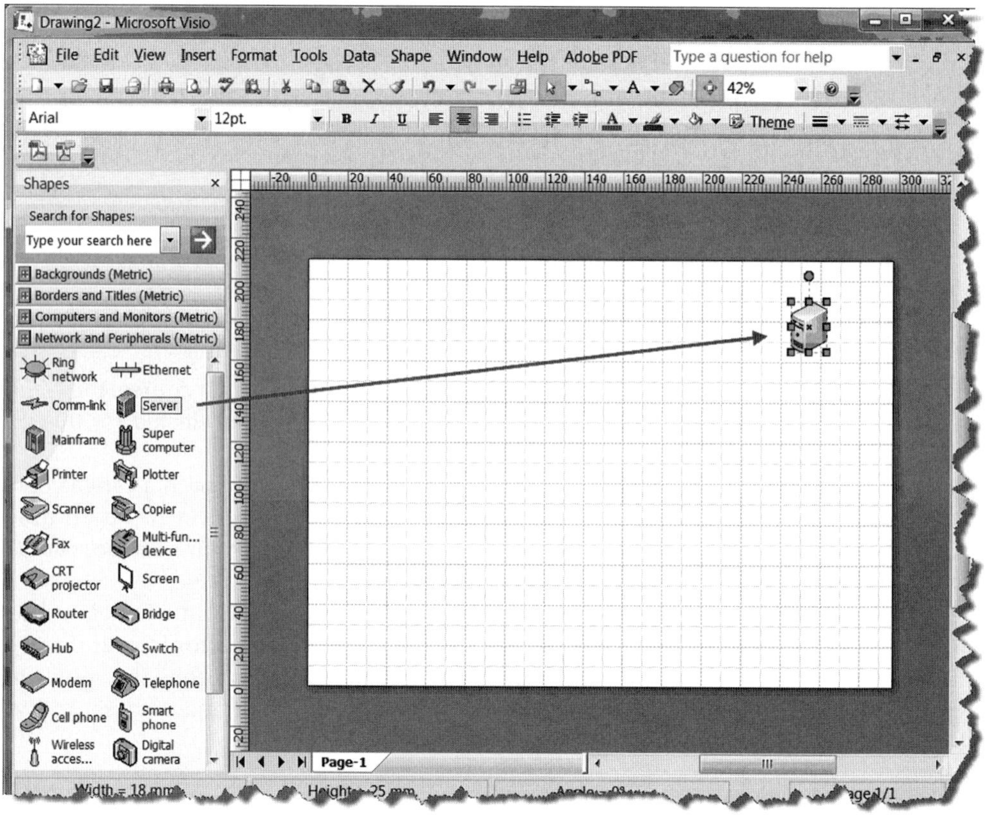

FIGURE 4a-2 Drawing Canvas with Icon Being Dragged

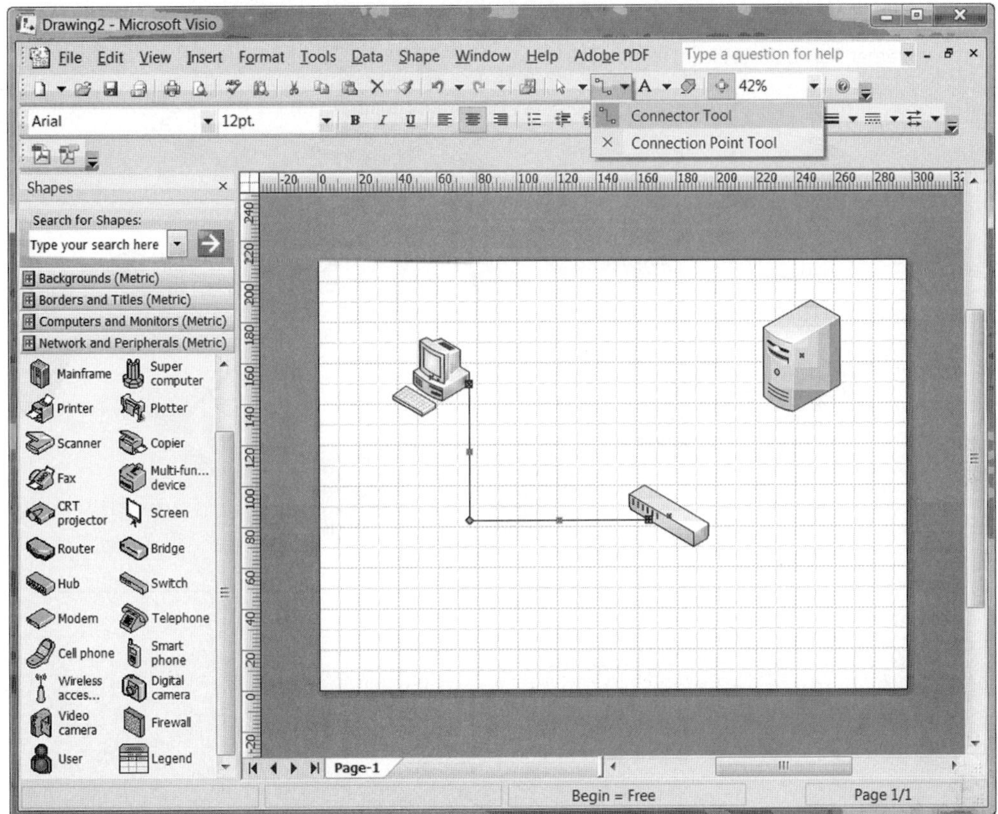

FIGURE 4a-3 Adding Connections

Not shown on the figure, you can double-click on an icon. This adds text below the icon. Visio is not fussy about preventing lines from overlapping text. Overall, Visio diagrams are easy to create but not extremely pretty.

EXERCISE

In Microsoft Office Visio, create something like the drawing in Figure 4a-4. The drawing has a print server. A print server is a device that allows several users in an office to share a printer. A print server plugs into a printer via a USB port. It also plugs into a switch via a UTP cord.

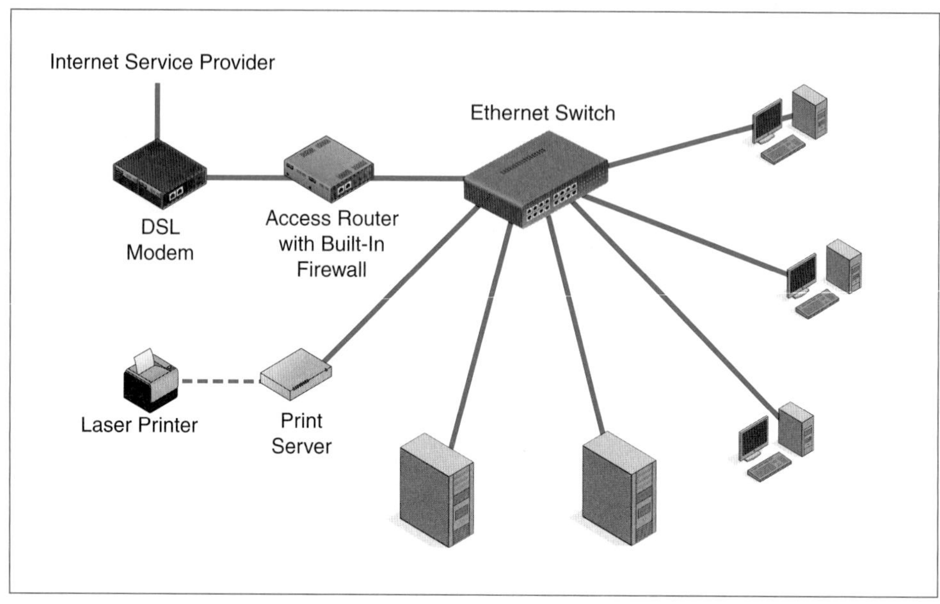

FIGURE 4a-4 Sample Drawing

CHAPTER 5

Physical Layer Propagation

LEARNING OBJECTIVES

By the end of this chapter, you should be able to:

- Describe signals and propagation effects.
- Explain unshielded twisted-pair (UTP) wiring, including relevant propagation effects that must be controlled by limiting cord length and by limiting the untwisting of pairs during connectorization.
- Describe the differences between serial and parallel transmission, including the speed advantage of parallel transmission.
- Describe optical fiber cabling, including relevant propagation effects and different types of optical fiber cabling and signaling.
- Describe Public Switched Telephone Network access lines (the local loop) and ADSL service.
- Describe cable modem service.

INTRODUCTION: PHYSICAL LAYER PROPAGATION

Chapters 1 through 4 gave you a broad introduction to networking. In the rest of the book, we will look at key topics in more detail. In this chapter, we will look at wired physical layer propagation. Chapter 6 will cover switched networks. Chapters 7 and 8 then backtrack to physical propagation and data link layer concerns, this time for wireless networks.

Characteristics of the Physical Layer

Chapter 2 was an overview of layered standards. Most of that chapter focused on the data link, internet, transport, and application layers. In this chapter, we will go down to Layer 1, the physical layer, which differs from upper layers in two ways:

- It is the only layer that does not deal with messages. It takes the bits of frames and turns them into signals.
- It alone deals with propagation effects, which change signals when they travel over transmission lines.

Test Your Understanding

1. In what two ways does the physical layer differ from higher layers?

The Physical Layer

- The only layer that does not deal with messages
  - Individual bits are converted directly into signals.
- It alone deals with propagation effects—changes in signals as they propagate.

Signals and Propagation

- A signal is a disturbance that propagates down a transmission medium to the other side, which reads the signal.
- Propagation: The traveling of a signal down a transmission medium
- Propagation effects: changes in the signal as it propagates
  - Attenuation
  - Distortion
  - Etc.
- If propagation effects are too large, the receiver cannot read the signal.

Signaling

- In analog signaling, the signal rises and falls smoothly in intensity.
  - Small propagation errors are not fixed.
- In binary signals, there are *two* states per clock cycle.
  - Small propagation effects do not create errors in reading the signal.
- In digital signals, there are *a few* states per clock cycle.
  - Can transmit multiple bits per clock cycle
  - Less error immunity to propagation errors than in binary transmission
- Today, most transmission is binary.

FIGURE 5-1 Physical Layer Propagation (Study Figure)

Signals and Propagation

PROPAGATION Suppose that you and a friend are standing a few meters apart. You stretch a rope tautly between you. Now you close your eyes. Your friend jiggles the rope. The disturbance **propagates** (travels) down the rope. When it arrives, you feel it.

SIGNALS Something like this happens in network transmission. As Figure 5-2 shows, the transmitter creates a disturbance in a transmission medium—wire, optical fiber, or radio. This disturbance is the signal. A **signal**, then, is a disturbance that propagates down a transmission medium to the receiver, which reads the signal.

> A signal is a disturbance that propagates down a transmission medium to a receiver, which reads the signal.

PROPAGATION EFFECTS Note that the received signal differs from the transmitted signal. The changes in this figure are due to **propagation effects**—that is, changes in the signal during propagation. In the figure, the signal has **attenuated** (weakened). It

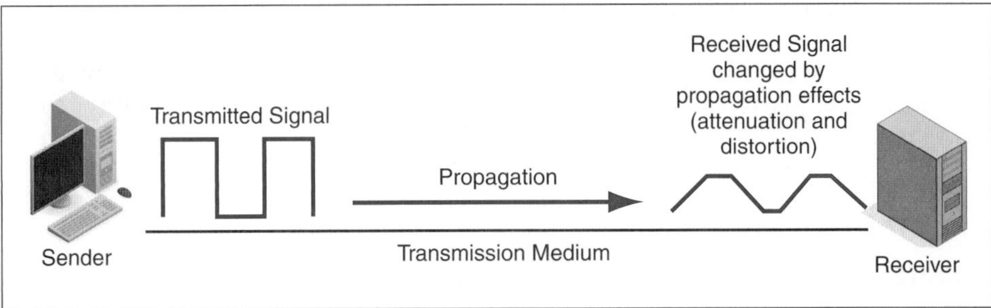

FIGURE 5-2 Signal and Propagation

also has been **distorted** (its shape has been changed). If propagation effects are too large, the receiver will not be able to interpret the signal correctly. In this chapter, we will see several propagation effects, most of which are important only in certain transmission media.

Test Your Understanding

2. a) What is a signal? b) What is propagation? c) What are propagation effects? d) Why are propagation effects bad?

Signaling

We looked at signaling in Chapter 1. We distinguished between analog signaling, binary signaling, and digital signaling.

- In analog signaling, the transmitted signal is analogous to the data input. As your voice rises and falls smoothly in intensity, the signal also rises and falls smoothly in intensity. Analog signals always introduce some degree of error.
- In binary signaling, there are *two* states that represent 1 and 0, respectively. In UTP wire transmission, there may be two voltage levels. In optical fiber, there is light for a 1 and no light for a 0. In any binary signaling, there are clock cycles. The signal is held constant during a clock cycle. At the end of the clock cycle, the signal may either change to the other state or remain the same. Binary signaling produces no reception error at all as long as propagation errors are moderate.
- Binary signaling is a special case of digital signaling, in which there are *a few* states. Unlike binary signaling, digital signaling with more than two states can send multiple bits per clock cycle. On the down side, digital signaling with more than two states is more error-prone than digital signaling with only two states, that is, binary signaling.

We will not look at analog signaling in this chapter. Except as otherwise noted, we will assume that signaling is binary rather than multistate digital signaling.

Test Your Understanding

3. a) What is the disadvantage of analog signaling? b) What is the advantage of binary signaling? c) What do we call signaling in which there are two states? d) What do we call signaling in which there are a few states?

UNSHIELDED TWISTED-PAIR (UTP) COPPER WIRING

Having looked at propagation effects and signaling in general, we will now see how these concepts apply to UTP transmission. Later, we will see how they apply to optical fiber transmission.

4-Pair UTP and RJ-45

THE 4-PAIR UTP CABLE The dominant standards family for local wired area networks is Ethernet. Ethernet LANs typically use **4-pair unshielded twisted-pair (UTP)** wiring to connect users to the nearest switch. They may also use 4-pair UTP to connect switches to other switches.[1] The name "4-pair unshielded twisted-pair (UTP)" sounds complicated,

4-Pair UTP Cable

- Ethernet is the dominant LAN standards family.
- Ethernet LANs typically use 4-pair unshielded twisted-pair wiring to connect hosts to the nearest switch.
- They may also use UTP to connect switches to other switches.

Cord Organization

- A length of UTP wiring is a cord.
- Each cord has eight copper wires.
  - Each wire is covered with dielectric (nonconducting) insulation.
- The wires are organized as four pairs.
  - Each pair's two wires are twisted around each other several times per inch.
- There is an outer plastic jacket that encloses the four pairs.

Connector

- RJ-45 connector is the standard connector.
- Plugs into an RJ-45 jack in a host, switch, or wall jack

Characteristics

- Inexpensive and easy to purchase and install
- Rugged: Can be run over with chairs, etc.
- Dominates media for access links

FIGURE 5-3 Unshielded Twisted-Pair (UTP) Wiring (Study Figure)

[1] "Unshielded" means that there is no metal shielding around the individual wire pairs or the entire jacket. Metal foil or mesh shielding reduces electromagnet interference effects, which we will see a little later in this section. However, shielded twisted-pair wiring is quite expensive, and unshielded twisted-pair is sufficient for office use and even for most factory use.

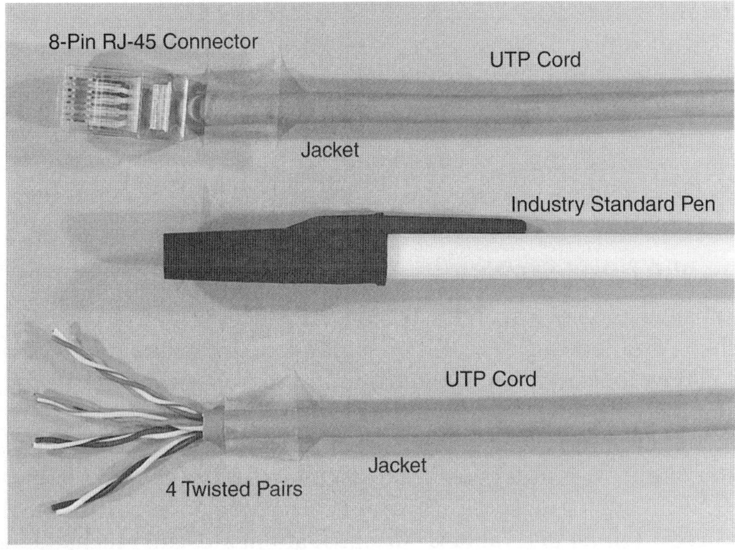

FIGURE 5-4 A 4-Pair Unshielded Twisted-Pair (UTP) Cord with RJ-45 Connector

but the medium is very simple. Figure 5-4 illustrates a 4-pair UTP cord with its four wire pairs showing.

- A length of UTP wiring is a **cord**.
- Each cord has eight copper wires.
- Each wire is covered with dielectric (nonconducting) **insulation**.[2] This prevents short circuits between the electrical signals traveling on different wires.
- The wires are organized as four pairs.
- Each pair's two wires are twisted around each other several times per inch.
- There is an outer plastic **jacket** that encloses the four pairs.

RJ-45 CONNECTORS At the two ends of a UTP cord, the wires must be separated and placed within an 8-pin **RJ-45 connector**, which also is shown in Figure 5-4. The RJ-45 connector at each end of a 4-pair UTP cord snaps into an **RJ-45 jack** (port) in the host, the switch, or the wall jack.[3]

EASY, INEXPENSIVE, AND RUGGED UTP is inexpensive to purchase, easy to **connectorize**[4] (add connectors to), and relatively easy to install. It is also rugged, so if a chair runs over it accidentally, it probably will survive undamaged. Four-pair UTP dominates corporate usage in access links from a host to the first switch because of UTP's low cost and durability.

---

[2] Benjamin Franklin coined the terms *conductor*, *insulation*, and many other terms in electricity (including *positive* and *negative*). He also created the theory of electricity flow, although he thought that electricity flowed from the positive end of a battery to the negative end. Today, we know that electrons flow from negative to positive.

[3] Home telephone connections use a thinner RJ-11 connector and jack. They were designed to terminate six wires but usually terminate only a single pair.

[4] Yes, *connectorize* is a really ugly term. Hey, I don't make up these names!

Test Your Understanding

4. a) What is a length of UTP wiring called? b) In 4-pair UTP, how many wires are there in a cord? c) How many wire pairs are there in a cord? d) What surrounds each wire? e) How are the two wires of each pair arranged? f) What is the outer covering called?
5. Why is 4-pair UTP dominant in LANs for the access line between a host and the switch that serves the host?

Attenuation and Noise Problems

UTP signals change as they travel down the wires. If they change too much, they will not be readable. As noted earlier in this chapter, there are several types of propagation effects. We will now look at the most three important propagation effects for UTP transmission, beginning with attenuation and noise.

ATTENUATION Figure 5-5 illustrates that when signals travel, they **attenuate** (grow weaker). To give an analogy, as you walk away from someone who is speaking, his or her voice will grow fainter and fainter. If a signal attenuates too much, the receiver will not be able to recognize it.

Attenuation is measured relative to the original strength of the signal. If a signal starts at the power level 20 milliwatts (mW) and falls to 5 mW, then signal had decreased to 25 percent (5/20) of its original value. Expressed another way, the signal has attenuated by 75 percent.

NOISE Electrons within a wire are constantly moving randomly, and these moving electrons generate random electromagnetic energy. This random electromagnetic energy is **noise**. The receiver actually gets the total of the signal plus the noise.

Random electromagnetic energy is noise.

The mean of the noise energy is the **noise floor**—despite the fact that it is an average and not a minimum, as the name *floor* would suggest. Figure 5-5 shows a noise floor.

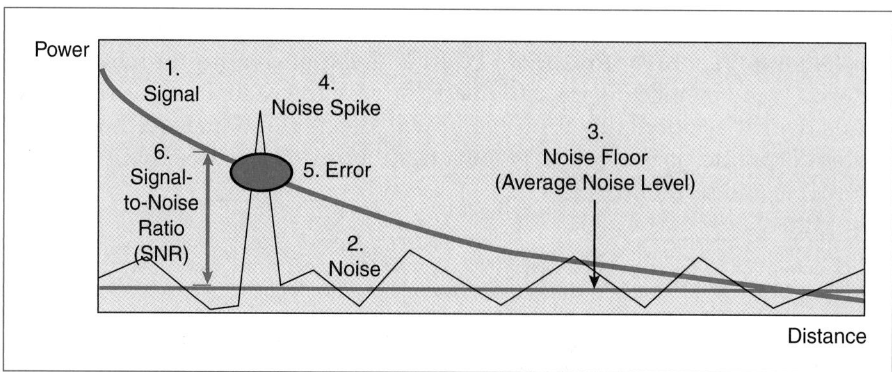

FIGURE 5-5 Attenuation and Noise

The mean of the noise energy is the noise floor.

As a consequence of noise being a random process, there are occasional **noise spikes** that are much higher or lower than the noise floor. As Figure 5-5 shows, if a noise spike is about as large as the signal, the combined signal and noise may be unrecognizable by the receiver.

NOISE, ATTENUATION, AND PROPAGATION DISTANCE If a signal is far larger than the noise floor, then we have a high **signal-to-noise ratio (SNR)**. With a high SNR, few random noise spikes will be large enough to cause errors. However, as a signal attenuates during propagation, it falls ever closer to the noise floor. Noise spikes will equal the signal's strength more frequently, so errors will become more frequent. In other words, even if the noise level is constant, longer propagation distances create attenuation that results in a lower SNR and therefore more noise errors.

Even if the noise level is constant, longer propagation distances result in a lower SNR and therefore more noise errors.

Note that noise problems do not begin when the signal reaches the noise floor. Noise spikes always cause some errors. They simply cause more errors as the SNR decreases. The number of errors becomes unacceptable well before the signal reaches the noise floor.

The number of errors becomes unacceptable well before the signal reaches the noise floor.

LIMITING UTP CORD DISTANCE TO LIMIT ATTENUATION AND NOISE PROBLEMS Fortunately, installers can control attenuation and noise by limiting the length of UTP cords. The Ethernet standard currently limits UTP propagation distances to 100 meters at all speeds up to 1 Gbps. If UTP cords are restricted to 100 meters, the signal still will be comfortably larger than the noise floor when it arrives at the receiver, so there will be few noise errors.

The Ethernet standard limits UTP cords to 100 meters at all speeds up to 1 Gbps.

Test Your Understanding

6. a) Describe the attenuation problem and why it is important. b) A signal is transmitted at 50 mW. When it reaches the receiver, its strength is 5 mW. On a percentage basis, how strong is the final signal relative to the initial signal? c) By what percentage has the signal declined in strength? d) Describe the noise problem. e) What is the signal-to-noise ratio, and why is it important? f) As a signal propagates down a UTP cord, the noise level is constant. Will greater propagation distance result in fewer noise errors, the same number of noise errors, or more noise errors? Explain. g) Does noise only cause errors when the signal strength falls to the noise floor or below it? Explain.
7. a) What problem (or problems) is (are) reduced to acceptable limits by limiting the length of a UTP cord in Ethernet? b) What is the typical limit on UTP cord length in Ethernet standards?

Electromagnetic Interference (EMI) in UTP Wiring

GENERAL INTERFERENCE Noise is unwanted electromagnetic energy *within* the propagation medium. In turn, **electromagnetic interference (EMI)**—or more simply, **interference**—is unwanted *external* electromagnetic energy coming from electrical motors, fluorescent lights, and even other nearby UTP cords (which always radiate some of their signal). Like noise energy, interference energy adds to the signal energy and can make the received signal unreadable.

USING TWISTED-PAIR WIRING TO REDUCE INTERFERENCE Fortunately, there is a simple way to reduce interference to an acceptable level. This is to twist each pair's wires around each other several times per inch, as Figure 5-6 illustrates.

Consider what happens over a full twist. Over the first half of the twist, the interference might add to the signal. Over the other half, however, this same interference would subtract from the signal. The interference on the two halves would cancel out, and the net interference would be zero.[5]

Does twisting really work this perfectly? No, of course not. However, twisting the wiring dramatically reduces interference, limiting it to an acceptable level. As a historical note, Alexander Graham Bell himself invented twisted-pair wiring as a way to reduce interference in telephone transmission.

CROSSTALK INTERFERENCE As Figure 5-7 shows, individual pairs in a cord will radiate some of their energy, producing electromagnetic interference in other pairs within the cord. This mutual interference among wire pairs in a UTP cord is **crosstalk interference**. It is always present in wire bundles and must be controlled. Fortunately,

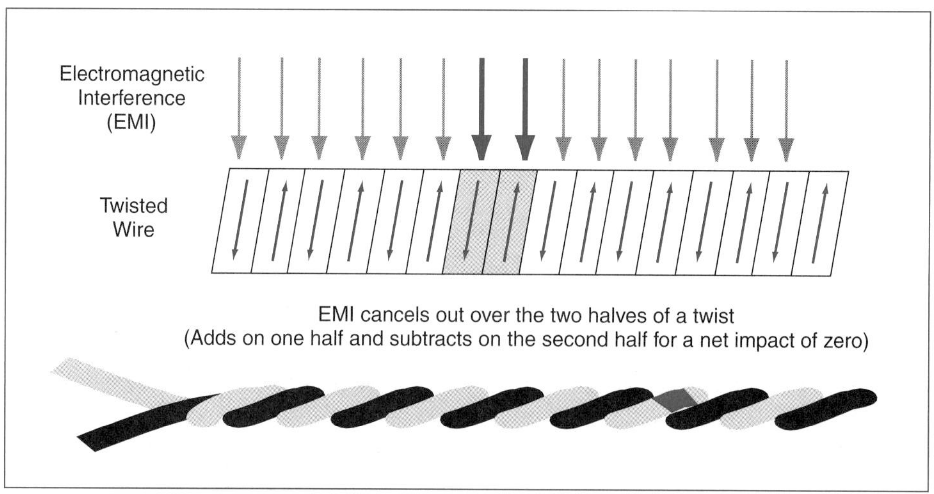

FIGURE 5-6 Electromagnetic Interference (EMI) and Twisting

[5] To give another example, suppose that an aircraft makes a round trip. In still air, it flies at 600 kph. Going one way, it faces a 50 kph head wind. Going back, it has a 50 kph tail wind. Across the two trips, the impact of the head wind is zero.

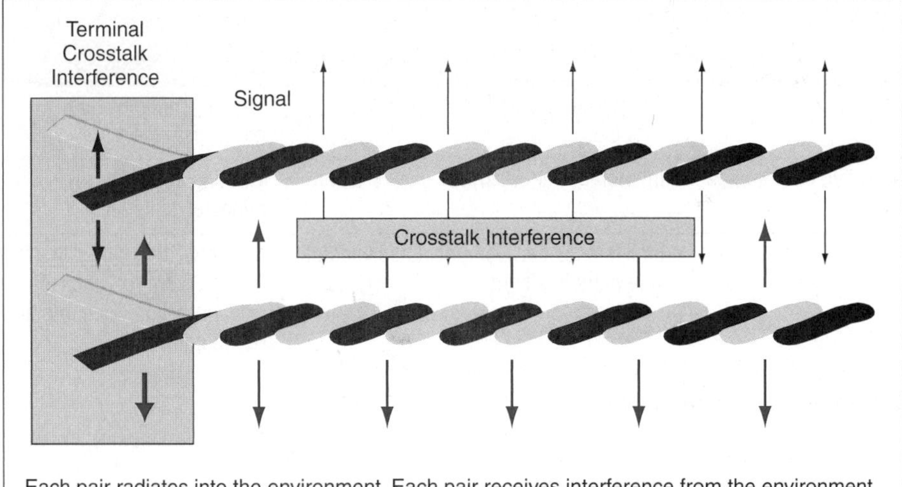

FIGURE 5-7 Crosstalk Interference and Terminal Crosstalk Interference

the twisting of each pair normally keeps crosstalk interference in a UTP cord to a negligible level.

TERMINAL CROSSTALK INTERFERENCE Unfortunately, when a UTP cord is connectorized, its wires must be untwisted to fit into the RJ-45 connector, as shown in Figure 5-7. The eight wires are now parallel, so there is no protection from crosstalk interference. Crosstalk interference at the ends of the UTP cord, which is **terminal crosstalk interference**, usually is much larger than the rest of the crosstalk interference over the entire rest of the cord.

Fortunately, there is a simple way to deal with terminal crosstalk interference. Installers must be careful not to untwist UTP wires more than 1.25 cm (half an inch) when adding connectors. This precaution will not completely eliminate terminal crosstalk interference, but it will limit terminal crosstalk interference to an acceptable level.

Installers must be careful not to untwist UTP wires more than 1.25 cm (half an inch) when adding connectors. This precaution will not completely eliminate terminal crosstalk interference, but it will limit crosstalk interference to an acceptable level.

HIERARCHY OF INTERFERENCE TYPES Note in Figure 5-8 that the three types of interference are related. Interference is the general term. Crosstalk interference is a particular type of interference. Terminal crosstalk interference is a particular type of crosstalk interference. The only type of interference that commonly is a threat is terminal crosstalk interference.

Type of Interference	Propagation Effect Reduced to Insignificance	Mitigation
Interference	All interference from outside a UTP wire pair	Twisting automatically mitigates the effect.
Cross-Talk Interference	Interference between wire pairs in a cord	Twisting automatically mitigates the effect.
Terminal Cross-Talk Interference	Interference between untwisted ends of wire pairs in a cord	Not untwisting wire pairs more than 1.25 cm (0.5 inches) when inserting them into an RJ-45 connector

FIGURE 5-8 Types of Interference

CONTROLLING PROPAGATION EFFECTS In this section on UTP, we have looked at the three main propagation effects for UTP and the two ways we can reduce them to unimportance. Figure 5-9 summarizes this information. It is important for you to understand this figure completely.

Test Your Understanding

8. a) Distinguish between interference, crosstalk interference, and terminal crosstalk interference. b) How is interference controlled? c) How is terminal crosstalk interference controlled in general? Explain. d) Does this precaution eliminate terminal crosstalk interference? e) What is the only type of interference that is frequently a problem?

Serial and Parallel Transmission

Figure 5-10 shows an important distinction in wire communication: serial versus parallel transmission.

SERIAL TRANSMISSION In slower versions of Ethernet that run at 10 Mbps and 100 Mbps, a single pair in a UTP cord transmits the signal in each direction. Two-way transmission, then, uses two of the four pairs—one in each direction. The other two pairs are not used, although the standard calls for them to be present. If one pair of wires is used to send a transmission, this is **serial transmission** because the transmitted bits must follow one another in series.

Precaution	Propagation Effect Reduced to Insignificance
Limiting UTP distance to 100 meters	Attenuation and Noise
Limiting the untwisting of wires to less than 1/2 inch at the connector	Terminal crosstalk interference

FIGURE 5-9 Controlling UTP Propagation Effects

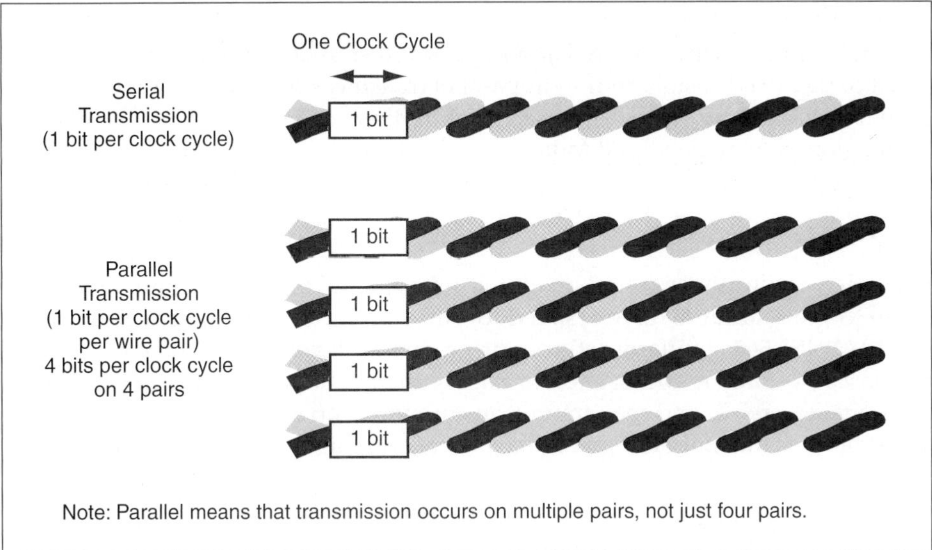

FIGURE 5-10 Serial versus Parallel Transmission

PARALLEL TRANSMISSION For gigabit Ethernet, however, when the host or switch transmits, it transmits on all four pairs in each direction.[6] This means that it can transmit four bits at a time instead of just one. Sending data simultaneously on two or more transmission lines in the same direction is **parallel transmission**. The benefit of parallel transmission is that it is faster than serial transmission for the same clock cycle duration. Speed is the key benefit of parallel transmission.

Parallel transmission is sending data simultaneously on two or more transmission lines in the same direction.

Note that parallel transmission does not mean the use of *four* transmitting pairs (or grounded wires). It means using *more than a single pair* of wires.[7] For example, your computer's internal bus, which carries signals among components in your system unit, has about 100 wires for high-density parallel transmission. To give another example, old parallel cables used eight wires to carry data in each direction.

TREND In general, serial transmission has largely replaced parallel transmission in practice. Reducing the clock cycle in serial transmission has the same benefit as using more transmission paths, and it is generally cheaper. Parallel transmission survives only in some short-distance transmission applications such as gigabit Ethernet and PC parallel busses.

---

[6] What if both sides transmit at the same time? Each knows what signal it sent. It subtracts this signal from the signal it hears on the wire. The remaining signal is the signal sent by the other side.

[7] You actually can have a single wire plus a ground.

Test Your Understanding

9. a) Distinguish between serial and parallel transmission. b) What is the main benefit of parallel transmission? c) In parallel transmission, how many pairs (or single wires and ground wires) are used in each direction? d) Is serial or parallel transmission more widely used today?

Wire Quality Categories

UTP cords vary in transmission quality. If wire quality is too low, signals will not be able to travel far at a given speed or may not be able to travel at all. The **TIA/EIA/ANSI-568** standard defines wiring quality levels as **category** numbers. The term *category* usually is abbreviated to **cat**. Figure 5-11 shows the relationship between wiring quality, speed, and maximum transmission propagation distances in Ethernet. Quite simply, as transmission speeds have increased, higher-quality wiring has become necessary.

Category is a measure of UTP quality.

CAT 5E AND CAT 6 UP TO 1 GBPS In most buildings today, the wiring is **Cat 5e** (enhanced) or **Cat 6**. Both categories of wire quality can carry Ethernet signals at 1 Gbps. Both can support these speeds over distances up to 100 meters. This is more than sufficient for most corporate needs for the foreseeable future for desktop access lines.[8]

CAT 6 AND CAT 6A UP TO 10 GBPS Within data centers and switching centers, however, and between central switches in a corporate network, speeds of 10 Gbps are becoming important. Although optical fiber can be used at these distances, UTP wiring is less expensive, so new categories have been defined for twisted-pair wiring. The first category intended for 10 Gbps was **Cat 6**. However, Cat 6 was a disappointment. It could only carry 10 Gbps Ethernet traffic up to 55 meters, not the usual 100 meters. Consequently,

Category	Technology	Maximum Speed	Maximum Ethernet Distance at this Speed
5e	UTP	1 Gbps	100 meters
6	UTP	10 Gbps	55 meters
6A	UTP	10 Gbps	100 meters

FIGURE 5-11 Wire Quality Standards

[8] Older buildings have some Cat 3 and Cat 4 UTP wiring. Cat 3 and Cat 4 wiring can carry 10 Mbps Ethernet signals up to 100 meters. The UTP wiring in the first author's building has been rated as "Cat 3 on a good day."

augmented category 6 (**Cat 6A**[9] or AC6) wiring was defined to allow UTP to carry 10 Gbps up to the usual 100 meters.[10]

Test Your Understanding

10. a) What is the maximum length of UTP cords in Ethernet standards up to and including 1 Gbps? b) What wiring characteristic do UTP categories standardize? c) What two UTP quality categories dominate sales in building wiring today? d) Cat 5e and Cat 6 are sufficient for Ethernet transmission up to what speed? e) What new categories of wiring were developed for 10 Gbps Ethernet transmission? f) How far can Cat 6 and Cat 6A carry 10 Gbps Ethernet signals?

OPTICAL FIBER

Light through Glass

In the 1840s, scientists discovered that light would follow water flowing out of a hose. Where the water stream bent, the light would follow. This raised the possibility that light signals could be sent through glass rods—a medium that allows more controlled transmission than water.

The Core and Cladding

Figure 5-12 illustrates optical fiber operation. Fiber uses binary signaling. During each clock cycle, the light is turned on for a 1, off for a 0.

To send signals, the **transceiver** (sender/receiver) injects light into a very thin glass rod called the **core**. Typical core diameters are 8.3, 50, or 62.5 microns (millionths of a meter). In comparison, an average human hair is about 75 microns thick.[11]

Surrounding the core is a thicker glass cylinder called the **cladding**. The cladding normally is 125 microns in diameter, regardless of the core diameter. Consequently, an optical fiber strand with a 50 micron core diameter is often called 50/125 fiber.

As Figure 5-12 shows, if a light ray enters at an angle, it hits the cladding and reflects back into the core with **perfect internal reflection** so that no light escapes into the cladding or beyond.[12] In contrast, UTP tends to radiate energy out of the wire bundle, causing rapid attenuation.

---

[9] Why is improved Cat 5 wiring called Cat 5e (enhanced), while improved Cat 6 wiring is Cat 6A (augmented) or AC6? The answer is that these two improved categories were created by different standards agencies that had different terminology.

[10] Another way to carry 10 Gbps traffic up to 100 meters is to use Category 7 wiring. In contrast to other wiring categories, Cat 7 wiring uses shielded twisted-pair (STP) wiring instead of UTP. STP wiring places a metal foil shield around each twisted pair and also a wire mesh around the four pairs. This almost completely eliminates electromagnetic interference. However, STP is more difficult and expensive to buy and install than UTP. In addition, Cat 7 STP wiring requires a new type of connector. Companies are likely to be very reluctant to use Cat 7 STP unless they have high-EMI environments.

[11] Actually, human hair varies in thickness from about 40–120 microns. Hair less than 60 microns thick is considered to be fine hair. Hair more than 80 microns thick is called thick or coarse. See, you learned something from reading this book.

[12] This is based on Snell's Law. The cladding has a slightly lower index of refraction than the core. This difference in index of refraction creates perfect internal reflection. By the way, if the fiber is bent too much, the angle between the core and cladding will become wrong for the light waves, and a great deal of light will be lost into the cladding. Of course, if optical fiber is bent even farther, it will crack. Glass is glass.

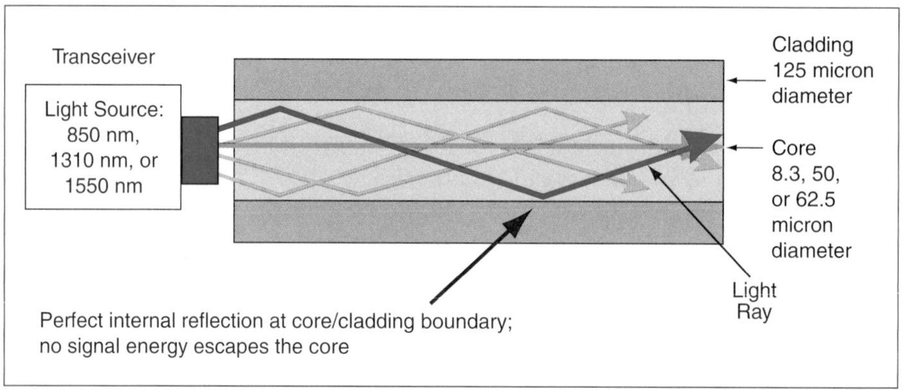

FIGURE 5-12 Optical Fiber Transmission

An opaque coating surrounds the cladding to keep out light and to strengthen the fiber. Around this opaque coating are strands of yellow Aramid (Kevlar[TM][13]) yarn to strengthen the fiber. The coating, Aramid strands, and outer jacket bring the outer diameter of a LAN fiber strand to about 900 microns (0.9 mm), regardless of the core diameter.

Test Your Understanding

11. a) How does optical fiber signaling represent 1s and 0s? b) In optical fiber, what are the roles of the core and the cladding in light transmission? c) How thick are fiber cores? d) How thick is the fiber cladding? e) How thick is the outer jacket?

Transceiver Frequency and Wavelength

In optical fiber, a **transceiver** (transmitter/receiver) injects light into the core and accepts light arriving from the core.

Figure 5-13 shows that light waves can be described in terms of **frequency**, which is the number of times the wave goes through a complete cycle per second. Frequency is measured in **hertz (Hz)**. One hertz is one cycle per second. High frequencies are expressed in metric notation.

The **wavelength,** in turn, is the distance between comparable parts on two successive cycles—peak to peak, trough to trough, and so on. In light, wavelengths are measured in **nanometers (nm)**. A nanometer is one billionth of a meter.

In optical fiber transmission, laser light is reported in wavelengths. In fiber, wavelengths come in three main "windows" centered at 850 nm, 1310 nm, and 1550 nm. These windows are approximately 50 nm wide. Within these windows, attenuation is very low compared with attenuation at other wavelengths.

In optical fiber transmission, laser light is reported in wavelengths.

[13] Yes, Kevlar is the stuff used to make bulletproof vests. If you cut optical fiber with a wire cutter built for UTP, you will dull the wire cutter very quickly. I have verified this the hard way.

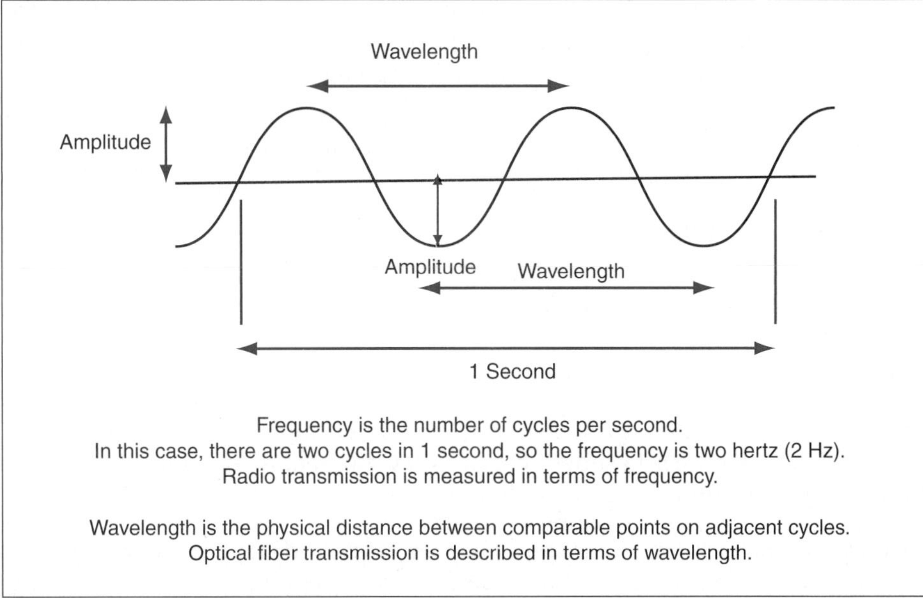

FIGURE 5-13 Frequency and Wavelength

Longer wavelengths bring longer propagation distances at higher speeds. However, longer wavelengths also increase the transceivers' costs. Consequently, the goal is to select the shortest wavelength that will provide the speed and distance needed. For LANs, this is almost always 850 nm. For WANs, which must span long distances, 1310 nm is the most widely used wavelength, although 1550 nm is also used.

Longer-wavelength light can travel farther and faster, but longer-wavelength transceivers are more expensive. For LAN fiber 850 nm transceivers usually are sufficient. WAN fiber normally uses 1310 nm or even 1550 nm transceivers.

The third characteristic of a wave is **amplitude**, which is the strength of the wave. Amplitude usually is measured as power.

Test Your Understanding

12. a) What are the three characteristics of light waves? b) Do we typically measure light in optical fiber using wavelength or frequency? c) What are typical light wavelengths? d) What two wavelengths of light can travel the farthest? e) What light wavelength is the least expensive to generate? f) What light wavelength dominates in LANs? Why does it dominate?

Farther Yes, but Not Necessarily Faster

Optical fiber can carry signals much *farther* than UTP at any given speed. While UTP transmission is limited to about 100 meters, optical fiber can easily span distances of 200 to 300 meters. This is optical fiber's main benefit today.

Some texts say that optical fiber can also carry signals faster than UTP. However, both UTP and fiber can carry signals up to 10 Gbps, which is the highest speed of Ethernet today. Optical fiber is normally used for going farther, not going faster.

Optical fiber is normally used to carry signals farther, not faster.

Test Your Understanding

13. a) What is the main benefit of optical fiber compared with UTP today? b) Which costs less to buy and install—UTP or optical fiber?

The Roles of Fiber and Copper

Although optical fiber use is growing rapidly, it is not completely replacing copper UTP wiring. Rather, as Figure 5-14 shows, the two play different roles in most LANs, so they are not direct competitors.

- UTP is dominant for access lines between hosts and the nearest switch because UTP is less expensive than fiber and because access line distances normally are well within UTP's usual 100 meter distance limit and speed limits. Also, UTP is somewhat more rugged than fiber, and this can be important in office areas, where optical fiber deals with its traditional natural enemy—moving chairs.
- In contrast, while UTP may be used for some trunk lines that connect switches to other switches, most trunk lines are optical fiber. Trunk lines tend to span longer distances and so can justify the higher cost of optical fiber.

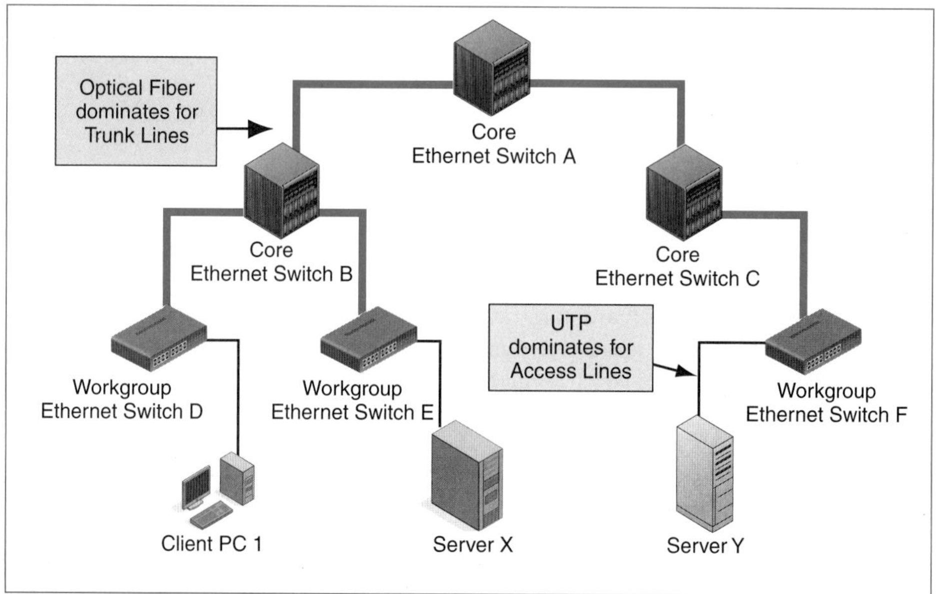

FIGURE 5-14 Roles of UTP and Optical Fiber in LANs

Test Your Understanding

14. a) Distinguish between access lines and trunk lines. b) In LAN transmission, what are the typical roles of UTP and optical fiber?

Full-Duplex Operation and Connectors

Figure 5-15 illustrates how an optical fiber strand is constructed.

TWO STRANDS FOR FULL-DUPLEX OPERATION Note in Figure 5-15 that an **optical fiber cord** normally has two **strands** of fiber—one for transmission in each direction. This gives **full-duplex communication** (simultaneous two-way communication).

CONNECTORS In UTP, there is only one type of connector—the RJ-45 connector. However, there are several types of optical fiber connectors. The most popular are the **SC** and **ST** connectors, which are shown in Figure 5-15. There also are several smaller connectors, which are called **small form factor (SFF)** connectors. These smaller connectors allow more ports to be placed on a switch of a given size. This diversity of connector types somewhat complicates fiber selection, but not too much, because it is possible to put different types of connectors at the two ends of a fiber cord so that a single cord can connect to ports on two different switches or routers with different types of plugs.

Test Your Understanding

15. a) What is the ability to transmit in both directions simultaneously called? b) Why does a fiber cord normally need two fiber strands? c) Does optical fiber have a single connector type? d) What are the two most common fiber connector types? e) What are small optical fiber connectors called?

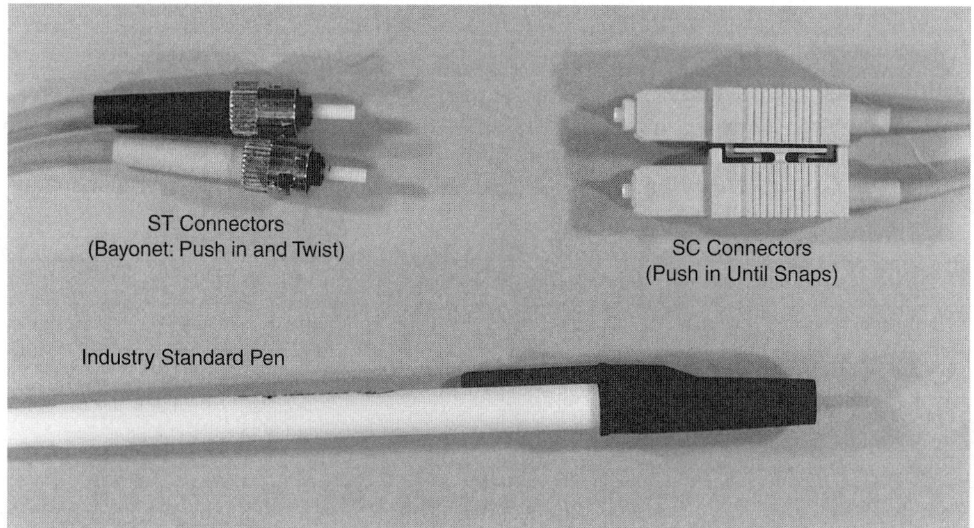

FIGURE 5-15 Full-Duplex Optical Fiber Cord with Two Strands and SC and ST Connectors

LAN Fiber versus Carrier WAN Fiber

Corporations use optical fiber to create LANs. Carriers use optical fiber to carry signals over long WAN distances. This is a book about corporate networking, so we will focus on the types of fiber that most corporations encounter, fiber for local area networks. Figure 5-16 illustrates the main differences between LAN and carrier WAN fiber.

WAVELENGTHS First, LAN fiber almost always uses 850 nm lasers. There is no need to use more expensive 1310 nm or 1550 nm lasers to carry signals over LAN distances. Carrier fiber, in contrast, normally uses 1310 or 1550 nm lasers because it needs longer-wavelength light for the distances carriers need to span (usually, tens of kilometers).

MULTIMODE FIBER LANs use **multimode fiber**, which has a relatively thick core, as Figure 5-17 shows. Initially, North American multimode fiber had a core diameter of 62.5 microns. In Europe, which developed fiber slightly later, the core diameter was a little thinner—only 50 microns in diameter. A thinner core can carry signals farther. Although 50-micron fiber is no more expensive than 62.5-micron fiber today, most U.S. organizations continue to standardize on 62.5-micron fiber, while European companies primarily use 50-micron fiber.

As Figure 5-17 shows, because multimode fiber has a thick core, light rays (called **modes**) can enter over a wide range of angles. Light traveling straight through the core arrives fastest. In contrast, modes entering at high angles will zigzag through the core many times. It will travel much farther and so will arrive later.

This difference in arrival times causes a problem called **modal dispersion**, in which some light modes from successive light pulses overlap. If the cord length is too great, light from successive light pulses will overlap so much that the signal will be unreadable. Modal dispersion is a limiting factor for distance in multimode fiber.

Modal dispersion is a limiting factor for distance in multimode fiber.

CARRIER SINGLE-MODE FIBER In contrast, carriers use single-mode fiber. This fiber has a core diameter of only 8.3 microns. With such a thin diameter, only a single mode can

Characteristic	LAN Fiber	Carrier WAN Fiber
Required Distance Span	200 meters to 300 meters	1 to 40 kilometers
Transceiver Wavelength	850 nm	1310 nm (and sometimes 1550 nm)
Type of Fiber	Multimode (thick core)	Single mode (thin core)
Core Diameter	50 microns or 62.5 microns	8.3 microns
Primary Distance Limitation	Modal dispersion	Absorptive attenuation
Quality Metric	Modal bandwidth (MHz.km)	NA

FIGURE 5-16 LAN Fiber versus Carrier WAN Fiber

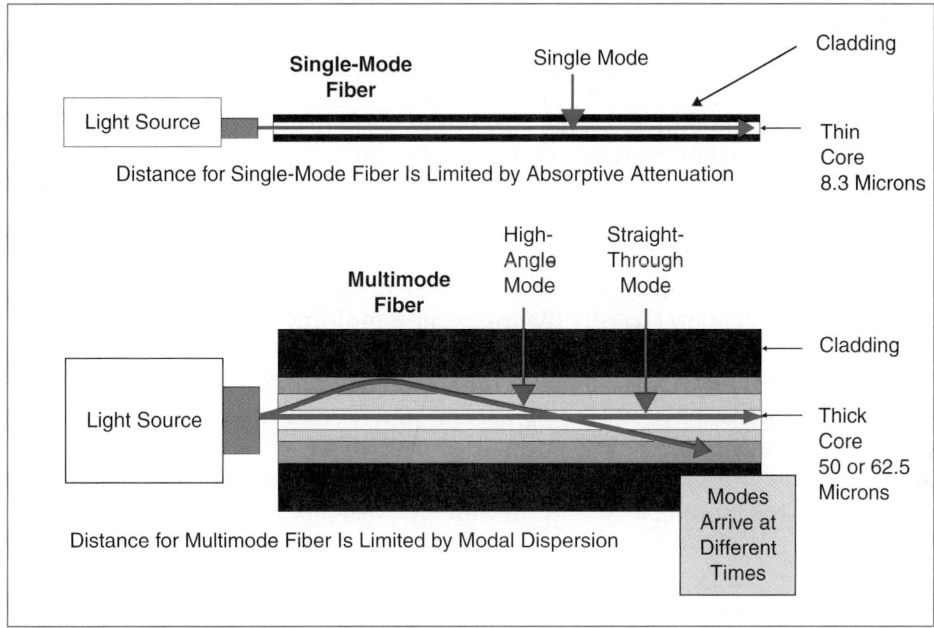

FIGURE 5-17 Multimode Fiber and Single-Mode Fiber

travel through the fiber. There is no modal dispersion to limit distance. For single-mode fiber, the only distance limitation is **absorptive attenuation**, with the light being absorbed by glass molecules. Fortunately, this attenuation is very low, especially at longer wavelengths. Consequently, single-mode fiber can carry signals much farther than multimode fiber. However, less expensive multimode fiber is sufficient for LANs and dominates LAN fiber usage.

The limiting factor in single-mode fiber is absorptive attenuation.

MULTIMODE FIBER QUALITY Just as UTP wiring has quality categories, multimode LAN fiber varies in quality. Simply put, higher-quality multimode fiber is better at limiting modal dispersion.[14] This gives greater propagation distance.

We saw that wire quality is given as a series of discrete quality levels (Cat 5e, Cat 6, etc.). Multimode optical fiber quality, in contrast, is measured as a *continuous* variable called **modal bandwidth**. Modal bandwidth is measured as **MHz.km**. Usually, two modal bandwidths are listed, as in 200/500. The first gives the modal bandwidth of the fiber at 850 nm. The second gives the modal bandwidth at 1310 nm.

---

[14] As one way to reduce modal dispersion, fiber manufacturers today only build graded-index multimode fiber, in which the index of refraction decreases from the center of the core to the core's outer edge. This slows light traveling down the center, compared with light farther out in the radius. Consequently, light going straight through along the center is slowed down, while light zigzagging through the core is speeded up during its time away from the center. This reduces the time lapse between the direct mode and other modes, thus reducing modal dispersion. I did not mention this in the main text because all multimode fiber today is graded-index multimode fiber.

Multimode fiber quality is measured by modal bandwidth, which is measured as MHz.km.

The modal bandwidth concept is too difficult to explain in an introductory network course (fortunately for you). Quite simply, more modal bandwidth is better for signal distance.

Test Your Understanding

16. a) Do LANs normally use multimode fiber or single-mode fiber? b) What is the main physical characteristic distinguishing multimode fiber from single-mode fiber? c) Why is multimode fiber used in LANs? (Give a complete explanation.) d) What are typical core diameters for multimode fiber? e) What is modal dispersion? f) What is the limiting propagation problem for multimode fiber transmission distance? g) Of what is modal bandwidth a measure? h) In what units is modal bandwidth expressed? i) What is the core diameter of single-mode fiber? j) What is the limiting factor for transmission distance in single-mode fiber?

Wavelength, Core Diameter, Modal Bandwidth, and Distance in Multimode Fiber

To increase LAN propagation distance, one can increase the light wavelength (rarely done), decrease the core diameter (not commonly done), and buy better-quality fiber having greater modal bandwidth (the most commonly done approach). In practice, however, LAN fiber uses only 850 nm light because longer wavelengths would be too expensive. Figure 5-18 gives examples of how modal bandwidth and core diameter can affect propagation distance.

1000BASE-SX WITH 62.5/125 FIBER For example, gigabit Ethernet, which carries signals at 1 Gbps, is dominated by the **1000BASE-SX** standard, which transmits light at 850 nm. (The S stands for short wavelength to denote 850 nm transceivers.) With a 62.5-micron core diameter and 160 MHz.km modal bandwidth, 1000BASE-SX can transmit gigabit Ethernet signals over a distance of 220 meters (the length of two football fields). With 200 MHz.km fiber, the distance rises to 275 meters. Fiber manufacturers produce fiber in a wide range of modal bandwidths.

Wavelength	Core Diameter	Modal Bandwidth	Maximum Propagation Distance
850 nm	62.5 microns	160 MHz.km	220 meters
850 nm	62.5 microns	200 MHz.km	275 meters
850 nm	50 microns	500 MHz.km	550 meters

FIGURE 5-18 Wavelength, Core Diameters, Modal Bandwidth, and Maximum Propagation Distance for Ethernet 1000BASE-SX

1000BASE-SX WITH 50/125 FIBER Few companies need longer distances than 62.5/125 multimode fiber can provide for their trunk lines. However, if they do, they can use 50/125 fiber rather than 62.5/125 fiber. Thinner fiber offers higher modal bandwidth. With a modal bandwidth of 500 MHz.km, 50/125 fiber can carry 1000BASE-SX signals up to 550 meters.

Test Your Understanding

17. In what two ways can businesses select multimode fiber to transmit Ethernet 1000BASE-SX LAN signals farther?

Noise and Electromagnetic Interference

In fiber, absorptive attenuation and modal dispersion are important limitations for different types of fiber. However, noise and electromagnetic interference, which are serious problems in UTP, are *not* problems for fiber.

There is no noise energy in fiber. Electrons randomly moving around within the transmission medium generate electromagnetic energy called noise. However, electrons do not generate light energy that adds to the light signal.

In UTP, interference is an important concern. In optical fiber, this concern vanishes. The only type of electromagnetic interference applicable to fiber is light coming into the fiber from the outside. Fiber prevents this completely by having an opaque coating around the cladding.

Test Your Understanding

18. Are noise and interference major propagation problems for optical fiber propagation? Explain.

THE PUBLIC SWITCHED TELEPHONE NETWORK (PSTN)

Up to this point, we have been looking at UTP and optical fiber in the context of local area networks. However, both technologies are used in the worldwide telephone network, which is technically the **Public Switched Telephone Network (PSTN)**. Figure 5-19 shows the main elements of the PSTN.

The Business Local Loop

The access link between the customer premises and the nearest telephone company switch (called the end office switch) is called the **local loop**. Figure 5-20 shows the three main technologies that dominate the local loop.

TWO-PAIR DATA-GRADE UTP Two of these technologies are used to provide leased lines to businesses. One of these is a type of UTP wiring we have not discussed before. This is **2-pair data-grade UTP**. As the name suggests, this type of UTP wiring uses *two* twisted pairs instead of four. Each pair is for transmission in one direction. *Data grade* means that this wire was designed to carry data instead of voice. Two-pair data-grade UTP is used in the slowest leased lines—those up to about 2 Mbps.

Local loops, of course, are longer than 100 meters. Often, there can be one or two kilometers between the customer premises and the nearest end office switch. The 100 meter limit on UTP cords that we saw earlier in this chapter is specific to 4-pair UTP and LANs.

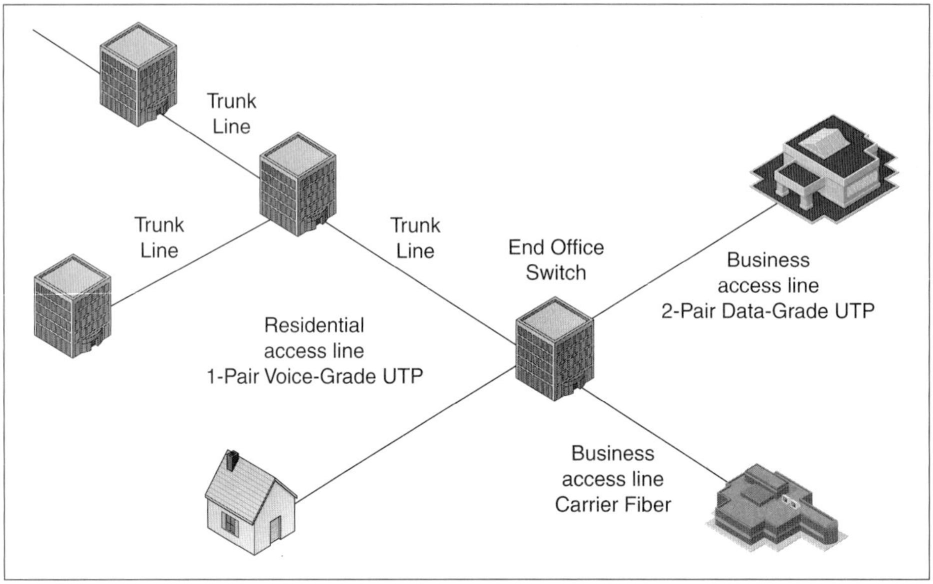

FIGURE 5-19 The Public Switched Telephone Network (PSTN)

Local Loop Technology	Business or Residential	Considerations
2-Pair Data-Grade UTP	Business	For leased lines up to 2 Mbps Must be pulled to the customer premises Not limited to 100 meters
Optical Fiber	Business	For leased lines faster than 2 Mbps Must be pulled to the customer premises
1-Pair Voice-Grade UTP	Residential	Designed only for voice transmission Not limited to 100 meters If a 1-pair VG UTP line carries data, the service is called digital subscriber line (DSL) service Already installed, so avoids the expense of pulling a new line to residences
Optical Fiber	Residential	Fiber to the Home (FTTH) New Installing FTTH in entire neighborhoods to reduce installation costs

FIGURE 5-20 Local Loop Technologies

The 100 meter limit on UTP cords that we saw earlier in this chapter is specific to 4-pair UTP and LANs. It does not apply to other types of UTP.

CARRIER FIBER Two-pair data-grade UTP is fine for leased lines up to about 2 Mbps. For all faster leased lines, carriers use carrier-grade optical fiber. This means single-mode fiber and higher wavelengths of light—1310 nm and 1550 nm.

Test Your Understanding

19. a) Compare the types of UTP used in corporate buildings and in low-speed PSTN leased lines. b) Is all UTP wiring limited to 100 meters? Explain.

The Residential Local Loop and Digital Subscriber Lines (DSLs)

ONE-PAIR VOICE-GRADE UTP While businesses needing leased lines may use 2-pair data-grade UTP, the lines running to all residential premises are **1-pair voice-grade UTP**. As its name suggests, this type of wiring was designed to carry voice, not high-speed data.

However, telephone companies have learned to send data over 1-pair voice-grade UTP wiring. They have done this because 2-pair data-grade UTP and carrier fiber usually have to be pulled from the end office to the customer premises whenever a business needs a leased line. However, every residence already has 1-pair voice-grade UTP running to it. This greatly reduces the cost of **provisioning** (providing service). Lower costs to set up a service translate into lower monthly service fees.

Whenever data is transmitted over 1-pair voice-grade UTP, the service is called **digital subscriber line (DSL)** service. The *subscriber line* is the line already running to the residential premises. DSL sends data over these voice-grade lines.[15]

Whenever data is transmitted over 1-pair voice-grade UTP, the service is called digital subscriber line (DSL) service.

Residential service is **asymmetric DSL (ADSL)** service with much higher download (to the home) speeds than upload (from the home) speeds. Asymmetry is attractive because many services that require high speeds require faster downstream speeds than upstream speeds. Two common examples are web service and video.

Figure 5-21 shows the main technical elements of ADSL service. The customer needs a DSL modem, which plugs into a wall jack via a home telephone cord. A computer can plug into the DSL modem for digital transmission service. At the same time, the customer can plug his or her telephones into wall jacks and use them while the line is used for data transmission.

To use DSL service effectively, the user must plug a **splitter** into each telephone jack. The splitter separates the voice signal from the data signal so that they cannot interfere with each other. Phones plug into the voice port of a splitter. A DSL modem, if one is present, plugs into the data port on the splitter.

---

[15] As we will see in Chapter 6, businesses often get *symmetric* DSL service, with equal speed in both directions. Symmetric service is needed for connecting sites to other sites, because traffic is likely to be about the same in both directions. Symmetric DSL service also offers service-level agreements because businesses need the predictability that SLAs offer.

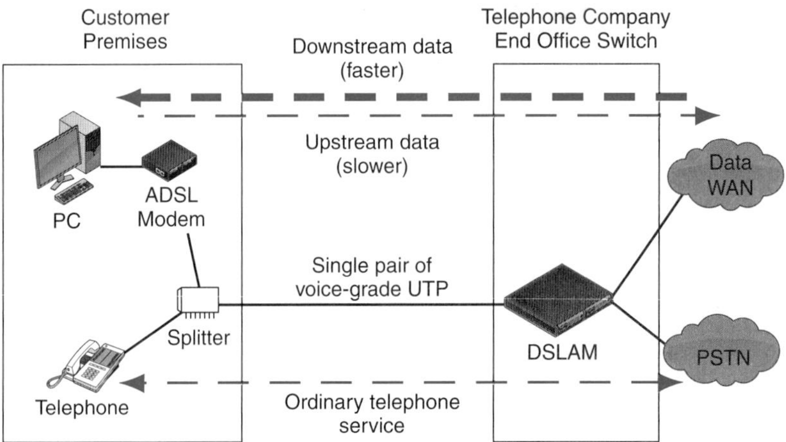

FIGURE 5-21 Asymmetric Digital Subscriber Line (DSL) Service

At the telephone company end office switch, the carrier needs to install a **DSL access multiplexer (DSLAM)**. This device separates voice and data traffic. It links voice traffic to the traditional voice part of the Public Switched Telephone Network. It links data traffic to a data network.

FIBER TO THE HOME (FTTH) Although DSL speeds today are quite fast, subscribers want to bring high-definition video into their homes, and they want multiple channels at a time. We cannot do this with digital subscriber lines using 1-pair voice-grade UTP copper wire. To provide extremely high speeds, a number of carriers are beginning to bring **fiber to the home (FTTH)** by running carrier-grade fiber from the end office switch to residential households.

Running new fiber to each household is very expensive, so implementation has been slow. However, by converting entire neighborhoods to FTTH, carriers have been able to lower their per-house installation costs.

Test Your Understanding

20. a) Compare the 4-pair UTP wiring used in corporate buildings and the UTP wiring in the residential local loop. b) Distinguish between the UTP wiring in the residential local loop and the UTP wiring used for lower-speed leased lines (under about 2 Mbps). c) What is provisioning? d) What type of transmission line technology do all DSL services use? e) Why is 1-pair voice-grade UTP attractive? f) Why is it all right that ADSL consumer service is asymmetric? g) What equipment does the residential customer need for DSL service? h) What equipment does the telephone carrier have to add to its end office switch? i) What is the purpose of this device?

CABLE MODEM SERVICE

Telephone Service and Cable TV

In the 1950s, **cable television** companies sprang up in the United States and several other countries, bringing television into the home. Initially, cable only brought over-the-air

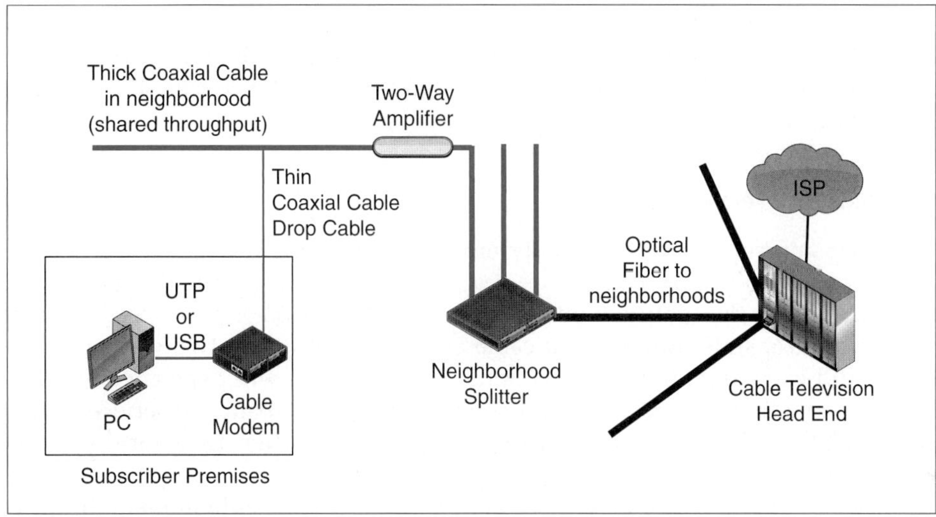

FIGURE 5-22 Cable Modem Service

TV to rural areas. Later, it began to penetrate urban areas by offering far more channels than urban subscribers could receive over the air. In the 1970s, many books and articles forecast a "wired nation" in which two-way cable and the advent of 40-channel cable systems would soon turn cable into an information superhighway. (After all, it would be impossible to fill 40 channels just with television, wouldn't it?) However, available services did not justify the heavy investment to make cable a two-way service[16] until many years later.

Figure 5-22 shows how cable television operates. The cable television operator has a central distribution point, called a **head end**. From the head end, signals travel out to neighborhoods via optical fiber.

From neighborhood splitters, signals travel through coaxial cable. The transmission of an electrical signal always requires *two* conductors. In UTP, the two conductors are the two wires in a pair. In coaxial cable, the first conductor is a wire running through the center of a coaxial cable. The second conductor is a mesh wire tube running along the cable. The two conductors have the same axis, so the technology is called **coaxial**. Before the advent of high-definition HDMI cables, you typically connected your VCR to your television with coaxial cable.

The cable television company runs signals through the neighborhood using *thick coaxial cable* that looks like a garden hose. The access line to individual homes is a thin coaxial cable **drop cable**. The resident connects the drop cable to his or her television.

---

[16] This was proven in the dissertation of a Stanford PhD student. The student received a contract from the White House to do the study. Unfortunately, when the study was finished, Richard Nixon was being impeached, and the Executive Office of the President of the United States refused to release the study—despite the fact that the results of the study were already widely known. The study was released a year later, and the student was able to get his PhD.

Test Your Understanding

21. a) What transmission medium do cable television companies primarily use? b) Why is coaxial cable called "coaxial"? c) Distinguish between the cable television trunk cable and drop cable.

Cable Modem Service[17]

Cable television companies eventually moved beyond one-way television service to two-way broadband (fast) data service. For television, the repeaters that boost signals periodically along the cable run only had to boost television signals travelling downstream. Data transmission required cable companies to install **two-way amplifiers**, which could carry data in both directions. Although this was expensive, it allowed cable companies to compete in the burgeoning market for broadband service. As in the case of ADSL, cable television service was asymmetric, offering faster downstream speeds than upstream speeds.

Instead of having a DSL modem, the subscriber has a **cable modem**. In general, this cable data service is called **cable modem service**. The coaxial cable drop line goes into the cable modem. The cable modem has a USB port and an Ethernet RJ-45 connector. The subscriber plugs a computer or access router into one of the two ports.

At the cable television head end, the cable television company connects to an Internet service provider. This allows subscribers to connect to hosts on the Internet.

Test Your Understanding

22. a) What types of amplifier are needed for cable data service? b) What device do customers need for cable modem service?

CONCLUSION

Synopsis

Chapter 2 discussed standards in general. It focused on the data link layer through the application layer because these layers all operate in the same general way—by sending messages. The physical layer, which was the focus of this chapter, is very different. It sends signals instead of messages, and it alone governs transmission media and propagation effects.

When signals propagate, they endure propagation effects such as attenuation. If propagation effects are too large, the receiver will not be able to read the signal. In this chapter, we looked at three transmission media and their major propagation effects: unshielded twisted-pair copper wiring (4-pair UTP), optical fiber, and, to some extent, coaxial cable.

---

[17] Which is better—ADSL service or cable modem service? ADSL services tout the fact that ADSL access lines are not shared, while coaxial cable trunk lines passing through neighborhood are shared. However, DSLAMs are shared in ADSL and often lack full capacity to serve all subscribers simultaneously. In both ADSL and cable modem service, the trunk line leading back to the data network is shared. More importantly, service providers often do not give the highest possible speed at lower monthly prices. In general, cable modem has tended to be somewhat faster and also somewhat more expensive than ADSL service.

In unshielded twisted-pair wiring, a cord consists of four twisted wire pairs. There is an RJ-45 connector at each end. UTP is rugged and inexpensive and so dominates the access links that connect computers to workgroup switches.

The two wires of each pair in a UTP cord are twisted around each other to reduce interference problems. Two simple installation expedients keep propagation problems to acceptable levels. First, restricting UTP cord lengths to 100 meters usually prevents serious attenuation and noise errors. Second, limiting the untwisting of wires to no more than 0.5 in. (1.25 cm) usually keeps terminal crosstalk interference to an acceptable level.

Earlier versions of Ethernet transmitted serially—sending on only one wire pair in each direction. Gigabit Ethernet transmits in parallel, sending on all four wire pairs when it transmits. Other forms of parallel transmission use more than four transmission paths. Parallel transmission is faster than serial transmission for a given clock cycle but is more expensive, so serial transmission now dominates in the marketplace.

There are different grades (called categories) of UTP quality. Almost all UTP sold today is Category 5e or Category 6 wiring. These are sufficient for up to 1 Gbps in Ethernet. To achieve 100-meter distances at 10 Gbps over UTP, companies need to install newer Cat 6A wiring, although if they only need a distance of up to 55 meters, they can send 10 Gbps Ethernet over Cat 6 wiring.

For trunk lines, optical fiber dominates in LAN transmission and will grow even more dominant as speed requirements increase. In fiber, a transceiver injects signals into a thin glass core that is 62.5 microns, 50 microns, or 8.3 microns in diameter. There are three major propagation windows centered around 850 nm, 1310 nm, and 1550 nm. Attenuation is lower at longer wavelengths, so carrier WAN fiber normally uses 1310 or 1550 nm light. However, 850 nm transceivers are much less expensive, and they can easily span typical LAN distances, so LAN fiber uses 850 nm light.

The transceiver uses simple binary on/off light signaling. An optical fiber cord has two strands for full-duplex communication. Fiber can use several connectors, including SC, ST, and newer small form factor (SFF) attenuation.

LAN fiber almost always uses multimode fiber, which has a "thick" diameter of 62.5 or 50 microns. This is less expensive than single-mode fiber, which has a core diameter of only 8.3 microns. Although multimode cannot carry signals as far as single-mode fiber, it can easily carry light 200 meters or more. This is more than adequate for nearly all LAN trunk lines. Multimode fiber propagation is limited by modal dispersion. Higher-quality multimode fiber, which is characterized by a higher modal bandwidth (measured as MHz.km), can carry signals farther than lower-quality multimode fiber. In addition, multimode fiber with a thinner core diameter (50 microns instead of 62.5 microns) can carry signals farther.

Carriers use single-mode fiber, which is very expensive but can carry high-speed signals over the very long distances needed by carriers (tens of kilometers). Single-mode fiber is limited by attenuation, which is very low but is the limiting propagation effect over long distances.

After looking at UTP and fiber in local area networks, we looked at how they are used in the Public Switched Telephone Network access lines (local loops). Of course, the PSTN always uses carrier fiber—single-mode fiber with 1310 or 1550 nm light. For low-speed leased lines, carriers run 2-pair data-grade UTP cords to each business. For residences, 1-pair voice-grade UTP already runs to each household. Carriers have learned to offer data services over 1-pair voice-grade UTP. These are called digital subscriber line (DSL) services. We also looked at how cable television companies offer competing cable modem service.

END-OF-CHAPTER QUESTIONS

Thought Questions

1. What type of interference is most likely to create problems in UTP transmission?
2. When a teacher lectures in class, is the classroom a full-duplex communication system or a half-duplex communication system?

Troubleshooting Questions

1. A tester shows that a UTP cord has too much interference. What might be causing the problem? Give at least two alternative hypotheses, and then describe how to test them.
2. What kinds of errors are you likely to encounter if you run a length of UTP cord 300 meters? (Recall that the standard calls for a 100 meter maximum distance.)

Perspective Questions

1. What was the most surprising material for you in this chapter?
2. What was the most difficult thing for you in this chapter?

CHAPTER 5a

Hands-On: Cutting and Connectorizing UTP[1]

INTRODUCTION

Chapter 5 discussed UTP wiring in general. This chapter discusses how to cut and connectorize (add connectors to) solid UTP wiring.

SOLID AND STRANDED WIRING

Solid-Wire UTP versus Stranded-Wire UTP

The TIA/EIA-568 standard requires that long runs to wall jacks use **solid-wire UTP**, in which each of the eight wires really is a single solid wire.

However, patch cords running from the wall outlet to a NIC usually are **stranded-wire UTP**, in which each of the eight "wires" really is a bundle of thinner wire strands. So stranded-wire UTP has eight bundles of wires, each bundle in its own insulation and acting like a single wire.

Relative Advantages

Solid wire is needed in long cords because it has lower attenuation than stranded wire. In contrast, stranded-wire UTP cords are more flexible than solid-wire cords, making them ideal for patch cords—especially the one running to the desktop—because they can be bent more and still function. They are more durable than solid-wire UTP cords.

Adding Connectors

It is relatively easy to add RJ-45 connectors to solid-wire UTP cords. However, it is very difficult to add RJ-45 connectors to stranded-wire cords. Stranded-wire patch cords should be purchased from the factory precut to desired lengths and preconnectorized.

[1] This material is based on the author's lab projects and on the lab project of Professor Harry Reif of James Madison University.

Solid-Wire UTP
- Each of the eight wires is a solid wire
- Low attenuation over long distances
- Easy to connectorize
- Inflexible and stiff—not good for runs to the desktop

Stranded-Wire UTP
- Each of the eight "wires" is itself several thin strands of wire within an insulation tube
- Flexible and durable—good for runs to the desktop
- Impossible to connectorize in the field (bought as patch cords)
- Higher attenuation than solid-wire UTP—Used only in short runs
  - From wall jack to desktop
  - Within a telecommunications closet (see Chapter 3)

FIGURE 5a-1 Solid-Wire and Stranded-Wire UTP (Study Figure)

In addition, when purchasing equipment to connectorize solid-wire UTP, it is important to purchase crimpers designed for solid wire.

CUTTING THE CORD

Solid-wire UTP normally comes in a box or spool containing 50 meters or more of wire. The first step is to cut a length of UTP cord that matches your need. It is good to be a little generous with the length. This way, bad connectorization can be fixed by cutting off the connector and adding a new connector to the shortened cord. Also, UTP cords should never be subjected to pulls (strain), and adding a little extra length creates some slack.

STRIPPING THE CORD

Now the cord must be stripped at each end using a **stripping tool** such as the one shown in Figure 5a-2. The installer rotates the stripper once around the cord, scoring (cutting into) the cord jacket (but not cutting through it). The installer then pulls off the scored end of the cord, exposing about 5 cm (about 2 inches) of the wire pairs.

It is critical not to score the cord too deeply, or the insulation around the individual wires may be cut. This creates short circuits. A really deep cut also will nick the wire, perhaps causing it to snap immediately or later.

WORKING WITH THE EXPOSED PAIRS

Pair Colors

The four pairs each have a color: orange, green, blue, or brown. One wire of the pair usually is a completely solid color. The other usually is white with stripes of the pair's color. For instance, the orange pair has an orange wire and a white wire with orange stripes.

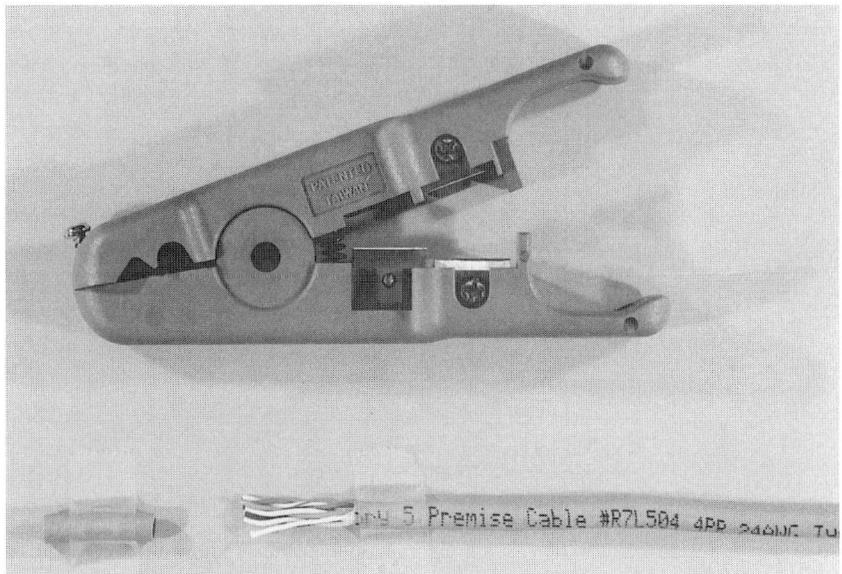

FIGURE 5a-2 Stripping Tool

Untwisting the Pairs

The wires of each pair are twisted around each other several times per inch. These must be untwisted after the end of the cord is stripped.

Ordering the Pairs

The wires now must be placed in their correct order, left to right. Figure 5a-3 shows the location of Pin 1 on the RJ-45 connector and on a wall jack or NIC.

Which color wire goes into which connector slot? The two standardized patterns are shown in Figure 5a-4. The T568B pattern is much more common in the United States.

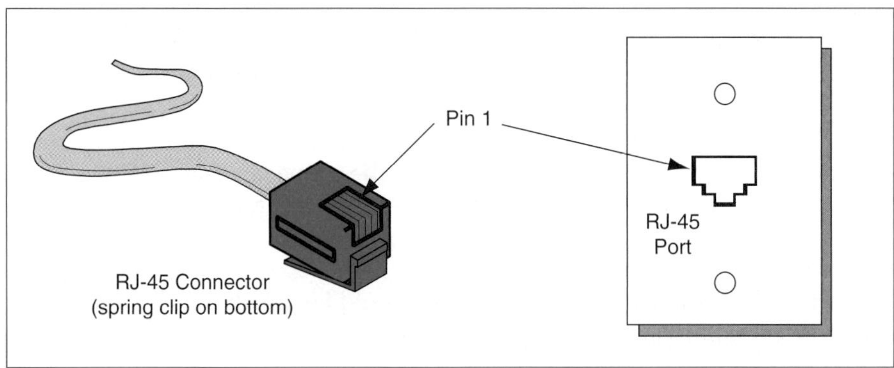

FIGURE 5a-3 Location of Pin 1 on an RJ-45 Connector and Wall Jack or NIC

Pin*	T568A	T568B
1	White-Green	White-Orange
2	Green	Orange
3	White-Orange	White-Green
4	Blue	Blue
5	White-Blue	White-Blue
6	Orange	Green
7	White-Brown	White-Brown
8	Brown	Brown

Note: Do not confuse T568A and T568B pin colors with the TIA/EIA-568 Standard.

FIGURE 5a-4 T568A and T568B Pin Colors

The connectors at both ends of the cord use the same pattern. If the white-orange wire goes into Pin 1 of the connector on one end of the cord, it also goes into Pin 1 of the connector at the other end.

Cutting the Wires

The length of the exposed wires must be limited to 1.25 cm (0.5 inch) or slightly less. After the wires have been arranged in the correct order, a cutter should cut across the wires to make them this length. The cut should be made straight across, so that all wires are of equal length. Otherwise, they will not all reach the end of the connector when they are inserted into it. Wires that do not reach the end will not make electrical contact.

ADDING THE CONNECTOR

Holding the Connector

The next step is to place the wires in the RJ-45 connector. In one hand, hold the connector, clip side down, with the opening in the back of the connector facing you.

Sliding in the Wires

Now, slide the wires into the connector, making sure that they are in the correct order (white-orange on your left). There are grooves in the connector that will help. Be sure to push the wires all the way to the end or proper electrical contact will not be made with the pins at the end.

Before you crimp the connector, look down at the top of the connector, holding the tip away from you. The first wire on your left should be mostly white. So should every second wire. If they are not, you have inserted your wires incorrectly.[2]

[2] Thanks to Jason Okumura, who suggested this way of checking the wires.

Some Jacket Inside the Connector

If you have shortened your wires properly, there will be a little bit of jacket inside the RJ-45 connector.

CRIMPING

Pressing Down

Get a really good **crimping tool** (see Figure 5a-5). Place the connector with the wires in it into the crimp and push down firmly. Good crimping tools have ratchets to reduce the chance of your pushing down too tightly.

Making Electrical Contact

The front of the connector has eight pins running from the top almost to the bottom (spring clip side). When you **crimp** the connector, you force these eight pins through the insulation around each wire and into the wire itself. This seems like a crude electrical connection, and it is. However, it normally works very well. Your wires are now connected to the connector's pins. By the way, this is called an **insulation displacement connection (IDC)** because it cuts through the insulation.

Strain Relief

When you crimp, the crimper also forces a ridge in the back of the RJ-45 connector into the jacket of the cord. This provides **strain relief**, meaning that if someone pulls on the cord (a bad idea), they will be pulling only to the point where the jacket has the ridge forced into it. There will be no strain where the wires connect to the pins.

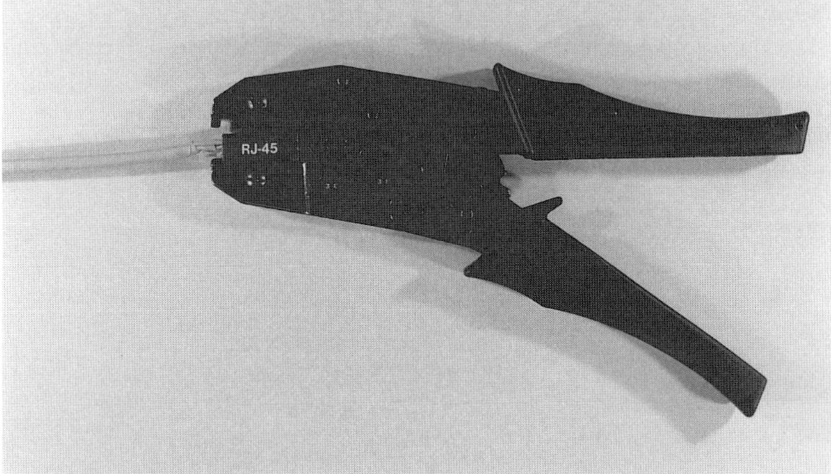

FIGURE 5a-5 Crimping Tool

TESTING

Purchasing the best UTP cabling means nothing unless you install it properly. Wiring errors are common in the field, so you need to test every cord after you install it. Testing is inexpensive compared to troubleshooting subtle wiring problems later.

Testing with Continuity Testers

The simplest testers are **continuity testers**, which merely test whether the wires are arranged in correct order within the two RJ-45 connectors and are making good electrical contact with the connector. They cost only about $100.

Testing for Signal Quality

Better testers cost $500–$2,000 but are worth the extra money. In addition to testing for continuity problems, they send **test signals** through the cord to determine whether the cord meets TIA/EIA-568 signal-quality requirements. Many include **time domain reflectometry (TDR)**, which sends a signal and listens for echoes in order to measure the length of the UTP cord or to find if and where breaks exist in the cord.

Test Your Understanding

1. a) Explain the technical difference between solid-wire UTP and stranded-wire UTP. b) In what way is solid-wire UTP better? c) In what way is stranded-wire UTP better? d) Where would you use each? e) Which should only be connectorized at the factory?
2. If you have a wire run of 50 meters, should you cut the cord to 50 meters? Explain.
3. Why do you score the jacket of the cord with the stripping tool instead of cutting all the way through the jacket?
4. a) What are the colors of the four pairs? b) If you are following T568B, which wire goes into Pin 3? c) At the other end of the cord, would the same wire go into Pin 3?
5. After you arrange the wires in their correct order and cut them across, how much of the wires should be exposed from the jacket?
6. a) Describe RJ-45's insulation displacement approach. b) Describe its strain relief approach.
7. a) Should you test every cord in the field after installation? b)For what do inexpensive testers test? c) For what do expensive testers test?

CHAPTER 6

Switched Wired Networks

LEARNING OBJECTIVES
By the end of this chapter, you should be able to:

- Describe Ethernet physical layer standards and how they affect network design.
- Describe the Ethernet data link layer and the Ethernet MAC layer frame.
- Explain basic Ethernet data link layer switch operation.
- Apply Ethernet switch purchasing criteria to specific situations.
- Explain Ethernet security.
- Describe LANs that are internets with subnets.
- Describe switched wired wide area networks (WANs).
- Describe leased line switched networks.
- Describe public switched data networks (PSDNs).

THE BANK OF PARADISE

The Bank of Paradise (real company, fake name) is a mid-size U.S. bank. It is located in an unnamed state in the middle of the Pacific Ocean. Its network administrator has just finished an inventory of the bank's network assets. His staff is now drawing the bank's network elements on a large map. The map needs to be very large because the bank has about 800 Ethernet switches, 400 routers, and over 300 wireless access points. Drawing them will be a large job. Managing them on a day-to-day basis is a far larger job.

SWITCHED WIRED NETWORKS

In the last chapter, we looked at physical layer transmission. In this chapter, we will look at standards for wired switched data networks. Switched data network standards cover both physical layer standards and data link layer (Layer 2) standards. Data link layer standards govern frame organization and switch operation.

We begin by looking at Ethernet, which is the dominant switched network standard for wired LANs. We look at both physical layer and data link layer Ethernet standards. We will also look at issues in the management of Ethernet networks.

Switched Network Standards

- Data Link layer standards
  - Switch operation
  - Frame organization
- Physical layer standards
  - Uses UTP and optical fiber
  - Adds signaling

Ethernet

- Dominant in wired LANs
- Became dominant because of its low cost and good performance

Switched WANs

- Leased line networks
  - Company leases lines to connect its sites
  - Installs switches to connect the leased lines
  - Manages the resulting networks
- Public Switched Data Networks (PSDNs)
  - PSDN vendor manages the switching cloud
  - Firm only needs to install a single leased line from each site to the vendor's nearest point of presence (POP)
  - Frame Relay is the dominant PSDN standard.
  - Metropolitan area Ethernet is growing.

FIGURE 6-1 Switched Networks (Study Figure)

Afterward, we look at switched WAN services. Companies can build their own switched WANs by connecting their sites with leased line circuits, adding their own switching, and then managing the network on an ongoing basis. An alternative to leased line data networks is the public switched data network (PSDN), in which a WAN carrier offers total service. The user organization only needs to connect each of its sites via a leased access line to a PSDN carrier point of presence.

In Chapters 7 and 8, we will look at wireless (radio) data networks. Like switched wired networks, wireless networks also require both physical and data link layer standards. Chapter 9 will take us above the data link layer to internetworking.

ETHERNET

The dominant standards family for wired LANs today is Ethernet. Once there were several competing wired LAN technologies, but Ethernet has eliminated the competition through good performance and low costs.

> Ethernet became the dominant LAN technology by providing good performance and low cost.

Test Your Understanding

1. a) What is the dominant LAN technology today? b) Why did it become dominant?

Creating Ethernet Standards

THE 802 COMMITTEE Nearly all LAN standards of all types today are created by the **802 LAN/MAN Standards Committee** of the **Institute for Electrical and Electronics Engineers (IEEE).** The "**802 Committee,**" as everybody calls it, is broadly responsible for creating local area network standards and metropolitan area network standards. (Metropolitan area networks are wide area networks that are restricted to a city and its surrounding areas.)

THE 802.3 ETHERNET WORKING GROUP The 802 Committee delegates the actual work of developing standards to specific working groups. The 802 Committee's **802.3 Working Group,** for example, creates Ethernet-specific standards. We will use the terms *Ethernet* and *802.3* interchangeably in this book.

We will use the terms *Ethernet* and *802.3* interchangeably in this book.

The 802 Committee

Committee of the Institute for Electrical and Electronics Engineers (IEEE)

IEEE created the 802 LAN/MAN Standards Committee for LAN standards.

This committee is usually called the 802 Committee.

The 802 Committee creates working groups for specific types of standards.

- 802.1 for general standards
- 802.3 for Ethernet standards
- 802.11 for wireless LAN standards
- 802.16 for WiMax wireless metropolitan area network standards

The 802.3 Working Group

This group is in charge of creating Ethernet standards.

The terms *802.3* and *Ethernet* are interchangeable today.

Figure 6-4 shows Ethernet physical layer standards.

Ethernet also has data link layer standards (frame organization, switch operation, etc.).

Ethernet Standards are OSI Standards

Layer 1 and Layer 2 standards are almost universally OSI standards.

Ethernet is no exception.

ISO must ratify them.

In practice, when the 802.3 Working Group finishes standards, vendors begin building compliant products.

FIGURE 6-2 Creating Ethernet Standards (Study Figure)

OTHER WORKING GROUPS The 802 Committee has several other working groups. For example, the 802.11 Working Group creates the wireless LAN standards that we will see in Chapters 7 and 8. The 802.1 Working Group creates general standards used in many different LAN technologies, most notably security standards.

ETHERNET STANDARDS ARE OSI STANDARDS As just noted, Ethernet standards are LAN standards, so they include both Layer 1 (physical) and Layer 2 (data link) standards. Recall from Chapter 2 that standards at the lowest two layers are always OSI standards. Although the 802.3 Working Group creates Ethernet standards, these standards are not official OSI standards until ISO ratifies them later. In practice, however, as soon as the 802.3 Working Group releases an 802.3 standard, vendors begin building products based on the specification.

Test Your Understanding

2. a) What working group creates Ethernet standards? b) To what committee does this working group report? c) In what organization is this committee? d) Are there other working groups? e) What types of standards does the 802.1 Working Group create? f) Does this book use *Ethernet* and *802.3* interchangeably? g) Why would you expect Ethernet standards to be OSI standards? h) When do vendors begin developing products based on Ethernet standards?

ETHERNET PHYSICAL LAYER STANDARDS

We will look first at Ethernet physical layer standards.

Network Interface Cards (NICs)

To connect to an Ethernet network, a host needs an Ethernet network interface card (NIC). Figure 6-3 shows a typical NIC installed as an expansion card. It has an RJ-45

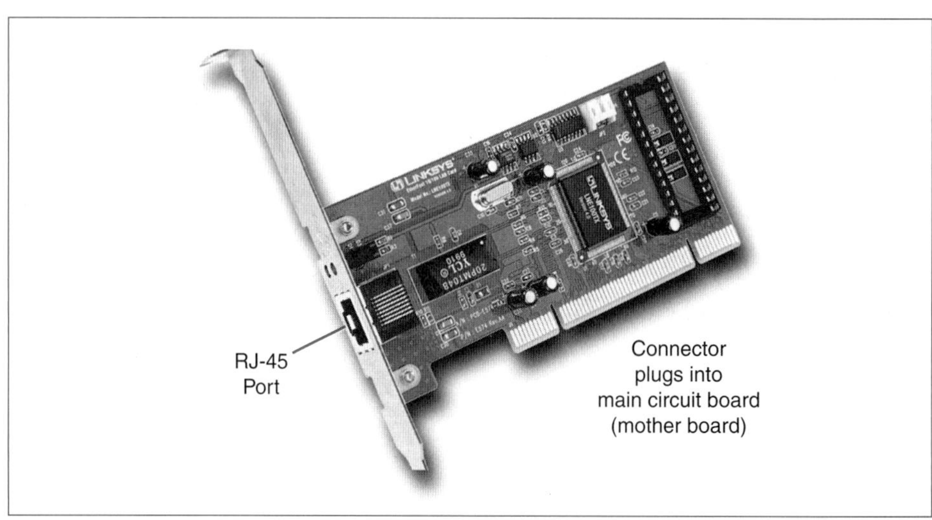

FIGURE 6-3 Network Interface Card (NIC)

port that is accessible from outside the computer. In many newer computers, and in all notebooks, NIC circuitry is built into the motherboard.

Major Ethernet Physical Layer Standards

We looked at 4-pair UTP and optical fiber standards in the last chapter. Why are we looking at UTP and fiber standards again in this chapter? The answer is that Ethernet physical layer standards go beyond physical media standards. They also specify standards for Ethernet-specific signaling standards. As Ethernet has grown in speed, Ethernet electrical signaling has grown steadily more sophisticated.

> Ethernet physical layer standards = UTP and fiber physical media standards + Ethernet-specific signaling standards.

When the 802.3 Working Group first created physical layer Ethernet standards, it used technologies that are no longer in use. Figure 6-4 shows some of the Ethernet physical layer standards that have been ratified more recently.

Vendors have stopped manufacturing products based on the oldest standards, and for higher speeds, vendors are manufacturing products for only some standards. In addition, some vendors provide optical fiber of a higher quality than standards specify. These can carry Ethernet signals over longer distances. If you are designing an Ethernet network, you will need to conduct timely research on what is available.

BASEBAND VERSUS BROADBAND TRANSMISSION As Figure 6-4 illustrates, all recent Ethernet physical layer standards have *BASE* in their names. This is short for *baseband*. In **baseband** transmission, the transmitter simply injects signals into the transmission medium, as Figure 6-5 illustrates. For UTP, signals consist of voltage changes that propagate down the wires. For fiber, signals are light pulses.

In contrast, in **broadband** transmission, the sender transmits signals in radio channels. However, radio-based broadband transmission is more expensive than baseband transmission. Although the 802.3 Working Group came up with early broadband LAN standards, these standards did not thrive because of their high cost.

The names of Ethernet physical layer standards also indicate transmission speeds—100 Mbps, 1000 Mbps, or more. These are the speeds of NICs and switch ports. For example, a 12-port gigabit Ethernet switch will be able to send and receive at 12 Gbps across all of its ports.

100 MBPS PHYSICAL LAYER 802.3 STANDARDS The 802.3 Working Group produced 100 Mbps standards for UTP (T) and optical fiber (F). The **100BASE-TX** UTP standard is the dominant Ethernet standard for connecting hosts to switches today. The 100 Mbps fiber standard is no longer used.[1]

> 100BASE-TX is the dominant Ethernet standard for connecting hosts to switches today.

[1] Why TX and FX instead of T and F? The 802.3 Working Group created other 100 Mbps Ethernet standards, but only the TX and FX standards saw market acceptance.

Ethernet Physical Layer Standards = UTP and Fiber Physical Media Standards
+ Ethernet-Specific Signaling Standards

Physical Layer Standard	Speed	Maximum Run Length	Medium
UTP			
100BASE-TX	100 Mbps	100 meters	6-pair Category 5e or higher
1000BASE-T	1,000 Mbps	100 meters	6-pair Category 5e or higher
10GBASE-T	10 Gbps	55 meters or 100 meters	6-Pair Category 6: Maximum length 55 meters 6-Pair Category 6A or Category 7: Maximum length 100 meters
Optical Fiber 1 Gbps			
1000BASE-SX	1 Gbps	220 meters	62.5/125 micron multimode, 850 nm, 160 MHz.km modal bandwidth
1000BASE-SX	1 Gbps	275 meters	62.5/125 micron multimode, 850 nm, 200 MHz.km modal bandwidth
1000BASE-SX	1 Gbps	500 meters	50/125 micron multimode, 850 nm, 400 MHz.km modal bandwidth
1000BASE-SX	1 Gbps	550 meters	50/125 micron multimode, 850 nm. 500 MHz.km modal bandwidth
1000BASE-LX	1 Gbps	550 meters	62.5/125 micron multimode, 1,310 nm
1000BASE-LX	1 Gbps	5 km	9/125 micron single mode, 1,310 nm
Faster Optical Fiber			
10GBASE-SR/SW	10 Gbps	65 meters	62.5/125 micron multimode, 850 nm
10GBASE-LX4	10 Gbps	300 meters	62.5/125 micron multimode, 1,310 nm, wave division multiplexing
10GBASE-LR/LW	10 Gbps	10 km	9/125 micron single mode, 1,310 nm
10GBASE-ER/EW	10 Gbps	40 km	9/125 micron single mode, 1,550 nm
40 Gbps Ethernet 100 Gbps Ethernet	40 Gbps or 100 Gbps	Under development	To be determined

Notes: S = 850 nm, L = 1310 nm, and E = 1550 nm

For 10GBASE-x, LAN versions (X and R) transmit at 10 Gbps. WAN versions (W) transmit at 9.95328 Gbps for carriage over SONET/SDH links.

The 40 Gbps and 100 Gbps Ethernet standards are still under development but are close to implementation.

Vendors do not produce products for all versions of all standards.

FIGURE 6-4 Ethernet Physical Layer Standards

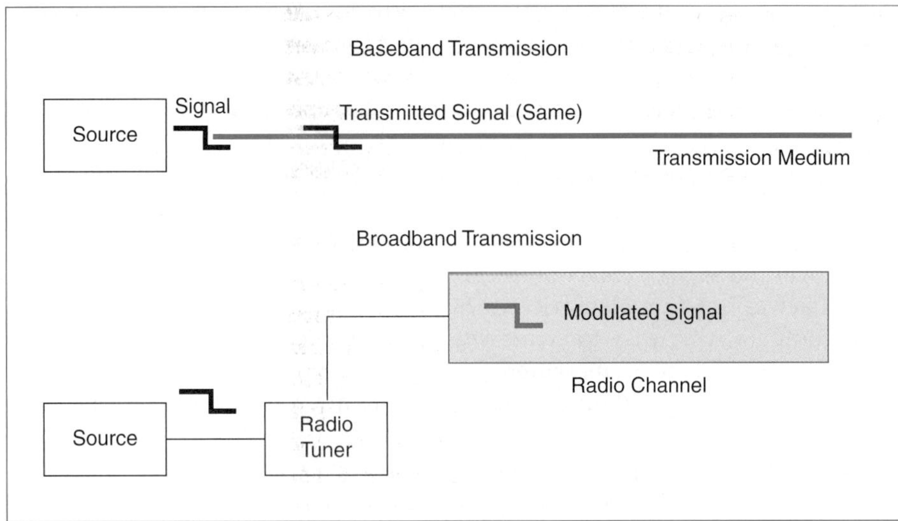

FIGURE 6-5 Baseband versus Broadband Transmission

GIGABIT ETHERNET (1000BASE-X) Advancing by another factor of 10, the 802.3 Working Group then produced **gigabit Ethernet (1000BASE-x).** One gigabit per second is the dominant speed for connecting switches to switches, switches to routers, and routers to routers. However, companies are increasingly using gigabit Ethernet to connect servers and some desktops to the switches that serve them.

A UTP version, **1000BASE-T,** can be used to bring gigabit Ethernet to the desktop. It can also be used to connect switches to other switches and switches to routers. Optical fiber connections are more expensive than UTP connections, so 1000BASE-T is attractive.

However, most firms use fiber versions of gigabit Ethernet. Although fiber is more expensive to install, fiber can carry signals farther than UTP. The **1000BASE-SX** (short wavelength) standard for gigabit Ethernet transmits at 850 nm, using inexpensive laser signaling and multimode fiber. This standard has a maximum transmission distance of 220 meters for older 62.5 micron fiber with a modal bandwidth of 160 MHz-km, and 275 meters for newer 62.5 micron fiber with a modal bandwidth of 200 MHz-km. In addition, some fiber vendors offer ultra-high modal bandwidth fiber that can extend 1000BASE-SX to nonstandard but reliable lengths of 300 to 500 meters. Overall, 1000BASE-SX is the dominant gigabit Ethernet standard and is the most widely used standard for connecting switches to other switches in organizations.[2]

Standard 1000BASE-SX is the most widely used standard for connecting switches to other switches in organizations.

TEN GIGABIT PER SECOND ETHERNET The next step is 10 Gbps. As you can see from Figure 6-4, the 802.3 Working Group has developed several standards for 10 Gbps

[2] The 1000BASE-LX (long wavelength) version uses more expensive lasers to transmit at 1,310 nm. It can send data twice as far as SX using multimode fiber. With single-mode fiber, it can send data 5 km. However, 1000BASE-SX is used far more widely than 1000BASE-LX.

Ethernet. As we saw in the last chapter, Cat 6 UTP can carry 10 Gbps Ethernet signals 55 meters, while Cat 6A UTP can carry traffic 100 meters.

ETHERNET AT 40 GBPS AND 100 GBPS The 802.3 Working Group is now working on the next generation of Ethernet speed standards. Traditional Ethernet experts wanted to continue the pattern of increasing speed by a factor of 10. This would give the next generation of Ethernet a speed of 100 Gbps.

However, wide area network specialists know that wide area transmission will take place over the SONET/SDH physical layer standard for carrier fiber. For 10 Gbps Ethernet, this was not a problem because SONET/SDH offers a speed of 9.95328 Gbps, so it was comparatively easy to create versions that ran at a slightly different speed. However, the SONET/SDH increment beyond 10 Gbps is about 40 Gbps instead of 100 Gbps. Consequently, Ethernet specialists who are focused on metropolitan area networking wanted the next version to run at 40 Gbps.

In the end, the 802.3 Working Group decided to do both, reasoning that the two standards are likely to be used very differently. Forty Gbps will be attractive for metro Ethernet, while 100 Mbps will be attractive for wired LANs.

TERABIT ETHERNET Although the 802.3 Working Group is still working on 40 Gbps and 100 Gbps Ethernet, much faster speeds may be coming. In fact, several companies have already demonstrated experimental Ethernet networks operating at a terabit per second or more.

Test Your Understanding

3. a) At what layer(s) is the 100BASE-TX standard? b) What can you infer from the name 100BASE-TX? c) Distinguish between baseband and broadband transmission. d) Why does baseband transmission dominate for LANs? e) What is the most widely used 802.3 physical layer standard for connecting hosts to switches today? f) What is the most widely used 802.3 physical layer fiber standard for connecting switches to other switches today? g) What are the two speeds for the next iteration of Ethernet? h) At what speed have experimental Ethernet systems been demonstrated?

Link Aggregation (Trunking)

Ethernet transmission capacity usually increases by a factor of 10. What should you do if you only need somewhat more speed than a certain standard specifies? For instance, suppose that you have gigabit Ethernet switches and need a switch-to-switch link of 1.5 Gbps instead of 10 Gbps.

Figure 6-6 illustrates that sometimes two or more trunk lines connect a single pair of switches. The IEEE calls this **link aggregation.** Networking professionals also call this **trunking** or **bonding.**

Link aggregation allows you to increase trunk speed incrementally, by a factor of two or three, instead of by a factor of ten. This incremental growth uses existing ports and usually is inexpensive compared to purchasing new faster switches.

However, after two or three aggregated links, the company should compare the cost of link aggregation with the cost of a tenfold increase in capacity by moving up to

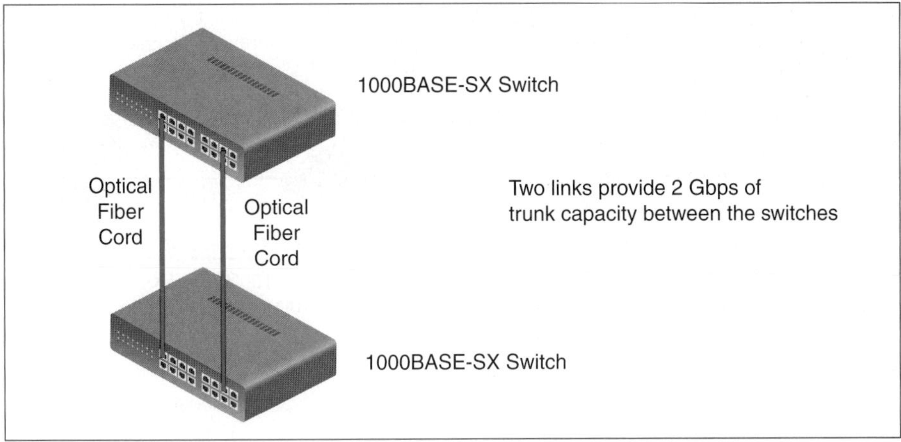

FIGURE 6-6 Link Aggregation (Trunking or Bonding)

the next Ethernet speed. Going to a much faster trunk line will also give more room for growth.

Test Your Understanding

4. a) What is link aggregation (trunking or bonding)? b) If you need to connect two 1000BASE-SX switches at 2.5 Gbps, what are your options? c) Why may link aggregation be more desirable than installing a single faster link? d) Why may link aggregation not be desirable if you will need several aggregated links to meet capacity requirements?

Ethernet Physical Layer Standards and Network Design

USING FIGURE 6-4 Note that if you know the speed you need (100 Mbps, 1 Gbps, and so forth) and if you know what distance you need to span, the information in Figure 6-4 will show you what type of transmission link you can use. Because link aggregation is available on all core switches, you have even more choices.

For instance, suppose that you need a speed of 2.5 Gbps between two switches that are 130 meters apart. You would need optical fiber to span this distance. For speed, 1 Gbps would not be sufficient, and 10 Gbps might be too expensive. Your best choice probably would be three bonded 1000BASE-SX links, although you would consider the cost of moving up to a 10 Gbps fiber standard.

Alternatively, if you are designing a network from scratch, say for a new facility, the options presented in Figure 6-4 will allow you to consider alternative placements for your switches. With longer physical links, you can place your switches farther apart on average, reducing the total number of switches. This can save money.

SWITCHES REGENERATE SIGNALS TO EXTEND DISTANCE The 100-meter Ethernet limit for UTP and the longer distance limits for fiber shown in Figure 6-4 are physical layer standards. Consequently, they only apply to connections *between a single pair of*

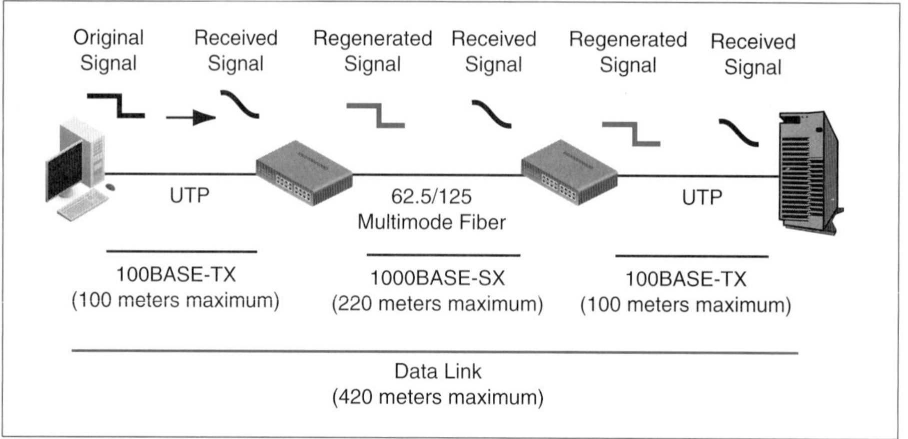

FIGURE 6-7 Data Link Using Multiple Switches

devices—for example, a host and a switch, between two switches, or a between a switch and a router.

The 100-meter Ethernet limit for UTP and the longer distance limits for fiber shown in Figure 6-4 only apply to physical links *between pairs of devices*, not to end-to-end data links between hosts across multiple switches.

What should you do if a longer distance separates the source host and the destination host? Figure 6-7 shows a data link with two intermediate switches. In addition to the two 100-meter maximum length UTP access links, there is a 220-meter maximum length 1000BASE-SX optical fiber link (using 62.5/125 micron 160 MHz-km modal bandwidth fiber) between the two switches. This setup can support a data link with a maximum span of 420 meters.

Each switch along the way **regenerates** the signal. If the signal sent by the source host begins as a 1, it is likely to be distorted before it reaches the first switch. The first switch recognizes it as a 1 and generates a clean new 1 signal to send to the second switch. The second switch regenerates the 1 as well.

The key point is that Figure 6-4 shows maximum distances between *pairs* of devices, not maximum end-to-end transmission distances. To deliver frames over long distances, intermediate switches regenerate the signal. There is no maximum end-to-end distance between pairs of hosts in an Ethernet network. Although cumulative delay might be a problem with a dozen or more intermediate switches, this rarely is a problem in real networks.

There is no maximum end-to-end distance between pairs of hosts in an Ethernet network.

Test Your Understanding

5. a) How could you use the information in Figure 6-4 in network design? b) If more than one type of Ethernet standard shown in Figure 6-4 can span the distance you need, what would determine which one you choose? c) In Figure 6-4, is the

maximum distance the maximum distance for a single physical link or for the data link between two hosts across multiple switches? d) At what layer or layers is the 802.3 100BASE-TX standard defined—physical, data link, or internet? e) How does regeneration allow a firm to create LANs that span very long distances? f) If you need to span 300 meters by using 1000BASE-SX, what options do you have? (Include the possibility of using an intermediate switch.) g) How would you decide which option to choose?

THE ETHERNET FRAME

So far, we have been looking at Ethernet physical layer standards. Now, we will look at Ethernet data link layer standards, beginning with frame organization.

Layering

THE LOGICAL LINK CONTROL LAYER When the 802 Committee assumed control over Ethernet standardization, it realized that it would have to standardize non-Ethernet LAN technology as well. Consequently, the 802 Committee divided the data link layer into two layers, as Figure 6-8 illustrates.

- The lower part of the standard—the **media access control (MAC)** layer—is specific to the particular LAN technology. For example, there are separate media access control layer standards for Ethernet and 802.11 wireless networks.
- The upper layer—the **logical link control (LLC)** layer—adds some functionality on top of technology-specific functionality. Unfortunately, time has proven the added functionality of the LLC layer to be of little value, so it is now largely ignored. As we will see a little later, the sender adds an LLC subheader to each 802.3 frame. There is only a single LLC layer standard, **802.2.**

THE 802.3 MAC LAYER STANDARD As just noted, the lower part of the data link layer is the MAC layer. The MAC layer defines functionality specific to a particular LAN technology.

Note in Figure 6-8 that while Ethernet (802.3) has many physical layer standards, it only has a single media access control layer standard, the **802.3 MAC Layer Standard.** This standard defines Ethernet frame organization and NIC and switch operation.

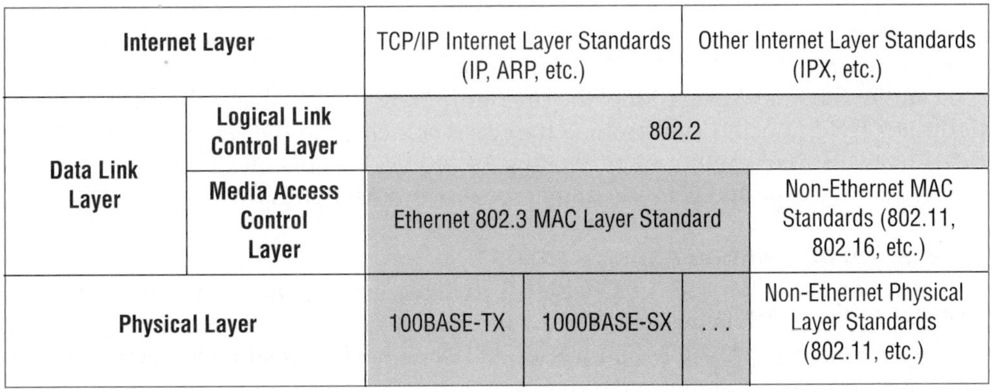

<table>
<tr><td colspan="2">Internet Layer</td><td colspan="2">TCP/IP Internet Layer Standards (IP, ARP, etc.)</td><td colspan="2">Other Internet Layer Standards (IPX, etc.)</td></tr>
<tr><td rowspan="2">Data Link Layer</td><td>Logical Link Control Layer</td><td colspan="4">802.2</td></tr>
<tr><td>Media Access Control Layer</td><td colspan="3">Ethernet 802.3 MAC Layer Standard</td><td>Non-Ethernet MAC Standards (802.11, 802.16, etc.)</td></tr>
<tr><td colspan="2">Physical Layer</td><td>100BASE-TX</td><td>1000BASE-SX</td><td>. . .</td><td>Non-Ethernet Physical Layer Standards (802.11, etc.)</td></tr>
</table>

FIGURE 6-8 Layering in 802 Networks

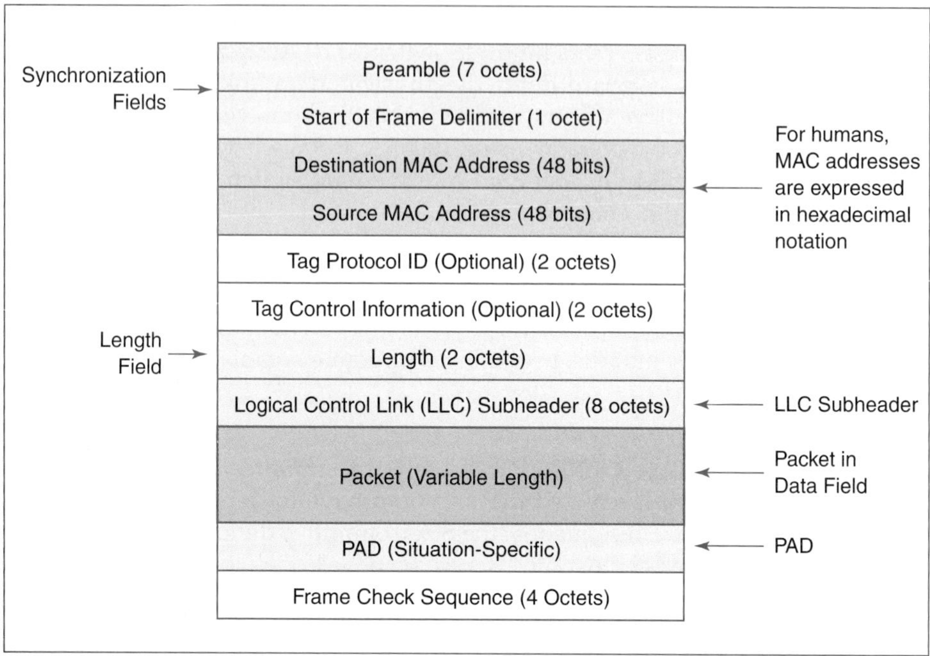

FIGURE 6-9 The Ethernet MAC-Layer Frame

Test Your Understanding

6. a) Distinguish between the MAC and LLC layers. b) Does Ethernet have multiple physical layer standards? c) Does Ethernet have multiple MAC layer standards? d) What is the name of Ethernet's single MAC standard?

The Ethernet Frame's Organization

Figure 6-9 shows the Ethernet MAC layer frame, which we saw briefly in Chapter 2. We will now look at the Ethernet frame in more depth. Recall that an *octet* is a byte.

PREAMBLE AND START OF FRAME DELIMITER FIELDS Before a play in American football, the quarterback calls out something like "Hut one, hut two, hut three, hike!" This cadence synchronizes all of the offensive players.

In the Ethernet MAC frame, the **preamble field** (7 octets) and the **start of frame delimiter field** (1 octet) synchronize the receiver's clock to the sender's clock. These fields have a strong rhythm of alternating 1s and 0s. The last bit in this sequence is a 1 instead of the expected 0, to signal that the synchronization is finished.

SOURCE AND DESTINATION ADDRESS FIELDS

Hex Notation. We saw in Chapter 2 that the source and destination Ethernet address fields are 48 bits long and that while computers work with this raw 48-bit form, humans normally express these addresses in Base 16 **hexadecimal (hex) notation**.

- First, divide the 48 bits into twelve 4-bit units, which computer scientists call nibbles.
- Second, convert each nibble into a hexadecimal symbol, using Figure 6-10.

4 Bits	Decimal (Base 10)	Hexadecimal (Base 16)
0000	0	0 hex
0001	1	1 hex
0010	2	2 hex
0011	3	3 hex
0100	4	4 hex
0101	5	5 hex
0110	6	6 hex
0111	7	7 hex

4 Bits*	Decimal (Base 10)	Hexadecimal (Base 16)
1000	8	8 hex
1001	9	9 hex
1010	10	A hex
1011	11	B hex
1100	12	C hex
1101	13	D hex
1110	14	E hex
1111	15	F hex

Divide a 48-bit Ethernet address into 12 four-bit "nibbles."
Convert each group of 4 bits into a Hex symbol.
Combine two hex symbols into pairs and place a dash between pairs.
For example, A1-36-CD-7B-DF hex begins with 10100001 for A1.

FIGURE 6-10 Hexadecimal Notation

- Third, write the symbols as six pairs with a dash between each pair—for instance, B2-CC-66-0D-5E-BA. (Each pair represents one octet.)

To convert a hex address back to binary, change each symbol back to its 6-bit pattern. For example, if the first hex pair is E2, E is 1110, and 2 is 0010. So E2 is equivalent to the octet 11100010. In this conversion, you must keep the two leading zeros in 0010.

MAC Layer Addresses. Ethernet addresses exist at the MAC layer, so Ethernet addresses are **MAC addresses.** They are also called **physical addresses** because physical devices (NICs) implement Ethernet at the physical, MAC, and LLC layers.

TAG FIELDS

We will look at the two tag fields later in this chapter, when we look at virtual LANs (VLANs). These tag fields are optional and are only used under certain circumstances.

LENGTH FIELD

The **length** field contains a binary number that gives the *length of the data field* (not of the entire frame) in octets. The maximum length of the data field is 1,500 octets. There is no minimum length for the data field.

THE DATA FIELD

The **data** field contains two subfields: the LLC subheader and the packet that the frame is delivering.[3]

[3] Why does the data field have two parts? The answer is that the data field of the MAC layer frame actually is an encapsulated LLC layer frame, which has a header (the LLC subheader) and a data field consisting of the packet being carried in the LLC frame. However, to avoid damaging neurons, it is best simply to think of the MAC layer data field as having two parts.

LLC Subheader. The **logical link control layer (LLC) subheader** is eight octets long.[4] The purpose of the LLC subheader is to describe the type of packet contained in the data field. For instance, if the LLC subheader ends with the code 08-00 hex (Base 16), then the data field contains an IP packet.[5]

The Packet. The data field also contains the packet that the MAC layer frame is delivering. The packet usually is far longer than all other fields combined. The packet encapsulated in the data field usually is an IP packet. However, it could also be a packet from another standards architecture, for instance, an IPX packet. As long as the source and destination hosts understand the packet format, there is no problem.

PAD FIELD The **PAD** field is unusual because it does not always exist. Although there is no minimum length for 802.3 MAC layer frame data fields, if the data field is less than 46 octets long, then the sender must add a PAD field so that the total length of the data field and the PAD field is exactly 46 octets long. For instance, if the data field is 26 octets long, the sender will add a 20-octet PAD field. If the data field is 46 octets long or longer, the sender will not add a PAD field.

There is no minimum length for the data field, but if the data field is less than 46 octets long, then a PAD field must be added to bring the total length of the data and pad fields to 46 octets.

FRAME CHECK SEQUENCE FIELD The last field in the Ethernet frame is the **frame check sequence** field, which permits error detection. This is a four-octet field. The sender does a calculation based on other bits in the frame and places the result in the frame check sequence field. The receiver redoes the calculation and compares its result with the contents of the frame check sequence field. If the two are different, there is an error in the frame. If there is an error, the receiver simply discards the frame. There is no retransmission of damaged frames.

Test Your Understanding

7. a) What is the purpose of the preamble and start of frame delimiter fields? b) Why are Ethernet addresses called MAC addresses or physical addresses? c) What are the steps in converting 48-bit MAC addresses into hex notation? d) The length field gives the length of what? e) What are the two components of the Ethernet data field? f) What is the purpose of the LLC subheader? g) What type of packet is usually carried in the data field? h) What is the maximum length of the data field? i) Who adds the PAD field—the sender or the receiver? j) Is there a minimum length for the data field? k) If the data field is 40 octets long, how long a PAD field must the sender add? l) If the data field is 400 octets long, how long a PAD field must the sender add? m) What is the purpose of the frame check sequence field? n) What happens if the receiver detects an error in a frame? o) Convert 11000010 to hex. p) Convert 7F hex to binary.

---

[4] There are some exceptions, but they are rare.

[5] The LLC subheader has several fields. In the SNAP version of LLC, which is almost always used, the first three octets are always AA-AA-03 hex. The next three octets are almost always 00-00-00 hex. The final two octets constitute the Ethertype field, which specifies the kind of packet in the data field. Common hexadecimal Ethertype values are 0800 (IP), 8137 (IPX), 809B (AppleTalk), 80D5 (SNA services), and 86DD (IP version 6).

BASIC ETHERNET DATA LINK LAYER SWITCH OPERATION

In this section, we will discuss the basic data link layer operation of Ethernet switches. This is also governed by the 802.3 MAC layer standard. In the section after this one, we will discuss other aspects of Ethernet switching that a firm may or may not use.

Frame Forwarding

Figure 6-11 shows an Ethernet LAN with three switches. Larger Ethernet LANs have dozens of switches, but the operation of individual switches is the same whether there are a few switches or many. Each individual switch makes a decision about which port to use to send the frame back out to the next switch.

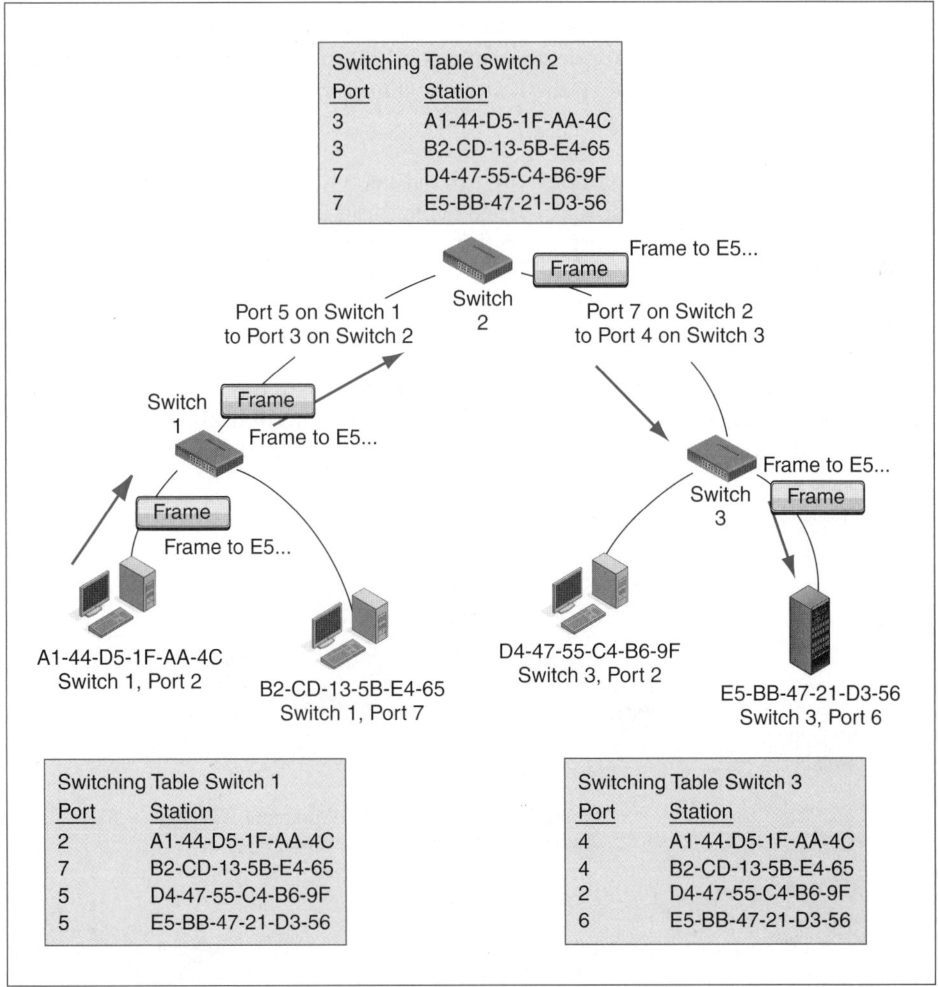

FIGURE 6-11 Multiswitch Ethernet LAN

In the figure, Host A1 wishes to send a frame to Host E5. This frame must go to Switch 1, then Switch 2, and then to Switch 3. Switch 3 will send the frame to Host E5.

- Host A1 puts E5 (later octets dropped for brevity) in the destination address field of the frame. It sends the frame to Switch 1, through Port 2.
- Switch 1 looks up the address E5 in its switching table. It sees that E5 is associated with Port 5, so it sends the frame out Port 5. This is a very simple process, so it requires little processing power. This means that Ethernet switches are inexpensive for the volume of traffic they carry.
- The frame going out Port 5 on Switch 1 goes into Port 3 on Switch 2. Switch 2 now looks up the address E5 in its switching table. This address is associated with Port 7, so Switch 2 sends the frame out Port 7.
- The frame arrives at Switch 3 though Port 4. Switch 3 now looks up the address E5 in the switching table. This time, the address is associated with Port 6. Switch 3 sends the frame out Port 6. This takes it to the destination Host E5.

Note that each switch only knows the information in its switching table. More specifically, it only knows what port to use to send the frame back out. Switches do not know the entire data link between the source host and the destination host.

Test Your Understanding

8. a) Do switches know the entire data link between the source and destination host? b) What does a switch know? c) Trace everything that will happen when Host E5 sends a frame to D4. d) Trace everything that will happen when Host E5 sends a frame to B2.

Hierarchical Switch Topology

HIERARCHICAL SWITCH ORGANIZATION Note that the switches in Figure 6-12 form a **hierarchy,** in which each switch has only one parent switch above it. In fact, the Ethernet standard *requires* a **hierarchical topology** for its switches. Otherwise, loops

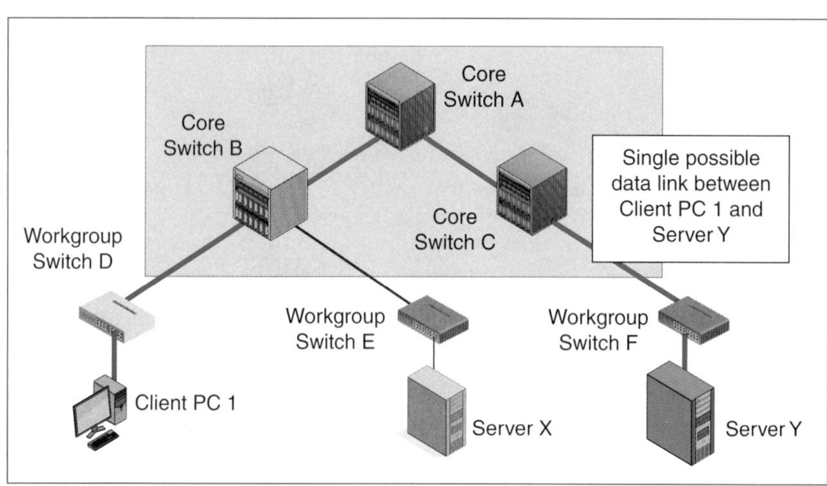

FIGURE 6-12 Hierarchical Ethernet LAN

would exist, causing frames to circulate endlessly from one switch to another around the loop or causing other problems. Figure 6-12 shows a larger switched Ethernet LAN organized in a hierarchy.

Ethernet *requires* a hierarchical switch hierarchy.

SINGLE POSSIBLE PATH BETWEEN END HOSTS In a hierarchy, there is only a single possible path between any two end hosts. (To see this, select any two hosts at the bottom of the hierarchy and trace a path between them. You will see that only one path is possible.)

In a hierarchy, there is only a single possible path between any two end hosts.

WORKGROUP VERSUS CORE SWITCHES In a hierarchy of Ethernet switches, there are workgroup switches and core switches. Figure 6-12 illustrates these two types of switches.

- **Workgroup switches.** Switches that connect hosts to the network via access lines are called workgroup switches (Switches D, E, and F in Figure 6-12).
- **Core switches.** Switches farther up the hierarchy (Switches A, B, and C in Figure 6-12) that carry traffic via trunk lines between pairs of switches, switches and routers, and pairs of routers are called core switches. The collection of all core switches plus the trunk lines that connect them is the network's **core.**

Workgroup switches connect hosts to the network.
Core switches connect switches to other switches.

Workgroup switches handle only the traffic of the hosts they serve. However, core switches must be able to carry the conversations of dozens, hundreds, or thousands of hosts. Consequently, core switches need to have much higher port speeds than workgroup switches. Their cost also is much higher.

The dominant port speed for workgroup switches today is 100 Mbps. In contrast, the dominant port speed for core switches today is 1 Gbps, and some core switches already use port speeds of 10 Gbps.

Test Your Understanding

9. a) How are switches in an Ethernet LAN organized? b) Because of this organization, how many possible paths can there be between any two hosts? c) In Figure 6-12, what is the single possible path between Client PC 1 and Server Y? d) Between Client PC 1 and Server X?
10. a) Distinguish between workgroup switches and core switches in terms of which devices they connect. b) How do they compare in terms of port speeds? Explain.

Only One Possible Path: Low Switching Cost

We have just seen that a hierarchy allows only one possible path between any two hosts. If there is only a single possible path between any two hosts, it follows that, in every switch along the path, the destination address in a frame will appear only once in the switching table—for the specific outgoing port needed to send the frame on its way.

This allows a simple table lookup operation that is very fast and therefore costs little per frame handled. This is what makes Ethernet switches inexpensive. As noted in the introduction, simple switching operation and therefore low cost has led to Ethernet's dominance in LAN technology.

The fact that there is only a single possible path between any two end hosts in an Ethernet hierarchy makes Ethernet switch forwarding simple and therefore inexpensive. This low cost has led to Ethernet's dominance in LAN technology.

In Chapter 8, we will see that routers have to do much more work when a packet arrives because routers are connected in a mesh, so there are multiple alternative routes between any two hosts. Each of these alternative routes appears as a row in the routing table. Therefore, when a packet arrives, a router must first identify all possible routes (rows) and then select the best one—instead of simply finding a single match. This additional work per forwarding decision makes routers very expensive for the traffic load they handle.

Test Your Understanding

11. a) What is the benefit of having a single possible path? Explain in detail. b) Why has Ethernet become the dominant LAN technology? c) Why are routers expensive for the traffic volume they handle?

ADVANCED ETHERNET SWITCH OPERATION

Now that we have discussed basic Ethernet switch operation involved in frame forwarding, we will begin looking at additional aspects of Ethernet switch operation that are important in larger Ethernet networks.

The Rapid Spanning Tree Protocol (RSTP)

SINGLE POINTS OF FAILURE We have just seen that having only a single possible path between any two hosts allows rapid frame forwarding and, therefore, low switch cost. Unfortunately, having only a single possible path between any two computers also makes Ethernet vulnerable to **single points of failure,** in which the failure of a single component (a switch or a trunk line between switches) can cause widespread disruption.

Having only a single possible path between end hosts in a switched Ethernet network reduces cost, but it creates single points of failure, meaning that a single failure can cause widespread disruption.

To understand this, suppose that Switch 2 in Figure 6-13 fails. Then the hosts connected to Switch 1 will not be able to communicate with hosts connected to Switch 2 or Switch 3. For a second example, suppose that the trunk line between Switch 1 and Switch 2 fails. In this case, too, the network also will be broken into two parts.

Although the two parts of the network might continue to function independently after a failure, many firms put most or all of their servers in a centralized server room. In such firms, clients on the other side of the broken network would lose most of their

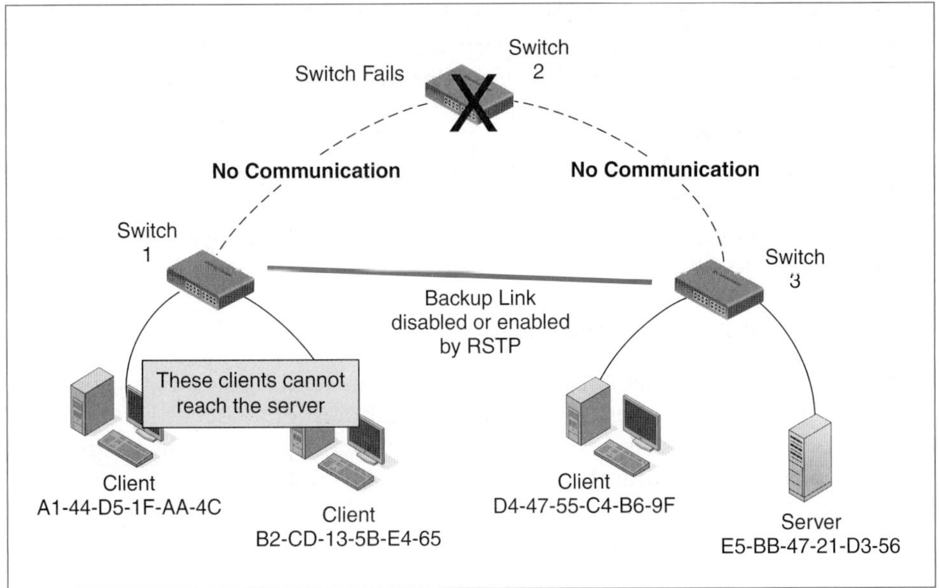

FIGURE 6-13 Single Point of Failure and the Rapid Spanning Tree Protocol

ability to continue working. For example, in the figure, Client A1-44-D5-1F-AA-4C, which connects to Switch 1, cannot reach Server E5-BB-47-21-D3-56, which connects to Switch 3. External connections also tend to be confined to a single network point for security reasons. Computers on the wrong side of the divide after a breakdown would lose external access.

THE RAPID SPANNING TREE PROTOCOL Fortunately, the 802.1 Working Group (not to be confused with the 802.3 Working Group) created a way to provide **backup links.** To see how this method works, suppose that Switch 2 in Figure 6-13 fails. Then the hosts connected to Switch 1 will not be able to communicate with the hosts attached to Switch 2 or Switch 3. For a second example, suppose that the link between Switch 1 and Switch 2 fails. Then the network will be broken in two parts.

As Figure 6-13 shows, a company can install a backup link—in this case, between Switch 1 and Switch 3. The backup link provides redundancy so that frames can take alternative paths if there is a failure.

Of course, the backup line will create a loop, which is forbidden in Ethernet. Fortunately, under the **Rapid Spanning Tree Protocol (RSTP),**[6] switches can detect a loop and disable it by turning off switch ports selectively to disable some links. If the switches are configured properly, the switches will specifically disable the backup link between Switches 1 and 3. Later, if Switch 2 fails, the RSTP protocol will reestablish the backup link between Switches 1 and 3. This will repair most parts of the network temporarily, until the failure point can be restored.

[6] There was an earlier standard, the Spanning Tree Protocol (802.1D), which is now deprecated because of its slow operation.

Unfortunately, the use of RSTP is anything but automatic. Although RSTP will always break loops, getting it to restore particular backup links when specific problems occur requires a great deal of planning and extensive switch configuration.

Test Your Understanding

12. a) Why is having a single possible path between any two hosts in an Ethernet network dangerous? b) What is a single point of failure? c) What standard allows redundancy in Ethernet networks? d) Is it easy or difficult to create backup links effectively in RSTP?

Virtual LANs and Ethernet Switches

VLANs In a normal Ethernet network, any client can send frames to any server, and any server can reach any client. However, many Ethernet switches can now create virtual LANs (VLANs). As Figure 6-14 shows, VLANs are groups of clients and servers that are allowed to communicate with each other but not with clients or servers on other VLANs. In the figure, clients and servers in VLAN 3 (indicated by rectangles) cannot communicate with clients and servers on VLAN 47 (indicated by ovals).

CONGESTION REDUCTION VLANs are used for two main reasons. First, some servers tend to **broadcast** frames to all clients. (One reason for the server to do this is to advertise its availability to its clients every 30 seconds or so.) In a large network, this broadcasting can create a great deal of congestion. With VLANs, however, the server will not flood the entire network with traffic; the frames will go only to the clients on the server's VLAN.

SECURITY A second reason for using VLANs is security. If clients on one VLAN cannot reach servers on other VLANs, they cannot attack these servers. In addition, if a client

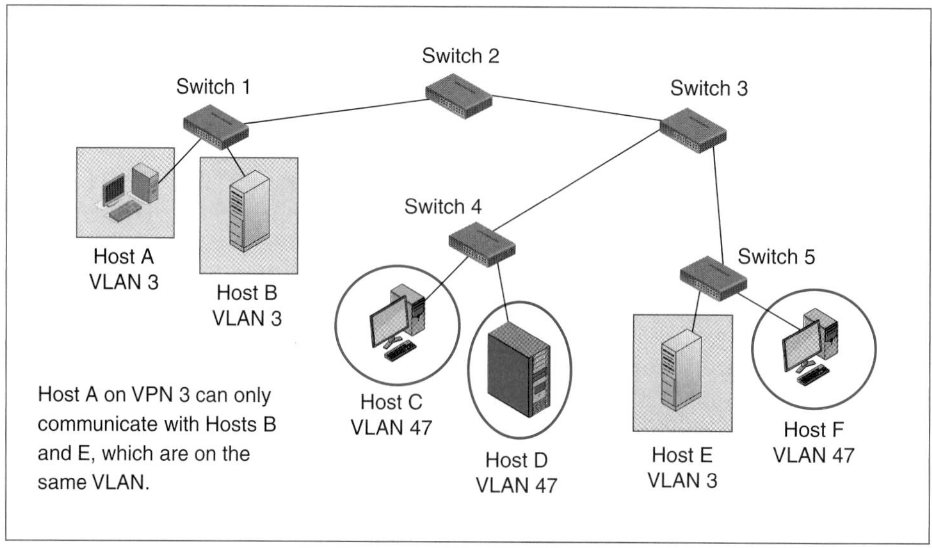

FIGURE 6-14 Ethernet Virtual LANs (VLANs)

becomes infected with a virus, it can only pass the virus on to other clients and servers on its own VLAN.

THE 802.1Q VLAN STANDARDS Until recently, there was no standard for VLANs, so if you used VLANs, you had to buy all of your Ethernet switches from the same vendor. However, as Figure 6-9 shows, the **802.1Q** standard extends the Ethernet MAC layer frame to include two optional **tag fields** after the address fields.

802.1Q is the standard for frame tagging.

The first tag field (the **Tag Protocol ID** field) simply indicates that this is a tagged frame. The second tag field (the **Tag Control Information** field) contains a 12-bit VLAN ID that the sender sets to 0 if the firm does not use VLANs. Consequently, there are $2^{12}-1$ (4,095) possible VLANs.

The TCI field also has three priority bits. This gives up to eight (2^3) priority levels. Frame tagging is necessary to do priority-based switching in Ethernet.

Test Your Understanding

13. a) What is a VLAN? b) What two benefits do VLANs bring? c) How do VLANs bring security? d) When VLANs or priority are used, what two fields does the 802.1Q standard add to Ethernet frames? e) When VLANs are used, what does the Tag Protocol ID field tell a receiving switch or NIC? f) What information does the tag control information field tell the switch or receiver?

Manageability

If there is an Ethernet switch problem, discovering which switch is malfunctioning can be very difficult. Fixing the problem, furthermore, may require traveling to the switch to change its configuration. Switch troubleshooting can be very expensive, especially if the network staff must travel to distant switches to do diagnostics or configuration.

MANAGED SWITCHES AND THE MANAGER As Figure 6-15 shows, a company can mitigate these problems by using **managed switches.** As the name suggests, these switches have sufficient intelligence to be managed from a central computer called the **manager.** In most cases, management communication uses the Simple Network Management Protocol (SNMP) discussed in Chapter 4.

POLLING AND PROBLEM DIAGNOSIS Every few seconds, the SNMP manager polls each managed switch and asks for a copy of the switch's configuration parameters. If a problem occurs, the manager can discover quickly which switches are not responding and so can narrow down the source of the problem. In many cases, the status data collected frequently from the switches can pinpoint the cause of a problem.

FIXING SWITCHES REMOTELY In some cases, the network administrator can use the manager to fix switch problems remotely by sending commands to the switch. For instance, the manager can command the switch to do a self-test diagnostic. To give another example, the manager can tell the switch to turn off a port suspected of causing problems.

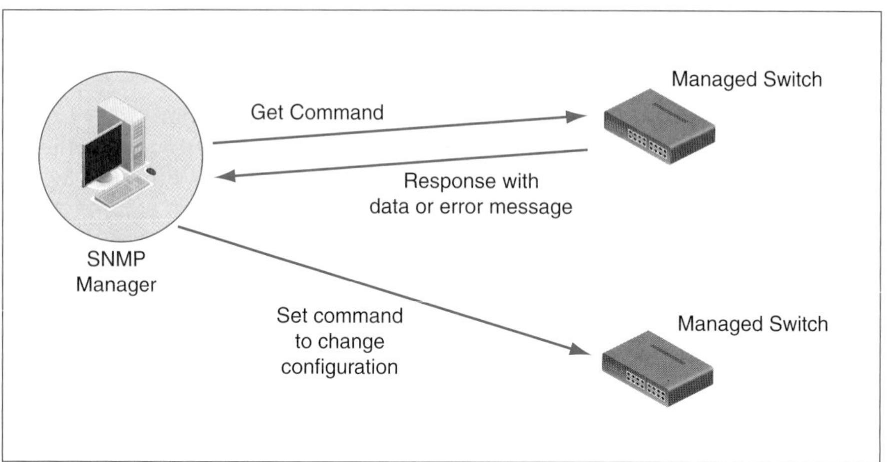

FIGURE 6-15 Managed Switches

PERFORMANCE SUMMARY DATA At the broadest level, software can search through the SNMP manager's data and can present the status data to the network administrator in summarized form, giving the administrator a good indication of how well the network is functioning and of whether changes will be needed to cope with expected traffic growth.

THE COST SAVINGS OF MANAGEABILITY Managed switches are much more expensive to buy than nonmanaged switches. However, central management slashes management labor, which is considerable. This labor cost reduction usually far offsets the higher switch purchase costs. The main benefit of network management, then, is to reduce overall costs.

> Managed switches are more expensive than nonmanaged switches, but they reduce management labor in large networks enough to more than offset managed switch purchase costs. Managed switches reduce overall costs.

Test Your Understanding

14. a) What are managed switches? b) What benefits do they bring? c) Do managed switches increase or decrease total costs?

Power over Ethernet (POE)

The telephone company wires that come into your home bring a small amount of power. You can plug a basic telephone into a telephone wall jack without having to plug it into a power outlet. USB cables also provide a small amount of electrical power to the devices they connect.

Similarly, the **power over Ethernet (POE)** standard can bring power to RJ-45 wall jacks. POE is important to corporations because it can greatly simplify electrical wiring for installing voice over IP (VoIP) telephones, wireless access points, and surveillance

Power over Ethernet (POE)

Switches can supply power to devices connected by UTP.

Latest POE Standard

Provide up to 25 watts to attached devices
Sufficient for most wireless access points
Sufficient for VoIP phones
Sufficient for surveillance cameras
Not sufficient for desktop or notebook PCs

POE Switches

New switches can be purchased with POE.
Companies can also add POE equipment to an existing non-POE switch.

FIGURE 6-16 Power over Ethernet (POE) (Study Figure)

cameras. Instead of having to provide power to each device via electrical wall jacks, the device can simply be plugged into the Ethernet wall jack.

The POE standard is limited to 25 watts of power.[7] Some nonstandard powered switches double this. However, both POE and nonstandard POE are only sufficient for low-power devices. POE does not provide enough power for desktop PCs or even laptop computers.

Companies that wish to supply power through their RJ-45 wall jacks will have to install either new switches compatible with the POE standard or modification kits that can add POE to existing switches.

Test Your Understanding

15. a) What is POE? b) Why is POE attractive to corporations? c) What maximum standard power does the POE standard specify? d) For what types of devices is POE sufficient? e) Is POE sufficient for desktop computers and most notebook computers?

ETHERNET SECURITY

Until recently, few organizations worried about the security of their wired Ethernet networks, presumably because only someone within the site could get access to the network, and security should be strong within the site. Unfortunately, experience has shown that attackers can easily get into sites, especially if a site has public areas.

Port-Based Access Control (802.1X)

A major security threat to Ethernet is that any attacker can plug his or her notebook PC into any Ethernet wall jack and have unfettered access to the network. To thwart this attack, companies can implement **802.1X,** which is a standard for **Port-Based Access**

[7] Technically, the standard that specifies 25 W is POE Plus. The original standard was simply POE.

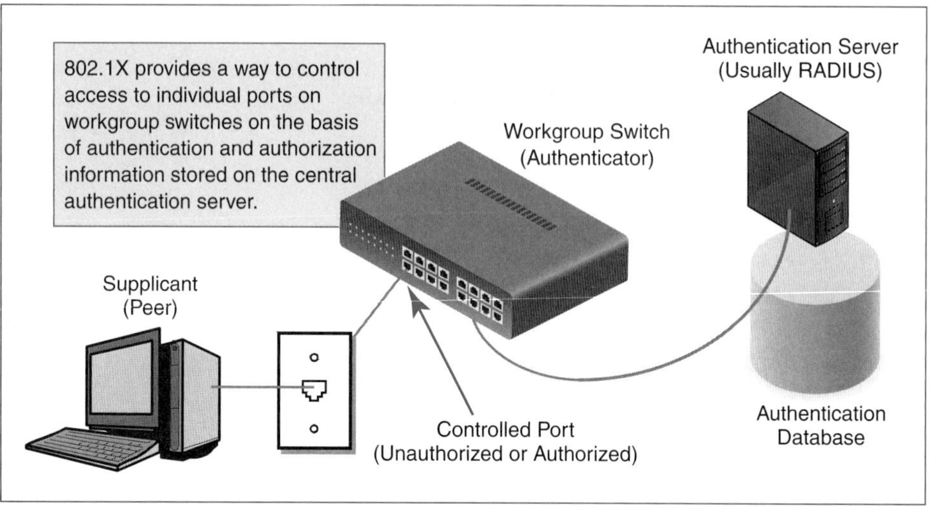

FIGURE 6-17 802.1X Ethernet Port-Based Access Control

Control on the workgroup switches that give users access to the network. Quite simply, a switch port will not allow the computer attached to the port to send traffic other than authentication traffic until the computer has authenticated itself.

Figure 6-17 illustrates 802.1X. The workgroup switch is called the **authenticator.** It gets this name because the workgroup switch provides authentication service to the supplicant computer. The 802.1X standard normally also uses a central authentication server to do the actual supplicant credentials checking. Typically, this is a **RADIUS** server.

When the supplicant host transmits its authentication credentials (password, etc.), the authenticator passes these credentials on to the authentication server. The authentication server checks these credentials against its authentication database. If the authentication server authenticates the credentials, it sends back a confirmation to the workgroup switch. The workgroup switch then allows the supplicant PC to send frames to other devices in the network.

Using a central authentication server provides four benefits.

- *Switch Cost*. First, having the central authentication server check credentials minimizes the processing power needed in the workgroup switch. Given the large number of workgroup switches, this can produce a major cost saving.
- *Consistency*. Second, having all credentials on the authentication server gives consistency in authentication. An attacker cannot succeed by trying many different workgroup switches until he or she finds one that is misconfigured and gives the attacker access.
- *Management Cost*. Third, management cost is reduced because credentials only need to be changed on the central authentication server when user authentication information is changed when a user joins the firm, leaves the firm, or needs other credential changes.
- *Rapid Changes*. Fourth, the credentials of individuals who are fired or suspended can be invalidated in seconds, on all workgroup switches.

Security Standards Come from the 802.1 Working Group

Note from its name that the 802.1X standard was created by the *802.1* Working Group, not by the *802.3* Working Group that creates Ethernet standards. The 802.1 Working Group produces standards that cut across all 802 network technologies. This includes security standards.

Test Your Understanding

16. a) What threat does 802.1X address? b) How does the standard address the threat? c) In 802.1X, what device is the authenticator? d) What are the benefits of using a central authentication server instead of having the individual authenticators do all authentication work?

LOCAL AREA NETWORKS THAT ARE INTERNETS WITH ETHERNET SUBNETS

For small LANs, Ethernet is sufficient for most firms. However, for very large LANs, like university campus LANs, it may be better to have a local internet (routed network). We have seen that the Rapid Spanning Tree Protocol is complex to implement in large networks. So are VLANs, priority, and a number of other capabilities of Ethernet management. Quite simply, Ethernet does not scale well to very large LANs.

Figure 6-18 shows a local internet, which adds routing to Ethernet switching. Routers divide the Ethernet LAN into a number of smaller **subnets.** Each subnet is an ordinary Ethernet LAN. This approach joins the simplicity and low cost of basic Ethernet operation within each subnet with the ability of internets to manage large networks well.

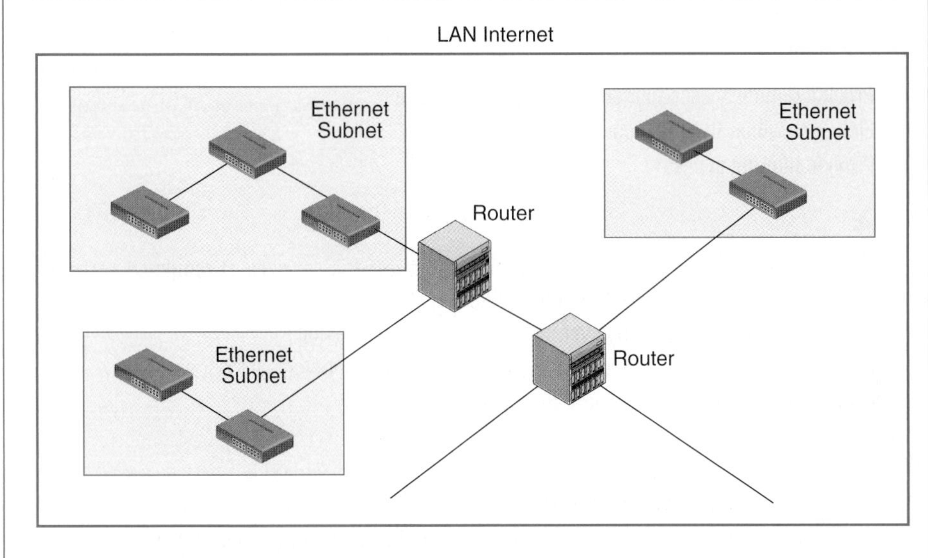

FIGURE 6-18 LAN Internet with Ethernet Subnets

Overall, a LAN can be a single switched (or wireless) network or an internet connecting multiple Ethernet or wireless subnets. The terms *LAN* and *switched* or *wireless network* are not synonymous.

Overall, a LAN can be a single switched (or wireless) network or an internet connecting multiple Ethernet or wireless subnets. The terms *LAN* and *switched* or *wireless network* are not synonymous.

Test Your Understanding

17. a) Why do many companies use LANs that are internets instead of pure Ethernet LANs? b) In LAN internets, what do we call individual Ethernet LANs?

SWITCHED WIRED WIDE AREA NETWORKS (WANs)

So far, we have been looking at switched Ethernet *LANs*. Switching is also used in wide area networks (WANs).

WANs

LANs are networks on the customer premises. WANs, in turn, connect different sites within the same company or different companies.

LANs are networks on the customer premises. WANs, in turn, connect different sites within the same company or between different companies.

Wide Area Networks (WANs)

Connect different sites

WAN Purposes

Provide remote access to individuals who are off site

Link sites within the same corporation

Provide Internet access

Carriers

Beyond their physical premises, companies must use the services of regulated carriers with rights of way for transmission in public areas.

Companies are limited to whatever services the carriers provide.

Prices for carrier services change abruptly and without technological reasons.

Prices and service availability vary from country to country.

High Costs and Low Speeds

High cost per bit transmitted, compared with LANs

Consequently, lower speeds (most commonly 256 kbps to about 50 megabits per second)

FIGURE 6-19 Wide Area Networks (WANs) (Study Figure)

THREE REASONS TO USE WANs There are three main purposes for WANs.

- The first is to provide remote access to customers or to individual employees who are working at home or traveling.
- The second is to link two or more sites within the same corporation. Given the large amount of site-to-site communication in most firms, this is the dominant WAN application.
- The third is to provide corporate access to the Internet.

CARRIERS A company can build its own LANs because they run through the company's own buildings and land. However, you cannot lay wires through your neighbor's yard, and neither can corporations. Transmission beyond the customer premises requires the use of regulated **carriers.** In exchange for regulation, carriers receive **rights of way** (permissions) to run their wires through public areas.

One shock that companies face when dealing with carriers is pricing. With LAN technology, prices closely follow costs, and prices change gradually as technology matures. However, with carriers, there often is little relationship between prices and costs. For instance, until recently, companies could purchase Frame Relay WAN service confident that it would be less expensive than leased line networking. Recently, however, many carriers abruptly and dramatically raised their Frame Relay prices and slashed their leased line prices. This created chaos in corporate WAN planning.

Another shock is *service limitations*. Usually, there are only a few competing carriers that a firm can use, and these carriers often offer only a few service options. There is nothing like the freedom companies have when they create LANs. (If you live in the United States and have a mobile phone, you get the idea.)

Global companies, furthermore, find that pricing and service options vary widely around the world. Options that are widely available in the United States and Europe often are rare in other parts of the world, and prices almost everywhere are higher than they are in the United States.

HIGH COSTS AND LOW SPEEDS Most LAN users are accustomed to at least 100 Mbps unshared speed to the desktop. In contrast, long-distance communication is much more expensive per bit transmitted, so companies usually content themselves with slower transmission speeds in WANs. Most WAN communication links operate at between 256 kbps and 50 Mbps, with most use coming at the lower end of this range.

Most WAN communication links operate at between 256 kbps and 50 Mbps, with most use coming at the lower end of this range.

Test Your Understanding

18. a) What are the three main purposes for WANs? b) What are carriers, and why must they be used? c) How are prices and costs related in carrier WAN services? d) Does a company have more service options with LANs or with WANs? e) Are service options and prices similar around the world? f) Compare LAN and WAN transmission speeds. g) Why are they different?

LEASED LINE NETWORKS FOR VOICE AND DATA

Leased Line Network Technology

When leased lines appeared in the 1960s, companies realized that they could build their own internal data networks to link their sites together. Figure 6-20 shows that companies need two pieces of technology.

- First, they needed leased lines to connect their sites to one another.
- Second, they needed a switch at each site.

Test Your Understanding

19. What two technologies are needed for leased line switched WANs?

Labor Costs

Designing a leased line network was a considerable task. As Chapter 4 discussed, a firm must first discover the traffic volumes between each pair of sites. In a leased line data WAN, the company must decide which sites to connect and what leased line speeds they need to connect the sites together. It must then contract with the telephone company for leased lines, install the switches, and test the network.

However, the real work begins after the leased line network is installed. Switched data networks in LANs often work with little day-to-day intervention. This is not true for switched WANs. The networking staff in the company must manage the network constantly.

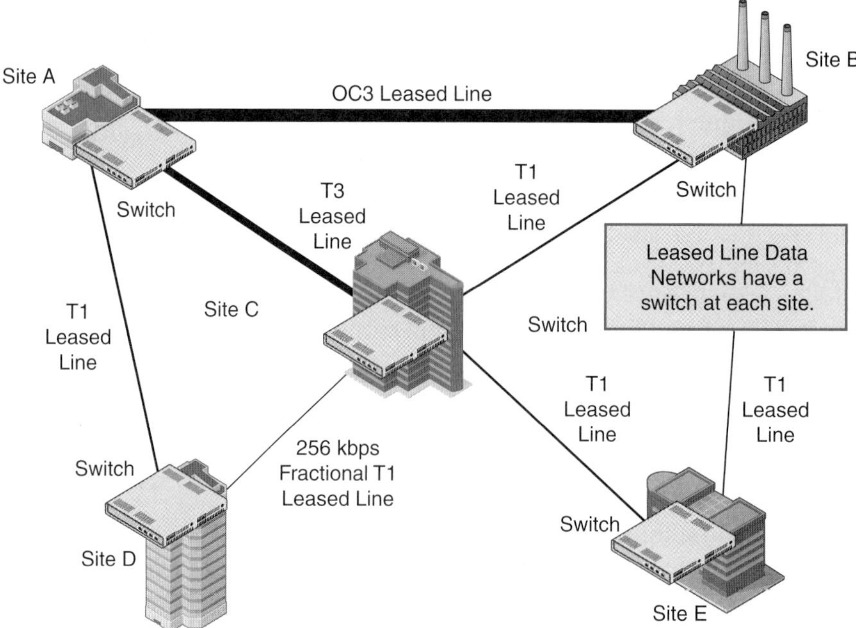

FIGURE 6-20 Leased Line Data Network

North American Digital Hierarchy		
Line	*Speed*	*Typical Transmission Medium*
56 kbps or 64 kbps (rarely offered)	56 kbps or 64 kbps	2-Pair Data-Grade UTP
T1	1.544 Mbps	2-Pair Data-Grade UTP
Fractional T1	128 kbps, 256 kbps, 384 kbps, 512 kbps, 768 kbps	2-Pair Data-Grade UTP
Bonded T1s (multiple T1s acting as a single line)	Small multiples of 1.544 Mbps	2-Pair Data-Grade UTP
T3	44.736 Mbps	Optical Fiber
CEPT Hierarchy (Europe)		
Line	*Speed*	*Typical Transmission Medium*
64 kbps	64 kbps	2-Pair Data-Grade UTP
E1	2.048 Mbps	2-Pair Data-Grade UTP
E3	34.368 Mbps	Optical Fiber
SONET/SDH Speeds		
Line	*Speed (Mbps)*	*Typical Transmission Medium*
OC3/STM1	155.52	Optical Fiber
OC12/STM4	622.08	Optical Fiber
OC48/STM16	2,488.32	Optical Fiber
OC192/STM64	9,953.28	Optical Fiber
OC768/STM256	39,813.12	Optical Fiber

FIGURE 6-21 Leased Line Speeds

Test Your Understanding

20. What are the cost elements in networks of leased lines?

Leased Line Speeds

Leased line speeds vary from 56 kbps to several gigabits per second. Figure 6-21 shows that different parts of the world use different standards for leased lines below 50 Mbps. The figure shows lower-speed leased lines in the United States and Europe. There also are differences in other countries.

56 KBPS AND 64 KBPS The lowest speed lines in these hierarchies operate at 56 or 64 kbps. This is barely higher than telephone modem speeds, so these leased lines are rarely sold today.

T1 AND E1 LEASED LINE At the next level of the hierarchy, the T1 line in the United States operates at 1.544 Mbps. The comparable European (CEPT) E1 line operates at 2.048 Mbps.

FRACTIONAL T1/E1 LEASED LINES The gap between 56 kbps/64 kbps and 1.544 Mbps/2.048 Mbps is large, so many U.S. carriers offer **fractional T1** leased lines operating at 128 kbps, 256 kbps, 384 kbps, 512 kbps, or 768 kbps. Fractional T1 lines provide intermediate speeds at intermediate prices. Similarly, carriers that offer E1 lines offer fractional lines.

T1/E1 and fractional T1/E1 lines provide speeds in the range of greatest corporate demand for WAN transmission—256 kbps to a few megabits per second. Consequently, T1/E1 and fractional T1/E1 lines are the most widely used leased lines.

T1/E1 and fractional T1/E1 lines are the most widely used leased lines.

BONDED T1S Sometimes, a firm needs somewhat more than a single T1 line but does not need the much higher speed of the T3 line (discussed next). Often, a company can **bond** a few T1s to get a few multiples of 1.544 Mbps. This is similar to link aggregation in Ethernet, which we saw earlier in this chapter. Bonding is also done with E1 lines.

T3 AND E3 LEASED LINES The next level of the hierarchy is the T3 line in the United States.[8] It operates at 44.736 Mbps. The comparable E3 line operates at 34.368 Mbps.

SONET/SDH Beyond T3/E3 lines, the world has nearly standardized on a single technology or, more correctly, on two compatible technologies. These are **SONET (Synchronous Optical Network)** in North America and **SDH (Synchronous Digital Hierarchy)** in Europe. Other parts of the world select one or the other.

Figure 6-21 shows that SONET/SDH speeds are multiples of 51.84 Mbps, which is close to the speed of a T3 line. SONET speeds are given by **OC (optical carrier)** numbers, while SDH speeds are given by **STM (synchronous transfer mode)** numbers.

The slowest offered SONET/SDH speed is 155.52 Mbps. Its speeds range up to several gigabits per second. SONET speed nearest to 10 Gbps is 9,953.28 Mbps. Ethernet uses this speed for WAN usage so that it can transmit data over physical layer SONET lines. (We will see later how Ethernet has begun to move outside the corporation.)

Test Your Understanding

21. a) Below what speed are there different leased line standards in different parts of the world? b) What is the exact speed of a T1 line? c) What are the speeds of comparable leased lines in Europe? d) Why are fractional T1 and E1 speeds desirable? e) List common fractional T1 speeds. f) What are the most widely used leased lines? g) What leased line standards are used above 50 Mbps?

---

[8] Although there are T2 and E2 standards, they are not offered commercially.

Digital Subscriber Lines (DSLs)

In the last chapter, we saw digital subscriber lines. While leased lines up to T1 and E1 use 2-pair data-grade UTP, DSLs send data over 1-pair voice-grade UTP.

One-pair voice-grade UTP is attractive for carrying data because these lines are already in place. The last chapter looked at asymmetric digital subscriber lines (ADSL lines), which provide high downstream (from the ISP) speeds but lower upstream (from the PC) speeds. This is fine for residential users, but businesses usually have traffic patterns that require high **symmetric** speeds—equal speeds in both directions. They also want service level agreements.

HDSL Fortunately, several business-oriented DSL standards are available, as Figure 6-22 indicates. The most popular business DSL is the **high-rate digital subscriber line (HDSL).** This standard allows symmetric transmission at 768 kbps (approximately half of a T1's speed) in both directions. A newer version, **HDSL2,** transmits at 1.544 Mbps in both directions. Like all DSLs, both use a single voice-grade twisted pair. Businesses find HDSL and HDSL2 attractively priced compared with T1 and fractional T1 lines.[9]

SHDSL The next step in business DSL is likely to be **SHDSL (super-high-rate DSL),** which can operate symmetrically over a single voice-grade twisted pair and over a speed range of 384 kbps–2.3 Mbps. In addition to offering a wide range of speeds and a higher top speed than HDSL2, SHDSL also can operate over somewhat longer distances.

	ADSL	HDSL	HDSL2	SHDSL
Uses existing 1-pair voice-grade UTP telephone access line to customer premises?*	Yes*	Yes*	Yes*	Yes*
Target market	Residences	Businesses	Businesses	Businesses
Downstream throughput	A few megabits per second	768 kbps	1.544 Mbps	384 kbps–2.3 Mbps
Upstream throughput	Slower than downstream	768 kbps	1.544 Mbps	384 kbps–2.3 Mbps
Symmetrical throughput?	No	Yes	Yes	Yes
QoS throughput guarantees?	No	Yes	Yes	Yes
*By definition, ALL DSLs use 1-pair voice-grade UTP residential access lines.				

FIGURE 6-22 ADSL versus Business-Class Symmetric Digital Subscriber Line (DSL) Services

[9] In fact, most T1 lines provided to businesses are double-bonded HDSL lines or HDSL2.

QUALITY-OF-SERVICE (QoS) GUARANTEES Generally, there are no hard guarantees for ADSL speeds, which are aimed at the service-tolerant but price-sensitive home market. However, throughputs for HDSL, HDSL2, and SHDSL generally come with strong quality-of-service guarantees because they are sold to businesses, which require predictable service. Meeting these guarantees requires more stringent engineering and management by the carrier and so increases carrier costs. This leads to higher prices for HDSL, HDSL2, and SHDSL than for ADSL services of comparable speed.

Test Your Understanding

22. a) How do the lowest-speed leased lines and DSL lines differ in terms of transmission media? b) Describe HDSL and HDSL2 in terms of speed. c) Describe SHDSL in terms of speed. d) Which DSL services usually offer QoS guarantees? e) What transmission medium do ADSL, HDSL, HDSL2, and SHDSL use?

PUBLIC SWITCHED DATA NETWORK (PSDN)

Companies that build their own leased line WANs must design, install, configure, and manage their leased lines. This is expensive. In contrast, many companies use **public switched data networks (PSDNs),** as Figure 6-24 illustrates. PSDNs allow a company to outsource the work of running a switched data network connecting its sites.

> PSDNs allow a company to outsource the work of running a switched data network connecting its sites.

Public Switched Data Network (PSDN) Access Lines

With a PSDN, the corporation needs only one leased line per site. This leased line has to run only from the site to the PSDN's nearest access point, called a **point of presence (POP).**

This means that if you have ten sites, you only need ten leased lines. Furthermore, most PSDN carriers have many POPs, so the few leased lines that are needed tend to span only short distances.

The PSDN Cloud

The PSDN's transport core usually is represented graphically as a **cloud.** This symbolizes the fact that although the PSDN has internal switches and trunk lines, the customer does not have to know how things work inside the PSDN cloud. The PSDN carrier handles almost all of the management work that customers have to do when running their own leased line networks. Customers merely have to send data to and receive data from the PSDN cloud, in the correct format. Although PSDN carrier prices reflect their management costs, there are strong **economies of scale** in managing very large PSDNs instead of individual corporate leased line networks. Quite simply, it is proportionally cheaper to manage the traffic of many firms than of one firm. There also are very large economies of scale in switching and leased line technologies. These economies of scale allow PSDN prices to remain relatively low.

Public Switched Data Network

- Outsource site-to-site networking to a PSDN carrier
- You need one leased line from each site to the nearest carrier point of presence (POP).
- PSDN core is shown as a cloud because user does not care what happens there.
- Economies of scale, so PSDNs usually are relatively inexpensive compared to switched leased line networks

X.25

- 1970s technology
- Slow and expensive
- Gone today

Frame Relay

- Started to grow in the 1990s
  - Inexpensive and fast compared to X.25
  - 256 kbps to about 40 Mbps
  - This is the range of greatest corporate demand for WAN speeds.
- Grew rapidly in the 1990s thanks to low prices
- Carriers have raised their prices to improve profit margins.
  - This has reduced growth.

ATM

- Created to replace the core of the Public Switched Telephone Network
  - Widely adopted for the Public Switched Telephone Network core
- Also provided as a PSDN service
  - Speeds of 1 Mbps to gigabits per second
  - Adoption for PSDN service has been limited

Metro Ethernet

- Metropolitan area network (MAN): a city and its environs
- MANs have smaller distances than national or international WANs, so lower prices and higher speeds
- Speeds of 1 Mbps to 100 Mbps and low prices
- No learning is needed because all firms are familiar with Ethernet.
- Carrier can provision or reprovision service speed rapidly, giving flexibility.
- The only PSDN service growing rapidly

FIGURE 6-23 Public Switched Data Networks (PSDNs) (Study Figure)

Test Your Understanding

23. a) Describe the physical components of PSDN technology. b) Do customers need leased lines if they use PSDNs? c) If a company has seven sites, how many leased lines will it need if it uses a PSDN? d) Why are PSDNs fairly inexpensive? e) Why is the PSDN transport core drawn as a cloud?

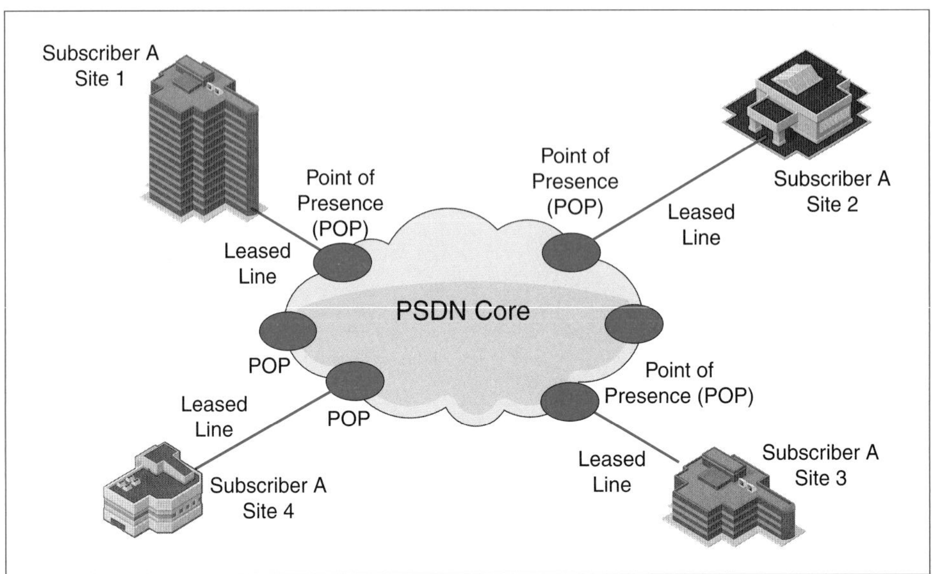

FIGURE 6-24 Public Switched Data Network (PSDN)

PSDN Standards

There are several standards for PSDN services. They have different speed ranges and also different prices.

X.25 The original PSDN standard was **X.25.** This was a 1970s technology that was slow, expensive, and never widely used. It is completely gone today, but it pioneered PSDN service.

FRAME RELAY In the 1990s, **Frame Relay** burst onto the scene. Frame Relay was much less expensive than X.25. Its speed range of 256 kbps to about 40 Mbps was not blindingly fast, but as noted earlier in this chapter, companies have lower speed requirements for WAN service than for LAN service. Frame Relay had exactly the speed range that corporations needed. In addition, only telephone companies offered leased lines, but new data networking companies could offer Frame Relay service. These new companies offered very low prices.

Frame Relay grew meteorically during the 1990s. Soon, it was used as widely as leased line networks. However, Frame Relay providers had very low profit margins because of their low prices. During the late 1990s, Frame Relay vendors began raising their prices while leased line prices fell precipitously because of improving technology.

ATM Another PSDN standard is **ATM.**[10] ATM actually was created as a new technology for the entire core of the worldwide Public Switched Telephone Network. In

[10] It stands for Asynchronous Transport Mode. Not very enlightening, is it? Fortunately, nobody ever spells it out.

fact, it has already replaced much of the traditional PSTN core, especially on international connections.

ATM offers higher speeds than X.25—from about 1 Mbps up to hundreds of gigabits per second. This makes ATM the fastest PSDN service. However, ATM is extremely expensive to install and manage, so ATM prices have also been very high. Consequently, ATM PSDN adoption has been very limited.

METROPOLITAN AREA ETHERNET Ethernet is a switched technology designed for LANs. However, many PSDN carriers have introduced metropolitan area Ethernet to extend Ethernet beyond the LAN. A **metropolitan area network (MAN)** is a WAN that is limited to a metropolitan area, such as a major city and its suburbs. MAN distances are shorter than national and international WAN distances, so MAN prices are lower, and typical speeds are higher. "Metro Ethernet" offers speeds of 1 Mbps to 100 Mbps at a very attractive price.

Metro Ethernet has several attractions beyond high speed and low price. Perhaps the most important is that corporations are already familiar with Ethernet. There is little learning with metro Ethernet compared to Frame Relay or ATM. Many people believe that in the future, companies will primarily use IP over Ethernet for nearly all connections.

Another advantage of metro Ethernet is provisioning. **Provisioning** is providing a service or changing a service. In metro Ethernet, service providers normally can reprovision the speed of a service within a day or so. This is much greater flexibility than companies have traditionally enjoyed with PSDNs.

Provisioning is providing a service or changing a service.

Given the promise of metro Ethernet, it is not surprising that metropolitan area Ethernet is the only PSDN service that is growing rapidly.

Test Your Understanding

24. a) What is the speed range of Frame Relay? b) Why is this speed range attractive? c) Why has ATM not been popular? d) What is metro Ethernet? e) For what reasons is it attractive?

Virtual Circuit Operation

Figure 6-25 shows the PSDN switches that sit inside the cloud in a mesh topology. In any mesh topology, whether partial or full, there are multiple alternative paths for frames to use to go from a source POP to a destination POP.

Selecting Best Possible Paths Through Meshes

Selecting the best possible path for each frame through a PSDN mesh would be complex and, therefore, expensive. In fact, if the best possible path had to be computed for each frame at

(continued)

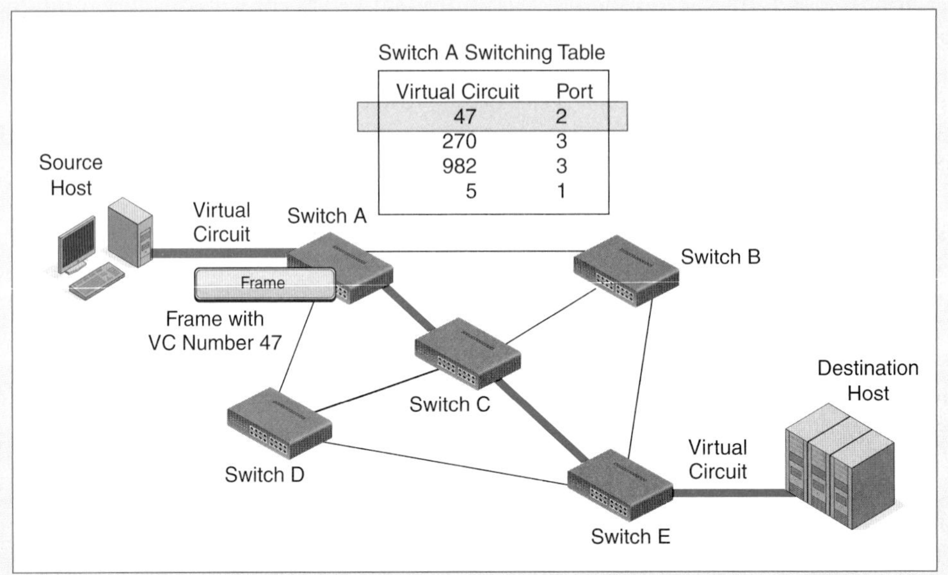

FIGURE 6-25 Virtual Circuit Operation

each switch along its path, PSDN switches would have to do so much work that they would be prohibitively expensive.

Virtual Circuits

Instead, most PSDNs select the best possible path between two sites *before transmission begins*. The actual transmission will flow along this path, called the **virtual circuit.** As Figure 6-25 shows, the switch merely makes a switching decision according to the virtual circuit number in the frame's header. This virtual circuit lookup is much faster than finding the best alternative path every time a frame arrives at the switch. The "heavy work" of selecting the best alternative path is only done once, before the beginning of communication.

PSDN Frame Headers Have Virtual Circuit Numbers Rather Than Destination Addresses

Note that PSDNs that use virtual circuits do not have destination addresses in their frame headers. Rather, each frame has a virtual circuit number in its header.

Frame Relay Virtual Circuits

In PSDNs that use virtual circuits, frame headers have virtual circuit numbers rather than destination addresses. This virtual circuit number in Frame Relay is the **Data Link Control Identifier** or **DLCI** (pronounced "DULL-see"). A DLCI typically is 10 bits long.[11] The switch

[11] Stations also have true Frame Relay addresses governed by the E.164 standard. These addresses are used to set up virtual circuits. Consequently, if an equipment failure renders a virtual circuit inoperable, a new virtual circuit can be set up using the E.164 addresses.

looks up this DLCI in its virtual circuit switching table and sends the frame out through the indicated port.

Test Your Understanding

25. a) Why are virtual circuits used? b) With virtual circuits, on what does a switch base its forwarding decision when a frame arrives? c) Do PSDN frames have destination addresses or virtual circuit numbers in their headers? d) What is the name of the Frame Relay virtual circuit number? e) How long, typically, is a DLCI? f) How many virtual circuits does this number of bits allow? (The answer is not specifically in the text.)

CONCLUSION

Synopsis

This chapter looked in some depth at Ethernet local area networking. Ethernet is the dominant technology for corporate LANs today. Ethernet's only serious competitor is wireless LANs, which we will see in the next chapter. We will learn that wireless LANs are not direct competitors to wired Ethernet LANs, but rather usually work in conjunction with wired Ethernet LANs.

The IEEE 802 LAN/MAN Standards Committee creates many LAN standards. The Committee's 802.3 Working Group specifically creates Ethernet standards. Like all networking standards, Ethernet standards exist at both the physical and data link layers. Therefore, they are OSI standards.

The 802.3 Working Group has created many physical layer Ethernet standards and is still creating better physical layer standards. Speeds range from 10 Mbps to 10 Gbps and are still moving higher. These standards use both 6-pair UTP and optical fiber.

The dominant Ethernet standard for access lines from hosts to switches is 100BASE-TX, while the dominant standard for switch-to-switch trunk lines is 1000BASE-SX using optical fiber. The newest and fastest standards (10GBASE-x and beyond) are being created first for metropolitan area networks (MANs) but are moving into LANs as well.

Although transmission links have maximum lengths, switches regenerate signals. Regeneration allows firms to send frames across many switches connected by trunk lines with little degradation.

The 802 Committee subdivided the data link layer into two layers. The media access control layer is specific to a particular technology, such as Ethernet or 802.11 wireless LANs. The logical link control layer deals with matters common to all LAN technologies. Ethernet has only a single MAC standard—the 802.3 Media Access Control standard. This standard specifies frame organization and switch operation.

The Ethernet frame has multiple fields. The preamble and start of frame delimiter fields synchronize the receiver's clock with the sender's clock. The destination and source MAC address fields are each 48 bits long, and NIC vendors assign Ethernet addresses to NICs at the factory. Because of human memory limitations (and to simplify writing), Ethernet MAC addresses usually are written in hexadecimal format, such as B2-CC-66-0D-5E-BA. The length field specifies the length of the data field (not the length of the frame as a whole). The data field has two parts: the LLC subheader, which describes the type of packet contained in the data field, and the packet itself. The PAD field is added if the data field is less than 46 octets long, in order to make the data field plus the PAD field exactly 46 octets in length. The receiving NIC uses the frame check sequence field to check for errors. If the receiver finds an error, it simply discards the frame.

Firms must organize their Ethernet switches in a hierarchy. This simplifies switching, making Ethernet switches inexpensive. Switches that connect hosts to the network are workgroup switches. Switches higher in the hierarchy are core switches. There must not be loops among switches because this would break the hierarchy. The Rapid Spanning Tree Protocol automatically detects and disables accidental loops. RSTP can also provide backup links in case of link or switch failures.

Most Ethernet switches can divide an Ethernet LAN into a number of VLANs that are groups of clients and servers that can talk to each other but cannot talk to clients and servers on different VLANs. Using VLANs reduces congestion when servers broadcast. It also provides security. To standardize VLANs (and priority), two tag fields are added to the Ethernet frame, right after the source address. The Tag Protocol ID field merely indicates that the frame is tagged. The Tag Control Information field has a 12-bit VLAN number to indicate to which VLAN a particular frame belongs.

Switches require care in purchasing. Managed switches are more expensive than other switches, but companies can manage them remotely. Using managed switches saves money overall by reducing management labor. Under Power over Ethernet (POE), some switches provide electrical power to the hosts they serve.

For Ethernet security, the 802.1X standard requires a computer that plugs into an RJ-45 wall jack to authenticate itself before it is allowed to use the network. The 802.1X standard uses a central authentication server.

Finally, for large LANs, Ethernet runs into serious management problems. Consequently, many companies use routers to turn their LANs into local internets. This divides the LAN into several smaller subnets that use Ethernet technologies.

Corporations build wide area networks (WANs) for individual remote access, site-to-site transmission, and Internet access. Among the technologies they use are networks of leased lines and public switched data networks (PSDNs).

Your personal experience probably has been limited primarily to LAN transmission, where low cost per bit transmitted allows companies to afford high speeds. You must adjust your thinking for wide area networks (WANs), where long distances make the price per transmitted bit very high, which in turn leads to companies limiting themselves primarily to low speeds—most typically, between 256 kbps and 50 Mbps. Most demand comes at the lower end of this range.

For site-to-site networking, companies have traditionally turned to networks of leased lines, which are reliable but expensive, and pure hub-and-spoke topologies, which are inexpensive but have many single points of failure.

Companies that use leased lines typically choose T1/E1 leased lines operating at 1.5 Mbps to 2 Mbps, fractional leased lines below T1/E1 speeds, and bonded T1/E1 lines. However, if they have some connections that require much higher speeds, they can choose T3/E3 lines operating at roughly 35 Mbps to 45 Mbps, or SONET/SDH lines operating at 156 Mbps to several gigabits per second. In addition, many firms use

HDSL, HDSL2, or SHDSL. These are digital subscriber line services, but they offer high symmetrical speeds and throughput guarantees.

With public switched data networks (PSDNs), the carrier does most of the transmission and management work. Companies merely need access devices (typically routers) at their sites and a single leased line from each site to the PSDN carrier's nearest point of presence. Frame Relay provides speeds of 256 kbps to 40 Mbps. This matches the range of greatest corporate demand, so Frame Relay has dominated the PSDN market. ATM is much faster but is very expensive and is only used modestly. Metropolitan area Ethernet offers speeds of 1 Mbps to 100 Mbps at attractive prices. In addition, metro Ethernet is familiar technology and can be provisioned very rapidly in useful increments.

(In a box.) Frame Relay and ATM use virtual circuits to simplify the operation of switches and, therefore, minimize switching costs. Virtual circuits are paths set up before communication begins. Switches base forwarding decisions on virtual circuit numbers rather than on destination addresses.

END-OF-CHAPTER QUESTIONS

Thought Questions

1. If the sender adds a PAD field to an Ethernet frame, the combined data field and PAD will be 46 octets long. How can the receiving NIC tell which part is the data field?
2. How do you think an Ethernet switch gets the information in its switching table? (Hint: The answer involves switch learning.)
3. Ethernet switches erase their switching tables frequently. Why do you think they do this?

Design Questions

1. Two switches are 47 meters apart. They need to communicate at 600 Mbps. What do you recommend? Justify your answer.
2. Two switches are 200 meters apart. They need to be able to communicate at 1.7 Gbps. What do you recommend? Justify your answer.
3. You will create a design for an Ethernet network connecting four buildings in an industrial park. Hand in a picture showing your network. There will be a core switch in each building.

   Building A is the headquarters building.

   Building B is 85 meters south and 90 meters east of the headquarters building. A line will run directly from Building A to Building B.

   Building C is 150 meters south of the headquarters building. A line will run directly from Building A to Building C.

   Building D is 60 meters west of Building C. A line will run directly from Building C to Building D.

   Computers in Building A need to communicate with computers in Building B at 600 Mbps.

   Computers in Building A need to be able to communicate with computers in Building C at 1 Gbps.

   Computers in Building B must communicate with computers in Building D at 500 Mbps.

   Computers in Building C must communicate with computers in Building D at 750 Mbps.

   a) Draw a picture of the situation.
   b) Determine the traffic volume on each transmission line. Explain your answers.
   c) Determine what leased line standard to use for each transmission line. Explain your answers.

Hands-On Exercises

BINARY AND HEXADECIMAL CONVERSIONS

If you have Microsoft Windows, the Calculator accessory shown in Chapter 1a can convert between binary and hexadecimal notation. Go to the *Start* button, then to *Programs* or *All Programs*, then to *Accessories*, and then click on *Calculator*. The Windows Calculator will pop up.

BINARY TO HEXADECIMAL

To convert eight binary bits to hexadecimal (hex), first choose *View* and click on *Scientific* to make the Calculator a more advanced scientific calculator. Click on the *Bin* (binary) radio button, and type in the 8-bit binary sequence you wish to convert. Then click on the *Hex* (hexadecimal) radio button. The hex value for that segment will appear.

HEXADECIMAL TO BINARY

To convert hex to binary, go to *View* and choose *Scientific*. Click on *Hex* to indicate that you are entering a hexadecimal number. Type the number. Now click on *Bin* to convert this number to binary.

One additional subtlety is that Calculator drops initial 0s. So if you convert 1 hex, you get 1. You must add three initial 0s to make this a 4-bit segment: 0001.

1. a) Convert 1100 to hexadecimal.
   b) Express the following MAC address in binary: B2-CC-66-0D-5E-BA, leaving a space after every eight bits. (Hint: B2 is 10110010.)
   c) Express the following MAC address in hex: 11000010 11001100 01100111 00001101 01011110 10111010. (Hint: 11000010 is C2.)

Perspective Questions

1. What was the most surprising thing you learned in this chapter?
2. What was the most difficult material for you in this chapter?

CHAPTER 6a

Hands-On: Ethernet Switching

LEARNING OBJECTIVES

By the end of this chapter, you should be able to:

- Set up a small Ethernet switched network.
- Observe what happens if you create a loop among Ethernet switches.

THE EXERCISE

This is a class exercise rather than an individual exercise. It is rather quick (taking 15 to 20 minutes), but it takes an investment in resources.

What You Will Need

- A number of Ethernet switches. In general, it is good to have one switch for every two to four students, with the low ratio being much better. These can be very cheap switches.
- Enough UTP cords to connect the switches to each other and to the wall jack that bring the campus network into the classroom. Each will need to be 3–6 m in length, depending on the layout of the classroom. Each student group should have sufficient room to work.
- Each Ethernet switch is powered. You may need to have some power cables so that all of the teams have power for their switches.
- Two notebooks to plug into the network.

Creating the Network

The students should create a network like the one in Figure 6a-1. There should be a top-level switch at the front of the classroom. It should plug into the wall jack that connects the classroom to the campus network.

Below the top-level switch, other switches should be arranged in a hierarchy. I find it is useful to have a simple hierarchy with two columns of switches as shown in the figure. It is important to keep a strict hierarchy among the switches.

After the switches are set up, attach PCs to switches at the end of each column. See if the PCs can connect to the Internet via the classroom wall jack. They should be able to do so.

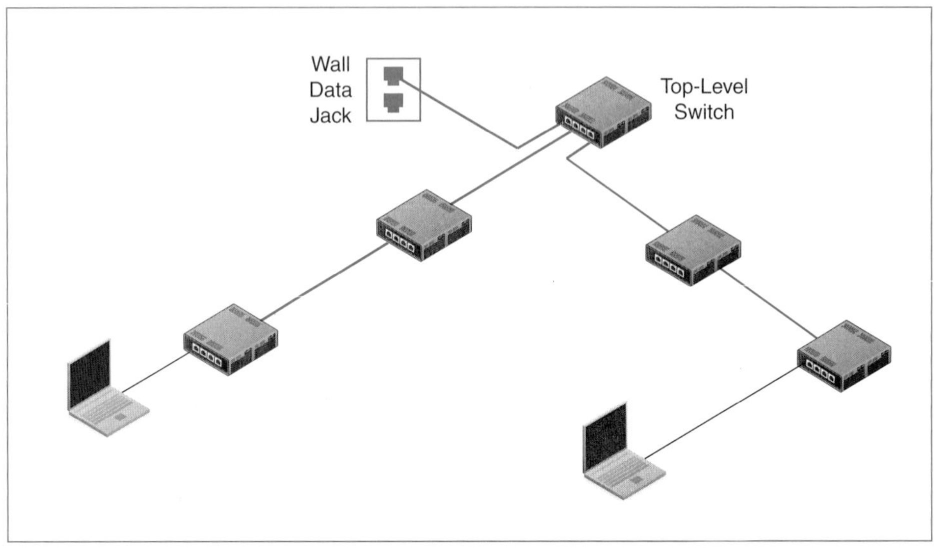

FIGURE 6a-1 The Network

At the end of this exercise, you can see how straightforward it is to set up a hierarchical Ethernet network. The switches are easy to power up, and RJ-45 connectors simply go "snap."

Creating a Loop

Now that the network is working, it is time to create a loop. Loops are not allowed in Ethernet, and you are about to see why. Connect two switches so that a loop is created, as Figure 6a-2 illustrates. Now see if the PCs can still access the Internet. They should not be able to do so.

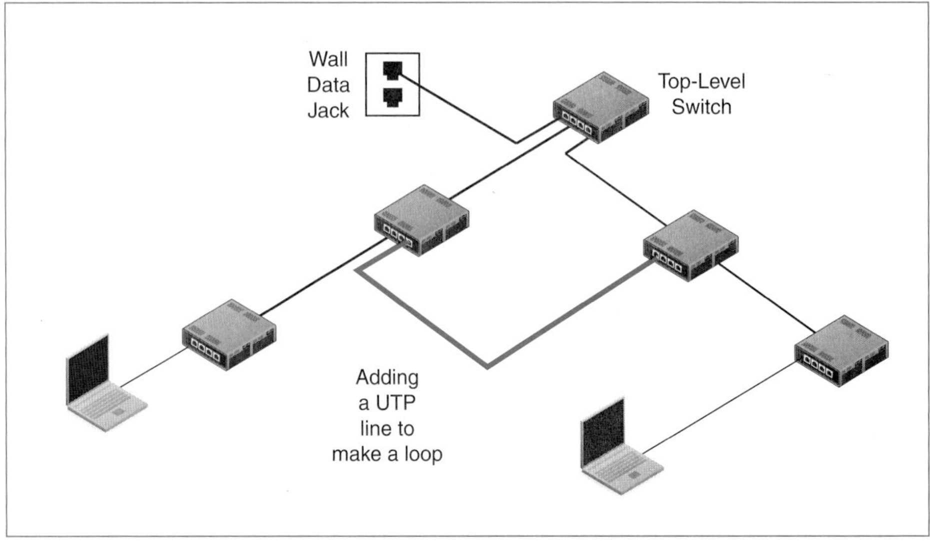

FIGURE 6a-2 Adding a Loop

CHAPTER 7

Wireless Networks I

LEARNING OBJECTIVES

By the end of this chapter, you should be able to:

- Explain radio signal propagation: frequencies, antennas, and wireless propagation problems.
- Describe wireless LAN technologies.
- Explain radio bands, bandwidth, and channels.
- Distinguish between normal and spread spectrum transmission.
- Describe 802.11 WLAN operation.
- Compare and contrast 802.11 transmission standards.

CHRISTOPHER LOREK

Christopher Lorek bought his wife a new notebook computer. Claire was pleased with her new computer, but she didn't want to run a new UTP line to it from their access router. She wanted mobility. So Christopher bought a new access router with a built-in access point. The new wireless router was inexpensive. It operated in the 2.4 GHz radio band, and it followed the 802.11g standard. Christopher also bought a wireless printer for them to use. It also operated in the 2.4 GHz band using the 802.11g standard.

Although Claire liked the mobility, she found Internet access painfully slow. Christopher downloaded an application to Claire's computer to identify nearby wireless access routers. When he used it, he found that there were only three available channels, and at least one of his neighbors had an access point operating on each channel. The problem, then, appeared to be interference.

Christopher returned the router to the store. He decided to get a router that would operate in the uncrowded 5 GHz band. He was pleased to find that routers that operated in the 5 GHz band also used the 802.11n standard, which is several times faster than 802.11g. The store had two 5 GHz routers. One operated in both the 2.4 GHz and 5 GHz band, but not simultaneously. The other operated in both bands simultaneously. The one that operated simultaneously in the two bands was more expensive.

Test Your Understanding

1. a) Which terms in this case were unfamiliar to you? b) What do you think Christopher should do?

INTRODUCTION

In the last chapter, we looked at wired switched networks. Technologies for those networks, for instance Ethernet, require both physical and data link layer standards. Consequently, they are OSI standards. In this chapter and the next, we will look at wireless communication in both LANs and WANs. Wireless networks are also single networks, which require physical and data link layer standards. So they too are OSI standards.

Although many people think of wireless transmission as something new and underdeveloped, businesses were already spending more on wireless networking than wired networking in 2008. Wireless transmission is the growth sector in networking today and will be for some time to come.

Test Your Understanding

2. a) At what layers do wireless networks operate? b) Are wireless network standards OSI standards or TCP/IP standards? Explain.

RADIO SIGNAL PROPAGATION

Chapter 5 discussed propagation effects in wired transmission media (UTP and optical fiber). Generally speaking, these effects can be well controlled by respecting cord distance limits and taking other installation precautions. This is possible because wired propagation is predictable. If you input a signal, you can estimate fairly precisely what it will be at the other end of a cord.

Chapter 6

Wired switched networks

Standards at Layer 1 (physical) and Layer 2 (data link)

Physical and data link layer standards are almost always OSI standards

Chapters 7 and 8

Wireless networking

Wireless LANs (WLANs)

Wireless WANs (WWANs)

Also require standards at Layers 1 and 2

So also are OSI standards

FIGURE 7-1 Perspective

In contrast, radio propagation is very unreliable. Radio signals bounce off obstacles, fail to pass through walls and filing cabinets, and have other problems we will look at in this section. Consequently, wireless networks, which use radio to deliver signals, are more complex to engineer than wired networks. Therefore we will spend more time on wireless propagation effects than we did on wired propagation effects.

Frequencies

Wireless radio signals propagate as waves, as we saw in Chapter 5. Figure 7-2 again notes that waves are characterized by amplitude, wavelength, and frequency. While optical fiber waves are given in terms of wavelength, radio waves are described in terms of *frequency*.

Frequency is used to describe the radio waves used in WLANs.

In waves, frequency is the number of complete cycles per second. One cycle per second is one **hertz (Hz).** Metric designations are used to describe frequencies. In the metric system, frequencies increase by a factor of 1,000, rather than 1,024. The most common radio frequencies for wireless data transmission are about 500 **megahertz (MHz)** to 10 **gigahertz (GHz).**

Test Your Understanding

3. a) Is wireless radio transmission usually expressed in terms of wavelength or frequency? b) What is a hertz? c) Convert 3.4 MHz to a number without a metric prefix. d) At what range of frequencies do most wireless systems operate?

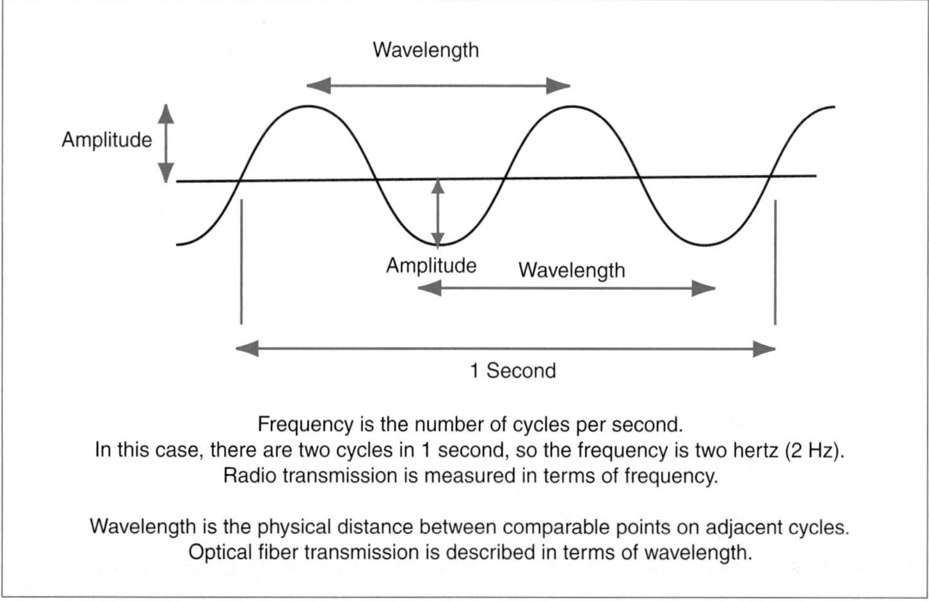

FIGURE 7-2 Electromagnetic Wave

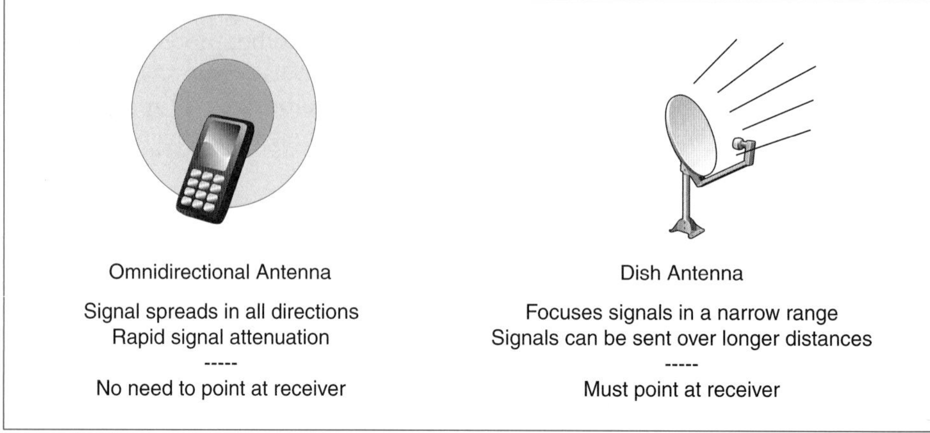

FIGURE 7-3 Omnidirectional and Dish Antennas

Antennas

Radio transmission requires an antenna. Figure 7-3 shows that there are two types of radio antennas: omnidirectional antennas and dish antennas.

- **Omnidirectional antennas** transmit signals equally strongly in all directions and receive incoming signals equally well from all directions. Consequently, the antenna does not need to point in the direction of the receiver. However, because the signal spreads in all three dimensions, only a small fraction of the energy transmitted by an omnidirectional antenna reaches the receiver. Omnidirectional antennas are best for short distances, such as those found in a wireless LAN (WLAN) or a cellular metropolitan area network.
- **Dish antennas,** in contrast, point in a particular direction, which allows them to focus stronger outgoing signals in that direction for the same power and to receive weaker incoming signals from that direction. (A dish antenna is like the reflector in a flashlight.) Dish antennas are good for longer distances because of their focusing ability, but users need to know the direction of the other antenna. Also, omnidirectional antennas are easier to use. (Imagine if you had to carry a dish with you whenever you carried your cellular phone. You would not even know where to point the dish!)

Test Your Understanding

4. a) Distinguish between omnidirectional and dish antennas in terms of operation. b) Under what circumstances would you use an omnidirectional antenna? c) Under what circumstances would you use a dish antenna? d) What type of antenna normally is used in WLANs? Why?

Wireless Propagation Problems

We have already noted that, although wireless communication gives mobility, wireless transmission is not very predictable, and there often are serious propagation problems. Figure 7-4 illustrates five common wireless propagation problems.

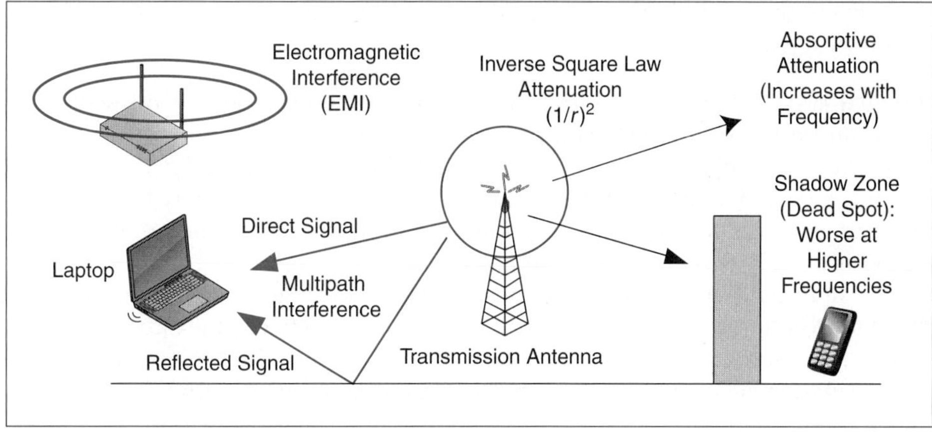

FIGURE 7-4 Wireless Propagation Problems

INVERSE SQUARE LAW ATTENUATION Compared with signals sent through wires and optical fiber, radio signals attenuate very rapidly. When a signal spreads out from any kind of antenna, its strength is spread over the area of a sphere. (In omnidirectional antennas, power is spread equally over the sphere, while in dish antennas, power is concentrated primarily in one direction on the sphere.)

The area of a sphere is proportional to the square of its radius, so signal strength in any direction weakens by an **inverse square law** ($1/r^2$), as Equation 5-1 illustrates. Here, S_1 is the signal strength at distance r_1, and S_2 is the signal strength at a farther distance, r_2.

$$S_2 = S_1{}^*(r_1/r_2)^2 \qquad \textbf{(Equation 5-1)}$$

To give an example, if you triple the distance ($r_1/r_2 = 1/3$), the signal strength (S_2) falls to only one-ninth ($1/3^2$) of its original strength (S_1). With radio propagation, you have to be relatively close to your communication partner unless the signal strength is very high, an omnidirectional antenna is used, or both.

To give a specific example, at 10 meters, the signal strength is 20 milliwatts (mW). How strong will the signal be at 20 meters?

- The distance doubles (so r_1/r_2 is 1/2).
- So we multiply the signal strength at 10 meters by 1/4 (1/2 squared)
- Twenty mW multiplied by 1/4 is 5 mW.
- So the strength of the signal at 10 meters will be 5 mW.

ABSORPTIVE ATTENUATION As a radio signal travels, it is partially absorbed by the air molecules, plants, and other things it passes through. This **absorptive attenuation** is especially bad in moist air, and office plants are the natural enemies of wireless transmission. Absorptive attenuation is especially bad for longer-distance outdoor propagation.

SHADOW ZONES (DEAD SPOTS) To some extent, radio signals can go through and bend around objects. However, if there is a large or dense object (such as a brick wall), blocking the direct path between the sender and receiver, the receiver may be in a **shadow zone (dead spot),** where the receiver cannot get the signal. If you have a cellular telephone and often try to use it within buildings, you probably are familiar with this problem.

MULTIPATH INTERFERENCE In addition, radio waves tend to bounce off walls, floors, ceilings, and other objects. As Figure 7-4 shows, this may mean that a receiver will receive two or more signals—a direct signal and one or more reflected signals. The direct and reflected signals will travel different distances and so may be out of phase when they reach the receiver. (For example, one may be at its highest amplitude while the other is at its lowest, giving an average of zero.)

This **multipath interference** may cause the signal to range from strong to nonexistent within a few centimeters (inches). If the difference in time between the direct and reflected signal is large, some reflected signals may even interfere with the next direct signal. Multipath interference is the most serious propagation problem at WLAN frequencies.

Multipath interference is the most serious propagation problem at WLAN frequencies.

ELECTROMAGNETIC INTERFERENCE (EMI) A final common propagation problem in wireless communication is *electromagnetic interference (EMI)*. Other devices produce EMI at frequencies used in wireless data communications. Among these devices are cordless telephones, microwaves, and especially devices in other nearby wireless networks.

FREQUENCY-DEPENDENT PROPAGATION PROBLEMS To complicate matters, two wireless propagation problems are affected by frequency.

- First, higher-frequency waves suffer more rapidly from absorptive attenuation than lower-frequency waves because they are absorbed more rapidly by moisture in the air, leafy vegetation, and other water-bearing obstacles. Consequently, as we will see in this chapter, WLAN signals around 5 GHz attenuate more rapidly than signals around 2.4 GHz.
- Second, shadow zone problems grow worse with frequency. As frequency increases, radio waves become less able to go through and bend around objects.

Test Your Understanding

5. a) Which offers more reliable transmission characteristics—UTP or radio transmission? b) Which attenuate more rapidly with distance—signals sent through wired media or radio signals? c) If the signal strength from an omnidirectional radio source is 8 mW at 30 meters, how strong will it be at 120 meters, ignoring absorptive attenuation? Show your work. d) How are shadow zones (dead spots) created? e) Why is multipath interference very sensitive to location? f) What is the most serious propagation problem in WLANs? g) List some sources of EMI. h) What propagation problems become worse as frequency increases?

WIRELESS LAN (WLAN) TECHNOLOGY

Having discussed wireless transmission briefly, we will look at wireless networking's widest application today, wireless local area networks. A **wireless local area network (WLAN),** like any type of LAN, operates on the customer premises.

Wireless LANs (WLANs) use radio for physical layer transmission on the customer premises.

802.11 WLAN Standards

The most important WLAN standards today are the **802.11** standards, which are created by the **IEEE 802.11 Working Group.** Recall that Ethernet standards are created by a different working group, the 802.3 Working Group.

Rather than being a competitor for wired Ethernet LANs, **802.11 WLANs** today primarily *supplement* wired LANs, but do not replace them. Figure 7-6 shows that mobile users typically connect by radio to devices called **wireless access points,** or simply, **access points.** These access points link the mobile user to the firm's wired Ethernet LAN.

Figure 7-7 shows a wireless access point and the wireless NICs that computers need to work with access points. The NICs in the figure are add-on devices. However, laptops and other portable devices typically come with built-in wireless NICs. Add-on wireless NICs are only needed if the user wishes to upgrade to a standard the built-in wireless NIC does not support.

Why is there normally a connection to the firm's main wired LAN? Quite simply, the servers that mobile host devices need, as well as the firm's Internet access router, usually are on the wired LAN. Wireless hosts need the wired LAN to reach the resources they need. A single 802.11 wireless access point can serve multiple hosts up to 30 to 100 meters away.

In a home, you are only likely to have a single access point. Businesses need far larger coverage areas. By placing wireless access points judiciously throughout a building, a company can construct a large 802.11 WLAN "cloud" that can serve mobile users anywhere in the building. We will spend most of this chapter looking at the technology of 802.11 WLANs.

802.11 Wireless LAN Technology

The dominant WLAN technology today

Standardized by the 802.11 Working Group

Wireless Computers Connect to Access Points (Figure 7-6 and Figure 7-7)

Supplement Wired LANs

Access points connect to the corporate LAN

So that wireless hosts can reach servers on the Ethernet LAN

So that wireless hosts can reach Internet access routers on the Ethernet LAN

Large 802.11 WLANs

Organizations can provide coverage throughout a building or a university campus

By the judicious installation of many access points

Speeds and Distances

Distances of 30 to 100 meters

FIGURE 7-5 802.11 Wireless LAN (WLAN) Standards (Study Figure)

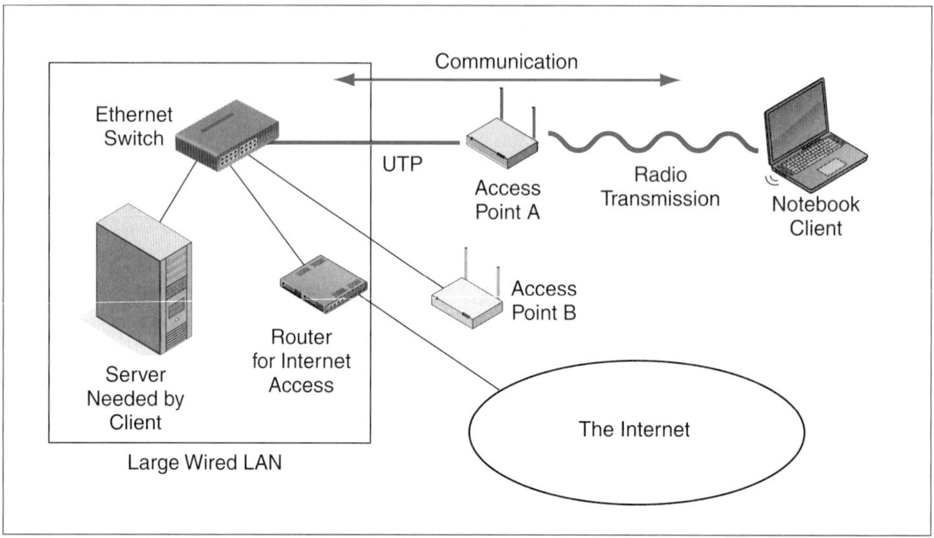

FIGURE 7-6 802.11 Wireless LAN (WLAN) Operation

Test Your Understanding

6. a) What 802 working group creates 802.11 standards? b) Why do wireless clients need access to the wired Ethernet LAN? c) How can firms provide WLAN coverage throughout a large building?

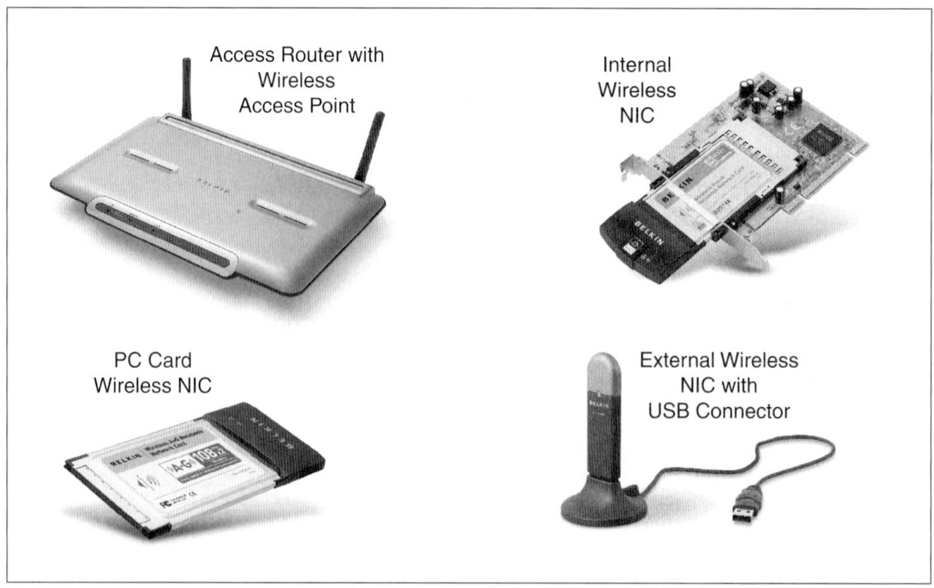

FIGURE 7-7 802.11 Wireless Access Point and Wireless NICs

RADIO BANDS, BANDWIDTH, AND SPREAD SPECTRUM TRANSMISSION

Bands and Bandwidth

Now we can begin looking at new information about radio transmission, beginning with the frequency spectrum, service bands, and channels.

THE FREQUENCY SPECTRUM AND SERVICE BANDS The **frequency spectrum** consists of all possible frequencies from zero hertz to infinity, as Figure 7-8 shows.

SERVICE BANDS The frequency spectrum is divided into contiguous spectrum ranges called **service bands** that are dedicated to specific services. For instance, in the United States, the AM radio service band lies between 535 kHz and 1,705 kHz. The FM radio service band, in turn, lies between 88 MHz and 108 MHz. The 2.4 GHz unlicensed band that we will see later for wireless LANs extends from 2.4000 GHz to 2.4835 GHz. There are also service bands for police and fire departments, amateur radio operators, communication satellites, and many other purposes.

CHANNELS Service bands are subdivided into smaller frequency ranges called **channels.** A different signal can be sent in each channel because signals in different channels do not interfere with one another. This is why you can receive different television channels successfully.

SIGNAL AND CHANNEL BANDWIDTH Figure 7-2 showed wave operating at a single frequency. In contrast, Figure 7-10 shows that signals do not operate at a single frequency. Rather, signals spread over a range of frequencies. This range is called the signal's **bandwidth.** Signal bandwidth is measured by subtracting the lowest frequency from the highest frequency.

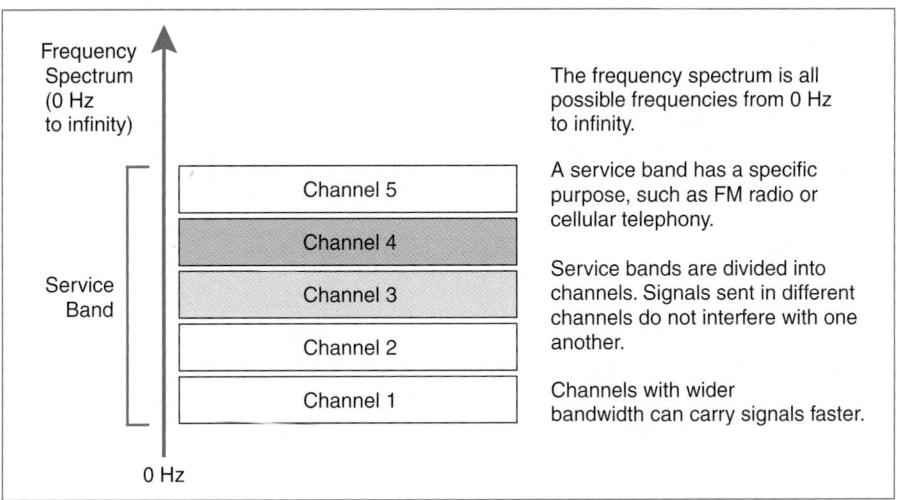

FIGURE 7-8 The Frequency Spectrum, Service Bands, and Channels

Signal Bandwidth

Figure 7-2 shows a wave operating at a single frequency
However, most signals are spread over a range of frequencies (Figure 7-10)
The range between the highest and lowest frequencies is the signal's bandwidth
The maximum possible transmission speed increases with bandwidth

Channel Bandwidth

Channel bandwidth is the highest frequency in a channel minus the lowest frequency
An 88.0 MHz to 88.2 MHz channel has a bandwidth of 0.2 MHz (200 kHz)
Higher-speed signals need wider channel bandwidths

Shannon Equation

$C = B\,[\mathrm{Log}_2\,(1+S/N)]$
- C = Maximum possible transmission speed in the channel (bps)
- B = Bandwidth (Hz)
- S/N = Signal-to-noise ratio measured as the power ratio

Note that doubling the bandwidth doubles the maximum possible transmission speed
Multiplying the bandwidth by X multiplies the maximum possible speed by X
Wide bandwidth is the key to fast transmission
Increasing S/N helps slightly, but usually cannot be done to any significant extent

Broadband and Narrowband Channels

Broadband means wide channel bandwidth and therefore high speed
Narrowband means narrow channel bandwidth and therefore low speed

The Golden Zone

Most organizational radio technologies operate in the golden zone in the 500 MHz to 10 GHz range
Golden zone frequencies are high enough for there to be large total bandwidth
- At higher frequencies, there is more available bandwidth

Golden zone frequencies are low enough to allow fairly good propagation characteristics
- At lower frequencies, signals propagate better

Growing demand creates intense competition for frequencies in the golden zone

FIGURE 7-9 Channel Bandwidth and Transmission Speed (Study Figure)

A channel also has a bandwidth. For instance, if the lowest frequency of an FM channel is 89.0 MHz and the highest frequency is 89.2 MHz, then the **channel bandwidth** is 0.2 MHz (200 kHz). AM radio channels are 10 kHz wide, FM channels have bandwidths of 200 kHz, and television channels are 6 MHz wide.

Why are there such large differences in channel bandwidth across service bands? The answer lies in the relationship between possible transmission speed in a channel and channel bandwidth. Shannon found that the maximum possible transmission speed (C)

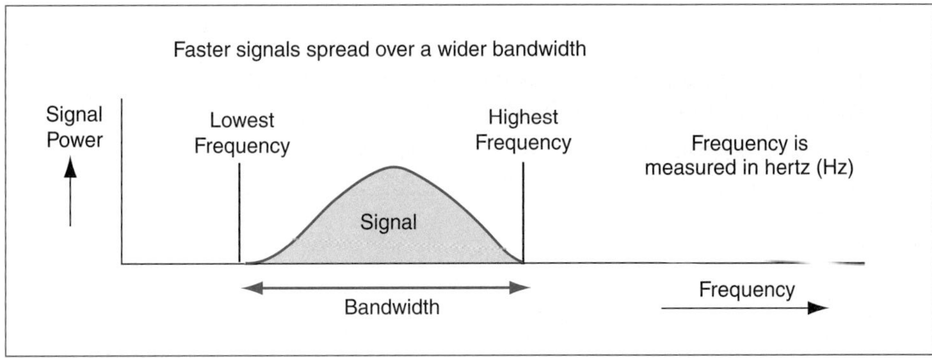

FIGURE 7-10 Signal Bandwidth

in bits per second when sending data through a channel is directly proportional to the channel's bandwidth (B) in hertz, as shown in the **Shannon Equation** (Equation 5-2).[1]

$$C = B[\mathrm{Log}_2(1 + S/N)] \qquad \textbf{(Equation 5-2)}$$

The maximum possible speed is directly proportional to bandwidth, so if you double the bandwidth, you can potentially transmit up to twice as fast. However, C is the *maximum* possible speed for a given bandwidth and signal-to-noise ratio. *Real transmission throughput* will always be less.

To transmit at a given speed, you need a channel wide enough to handle that speed. For example, video signals produce many more bits per second than audio signals, so television uses much wider channels than AM radio (6 MHz versus 10 kHz in AM radio transmission).

The signal-to-noise (S/N) ratio discussed in Chapter 5 is also important, but it is difficult to modify in practice. Radio signal strengths tend to be limited by law, and reducing noise is very difficult without going to super-cooled senders and receivers.

Channels with large bandwidths are called **broadband** channels. They can carry data very quickly. In contrast, channels with small bandwidths, called **narrowband** channels, can only carry data slowly. Although the terms *broadband* and *narrowband* technically refer only to the width of a channel, broadband has come to mean "fast," while narrowband has come to mean "slow."

Transmission systems that are very fast are usually called broadband systems even when they do not use channels.

THE GOLDEN ZONE Commercial mobile services operate in the high megahertz to low gigahertz range (approximately 500 MHz to 10 GHz). This is the **golden zone.** At lower frequencies, the spectrum is limited and has been almost entirely assigned. At higher frequencies, radio waves attenuate very rapidly with distance because of absorptive attenuation and cannot flow through or around objects as they do at lower frequencies. Consequently, the sender and receiver must have a **clear line of sight** (unobstructed

---

[1] Claude Shannon, "A Mathematical Theory of Communication," *Bell System Technical Journal* (July 1938), pp. 379–423, and (October 28, 1938), pp. 623–56.

direct path) between them. Even at the high end of the golden zone, absorption and shadow zone propagation problems are large. The golden zone is limited, and demand for channels and service bands in the golden zone is increasing rapidly. Consequently, there is strong competition for bandwidth in the golden zone.

The golden zone for commercial mobile services is 500 MHz to 10 GHz.

Test Your Understanding

7. a) Distinguish among the frequency spectrum, service bands, and channels. b) In radio, how can you send multiple signals without the signals interfering with one another?
8. a) Does a signal usually travel at a single frequency, or does it spread over a range of frequencies? b) What is channel bandwidth? c) If the lowest frequency in a channel is 1.22 MHz and the highest frequency is 1.25 MHz, what is the channel bandwidth? (Use proper metric notation.) d) Why is large channel bandwidth desirable? e) What do we call a system whose channels have large bandwidth? f) What other types of system do we call *broadband*?
9. a) Write the Shannon Equation. List what each letter is in the equation. b) What information does *C* give you? c) What happens to the maximum possible propagation speed in a channel if the bandwidth is tripled while the signal-to-noise ratio remains the same? d) Given their relative bandwidths, about how many times as much data is sent per second in television than in AM radio? (The information to answer this question is in the text.)
10. a) What is the golden zone in commercial mobile radio transmission? b) Why is the golden zone important? c) What is a clear line-of-sight limitation?

Licensed and Unlicensed Radio Bands

If two radio hosts transmit at the same frequency, their signals will interfere with each other. In the terminology of Chapter 5, this is electromagnetic interference. To prevent such chaos, governments regulate how radio transmission is used. The International Telecommunications Union, which is a branch of the United Nations, creates worldwide rules that define service bands and specify how individual radio service bands are to be used. Individual countries enforce these rules but are given discretion over how to implement controls.

LICENSED RADIO BANDS In **licensed radio bands,** stations must have a license to operate. They also need a license change if their antennas are moved. Television bands are licensed bands, as are AM and FM radio bands. Government agencies control who may have licenses. By doing so, the government limits interference to an acceptable level. In some licensed bands, the rules allow *mobile hosts* to move about while only central antennas are regulated. This is the case for mobile telephones.

UNLICENSED RADIO BANDS However, for companies that have wireless access points and mobile computers, even the requirement to license central antennas (in this case, access points) is an impossible burden. Consequently, the government has created a few **unlicensed radio bands.** In these bands, any wireless host can be turned on or moved around without the need for any government approval.

Licensed Radio Bands

If two nearby radio hosts transmit in the same channel, their signals will interfere

Most radio bands are licensed bands, in which hosts need a license to transmit

The government limits licenses to reduce interference

Television bands, AM radio bands, etc., are licensed

In cellular telephone bands, which are licensed, only the central transceivers are licensed, not the mobile phones

Unlicensed Radio Bands

Some bands are set aside as unlicensed bands

Hosts do not need to be licensed to be turned on or moved

802.11 operates in unlicensed radio bands

This allows access points and hosts to be moved freely

However, there is no way to stop interference from other nearby users

Your only recourse is to negotiate with others

At the same time, you may not cause unreasonable interference—for instance, by transmitting at excessive power

FIGURE 7-11 Licensed and Unlicensed Radio Bands (Study Figure)

The problem with unlicensed radio bands is that users of unlicensed radio bands must tolerate interference from others. If your neighbor sets up a wireless LAN next door to yours, you have no recourse but to negotiate with him or her over such matters as which channels each of you will use. At the same time, the law prevents you from creating unreasonable interference—for instance, by using illegally high transmission power.

Test Your Understanding

11. a) Do WLANs today use licensed or unlicensed bands? b) What is the advantage of using unlicensed bands? c) What is the disadvantage?

802.11 in the 2.4 GHz and 5 GHz Unlicensed Bands

It would be impossible for a company to have licenses for all of its access points and wireless hosts, so 802.11 operates in unlicensed radio bands. More specifically, WLANs today use two unlicensed bands. One is the 2.4 GHz band. The other is the 5 GHz band.

THE 2.4 GHZ UNLICENSED BAND The 2.4 GHz unlicensed band is the same in most countries in the world, stretching from 2.40 GHz to 2.4835 GHz. This commonality allows companies to sell generic 2.4 GHz radios, driving down the price of radios. In addition, radio propagation is better in the 2.4 GHz unlicensed band than in the higher-frequency 5 GHz band.

Unfortunately, the 2.4 GHz band is very limited. It has only 83.5 MHz of bandwidth. Traditionally, each 802.11 channel was 20 MHz wide, although 40 MHz bandwidth channels were introduced in 802.11n. Furthermore, due to the way channels are

The 2.4 GHz Unlicensed Band

Defined the same in almost all countries (2.400 GHz to 2.485 GHz)

Commonality reduces radio costs

Propagation characteristics are good

For 20 MHz 802.11 channels, only three nonoverlapping channels are possible

Channels 1, 6, and 11

This creates mutual channel interference between nearby access points transmitting in the same 20 MHz channel

Difficult or impossible to put nearby access points on different channels (Figure 7-5)

Also, potential problems from microwave ovens, cordless telephones, etc.

The 5 GHz Unlicensed Band

Radios are expensive because frequencies in different countries are different

Shorter propagation distance because of higher frequencies

Deader shadow zones because of higher frequencies

More bandwidth, so between 11 and 24 nonoverlapping channels

Allows different access points to operate on nonoverlapping channels

Some access points can operate on two channels to provide faster service

FIGURE 7-12 802.11 in the 2.4 GHz and 5 GHz Unlicensed Bands (Study Figure)

allocated, there are only three possible nonoverlapping 20 MHz 802.11 channels, which are centered at Channels 1, 6, and 11.[2] If nearby access points operate in the same channel, their signals will interfere with each other unless the access points are far apart. This interference is called **mutual channel interference.** If an 802.11n station finds itself in a crowded area, it will drop back to 20 MHz to reduce the interference it causes.

If you have only three access points that can all hear each other, there is no problem with having only three channels. You simply run each on a different channel, and there will be no interference. However, when you have multiple access points that can all hear each other, Figure 7-13 shows that there is no way to avoid having some mutual channel interference. You can minimize mutual channel interference somewhat by giving the shared channel to the two access points that are farthest apart, but this will only reduce interference somewhat.

In addition, the frequencies used in the 2.4 GHz band overlap the frequencies used in microwave ovens, cordless telephones, and Bluetooth equipment. This results in occasional interference that is difficult to diagnose.

THE 5 GHz UNLICENSED BAND The 802.11 standard can also operate in the 5 GHz band. There have been two problems with this band. The first is the cost of radios. In the 2.4

[2] Channel numbers were defined for the 2.4 GHz band when channels were narrower. A 20 MHz 802.11 channel overlaps several defined channels. Channels 1, 6, and 11 operate in the 2402 MHz to 2422 MHz, 2427 MHz to 2447 MHz, and 2452 MHz to 2472 MHz frequency ranges, respectively. Note that there are 5 MHz unused "guard bands" between the channels to prevent inter-channel interference.

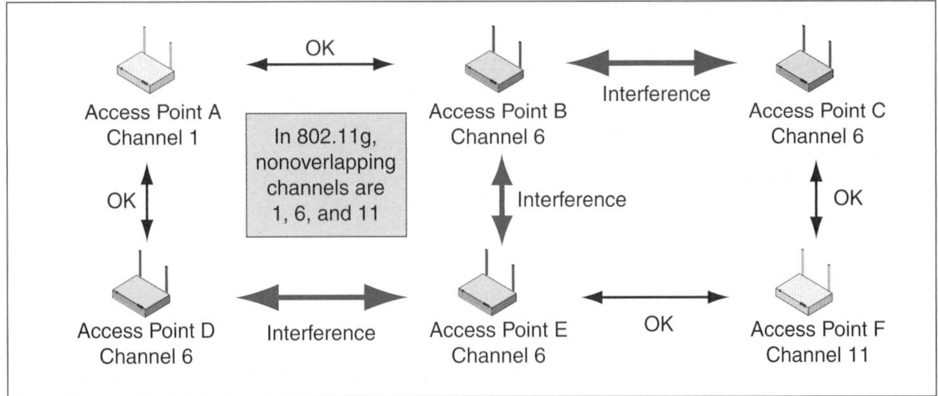

FIGURE 7-13 Mutual Interference in the 2.4 GHz Unlicensed Band

GHz band, high sales have allowed manufacturers to ride the learning curve down to lower costs. In addition, while the 2.4 GHz band is standardized throughout most of the world, different countries use different parts of the 5 GHz band. This also makes 5 GHz radios more expensive due to the need to use only those channels permitted within a given country.

Second, because of the 5 GHz band's higher frequencies, signals do not travel as far, and shadow zones are darker. This means that access points have to be placed closer together. It also means that siting access points to avoid dead spots is more difficult.

The big advantage of the 5 GHz band is that it is much wider than the 2.4 GHz band. In contrast to the 2.4 GHz band's mere three channels, the 5 GHz band provides between 11 and 24 nonoverlapping 20 MHz channels, depending on the frequencies allocated to unlicensed operation in a country.

Having many channels eliminates the mutual channel interference problem because it is easy to assign noninterfering channels to access points even in multi-floor buildings (which introduce interference in three dimensions).

In addition, in the sparsely used 5 GHz service bands, some access points can operate simultaneously on two different channels. This doubles the amount of bandwidth available to devices.

The disadvantage of operating in the 5 GHz band is higher radio cost because 5 GHz radios are inherently more expensive and because fewer of them are being produced, raising the cost per unit. However, this cost difference is rather small, and 5 GHz access points and clients are now available on retail shelves at only a modest premium.

Test Your Understanding

12. a) In what two unlicensed bands does 802.11 operate? b) How wide are 802.11 channels usually? c) Which licensed band is defined the same way in most countries around the world? d) Does the 2.4 GHz band or the 5 GHz band allow longer propagation distances for a given level of power? Justify your answer. e) How many nonoverlapping channels does the 2.4 GHz band support? f) Why is the number of nonoverlapping channels that can be used important? g) How many nonoverlapping channels does the 5 GHz band support?

NORMAL AND SPREAD SPECTRUM TRANSMISSION

Why Spread Spectrum Transmission?

At the frequencies used by WLANs, there are numerous and severe propagation problems. In these unlicensed bands, regulators mandate the use of a form of transmission called spread spectrum transmission. Spread spectrum transmission is transmission that uses far wider channels than transmission speed requires.

> Spread spectrum transmission is transmission that uses far wider channels than transmission speed requires.

Regulators mandate the use of spread spectrum transmission primarily to minimize propagation problems—especially multipath interference. (If the direct and reflected signals cancel out at some frequencies within the range, they will be double at other frequencies.)

In commercial transmission, security is *not* a reason for doing spread spectrum transmission. The military uses spread spectrum transmission for security, but it does so by keeping certain parameters of its spread spectrum transmission secret. Commercial spread spectrum transmission methods must make these parameters publicly known in order for two parties to communicate easily.

> In wireless LANs, spread spectrum transmission is used to reduce propagation problems and to reduce mutual interference between nearby hosts transmitting in the same channel, not to provide security.

Spread Spectrum Transmission

- You are required by law to use spread spectrum transmission in unlicensed bands
- Spread spectrum transmission reduces propagation problems
    - Especially multipath interference
- Spread spectrum transmission is NOT used for security in WLANs

Normal Transmission versus Spread Spectrum Transmission (See Figure 7-15)

- Normal transmission uses only the channel bandwidth required by your signaling speed
- Spread spectrum transmission: uses channels much wider than signaling speed requires

Orthogonal Frequency Division Multiplexing (See Figure 7-16)

- OFDM is the dominant spread spectrum transmission method today
- The sender divides the channel into multiple subchannels called carriers
- Part of the frame is sent in each carrier
- The frame is sent redundantly, so if some carriers are lost, the frame is still likely to get through
- It is difficult to transmit in a very wide channel
- It is much easier to transmit in many smaller-bandwidth subcarriers

FIGURE 7-14 Spread Spectrum Transmission (Study Figure)

How wide are spread spectrum channels? Earlier, we saw that 802.11 channel bandwidth was traditionally 20 MHz and may be twice as wide for the 802.11n standard.

Test Your Understanding

13. a) In unlicensed bands, what type of transmission method is required by regulators? b) What is the benefit of spread spectrum transmission for business communication? c) Is spread spectrum transmission done for security reasons in commercial WLANs?

Spread Spectrum Transmission Methods

NORMAL VERSUS SPREAD SPECTRUM TRANSMISSION As noted earlier in our discussion of the Shannon Equation, if you need to transmit at a given speed, you must have a channel whose bandwidth is sufficiently wide.

To allow as many channels as possible, channel bandwidths in *normal radio transmission* are limited to the speed requirements of the user's signal, as Figure 7-15 illustrates. For a service that operates at 10 kbps, regulators would permit only enough channel bandwidth to handle this speed.

In contrast to normal radio transmission, which uses channels just wide enough for transmission speed requirements, **spread spectrum transmission** takes the original signal, called a **baseband signal,** and spreads the signal energy over a much broader channel than is required.

ORTHOGONAL FREQUENCY DIVISION MULTIPLEXING (OFDM) There are several spread spectrum transmission methods. The 801.11 Working Group's most recent standards, 802.11g and 802.11n, use **orthogonal frequency division multiplexing (OFDM),** which Figure 7-16 illustrates.

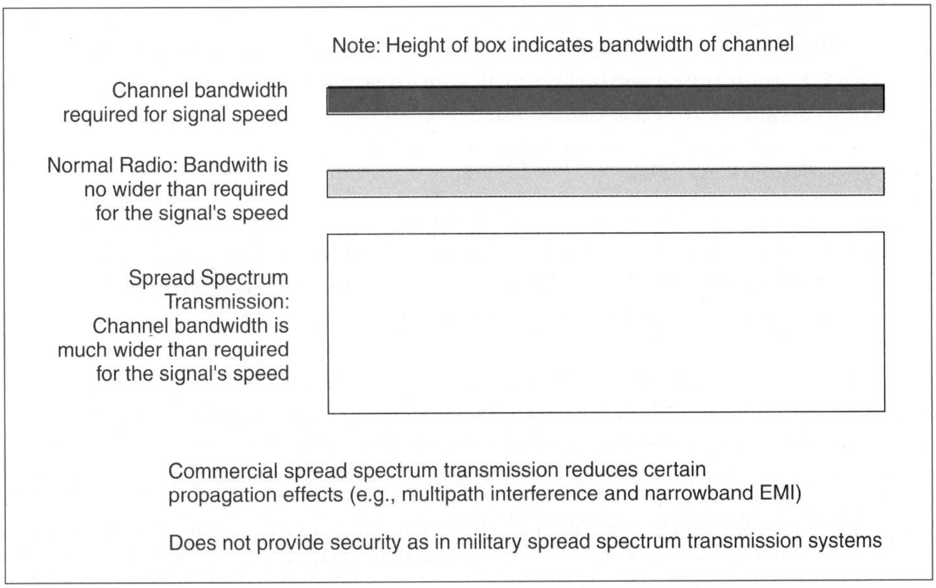

Note: Height of box indicates bandwidth of channel

Channel bandwidth required for signal speed

Normal Radio: Bandwith is no wider than required for the signal's speed

Spread Spectrum Transmission: Channel bandwidth is much wider than required for the signal's speed

Commercial spread spectrum transmission reduces certain propagation effects (e.g., multipath interference and narrowband EMI)

Does not provide security as in military spread spectrum transmission systems

FIGURE 7-15 Normal Radio Transmission and Spread Spectrum Transmission

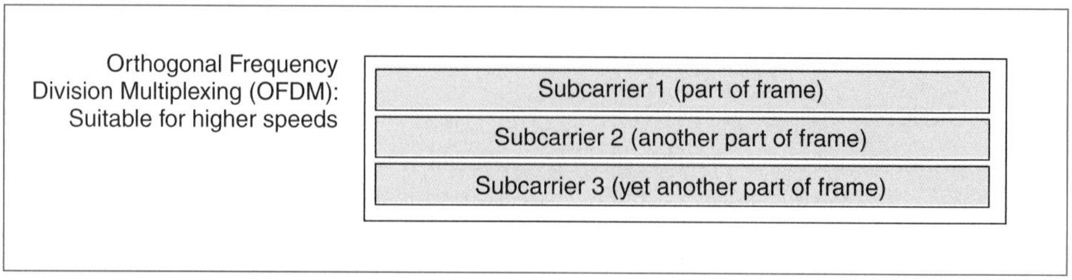

Orthogonal Frequency Division Multiplexing (OFDM): Suitable for higher speeds	Subcarrier 1 (part of frame)
	Subcarrier 2 (another part of frame)
	Subcarrier 3 (yet another part of frame)

FIGURE 7-16 Orthogonal Frequency Division Multiplexing (OFDM)

In OFDM, each broadband channel is divided into many smaller subchannels called **subcarriers.** Parts of each frame are transmitted in each subcarrier.[3] OFDM sends data redundantly across the subcarriers, so if there is impairment in one or even a few subcarriers, all of the data usually will still get through.

OFDM is complex and therefore expensive. However, sending data over a single very large channel reliably is difficult. In contrast, OFDM can be used at very high speeds because it is easier to send many slow signals reliably in many small subcarriers than it is to send one signal rapidly over a very wide-bandwidth channel.[4]

Test Your Understanding

14. a) In normal radio operation, how does channel bandwidth usually relate to the bandwidth required to transmit a data stream of a given speed? b) How does this change in spread spectrum transmission?

15. a) What spread spectrum transmission method is used for the most recent 802.11 standards? b) Describe it.

802.11 WLAN OPERATION

As noted at the beginning of this chapter, wireless LANs replace signals in copper wires or optical fiber with radio waves. WLANs allow mobile workers to stay connected to the network as they move through a building. In some cases, wireless LANs are less expensive to install than wired LANs, but this certainly is not always the case.

Extending the Wired LAN

As noted at the start of the chapter, and as Figure 7-17 shows, an 802.11 wireless LAN typically is used to connect a small number of mobile devices to a large wired LAN—typically, an Ethernet LAN—because the servers and Internet access routers that mobile hosts need to use usually are on the wired LAN.[5]

---

[3] In the 802.11g wireless LAN standard discussed later, each 20 MHz channel is divided into 52 subcarriers, each 312.5 kHz wide. Of the 52 subcarriers, 48 are used to send data and 4 are used to control the transmission.

[4] The ADSL services discussed in Chapter 6 generally also use OFDM, although in ADSL service, OFDM is called discrete multitone (DMT) service.

[5] There is a rarely used 802.11 *ad hoc mode,* in which no wireless access point is used. In ad hoc mode, computers communicate directly with other computers. (In contrast, when an access point is used, this is called 802.11 infrastructure mode.) In addition, 802.11 can create point-to-point transmission over longer distances than 802.11 normally supports. This approach, which normally is used to connect nearby buildings, uses dish antennas and higher power levels authorized for this purpose.

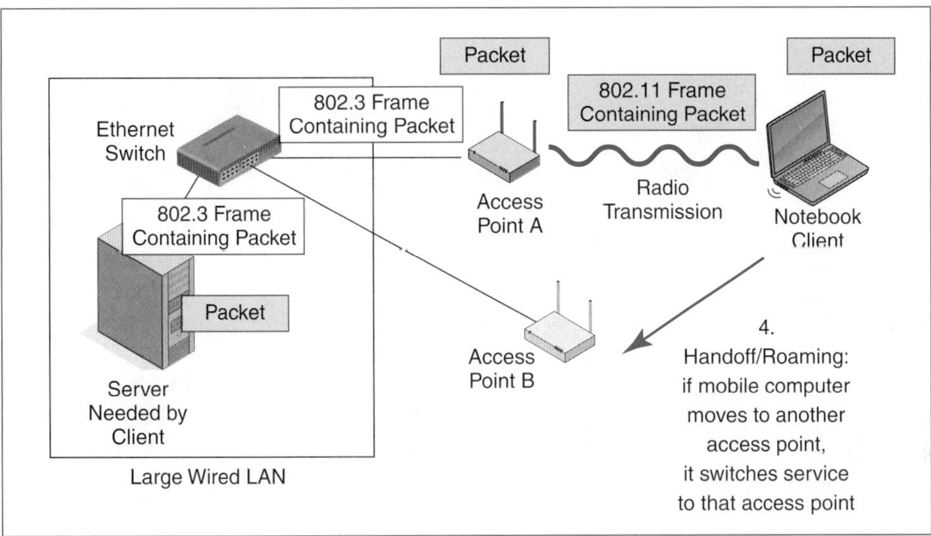

FIGURE 7-17 Typical 802.11 Wireless LAN Operation with Wireless Access Points

Wireless Access Points

When a wireless host wishes to send a frame to a server, it transmits the frame to a wireless access point.

As Figure 7-17 shows, when a wireless host transmits to a server on the wired LAN, it places the packet into an 802.11 frame.[6] The wireless access point removes the packet from the 802.11 frame and places the packet in an 802.3 frame. The access point sends this 802.3 frame to the server, via the wired Ethernet LAN. When the server replies, the wireless access point receives the 802.3 frame, removes the packet from the frame, and forwards the packet to the wireless host in an 802.11 frame.[7]

> The packet goes all the way from the wireless host to a server. The 802.11 frame only travels between the wireless host and the wireless access point. The 802.3 frame only travels between the wireless host and the server.

The wireless access point also controls hosts. It assigns transmission power levels to hosts within its range and performs a number of other supervisory chores.

Handoff/Roaming

When a mobile host travels too far from a wireless access point, the signal will be too weak to reach the access point. However, if there is a closer access point, the host can be **handed off** to that access point for service. In WLANs, the ability to use handoffs is also called **roaming.**[8]

---

[6] Note that 802.11 frames are much more complex than 802.3 Ethernet frames. Much of this complexity is needed to counter wireless propagation problems.

[7] This sounds like what a router does. However, a router can connect any two single networks. Access points are limited to connecting 802.3 and 802.11 networks.

[8] In cellular telephony, which we will see in the next chapter, the terms *handoff* and *roaming* mean different things.

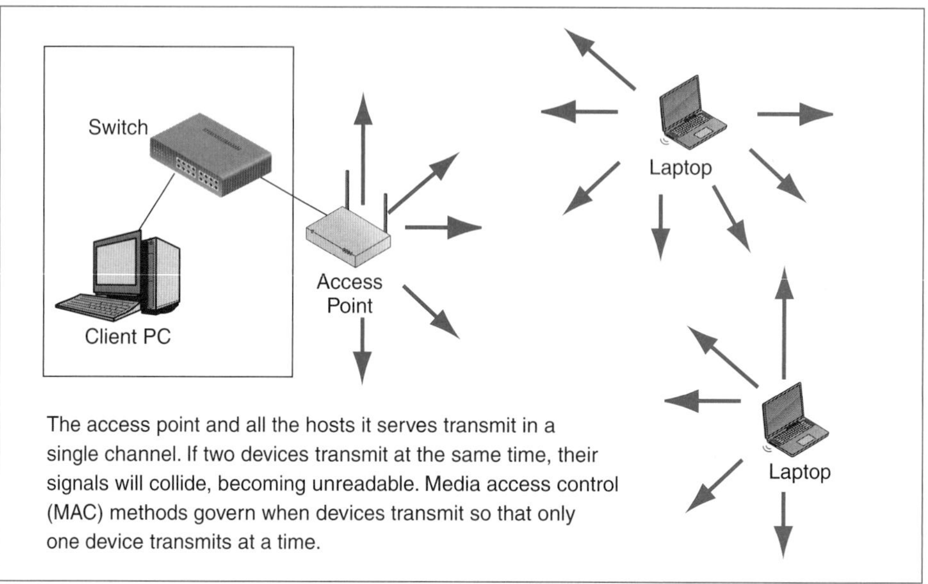

FIGURE 7-18 Hosts and Access Points Transmit in a Single Channel

Sharing a Single Channel

As Figure 7-18 shows, the access point and all of the wireless hosts it serves transmit and receive in a single channel. When a host or the access point transmits, all other devices must wait. (If two devices transmit in the same channel at the same time, their signals will interfere with each other.) As the number of hosts served by an access point increases, individual throughput falls because of this waiting. The box "Controlling 802.11 Transmission" discusses how **media access control** methods govern when hosts and access points may transmit so that collisions can be avoided.

> The access point and all of the wireless hosts it serves transmit and receive in a single channel. When a host or the access point transmits, all other devices must wait.

Test Your Understanding

16. a) List the elements in a typical 802.11 LAN today. b) Why is a wired LAN usually still needed if you have a wireless LAN? c) Why must the access point remove an arriving packet from the frame in which the packet arrives and place the packet in a different frame when it sends the packet back out? d) What is a handoff in 802.11? e) What is the relationship between handoffs and roaming in WLANs? f) When there is an access point and several wireless hosts, why may only one device transmit at a time? g) All wireless hosts and the access point that serves them transmit on the same channel. Why does this cause throughput to fall as the number of wireless hosts increases?

Controlling 802.11 Transmission

Media Access Control

As noted in the body of the text, the access point and the hosts it serves all transmit in the same channel. If two 802.11 devices (hosts or wireless access points) transmit at the same time, their signals will be jumbled together and will be unreadable. This is called a **collision.**

The 802.11 standard has two mechanisms for **media access control**—ensuring that hosts and the access point do not transmit simultaneously. The first, CSMA/CA+ACK, is mandatory. Access points and wireless hosts *must* support it. The second, RTS/CTS, is optional.[9]

Test Your Understanding

17. a) What is a collision? b) Why is it bad? c) What is the purpose of media access control? d) Does media access control limit the actions of wireless hosts, the access point, or both?

CSMA/CA+ACK Media Access Control

CSMA/CA

To reduce the number of collisions, wireless access points and wireless hosts use **carrier sense multiple access with collision avoidance (CSMA/CA).** Note the focus on collision *avoidance.* Figure 7-19 illustrates CSMA/CA.

CSMA requires that a host refrain from transmitting if it hears traffic. This is a very simple rule. Carrier sensing means listening for traffic. Multiple access means that this is a way of controlling how multiple hosts can access the network to transmit.

CSMA/CA (Carrier Sense Multiple Access with Collision Avoidance)

Sender listens for traffic

1. If there is traffic, waits
2. If there is no traffic:
   - 2a. If there has been no traffic for less than the critical time value, waits a random amount of time, then returns to Step 1.
   - 2b. If there has been no traffic for more than the critical value for time, sends without waiting.
   - This avoids collision that would result if hosts could transmit as soon as one host finishes transmitting.

ACK (Acknowledgment)

Receiver immediately sends back an acknowledgment.

If sender does not receive the acknowledgment, retransmits using CSMA.

CSMA/CA plus ACK is a reliable protocol.

FIGURE 7-19 CSMA/CA+ACK in 802.11 Wireless LANs

(*continued*)

---

[9] Actually, if you have even a single host with 802.11b equipment connected to an access point, RTS/CTS becomes mandatory. However, 802.11b wireless hosts are almost never encountered anymore.

CSMA requires not transmitting if a device hears traffic. Collision avoidance (CA) is a set of two rules that determine what a host or the access point does if it does *not* hear traffic.

- If the host does not hear traffic, it considers the last time it heard traffic. If the time since the last transmission exceeds a critical value, the host may transmit immediately.
- However, if the time is less than the critical value, the host sets a random timer and waits. If there still is no traffic after the random wait, the host may send.

If the last two points seem odd, note that the goal is to avoid collisions as much as possible. Two hosts are most likely to transmit at the same time if they both have been waiting for another host to finish transmitting. Without the random delay, both will transmit at the same time, causing a collision.

ACK

More specifically, 802.11 uses **CSMA/CA+ACK.** Collisions and other types of signal loss are still possible with CSMA/CA. When a wireless access point receives a frame from a host, or when a host receives a frame from an access point, the receiver *immediately* sends an acknowledgment frame, an **ACK.** A frame that is not acknowledged is retransmitted. Note that there is no wait when transmitting an ACK. This ensures that ACKs get through while other hosts are waiting.

Note also that retransmission makes CSMA/CA+ACK a reliable protocol. We saw in Chapter 2 that very few protocols are reliable because reliability usually costs more than it brings in benefits. The low error rates in wired media simply do not justify implementing reliability in Ethernet and other wired LAN protocols. However, wireless transmission has many errors, so a reliable protocol is required for reasonably good operation.[10]

Thanks to CSMA/CA+ACK, 802.11 is a reliable protocol.

Inefficient Operation

CSMA/CA+ACK works well, but it is inefficient. Waiting before transmission wastes valuable time. Sending ACKs also is time consuming. Overall, an 802.11 LAN can only deliver throughput (actual speed) of about half the rated speed of its standard—that is, the speed published in the standard.

This throughput, furthermore, is aggregate throughput shared by the wireless access point and all of the hosts sharing the channel. Individual host throughput will be substantially lower.

Test Your Understanding

18. a) Describe CSMA/CA+ACK. Do not go into detail about how long a host must wait to transmit if there is no traffic. b) Is CSMA/CA+ACK transmission reliable or unreliable? Explain. c) Why is CSMA/CA+ACK inefficient?

Request to Send/Clear to Send (RTS/CTS)

Although CSMA/CA+ACK is mandatory, there is another control mechanism called **request to send/clear to send (RTS/CTS).** Figure 7-20 illustrates RTS/CTS. As noted earlier, the RTS/CTS protocol is optional except in one rare case. Avoiding RTS/CTS whenever possible is wise because RTS/CTS is much less efficient, and therefore slower, than CSMA/CA+ACK.

---

[10] In addition, 802.11 uses forward error correction. It adds many redundant (extra) bits to each frame. If there is a small error, the receiver can use these redundant bits to fix the frame. If the receiver can make the repair, it does so and then sends back an ACK. This process makes wireless NICs more expensive than Ethernet NICs, but wireless transmission errors are so common that it makes economic sense to correct errors at the receiver in order to minimize retransmissions.

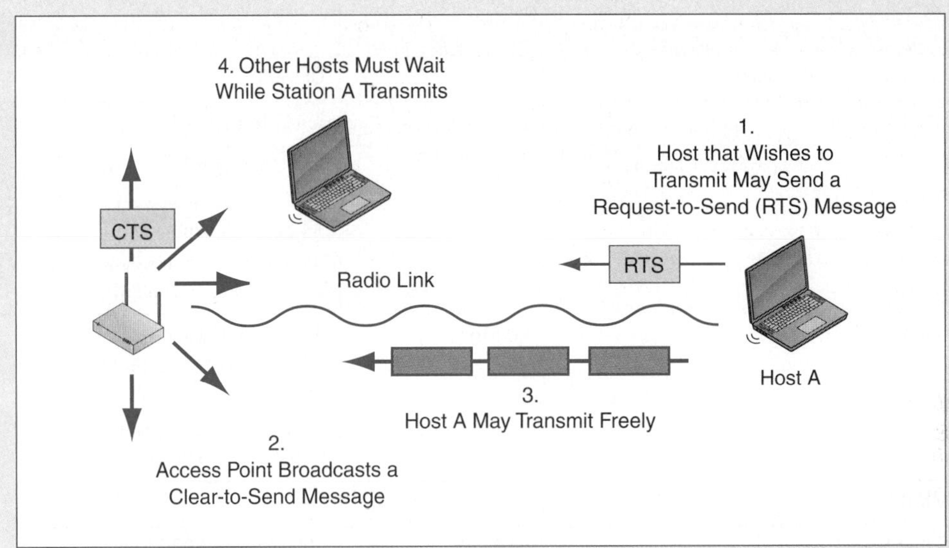

FIGURE 7-20 Request to Send/Clear to Send (RTS/CTS)

When a host wishes to send and is able to send because of CSMA/CA, the host may send a **request-to-send (RTS)** message to the wireless access point. This message asks the access point for permission to send messages.

If the access point responds by broadcasting a **clear-to-send (CTS)** message, then other hosts must wait. The host sending the RTS may then transmit, ignoring CSMA/CA.

Although RTS/CTS is widely used, keep in mind that it is only an option, while CSMA/CA is mandatory. Also, tests have shown that RTS/CTS reduces throughput when it is used.

RTS/CTS makes sense primarily when two wireless clients can both hear the access point but cannot hear each other. If CSMA/CD+ACK is used, the two stations may transmit at the same time.

Test Your Understanding

19. a) Describe RTS/CTS. b) Is CSMA/CA+ACK required or optional? c) Is RTS/CTS required or optional? d) Which is more efficient, RTS/CTS or CSMA/CA+ACK? e) When does RTS/CTS make sense to use?

802.11 TRANSMISSION STANDARDS

The 802.11 Working Group has created several WLAN transmission standards since 1997. However, 802.11g products and the newer 802.11n products dominate in today's marketplace. The **802.11g** standard dominates the installed base due to its past sales. The faster **802.11n** standard dominates sales today. Given its large installed base of devices, however, 802.11g will be important for many years to come.

Test Your Understanding

20. a) Which 802.11 standard dominates the largest installed base? b) Which 802.11 standard dominates sales today?

Characteristic	802.11g	802.11n
Spread Spectrum Method	OFDM	OFDM
Unlicensed Band	2.4 GHz	2.4 GHz and 5 GHz
Number of Nonoverlapping Channels (varies by country)**	3	3 in 2.4 GHz band and 12 to 24 in most countries in 5 GHz band So 802.11n has far more channels to operate on, especially in the uncrowded 5 GHz band
Channel bandwidth	20 MHz	40 MHz, but will drop back to 20 MHz if there is interference, on the two selected channels This approximately doubles transmission speed.
MIMO?	No	Yes
Rated Speed	54 Mbps	300 Mbps to 600 Mbps
Actual Throughput, 3 meters	25 Mbps	100 Mbps or more
Actual Throughput, 30 meters	20 Mbps	Less reduction at longer distances
Provides lower speed modes for stations that are far away	Yes	Yes
Distant stations transmit more slowly so decrease the individual throughput of all other devices using the access point.	Yes	Yes
Is throughput shared by all stations using an access point?	Yes	Yes
Backward-compatible with 802.11g?	NA	Yes
Remarks	Dominates today's installed base	Dominates sales today

FIGURE 7-21 The 802.11g and 802.11n Standards

Characteristics of 802.11g and 802.11n

Figure 7-21 compares the 802.11g and 802.11n standards across a number of important dimensions.

SPREAD SPECTRUM TRANSMISSION STANDARDS The figure shows the technology used in each standard. Both use OFDM.

SPREAD SPECTRUM TRANSMISSION IN UNLICENSED RADIO BANDS The 802.11g standard can only be used in the crowded 2.4 GHz band, which only has three channels.

In contrast, 802.11n can be used in both the 2.4 GHz band and the 5 GHz band. Operation in the 5 GHz band gives several times the number of channels that transmission in the lower band does. In addition, the 5 GHz band is less crowded, so it is usually possible to pick a channel that is not being used by a nearby access point.

SPEED

Bandwidth and MIMO. The 802.11n standard provides much faster service than 802.11g for two reasons. First, as noted earlier, its channel bandwidth is twice as wide (40 MHz as opposed to 20 MHz). This should provide a rough doubling in speed. Also, 802.11n uses MIMO. As we will see later, MIMO can bring an even larger increase in speed that doubling the bandwidth does.

Rated Speeds and Throughput. Figure 7-21 shows rated speed and throughput for 802.11 transmission standards.

- The figure shows that the *rated speeds* of 802.11g and 802.11n lie between 11 Mbps and 54 Mbps for 802.11g and up to 600 Mbps for 802.11n.
- As Figure 7-21 also shows, actual *throughput* usually is considerably lower than rated speeds and falls off rapidly with distance for 802.11g.
- Furthermore, this 802.11 throughput is *aggregate throughput* shared by all hosts that wish to transmit at the same time. For instance, if the aggregate shared throughput is 20 Mbps, hosts using a wireless access point serving 10 to 20 hosts might see individual throughput of only 2 to 5 Mbps, despite the fact that only a few hosts are likely to be transmitting at any given moment.

Transmission Modes and Distance. Each 802.11 transmission standard has multiple modes that operate at different speeds. Figure 7-21 lists the highest-speed mode for each 802.11 standard. When a host is near an access point, it normally can use the highest-speed mode. However, as a computer moves away from an access point, the highest-speed mode begins producing errors. The computer and access point will both switch to a lower-speed mode to reduce transmission errors. As a computer moves even farther from an access point, further drops in transmission mode will occur.

Of course, if a distant station transmits more slowly, it will take longer to send its transmissions. Other stations will have to wait longer, so they will experience greater loss in throughput.

BACKWARD COMPATIBILITY WITH 802.11G Due to the large installed base of 802.11g devices in business, the 802.11 Working Group ensured that 802.11n access points and network interface cards were created to be backwardly compatible with 802.11g equipment. This means that when an 802.11n wireless host nears an 802.11g access point, the host will be able to work with the access points, although it will only operate at 802.11g speeds and distances.

Test Your Understanding

21. a) In what radio bands can 802.11g and 802.11n operate? b) Why is the ability to operate in the 5 GHz band good?

22. a) Compare the rated speeds of 802.11g and 802.11n. b) For what two reasons is 802.11n faster than 802.11g? c) Distinguish between rated speed, aggregate throughput, and individual throughput in 802.11 WLANs. d) Why does transmission speed drop as a computer moves farther from an access point? e) How does the presence of a distant station harm all users of an access point?

23. a) Can an 802.11n wireless host use an 802.11g access point? b) If it does, will it get 802.11n speeds? Explain. c) Can an 802.11g wireless host use an 802.11n access point? The question is not in the text, but you should be able to answer this. d) If it does, will it get 802.11n speeds? Explain.

802.11n and MIMO

Figure 7-21 notes that 802.11n uses something called MIMO to achieve speeds far beyond the doubling that doubling the bandwidth brings. Figure 7-22 illustrates **multiple input/multiple output (MIMO)** transmission.

The key to higher throughput in MIMO is that the host or access point sends two or more **spatial streams** (radio signals) in the same channel between two or more different antennas on access points and wireless hosts.

In the figure, the two spatial streams, being in the same channel, should interfere with each other. However, the two spatial streams sent by different antennas will arrive at the two receiving antennas at slightly different times. Using special detection and separation methods based on differences in arrival times for the two spatial streams, the receiver can separate the two spatial streams in the same channel and so can read them individually.

Even with only spatial streams and two antennas each on the sender and receiver, MIMO can substantially increase throughput. Using more spatial streams and more antennas can increase throughput even more. The MIMO standard will permit two, three, or four antennas on each device and up to four data streams.

With two spatial streams, the maximum speed is 300 Mbps, and effective throughput is about 100 Mbps, depending on conditions. Three spatial streams can raise the maximum speed to 450 Mbps, and four can raise it to 600 Mbps. Today, vendors only support two spatial streams, but this gives the throughput that users typically receive from UTP to the desktop.

As the number of spatial streams increases, the number of sending and receiving antennas must go up. However, there is no simple relationship between the number of spatial streams and the required number of antennas.

Of equal importance, MIMO substantially increases propagation distance. It will provide better coverage at current 802.11 distances. It will also allow access points to be

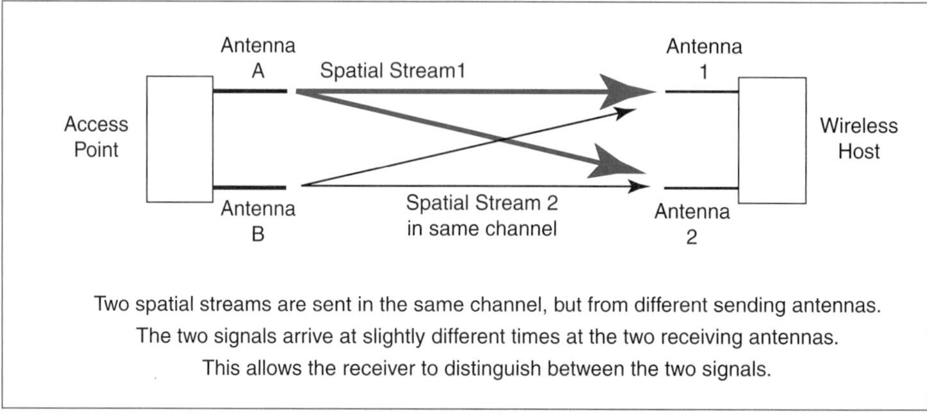

FIGURE 7-22 Multiple Input/Multiple Output (MIMO) Transmission

placed farther apart, reducing the number of access points needed to serve a building and therefore reducing costs.

MIMO brings both higher speeds and longer transmission distances.

In Figure 7-22, there are two antennas on each device. The 802.11n standard allows up to four antennas on each device. Most products today have two or three antennas. Even with only two data streams, reception improves with the number of receiving antennas.

Test Your Understanding

24. a) How does MIMO work? b) With two spatial streams, what is the maximum transmission speed in 802.11n? c) Under these conditions, what is typical throughput? d) What are the two benefits of MIMO? e) If you have two spatial streams, is it better to have three receiving antennas than two?

Faster Speeds

Even faster 802.11 speeds are coming. The **802.11ac** standard now under development will build on 802.11n in the 5 GHz band by improving modulation schemes and having even wider channels of 80 MHz to 160 MHz. This will at least quadruple 802.11n speeds.

In turn, the **802.11ad** standard now under development will operate in the 60 GHz band rather than the 2.4 GHz or 5 GHz band. The amount of spectrum available at 60 GHz will allow speeds of a few gigabit per second or more. However, absorptive attenuation is extremely high at 60 GHz, so 802.11ad will only be able to provide these speeds over short distances. A typical application may be HDTV transmission in the home.

Test Your Understanding

25. a) How will 802.11ac give higher speeds than 802.11n? b) How will 802.11ad give higher speeds than 802.11n? c) What will be the main limitation of 802.11ad?

Advanced Operation

Two other 802.11 developments are beginning to appear in products in very limited ways. In the future, they may become much more important.

MESH NETWORKING As we saw at the beginning of this chapter, the main purpose of access points today is to connect wireless hosts to the main wired LAN. However, it is possible to have all-wireless networks. As Figure 7-23 shows, it is possible for wireless access points to organize themselves into a mesh, routing frames from one to another to deliver frames to the right wireless hosts. Nonstandard 802.11n mesh networking products exist, and the **802.11s** standard for mesh networks is under development.

However, significant problems face mesh networking. Meshes must be self-organizing. If hosts and access points enter and leave the mesh frequently, the amount of processing power consumed in maintaining the access points' routing tables could make mesh networking prohibitively expensive.

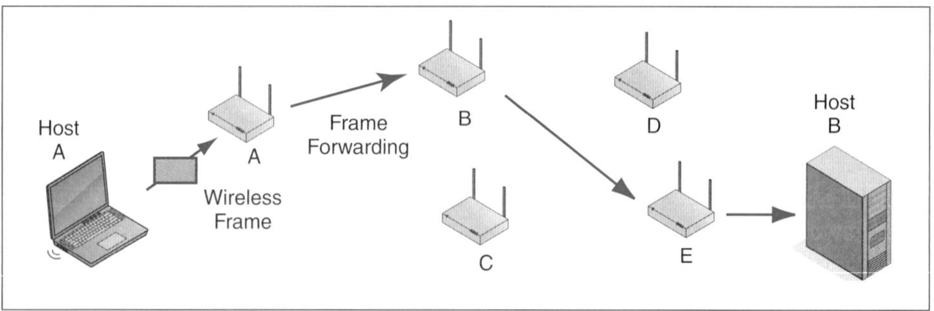

FIGURE 7-23 Wireless Mesh Network

In addition, it will be difficult to avoid access points near the geographical center of the mesh from being overloaded by the need to route many packets. (Think of sitting in the middle seat at a table during a Christmas dinner and constantly having to pass food back and forth.) If mesh networking works, but does not work well, it will have little value.

SMART ANTENNAS Another development is **smart antennas.** By having multiple antennas and changing the phase of waves coming from different antennas, an access point can focus signals toward individual hosts instead of broadcasting them, as Figure 7-24 indicates. This gives more power and therefore a stronger signal to each wireless host. Access points with nonstandard smart antennas are already being sold.

Test Your Understanding

26. a) Describe how mesh networking would work in 802.11 WLANs. b) What two problems mentioned in the text would 802.11 mesh networking designers have to overcome? c) What benefit do smart antennas bring?

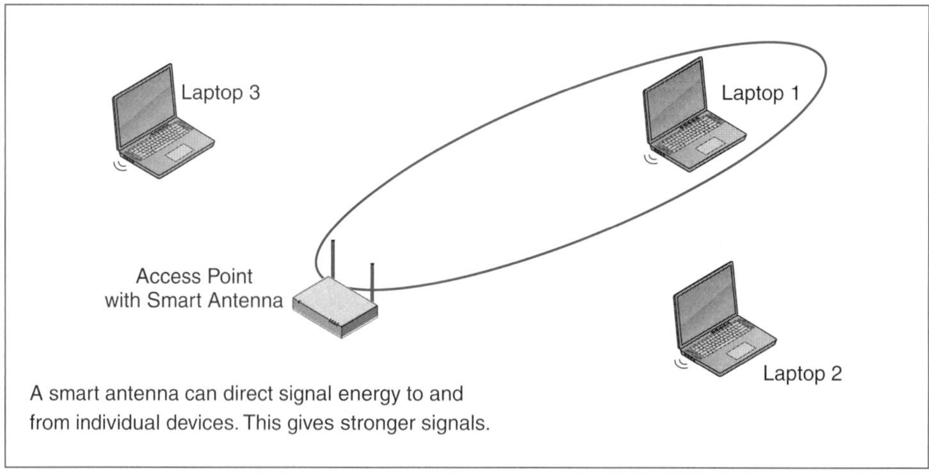

FIGURE 7-24 Smart Antenna

CONCLUSION

Synopsis

Chapters 5 and 6 looked at switched single networks. This chapter and the next look at wireless single networks. All single networks operate at Layers 1 and 2. OSI standards dominate at those layers, so we can expect all wireless network standards to be OSI standards.

This chapter spent a long time on physical layer propagation. This detail is needed because wireless propagation effects are complex. While we can predict what will happen as a signal travels down copper wire or optical fiber, predicting how strong a radio signal will be where a user is located is far more difficult. We looked at five wireless propagation problems: absorptive attenuation, inverse-square-law attenuation (yes, there are two types of attenuation), interference, shadow zones, and multipath interference. Multipath interference occurs when the destination device receives multiple signals, with some coming directly from the radio source and some bouncing off walls, ceilings, and other objects. The direct and reflected signals may interfere with each other, making the direct signal unreadable. Multipath interference is the biggest propagation problem in wireless LANs. Absorptive attenuation and shadow zones become worse at higher frequencies.

We looked at two types of antennas—omnidirectional antennas and dish antennas. Wireless LANs use omnidirectional antennas because users would not know where to point a dish antenna and certainly do not want to carry a dish around. Fixed users may use dishes pointing at a distant radio source to have stronger transmission and reception.

We looked at basic radio concepts, including the frequency spectrum that is all frequencies from 0 Hz to infinity. (Radio propagation is described by frequency, which is measured in hertz.) Service bands are contiguous chunks of the frequency spectrum that are reserved for particular purposes, such as FM radio, television, or police communication. Service bands are divided into channels. Signals are sent in a single channel, and signals in different channels do not interfere with each other. The golden zone falls between 500 MHz and 10 GHz. Most commercial wireless services and corporate WLANs operate in the golden zone.

Radio signals do not propagate at a single frequency. They spread over a range of frequencies, and the spread increases as signal speed increases. Consequently, to carry fast signals, channels must have wide bandwidth. According to the Shannon Equation, doubling bandwidth should double possible signal speed.

Most radio bands are licensed, meaning that you need a government license to operate and a new license every time you move an antenna. Obviously, that would not work with wireless LAN. Consequently, wireless LANs operate in unlicensed bands. In an unlicensed band, you can set up your network any way you wish. However you must tolerate interferences from nearby WLANs built by others.

Most WLAN technology operates in the 2.4 GHz band, in which radio prices are low and reception is good. However, there are only three nonoverlapping channels in this band, so nearby access points often interfere with one other. Some WLAN equipment operates in the 5 GHz unlicensed band, in which there are one or two dozen channels, depending on the country. The 5 GHz band is uncrowded, and the gap between 2.4 GHz prices and 5 GHz prices is narrowing. Consequently, use of the 5 GHz band is beginning to grow rapidly.

In the 2.4 GHz and 5 GHz bands, propagation effects tend to be worse at certain frequencies. This is especially true of multipath interference. Consequently, the government requires the use of spread spectrum transmission, in which the signal is spread far more than it needs to be for its speed. By sending a signal over a much wider range of frequencies, frequency-specific problems tend to be washed out. The two dominant WLAN standards, 802.11g and 802.11n, both use orthogonal frequency division multiplexing (OFDM), in which the channel is broken into much smaller channels called subcarriers. The frame is transmitted redundantly within the subcarriers. WLAN spread spectrum techniques, unlike military spread spectrum techniques, do not provide security.

In 802.11 WLAN operation, access points normally attach to the firm's main wired Ethernet LAN so that wireless clients can access servers and Internet access routers on the wired LAN. When a wireless host transmits, it sends its packet into an 802.11 frame. The access point removes the packet from the 802.11 frame, puts it in an 802.3 frame, and sends the frame to the server or Internet access router. The packet travels all the way; the 802.11 frame does not. We saw that when users move between access points, they can be transferred automatically to the new access point. This is called handoff or roaming.

The access point and the stations it serves all transmit in a single channel. Media access control is needed to ensure that they take turns transmitting so that their signals do not interfere. Wireless hosts that want to transmit often have to wait their turns. This reduces individual throughput. A box described the two main media access control protocols. CSMA/CA+ACK is mandatory on access points. Request to send/clear to send is optional but sometimes useful.

WLAN products on the market follow one of two 802.11 standards. The 802.11g standard allows equipment to be less expensive but has lower speed and distance. The 802.11n standard is much more advanced. Products using 802.11 are more expensive than 802.11g products, but have higher speeds and longer propagation distances. Price differences between 802.11g and 802.11n products are shrinking rapidly. The 802.11n standard now dominates in terms of sales, but there is a much larger installed base of 802.11g products. All 802.11n products are backward-compatible with 802.11g, so there is no problem mixing products from the two standards together, although they will all operate with 802.11g performance.

Figure 7-21 compares 802.11g and 802.11n in detail to show why 802.11n is faster and permits longer distances than 802.11g. It also notes that while 802.11g only operates in the 2.4 GHz band, 802.11n may use both the 2.4 GHz band and the 5 GHz band. However, inexpensive 802.11n products only use the 2.4 GHz band or can only operate in one band at a time.

The 802.11n standard introduces MIMO, which allows a sender to transmit two or more signals simultaneously in a single channel. The receiver can distinguish the two signals by their slightly different arrival times at its two or more antennas. In the future, we will see other advanced features. One will be mesh networking, in which wireless access points route wireless frames by themselves, with no central wired network. Another will be smart antennas, which focus their signal power at a particular wireless customer without needing a dish antenna.

In the next chapter, we will continue to look at 802.11 wireless LANs, focusing on security and management. We will then look at other local wireless technologies, including Bluetooth. We will finish with wireless wide area networks, which span a metropolitan area. The main technologies for these networks are 3G and 4G cellular service and WiMAX.

END-OF-CHAPTER QUESTIONS

Thought Questions

1. Reread the Christopher Lorek case. Are any terms still unfamiliar to you?
2. Telephone channels have a bandwidth of about 3.1 kHz. Do the following in Excel. Cut and paste your analyses into your homework. a) If a telephone channel's signal-to-noise ratio is 1,000 (the signal strength is 1,000 times larger than the noise strength), how fast can a telephone channel carry data? (Check figure: Telephone modems operate at about 30 kbps, so your answer should be roughly this speed.) b) How fast could a telephone channel carry data if the SNR were increased massively, from 1,000 to 10,000? (This would not be realistic in practice.) c) With an SNR of 1,000, how fast could a telephone channel carry data if the bandwidth were increased to 3.5 kHz? Show your work or no credit. d) What did you learn from these three analyses?
3. A friend wants to install a wireless home network. The friend wants to use either 802.11g or 802.11n. What advice would you give your friend?
4. A building has four sides that are each 100 meters, and the building is 100 meters tall. a) If access points have a range of 25 meters, how many access points would you need? b) If access points have a range of 33 meters, how many access points would you need. c) What did you learn from your answers?

Design Question

1. Consider a one-story building that is a square. It will have an access point in each corner and one in the center of the square. *All access points can hear one another*. a) Assign access point channels to the five access points if you are using 802.11g. Try not to have any access points that can hear each other use the same channel. Available channels are 1, 6, and 11. b) Were you able to eliminate interference between access points? c) Repeat the first two parts of the question, this time using 802.11a. Available channels are 36, 40, 44, 48, 53, 56, 60, 64, 149, 153, 157, and 161, but many NICs and access points only support channels below 100.

Troubleshooting Question

1. When you set up an 802.11g wireless access point in your small business, your aggregate throughput is only about 6 Mbps. a) List at least two possible reasons for this low throughput. b) Describe how you would test each. c) Describe what you would do if each proved to be the problem.

Perspective Questions

1. What was the most surprising thing you learned in this chapter?
2. What was the most difficult part of this chapter for you?

CHAPTER 8

Wireless Networks II

LEARNING OBJECTIVES

By the end of this chapter, you should be able to:

- Explain 802.11 WLAN security.
- Explain 802.11 wireless LAN management.
- Describe other local wireless technologies, including Bluetooth, ultrawideband (UWB) transmission, Zigbee, RFIDs, and software-defined radio.
- Differentiate between WLANs and wireless wide area networks.
- Explain cellular data service, including generations of cellular technology.
- Explain WiMAX wireless WAN networks.

TJX

The TJX Companies, Inc. (TJX) is a group of over 2,500 retail stores operating in the United States, Canada, England, Ireland, and several other countries. These companies do business under such names as TJ Maxx and Marshalls. In its literature, TJX describes itself as "the leading off-price retailer of apparel and home fashions in the U.S. and worldwide." With this mission statement, there is strong pressure to minimize costs.

On December 18, 2006, TJX detected "suspicious software" on its computer systems. On December 21, consultants confirmed that an intrusion had actually occurred. The next day, the company informed law enforcement authorities in the United States and Canada. Five days later, the security consultants determined that customer data had been stolen.

The consultants initially determined that the intrusion software had been working for seven months when it was discovered. A few weeks later, the consultants discovered that the company had also been breached several times in 2005. All told, the consultants estimated that 45.7 million customer records had been stolen. This was by far the largest number of personal customer records stolen from any company.

The thieves did not steal these records for the thrill of breaking in or to enhance their reputations among other hackers. They did it to steal information to make fraudulent credit card purchases in the names of the customers whose information had been stolen. These fraudulent purchases based on the stolen information did, in fact, take place.

TJX did not inform customers about the data breach until nearly a month later. The company said that it needed time to beef up its security. The company also said that law enforcement officials had told TJX not to release information about the breach immediately to avoid tipping off the data thieves about the investigation. Of course, the delay also left the customers ignorant of the danger they faced.

How did the breaches occur? When the attackers were caught, it was determined that the data thieves broke into poorly protected wireless networks in some Florida stores to get into the central TJX credit and debit card processing system in Massachusetts.[1] These stores protected their wireless network, but they used the obsolete Wired Equivalent Privacy (WEP) standard instead of newer and better security standards. In Massachusetts, poor firewall protection[2] allowed the data thieves to enter several systems and to install a sniffer that listened to the company's poorly encrypted traffic passing into and out of the processing center. Another problem was that TJX retained some sensitive credit card information that should not have been retained; it is this improperly retained information that the data thieves found valuable.[3]

A number of earlier (and smaller) data breaches had prompted the major credit card companies to create the **Payment Card Industry–Data Security Standard (PCI–DSS)**. Failure to implement PCI–DSS control objectives can result in fines and even the revocation of a company's ability to accept credit card payments.

At the time the data breach was discovered, TJX was far behind in its PCI-DSS compliance program. The company only complied with three of the twelve required control objectives. Internal memos[4] revealed that the company knew that it was in violation of the PCI–DSS requirements, particularly with respect to its weak encryption in retail store wireless networks. However, the company deliberately decided not to move rapidly to fix this problem. In November 2005, a staff member noted prophetically that "saving money and being PCI-compliant is important to us, but equally important is protecting ourselves against intruders. Even though we have some breathing room with PCI, we are still vulnerable with WEP as our security key. It must be a risk we are willing to take for the sake of saving money and hoping we do not get compromised."

When the staff member noted that "we have some breathing room with PCI," he probably was referring to the fact that TJX had been given an extension allowing it to be compliant beyond the standard's specified compliance date.[5] This additional time, ironically, was given *after* the data breaches had already begun. This extension was dependent upon evaluation of a TJX report on its compliance project by June 2006. It is unknown whether TJX complied with this requirement. The letter that authorized the extension was

[1] Mark Jewel, "Encryption faulted in TJX hacking," *MSNBC.com*, September 25, 2007. www.msnbc.msn.com/id/20979359/

[2] Ross Kerber, "Details Emerge on TJX Breach," *The Boston Globe*, October 25, 2007.

[3] Mark Jewel, op. cit. www.msnbc.msn.com/id/20979359/

[4] Schuman, Evan, "VISA Fined TJX Processor for Security Breach," *Eweek.com*, October 28, 2007. www.eweek.com/article2/0,1895,2208615,00.asp

[5] Evan Schuman, "In 2005, Visa Agreed To Give TJX Until 2009 To Get PCI Compliant," *StorefrontBacktalk*, November 9, 2007. storefrontbacktalk.com/story/110907visaletter

sent by a fraud control vice president for Visa. It ended with, "I appreciate your continued support and commitment to safeguarding the payment industry."

The company quickly became embroiled in commercial lawsuits and government investigations. These lawsuits involved the filing of briefs that shed additional light on the break-ins. For instance, sealed evidence from Visa and MasterCard placed the number of account records stolen at 94 million—roughly double TJX's estimates.[6]

TJX was sued by seven individual banks and bank associations. In December 2007, TJX settled with all but one of these banks, agreeing to pay up to $40.9 million. This would reimburse the banks for the cost of reissuing credit cards and other expenses.

The company also received a large fine from Visa. Actually, Visa could not fine TJX directly but could only fine TJX's merchant bank, the Fifth Third Bank of Ohio. (Merchant banks are financial institutions that serve—and should control—retail organizations that accept credit card payments.) However, merchant banks typically pass on the fine to the retailer, in this case TJX. During the summer of 2007, Visa fined TJX's merchant bank $880,000 and announced that it would continue to impose fines at $100,000/month until TJX had fixed its security problem. However, in the TJX settlement with the seven banks and banking associations, this fine was to be reduced by an undisclosed amount.

In this battle of corporate giants, consumers were handled last. At the time of this writing, TJX has proposed a settlement that would only involve very limited measures such as help with ID theft through insurance and other measures for the roughly 455,000 victims who had given personally identifiable information when they returned goods without a receipt. Other victims would be given a modest voucher or the opportunity to buy TJX merchandise at sale prices.[7]

On August 25, 2008, the Department of Justice charged 11 individuals with conducting the TJX break-in and the subsequent use of the stolen information.[8] Three were Americans, and they were jailed rapidly. Two more were in China. The other six lived in Eastern Europe. Although the three Americans conducted the actual data theft, they fenced the stolen information overseas. Two of the American defendants rapidly entered plea deals to testify against the alleged ringleader, Albert Gonzalez of Miami, Florida. Gonzalez was sentenced in 2010 to 20 years in prison.

One surprising thing about the indictment is that this criminal gang had not only plundered TJX of customer information. Subsequent investigation found that the attackers had repeated the crime with at least a half dozen other major companies.[9] These break-ins also began with the exploitation of WEP security.

Test Your Understanding

1. a) What did the attackers do first to break into TJX and other companies? b) Why do you think TJX failed to upgrade to stronger security than WEP? (There may be more than one consideration.) c) How was the TJX break-in an international crime?

---

[6] Ross Kerber, "Court Filling in TJX Breach Doubles Toll," *The Boston Globe*, October 24, 2007.

[7] John Leyden, "TJX Consumer Settlement Sale Offer Draws Scorn," *TheRegister.com*, November 20, 2007. www.theregister.co.uk/2007/11/20/tjx_settlement_offer_kerfuffle/

[8] U.S. Department of Justice, "Retail Hacking Ring Charged for Stealing and Distributing Credit and Debit Card Numbers from Major U.S. Retailers," August 5, 2008. www.usdoj.gov/criminal/cybercrime/gonzalezIndict.pdf

[9] Ibid.

INTRODUCTION

In Chapter 7, we focused on how 802.11 wireless LANs operate. This chapter continues to look at wireless networks. We will look at 802.11 security and management. We will then turn to other local wireless technologies, including Bluetooth. Finally, we will look at wireless metropolitan area network technologies, including 3G and 4G mobile telephone service.

Like switched networks, wireless networks are single networks. They are defined by standards at the physical and data link layers. Therefore, they are OSI standards, even if they are created by other organizations, such as the IEEE.

802.11 WLAN SECURITY

WLAN Security Threats

For most companies, security is the biggest problem with 802.11 wireless LANs. There are several major security threats to 802.11 WLANs:

- Most seriously, **drive-by hackers** park just outside a company's premises and intercept the firm's data transmissions. They can also mount denial-of-service attacks, send malware into the network, and do other mischief. Drive-by hacking software is readily downloadable from the Internet.
- Less importantly, **war drivers** are people who drive around a city looking for working access points that are unprotected. War driving is not illegal. Only if they try to break in do they become drive-by hackers.
- To guard against drive-by hackers, companies can secure their access points using techniques we will see a little later. However, departments or individual employees may set up unauthorized access points. These **rogue access points** often have no

Drive-By Hackers

Sit outside the corporate premises and read network traffic

Can send malicious traffic into the network

Easily done with readily available downloadable software

War Drivers

Merely discover unprotected access points—become drive-by hackers only if they break in

War driving per se is not illegal

Unprotected Access Points

Drive by hackers can associate with any unprotected access point

They gain access to the local area network without going through the site firewall

Rogue Access Points

Unauthorized access points that are set up by a department or an individual

Often have very poor security, making drive-by hacking easier

FIGURE 8-1 WLAN Security Threats (Study Figure)

security, so provide entry points for drive-by hackers who could not otherwise break into the network. Even if a company protects all of its official access points very well, a single rogue access point will make the security effort worthless.

Test Your Understanding

2. a) Distinguish between war drivers and drive-by hackers. b) What is a rogue access point?

802.11 Core Security Standards

Realizing the danger of drive-by hackers, the 802.11 Working Group decided to create a **core security standard** for wireless LANs. As Figure 8-2 shows, 802.11 core security standards provide protection between the wireless access point and the wireless host. This protection includes confidentiality, authentication, and message integrity. A drive-by hacker cannot intercept traffic to read it or send its own messages to the access point.

However, note in the figure that protection *only* extends between the wireless access point and the wireless host. It does not provide protection end-to-end between the wireless client host and the server host on the wired LAN (or the server host on the Internet). The core security standard has a very limited objective.

The protection provided by 802.11 core security standards *only* extends between the wireless access point and the wireless host.

Wired Equivalent Privacy (WEP) Security

When the 802.11 Working Group created its first standards in 1997, it only included a very rudimentary core security standard called **wired equivalent privacy (WEP)**.[10] This

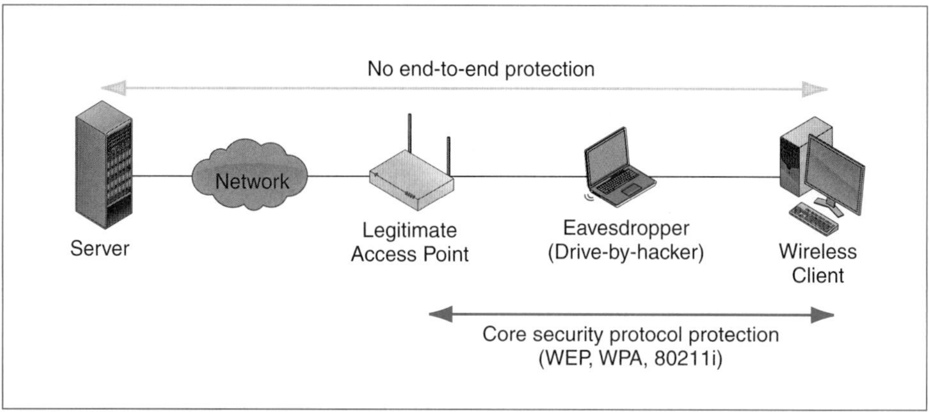

FIGURE 8-2 Core 802.11 Security Standards

[10] The WEP specification was only 10 pages long. In contrast, the specification for 802.11i security (discussed later) is 200 pages long.

Provide Security between the Wireless Station and the Wireless Access Point

- Client (and perhaps access point) authentication
- Encryption of messages for confidentiality

Wired Equivalent Privacy (WEP)

- Initial rudimentary core security standard provided with 802.11 in 1997
- Everyone shared the same secret encryption key, and this key could not be changed automatically
- Because secret key was shared, it does not seem to be secret
    - Users often give out freely
- Key initially could be cracked in 1–2 hours; now can be cracked in 3–10 minutes using readily available software

Wireless Protected Access (WPA)

- The Wi-Fi Alliance
    - Normally certifies interoperability of 802.11 equipment
    - Created WPA as a stop-gap security standard in 2002 until 802.11i was finished
- Designed for upgrading old equipment
    - WPA uses a subset of 802.11i that can run on older wireless NICs and access points
    - WPA added simpler security algorithms for functions that could not run on older machines
- Equipment that cannot be upgraded to WPA should be discarded

802.11i (WPA2)

- Uses AES-CCMP with 128-bit keys for confidentiality and key management
- Gold standard in 802.11 security
- But companies have large installed bases of WPA-configured equipment

FIGURE 8-3 802.11 Security Standards (Study Figure)

is the weak security standard that was used by TJX, as we saw at the beginning of this chapter.

With WEP, everyone sharing an access point had to use the same key, and WEP had no automated mechanism for changing this key. In nearly all cases, a firm with many access points and users normally gave all access points the same WEP key. Everybody in the firm knew the WEP key, so many employees thought it was not really secret and often were willing to share the key with unauthorized users. In addition, because there was no automated way to change all of the access point keys when an employee was fired, companies continued to use the same key even when there were clear security threats.

WEP had several other serious design problems. By 2001, software that would crack WEP keys quickly was readily available. Initially, cracking a WEP key with one of these programs took one or two hours. Today, it often takes only about 10 minutes.

Many companies began putting the brakes on WLAN implementation, and many pulled out their existing access points.[11]

Despite problems with WEP, companies continued to use it long after better 802.11 core security standards became available. One company that did so was TJ-MAX in the United States. As we saw at the beginning of this chapter, TJ-MAX made a conscious decision not to upgrade from WEP to a better core security protocol. The result was a disaster. Using a weak core security protocol is a terrible security policy. Although access points and clients usually continue to offer WEP as well as newer security protocols, only the newer security protocols should be used.

Test Your Understanding

3. a) When 802.11 was created, what security protocol did it offer? b) How long does it take to crack WEP today?

WPA (Wireless Protected Access)

The **Wi-Fi Alliance** is an industry trade group that certifies 802.11 products for interoperability. Normally, the Wi-Fi Alliance leaves standards creation to the 802.11 Working Group. However, when WEP's fatal flaws were discovered in 2000 and 2001, the market for wireless products began to falter. Furthermore, the 802.11i security standard being developed by the 802.11 Working Group was going to take years to be ratified.

As a stop-gap measure until 802.11i could be developed, the Wi-Fi Alliance created an interim security standard, **Wireless Protected Access (WPA)**, based on an early draft of the 802.11i core security standard. So that WPA could be used older wireless access points, the Wi-Fi Alliance watered down the 802.11i draft standard, using weaker security protocols. It announced this standard in 2002 and began certifying products for compliance in early 2003. Despite its weaknesses, WPA was much stronger than WEP. Companies flocked to it.

Test Your Understanding

4. a) Who created WPA? b) What is WPA's disadvantage compared with 802.11i?

802.11i (WPA2)

In 2004, the 802.11 Working Group finally ratified the **802.11i** standard. Most important, the 802.11i standard uses extremely strong **AES-CCMP** encryption. AES-CCMP has 128-bit keys and a key management method for automatically changing keys. In general, for every aspect of security, it uses the strongest security algorithms. Confusingly, the Wi-Fi Alliance refers to the 802.11i standard as **WPA2**.

---

[11] Some companies began taking other steps, like hiding the SSID (Service Set Identifier) of the access point. Users need to know this SSID to use an access point even if WEP is not used. Another common step was to only accept computers whose wireless NICs had registered MAC addresses. (All 802 LANs have 48-bit MAC addresses.) Unfortunately, these measures take a great deal of work, and they are easily cracked by readily available hacking software. They make sense if you are only concerned about unsophisticated but nosy neighbors at home.

However, before 802.11i appeared, most companies had already implemented WPA on their access points and wireless hosts. Reconfiguring all of these devices to work with 802.11i can be expensive, and until there were known cracks for WPA, companies were reluctant to make this investment. This changed in 2009, when a partial crack appeared for WPA. Companies that care strongly about WLAN security have transitioned to or are moving rapidly to 802.11i.

Test Your Understanding

5. a) What is the strongest security protocol for 802.11 today? b) What does the Wi-Fi Alliance call 802.11i? c) What encryption method does 802.11i use? d) What is deterring companies from converting from WPA to 802.11i? e) Why is WPA no longer viewed as a safe solution?

802.1X Mode Operation with Added Client–Access Point Security

WPA and 802.11i have two modes of operation. For large firms, the only mode that makes sense is **802.1X mode**. (The Wi-Fi Alliance calls it **enterprise mode**.) In Chapter 6, we saw that 802.1X was created for Ethernet wired networks. Its goal is to prevent attackers from simply walking into a building and plugging a computer into any wall jack or directly into a switch.

Figure 8-5 shows that each workgroup switch acts as an authenticator. The computer wishing access to the Ethernet network connects to the authenticator via UTP. Actual authentication is done by a central authentication server that normally uses the RADIUS protocol.

802.1X Mode (See Figure 8-5)

- Uses a central authentication server for consistency
- Authentication server also provides key management
- Wi-Fi Alliance calls this *enterprise mode*
- 802.1X standard protects communication with an extensible authentication protocol
  - Several EAP versions exist with different security protections
  - Firm implementing 802.1X must choose one
  - Protected EAP (PEAP) is popular because Microsoft favors it

Pre-Shared Key (PSK) Mode: Stations Share a Key with a Single Access Point

- For networks with a single access point
- Access point does all authentication and key management
- All users must know an initial pre-shared key (PSK)
- Each, however, is later given a unique key
- If the pre-shared key is weak, it is easily cracked
- Pass phrases that generate key must be at least 20 characters long
- Wi-Fi Alliance calls this *personal mode*

FIGURE 8-4 802.11 Security in 802.1X and PSK Modes (Study Figure)

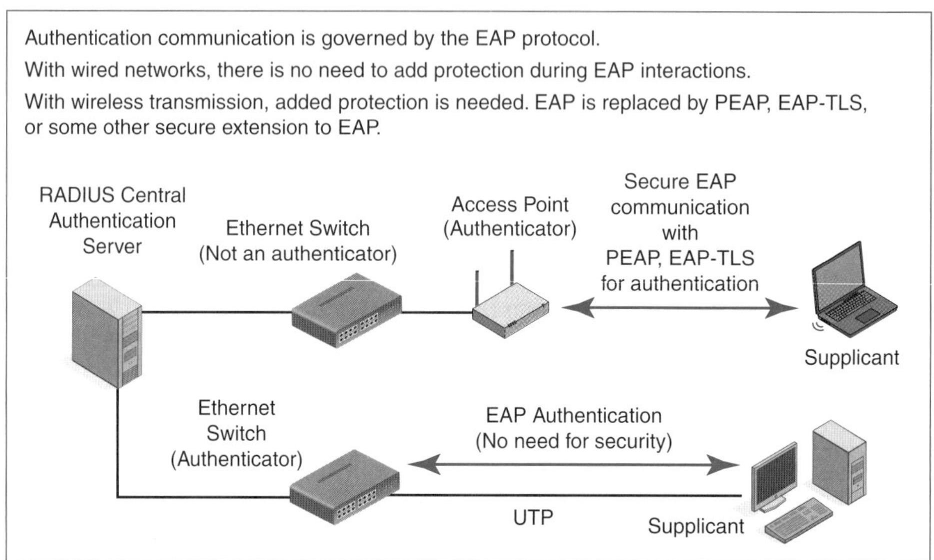

FIGURE 8-5 802.1X Mode for 802.11i (and WPA)

For Ethernet access, there is no need to have security between the computer seeking access and the workgroup switch that controls access. It is difficult for another person to tap the wired access line between the computer and the switch, and there are easier ways to break into a network. Consequently, as Figure 8-5 shows, there is no security between the wired hosts and the switch that is the 802.1X network access server.

With access points, however, transmissions between a wireless host and the access point are easy to intercept and mimic. The path between the host and the access point needs to be secure. To address this problem, the 802.11 Working Group extended the 802.1X standard by adding security between the access point and the wireless host before 802.1X mode authentication is done. The most common standard for adding this protection is the Protected Extensible Authentication Protocol (PEAP). Other standards for adding this security include EAP-TLS and EAP-TTLS. All of these standards create an SSL/TLS connection between the wireless access point and the wireless client before 802.1X is applied.

Test Your Understanding

6. a) In what mode of 802.11i and WPA operation is a central authentication server used? b) What does the Wi-Fi Alliance call 802.1X mode? c) Why does 802.1X not need security between the authenticator and the computer in Ethernet? d) Why does 802.1X need security between the authenticator and the computer in 802.11 wireless access? e) Why is extra EAP security necessary in 802.11 networks using 802.1X? f) What is the most popular standard for extended 802.1X security?

Pre-Shared Key (PSK) Mode

For homes and small businesses, adding a separate authentication server would be absurd. Consequently, the 802.11 Working Group created a simpler mode of operation beyond 802.1X mode. In this simpler mode, called **pre-shared key** mode, a single access point does all the work to provide a core security protocol. The Wi-Fi Alliance calls this **personal mode**.

PSK mode in 802.11i and WPA is for homes and small businesses with a single access point.

Figure 8-6 shows that the access points and wireless hosts begin with a shared 64-bit key. Everybody allowed to use the access point is told the shared key. This sounds like WEP, but this shared key is only used in initial authentication. After initial authentication, the wireless access point gives each authenticated user a new unique key to use while on the Internet. It also changes this key frequently. With this approach, too little traffic is created with each key to allow a drive-by hacker to crack the key.

In a home or very small business, having a handful of people know the shared key is not too dangerous. The larger danger is that the household or small business will select a weak key. To create the pre-shared key, the household or company creates a long **pass phrase** (which is much longer than a password) that the access point and each host use to generate the key. This pass phrase must be at least 20 characters long to provide adequate security. If short pass phrases are used, 802.11i and WPA in PSK mode are almost as easy to crack as WEP.

Test Your Understanding

7. a) How does PSK mode differ from 802.1X mode? b) What is a potential weakness of PSK mode? c) How long should pass phrases be with PSK?

Evil Twin Access Points and VPNs

Strong core security standards that protect communication between the wireless access point and wireless clients can greatly reduce risks. However, there is one type of attack that can break core security standards. This is the evil twin access point.

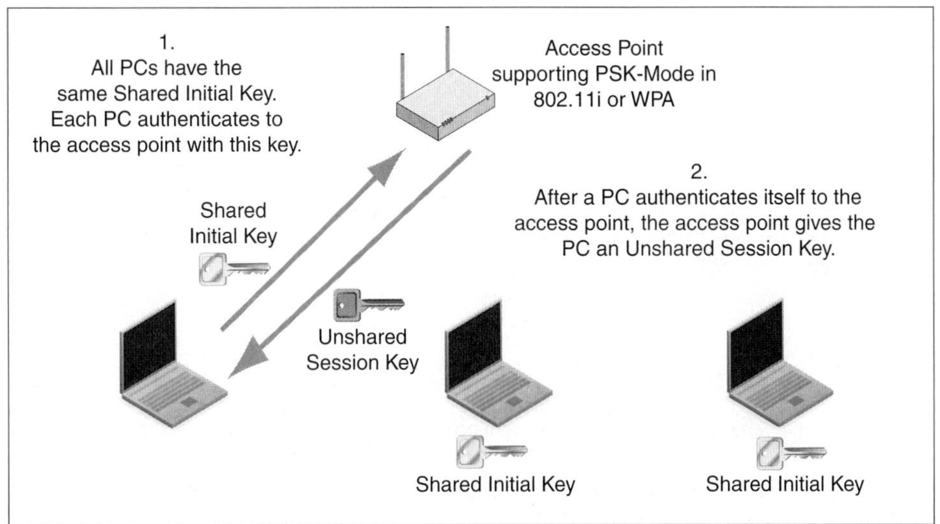

FIGURE 8-6 802.11i and WPA in Pre-Shared Key Mode

EVIL TWIN ACCESS POINTS Figure 8-7 illustrates an evil twin access point attack. Normally, the wireless client shown in the figure will associate with its legitimate access point. The two will establish a core security standard connection between them.

An **evil twin access point** is a notebook computer configured to act like a real access point. The evil twin operates at very high power. If the wireless host is configured to choose the highest-power access point, it will associate with the evil twin access point instead of the legitimate access point. The evil twin will establish a secure connection with the wireless client. This is Core Security Connection 1. Next, the evil twin associates with the legitimate access point, creating Core Security Connection 2.

An evil twin access point is a notebook computer configured to act like a real access point.

The evil twin can now read all traffic flowing between the wireless client and the legitimate wireless access point. When the wireless client sends encrypted frames, the evil twin decrypts them and then re-encrypts them before sending them on to the wireless access point. It does the same with traffic going in the other direction. Neither the wireless client nor the wireless access point knows that this is happening. Both seem to experience a normal secure connection. The wireless client has no problem getting to servers on the main wired LAN or getting to Internet servers.

What damage can an evil twin access point do? Most obviously, it can eavesdrop on the communication between the wireless client and the servers it uses. This allows the evil twin to steal important keys, corporate trade secrets, or personal information transmitted by the client.

The evil twin can also launch attacks against the client or any server attached to the network. On the client, the evil twin may be able to read information on the user's hard drive or plant malware for ongoing exploitation. The evil twin can launch attacks directly against servers, bypassing the company's firewall.

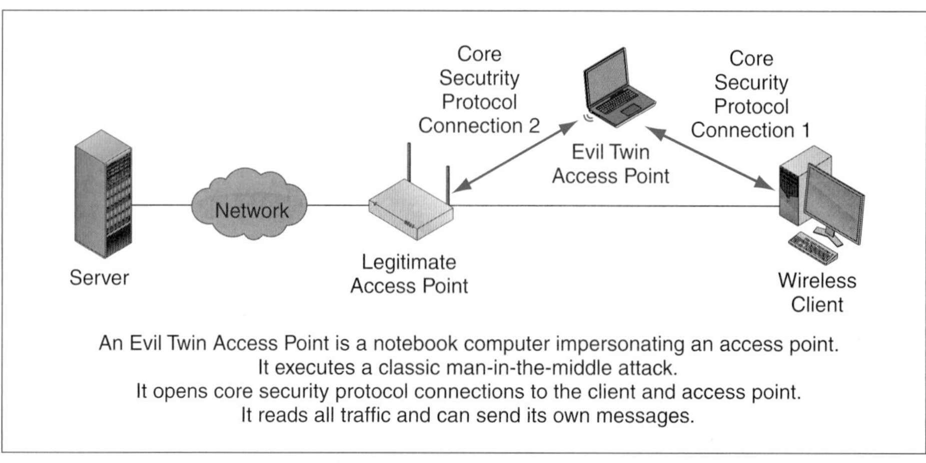

FIGURE 8-7 Evil Twin Access Point

VIRTUAL PRIVATE NETWORKS (VPNs) The evil twin attack is an example of a general class of attacks called **man-in-the-middle attacks**, in which an attacker intercepts messages and then passes them on. Man-in-the-middle attacks typically are very difficult to defeat. The main way to defeat them is to establish a virtual private network (VPN) connection between the wireless client and the server host it will use, as Figure 8-8 illustrates.

A **virtual private network (VPN)** is nothing more than a cryptographic system connection between a client and a server. A VPN gets its name from the fact that, as far as security is concerned, the client and server seem to have their own private network.

VPNs provide end-to-end protection between the client and server, including end-to-end authentication. If VPN authentication is based on a *pre-shared* secret that the client and server know ahead of time, an evil twin will not be able to conduct a man-in-the-middle attack against the client and server. Although the evil twin can read all wireless transmissions, it cannot intercept the pre-shared secret because this secret is never transmitted. So it will not be able to decrypt the messages passing between the client and the server, and it will not be able to send authenticated attack messages to either.

PERSPECTIVE Evil twin attacks are not theoretical concerns. They are commonplace, especially in wireless hot spots. Companies should use VPNs whenever clients and servers exchange sensitive communication.

Test Your Understanding

8. a) What is an evil twin access point? b) A company uses 802.11i for its core security protocol. How many 802.11i connections will the evil twin access point set up? c) What does the evil twin do when the client transmits to the legitimate access point? d) Distinguish between evil twin access points and rogue access points. e) How can VPNs defeat evil twin attacks? f) Why must the VPN secret be pre-shared to thwart a VPN attack?

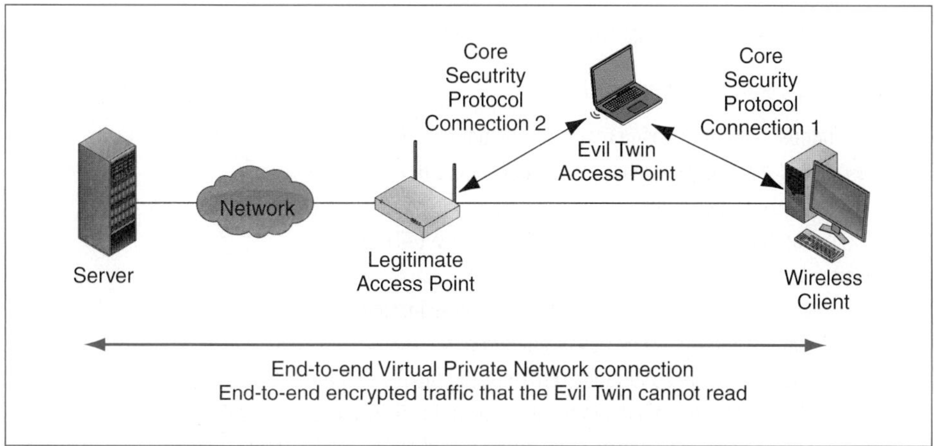

FIGURE 8-8 Using a Virtual Private Network to Counter an Evil Twin Attack

802.11 WIRELESS LAN MANAGEMENT

Until recently, the term *WLAN management* was almost an oxymoron. Large WLANs were like major airports without control towers. Companies quickly realized that they needed tools for centralized WLAN management.

Access Point Placement

The first management issue is where to place access points. If placement is not done systematically, there will be many dead spots, interference between many access points, or both.

INITIAL PLANNING The first step is to determine how far signals should travel. In many firms, a good radius is about 30 meters. If the radius is too great, many hosts will be far from their access points. Hosts far from the access point must drop down to lower transmission speeds, and their frames will take longer to send. This will reduce the access point's effective capacity. If the radius is too small, however, the firm will need many more access points to cover the space to be served.

Once an appropriate radius is selected (say 30 meters), the company gets out its building architecture drawings and begins to lay out 30 meter circles with as little overlap as possible. Where there are thick walls or other obstructions, shorter propagation distances must be assumed. And, of course, in a multistory building, this planning must be done in three dimensions.

Access Point Placement in a Building

- Must be done carefully for good coverage and to minimize interference between access points
- Lay out 30-meter or so radius circles on blueprints
- Adjust for obvious potential problems such as brick walls
- In multistory buildings, must consider interference in three dimensions
- Install access points and do site surveys to determine signal quality
- Adjust placement and signal strength as needed

Remote Access Point Management

- The manual labor to manage many access points can be very high
- Centralized management alternatives (See Figure 8-10)
  - Smart access points
  - Dumb access points, with intelligence in WLAN switches
- Desired functionality
  - Notify the WLAN administrators of failures immediately
  - Support remote access point adjustment
  - Should provide continuous transmission quality monitoring
  - Allow software updates to be pushed out to all access points or WLAN switches
  - Should work automatically whenever possible

FIGURE 8-9 Wireless LAN Management (Study Figure)

Finally, planners assign channels to access point positions. They attempt to minimize interference while doing so.

INSTALLATION AND INITIAL SITE SURVEYS Next, the access points are installed provisionally in the indicated places. However, the implementation work has just begun. When each access point is installed, an **initial site survey** must be done of the area around the access point to discover whether there are any dead spots or other problems. This requires signal analyzer software, which can run on a notebook computer or even a smart phone.

When areas with poor signal strength are found, surrounding access points must be moved appropriately, or their signal strengths must be adjusted until all areas have good signal strength.

ONGOING SITE SURVEYS Although the initial site survey should result in good service, conditions will change constantly. More people may be given desks in a given access point's range, signal obstructions may be put up for business purposes, and other changes must occur. Site surveys must be done periodically to ensure good service. They may also be done in response to specific reports of problems. To address problems, the network staff can adjust the power settings of nearby access points or take other actions.

Test Your Understanding

9. a) Describe the process by which access point locations are selected. b) After access points are installed provisionally, what must be done next?

Remote Management: Smart Access Points versus Wireless Switches

Large organizations have hundreds or even thousands of 802.11 wireless access points. Traveling to each one for manual configuration and troubleshooting would be extremely expensive. To keep management labor costs under control, organizations need to be able to manage access points remotely, from a central management console. Figure 8-10 illustrates two approaches to centralized wireless access point management.

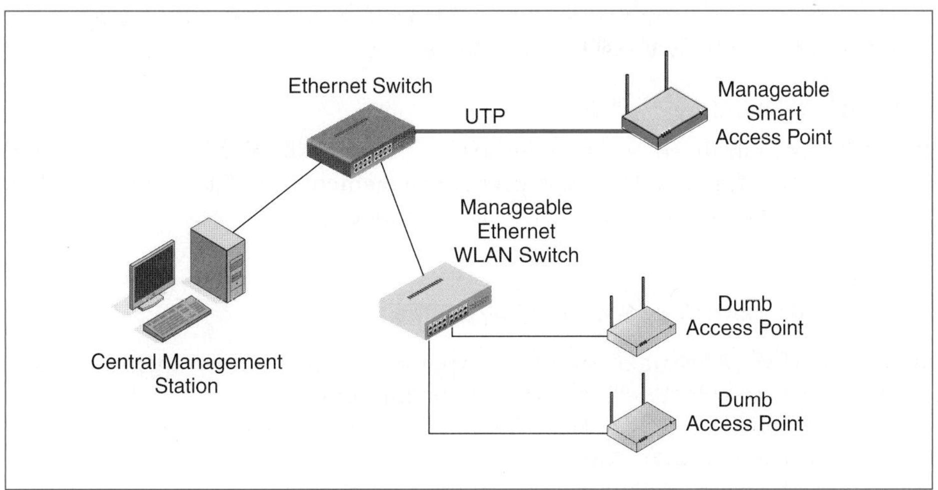

FIGURE 8-10 Wireless Access Point Management Alternatives

SMART ACCESS POINTS The simplest approach architecturally is to add intelligence to every access point. The central management console can then communicate directly with each of these **smart access points**[12] via the firm's Ethernet wired LAN. However, adding management capacity raises the price of access points considerably. Using smart access points is an expensive strategy.

WIRELESS LAN SWITCHES A second approach illustrated in Figure 8-10 is to use **WLAN switches**. As the figure shows, multiple access points connect to each wireless LAN switch. The management intelligence is placed in the WLAN switch rather than in the access points themselves. Vendors who sell WLAN switches claim that this approach reduces total cost because only inexpensive **dumb access points** are needed. Of course, smart access point vendors dispute these cost comparisons.

WIRELESS LAN MANAGEMENT FUNCTIONALITY Although technological approaches to centralized WLAN management vary, vendors generally agree on the types of functionality these systems should provide.

- These systems should notify the WLAN administrators of failures immediately so that malfunctioning access points can be fixed or replaced rapidly.
- They should provide continuous transmission-quality monitoring at all access points. In effect, they should provide continuous site surveys.
- They should help provide security by detecting rogue access points, evil twin access points, or legitimate access points that have improperly configured security.
- They should allow remote adjustment—for instance, telling nearby access points to increase their power to compensate for an access point failure. Such adjustments are also needed over time as furniture is moved (creating different shadow zones) or as the number of users in an area changes.
- They should allow software updates to be pushed out to all access points or WLAN switches, bypassing the need to install updates manually.
- The management software should be able to work automatically, taking as many actions as possible without human intervention.

Test Your Understanding

10. a) Why is centralized access point management desirable? b) What are the two technologies for remote access point management? c) What functions should remote access point management systems provide?

OTHER LOCAL WIRELESS TECHNOLOGIES

Although 802.11 WLANs are dominating corporate attention today, other local wireless technologies are important or may soon become important.

---

[12] Smart access points are also called fat access points.

Bluetooth Personal Area Networks (PANs)

PERSONAL AREA NETWORKS (PANS) FOR CABLE REPLACEMENT WLANs based on 802.11 can be large, especially if they use multiple access points. Another wireless networking standard, **Bluetooth**,[13] was created for wireless **personal area networks (PANs)**, which are intended to connect devices used by a single person. Bluetooth basically offers *cable replacement*—a way to get rid of cables between devices. It is not designed for full WLANs.

If you use a wireless mouse or have wireless headphones for your mobile phone, you are already using Bluetooth. Beyond these simple uses, Bluetooth can allow you to bring a notebook computer near a printer you have never used before. If the notebook and printer are both Bluetooth-enabled, they will be able to synchronize themselves and allow you to print with no extra configuration.

For Personal Area Networks (PANs)

- Devices on a person's body and nearby (mobile phone, PDA, notebook computer, etc.)
- Devices around a desk (computer, mouse, keyboard, printer)

Cable Replacement Technology

- For example, with a Bluetooth notebook, print wirelessly to a nearby Bluetooth-enabled printer
- Does not use access points
    - Uses direct device-to-device communication

Peer-to-Peer Connections

- Bluetooth does not use access points
- Devices do peer-to-peer pair connections
- A Bluetooth device can connect to eight others with peer connections

Disadvantages Compared with 802.11

- Short distance (10 meters)
- Low speed (3 Mbps today with a slower reverse channel)

Advantages Compared with 802.11

- Low battery power drain, so long battery life between recharges
- Bluetooth profiles (printing, etc.)
    - Somewhat rudimentary
    - Devices typically automate only a few

Bluetooth 3.0

- Can switch to 802.11 radio transmission for higher speeds than Bluetooth can provide

FIGURE 8-11 Bluetooth Personal Area Networks (PANs) (Study Figure)

[13] Bluetooth is named after King Harald Bluetooth, a Scandinavian king of the tenth century. As you might guess, Bluetooth was developed in Sweden, although it is now under the control of an international consortium.

PEER-TO-PEER CONNECTIONS While 802.11 uses access points, Bluetooth does not. Bluetooth uses peer-to-peer connections between pairs of devices. A Bluetooth device can set up peer-to-peer connections with up to eight other devices.

DISADVANTAGES COMPARED WITH 802.11

Limited Distance. While 802.11i normally has propagation distances up to 100 meters, 30–50 meters is a more realistic range in practice. Bluetooth, however, is normally limited to 10 meters or less.[14] For cable replacement around a desk or among the devices carried by a person, there is no need for longer distances. For a home or office WLAN, however, 10 meters is insufficient.

Low Speed. Bluetooth was not designed to handle heavy transmission loads. It currently offers a speed of only 3 Mbps with a slower reverse channel. This is sufficient for printing and most other Bluetooth applications, but it is not fast enough for WLANs.

ADVANTAGES COMPARED WITH 802.11

Long Battery Life. Although Bluetooth offers only low speeds and short distances, these limitations mean that radio transmission power is low, so battery life is quite long. This is very important for small portable devices.

Bluetooth Profiles. Bluetooth has one very important capability that 802.11 does not—**Bluetooth profiles**, which are application-layer standards designed to allow devices to work together automatically, with little or no user intervention. The most widely used profile is the Human Interface Device profile for mice, joysticks, keyboards, and other input devices. The Headset Profile, as the name suggests, is for wireless headsets. The Advanced Audio Distribution Profile is designed for one device to send music to another device or to speakers.

BLUETOOTH 3.0 Bluetooth 3.0 was created in 2009. Among other improvements, Bluetooth 3.0 was not limited to operating over Bluetooth's traditionally slow radio transmission speed. Although devices continue to pair using traditional Bluetooth radio, they can switch to 802.11 radio transmission and enjoy the much higher speeds of 802.11.

Test Your Understanding

11. a) Contrast how 802.11 and Bluetooth are likely to be used in organizations. b) What is a PAN? c) List a common use of Bluetooth. d) Does Bluetooth use access points? e) What are the speeds of Bluetooth transmission today? f) What is the normal maximum distance for Bluetooth propagation? g) What benefit do low speeds and short distances bring? h) Why are Bluetooth profiles attractive? i) How can a pair of Bluetooth devices transmit at tens of megabits per second?

Emerging Local Wireless Technologies

ULTRAWIDEBAND (UWB) TRANSMISSION We saw in Chapter 7 that maximum speed in a channel is governed primarily by a channel's bandwidth. However, spread spectrum transmission uses much wider channel bandwidths than it needs for its transmission

---

[14] This is with the standard 2.5 mW of power. There is an option for 100 mW power, which can raise the maximum propagation distance to 100 meters. This option is rarely built into Bluetooth products.

Ultrawideband (UWB)

Uses channels several gigahertz wide (spans multiple frequency bands)
Low power per hertz to avoid interference still gives very high speeds
But limited to short distance
Wireless USB provides 480 Mbps up to 3 meters, 110 Mbps up to 10 meters

ZigBee for almost-always-off sensor networks at low speeds

Very long battery life
250 kbps maximum

RFIDs: Like bar code tags but readable remotely

Software-Defined Radio

Can implement multiple wireless protocols
No need to have separate radio circuits for each protocol
Reduces the cost of multi-protocol devices

FIGURE 8-12 Emerging Local Wireless Technologies (Study Figure)

speeds. It does this because regulators require spread spectrum transmission in the unlicensed 2.4 GHz and 5 GHz bands in order to improve transmission.

An expanded form of the spread spectrum transmission is **ultrawideband (UWB)** transmission. While 802.11 spread spectrum transmission uses channels that are 22 MHz to 44 MHz wide, UWB uses channels that may be several *gigahertz* wide. These UWB channels actually span several entire service bands.

However, the amount of energy transmitted per hertz of bandwidth is very small in UWB. This prevents it from interfering with other transmissions within its huge range of frequencies. However, small energy per hertz times huge bandwidths is capable of transmitting data at high speeds, although only over short distances. This makes UWB capable of delivering video (even in high definition) and other bandwidth-hungry applications.

ZIGBEE The oddly named **ZigBee** technology is designed for wireless monitoring and control systems in businesses and homes. ZigBee is an almost-always-off technology for sensors that rarely send signals. In addition, when sensors and other ZigBee devices do transmit signals, they only need to send them at very slow speeds (up to 250 kbps, but usually slower). Such low performance requirements actually have a benefit: They lead to extremely long battery lives—months or even years. Although ZigBee was designed for scientific sensors, it has many possible everyday uses. For instance, a wall switch can communicate with a light via ZigBee.

RADIO FREQUENCY IDs (RFIDs) Today, bar-coded products must be run carefully over a laser scanner. In contrast, new **radio frequency ID (RFID)** tags can be used in place of bar code tags and only have to be brought *near* an RFID scanner to be read.

There are several RFID technologies. **Active RFID** tags have batteries and can be read tens of feet away. They can be used to record such things as when a firm's delivery trucks arrive and leave.

Less expensive **Passive RFID** tags derive their power from the radio signal sent by the reader. Passive RFID tags can only absorb a little power from the signal, so when they respond, their signals will not travel more than a few inches to about 4 feet. Passive ID tags are placed on pallets and boxes, but they are still much too expensive to be placed on most individual products at this time.

SOFTWARE-DEFINED RADIO Another trend to watch is **software-defined radio**, which will enable a wireless device to switch between 802.11, Bluetooth, ZigBee, and other wireless standards by switching software programs instead of having the functionality built entirely into hardware. In contrast, today's hardware-only approach to implementing multiple technologies is expensive because it requires multiple circuits to be built into devices.

Test Your Understanding

12. a) Compare the speeds of 802.11, Bluetooth, UWB, and ZigBee. b) Compare the distance limits of 802.11, Bluetooth, and UWB. c) What technology could replace universal product code (bar code) tags on products? d) What is the major promise of software-defined radio?

WIRELESS WIDE AREA NETWORKS (WWANS)

So far, we have been looking at wireless transmission in local area networks. WLANs today are well established. Now we will look at wireless transmission in the WAN environment. **Wireless WAN (WWAN)** transmission is less developed than wireless LAN transmission but holds enormous promise.

In wired LANs, we take very high speeds for granted. With WLANs, we tend to have to become accustomed to somewhat slower speeds, but these speeds are still sufficient for most needs. In wireless WANs, propagation distances are farther, so WWAN service costs more per bit transmitted than WLAN service. Consequently, prices are higher and transmission speeds are lower in WWANs than WLANs.

Most wireless WANs really are limited in distance. They typically are wireless metropolitan area networks. A metropolitan area network (MAN) serves a city and its suburbs. Over this limited distance, a single wireless WAN network can provide comprehensive coverage. Thanks to their limited coverage area, wireless metropolitan area networks can offer much higher speeds than they could if they had regional, national, or international scope.

Test Your Understanding

13. a) Compare wireless LANs and wireless WANs in terms of relative transmission speeds and the costs per bit transmitted. b) What is the coverage area for most wireless WANs? c) What is a metropolitan area network (MAN)?

Higher cost per bit transmitted than wireless LANs
Lower speeds than wireless LANs
Usually cover a metropolitan area rather than a regional, national, or international wireless network

FIGURE 8-13 Wireless Wide Area Networks (WWANs) (Study Figure)

CELLULAR DATA SERVICE

Cellular Service

Nearly everybody today is familiar with cellular telephony. In most industrialized countries, well over half all households now have a cellular telephone.[15] Many people now have *only* a cellular telephone and no landline.

Cells

CELLS AND CELLSITES Figure 8-14 shows that cellular telephony divides a metropolitan service area into smaller geographical areas called **cells**.

The user has a cellular telephone (also called a **mobile phone**, **mobile**, or **cellphone**). Near the middle of each cell is a **cellsite**, which contains a **transceiver** (transmitter/receiver) to receive mobile phone signals and to send signals out to the mobiles. The cellsite also supervises each mobile phone's operation (setting its power level, initiating calls, terminating calls, and so forth).

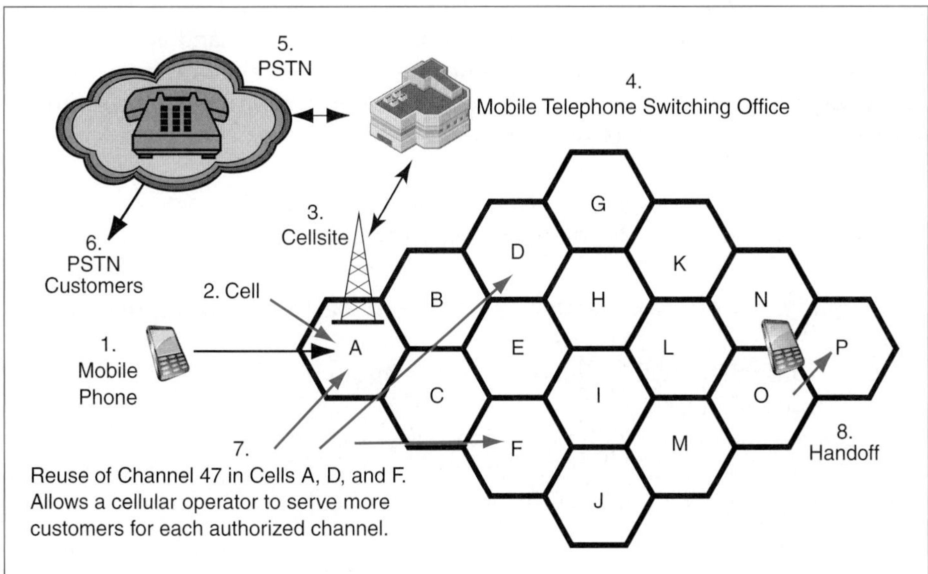

FIGURE 8-14 Cellular Technology

[15] Although cellular telephony was first developed in the United States, the United States has slightly lower market penetration than most other countries. One reason is that normal telephony is inexpensive in the United States, so moving to cellular service is an expensive choice. Another reason is that when someone calls a cellular phone in the United States, the cellular owner receiving the call pays; in most other countries, the caller pays. These two factors increase the relative price of using a cellular phone compared with using a landline phone in the United States. A third factor is that U.S. cellular carriers give inadequate coverage, even in large metropolitan areas. In most other countries, dropped calls and dead spots are rare.

MOBILE TELEPHONE SWITCHING OFFICE (MTSO) All of the cellsites in a cellular system connect to a **mobile telephone switching office (MTSO)**, which connects cellular customers to one another and to wired telephone users.

The MTSO also controls what happens at each of the cellsites. It determines what to do when people move from one cell to another, including deciding which cellsite should handle the transmission when the caller wishes to place a call. (Several cellsites may hear the initial request at different loudness levels; if so, the MTSO selects a service cellsite on the basis of signal loudness—not necessarily on the basis of physical proximity.)

Test Your Understanding

14. a) In cellular technology, what is a cell? b) What is a cellsite? c) What are the two functions of the MTSO?

Why Cells?

Why not use just one central transmitter/receiver in the middle of a metropolitan area instead of dividing the area into cells and dealing with the complexity of cellsites?

CHANNEL REUSE FOR MORE SUBSCRIBERS The answer is **channel reuse**. The number of channels permitted by regulators is limited, and subscriber demand is heavy. Cellular telephony uses each channel multiple times, in different cells in the network. This multiplies the effective channel capacity, allowing more subscribers to be served with the limited number of channels available.

> Cellular technology is used because it provides channel reuse—the ability to use the same channel in different cells. This allows cellular systems to support more subscribers.

TRADITIONALLY, NO CHANNEL REUSE IN ADJACENT CELLS With traditional cellular technologies, such as the **GSM** technology, you cannot reuse the same channel in adjacent cells, because there will be interference. For instance, in Figure 8-14, suppose that you use Channel 47 in Cell A. You cannot use it in Cells B or C. This reduces channel reuse. In general, the number of times you can reuse a channel is only about the number of cells, divided by seven. In other words, if you have 20 cells, you can reuse each channel only about three (20/7) times. (This is a rough calculation, so "about 2.857 timed" makes no sense.)

CHANNEL REUSE IN ADJACENT CHANNELS WITH CDMA Some cellular systems in the United States use a new cellular technology, **code division multiple access (CDMA)**. CDMA is a form of spread spectrum transmission. However, in contrast to the types of spread spectrum transmission used in 802.11 wireless LANs, which allow only one station to transmit at a time in a channel, CDMA allows multiple stations to transmit at the same time in the same channel.

In addition, CDMA permits stations in adjacent cells to use the same channel, without serious interference. In other words, if you have 20 cells, with CDMA you can reuse each channel 20 times. This allows you to serve far more customers with CDMA than you could with older forms of cellular telephony.

If CDMA is so good, why do only some systems use it? The answer is that it came out a few years later than GSM. The first CDMA cellular systems were not built until 1993, and even then, they represented a technological and economic risk. However, CDMA quickly proved itself, giving CDMA vendors a cost advantage.

CELLS AND WIRELESS LAN ACCESS POINTS In a sense, enterprise wireless LANs with many access points are like cellular technologies. They allow users to employ the limited number of frequencies available in WLANs many times within a building.

Test Your Understanding

15. a) Why does cellular telephony use cells? b) What is the benefit of channel reuse? c) If I use Channel 3 in a cell, can I reuse that same channel in an adjacent cell with traditional cellular technology? d) Can I reuse Channel 3 in adjacent cells if the cellular system uses CDMA transmission?

Handoffs versus Roaming

HANDOFFS If a subscriber moves from one cell to another within a city, the MTSO will implement a **handoff** from one cellsite to another. For instance, Figure 8-14 shows a handoff from Cell O to Cell P. The mobile phone will change its sending and receiving channels during the handoff, but this occurs too rapidly for users to notice.

ROAMING In contrast, if a subscriber leaves a metropolitan cellular system and goes to another city or country, this is called **roaming**. Roaming requires the destination cellular system to be technologically compatible with the subscriber's mobile. It also requires administration permission from the destination cellular system. Roaming is as much a business and administrative problem as it is a technical problem.

> In cellular telephony, handoffs occur when a subscriber moves between cells in a local cellular system. Roaming occurs when a subscriber moves between cellular systems in different cities or countries.

HANDOFF AND ROAMING IN 802.11 WLANs Recall from Chapter 7 that *handoff* and *roaming* mean the same thing in 802.11 WLANs. They both mean moving from one access point to another within the same WLAN. In other words, the terms *handoff* and *roaming* are used differently in cellular telephony from the way they are used in WLANs.[16] Figure 8-15 shows how the terms *handoff* and *roaming* are used differently in WLANs and cellular telephony.

Test Your Understanding

16. a) Distinguish between handoffs and roaming in cellular telephony. b) Distinguish between handoffs and roaming in 802.11 wireless LANs.

---

[16] Wouldn't it be nice if there were a networking terminology court that could punish this sort of thing?

	802.11 WLANs	Cellular Telephony
Relationship	Handoff and roaming mean the *same thing*	Handoff and roaming mean *different things*
Handoffs (means the same in both)	Wireless host travels between access points in an organization	Mobile phone travels between cellsites in the *same* cellular system
Roaming (means different things)	Wireless host travels between access points in an organization	Mobile phone travels to a *different* cellular system

FIGURE 8-15 Handoff and Roaming in 802.11 Wireless Networking and Cellular Telephony

Using Cellular Telephony for Data Transmission

SMART PHONES Originally, mobile phones were just that—telephones that you could carry with you. Today, however, many new mobile phones are **smart phones** that do far more. Most have cameras, and some even have video cameras. Most importantly, all have web browsers to support web access. Smart phones are indicative of a trend in cellular service from voice service to voice and data service. Future cellular technologies will primarily be data transmission technology, with voice transmission being a secondary consideration.

FIRST-GENERATION (1G) CELLULAR The early cellular telephones could only send and receive data at extremely low speeds. **First-generation (1G)** cellular technology used analog radio transmission and could only send and receive data at speeds below 10 kbps—far slower than even a telephone modem.

SECOND-GENERATION (2G) CELLULAR **Second-generation (2G)** technology introduced all-digital transmission, but it too was limited to a painfully slow 10 kbps. This was completely inadequate for web access, but it was good enough to introduce texting and the exchange of low-quality photographs.

THIRD-GENERATION (3G) CELLULAR **Third-generation (3G)** technology arrived around the year 2000. Third-generation services typically provide speeds of 300 to 500 kbps. This permits somewhat sluggish web access as well as the downloading of low-resolution video and the exchanging of high-quality photographs. This is the type of service that Apple 3G and 3Gs iPhone users get.

FOURTH-GENERATION (4G) CELLULAR The next step will be **fourth-generation (4G)** cellular technology. This technology will eventually bring speeds of 1 Gbps for stationary users and 100 Mbps for moving users. The 4G test systems that have appeared at the time of this writing (2010), however, are limited to typical speeds of about 3 Mbps and 5 Mbps. This is still an improvement over 3G download speeds by a factor of ten. Even current 4G systems are sufficient for downloading high-definition video and for providing Internet access at cable modem and ADSL speeds. Fourth-generation systems are advanced in many other ways. They are all-IP services, and they will also have excellent quality-of-service management.

Smart Phones

Combine cameras, web browsers, and other non-talk capabilities
Data capabilities are becoming more important than voice capabilities

First-Generation (1G) Cellular Technology

Introduced around 1980
Analog only
Can only send and receive below 10 kbps

Second-Generation (2G) Cellular Technology

Introduced around 1990
All-digital service
Still limited to 10 kbps
Sufficient for texting and the exchange of low-quality photographs

Third-Generation (3G) Cellular Technology

Introduced around 2000
Typical speeds of 300 kbps to 500 kbps
Sufficient for somewhat sluggish web access
Sufficient for low-quality video
Sufficient for exchanging high-quality photographs

Fourth-Generation (4G) Cellular Technology

Introduced around 2010
Typically 3 Mbps to 5 Mbps initially (X10 improvement from 3G)
Eventually, 100 Mbps to mobile users and 1 Gbps to stationary users
Sufficient for high-definition video
Runs over IP

WiMAX

Based on 802.16 standards
WiMAX Forum promotes WiMAX and conducts product interoperability tests.
802.16m will eventually provide 100 Mbps to mobile users and 1 Gbps to stationary users.

Long Term Evolution (LTE)

The 4G technology that most current carriers will use
Not a full 4G technology
LTE Advanced will eventually provide 100 Mbps to mobile users and 1 Gbps to stationary users.

Lies, Damned Lies, and Mobile Service Speeds

Mobile speeds vary widely.
All users share the capacity of a cellsite.
By time of day, usage increases and decreases, and throughput varies as a result.
Distance from a transmitter means slower speeds.
In some parts of a city, there may be insufficient capacity or poor reception.
Buildings and other obstructions may create local areas of poor service.

FIGURE 8-16 Generations of Cellular Service (Study Figure)

Two standards are vying for 4G service. The first is **WiMAX**, which is based on the IEEE 802 LAN/MAN Committee's **802.16** standard. The WiMAX Forum was created to promote this technology and to certify interoperability among WiMAX products. The first serious WiMAX systems were created in 2010. The first operational WiMAX systems are providing typical speeds of 3 Mbps to 5 Mbps. In the United States, Sprint is heavily committed to WiMAX. The 802.16m standard, which is now under development, is being designed to bring a full 1 Gbps to stationary users and 100 Mbps to rapidly moving users.

Most cellular carriers, however, are committed to **Long Term Evolution (LTE)** for their 4G technology. Service tests indicate that LTE will initially be comparable to early WiMAX implementations in speed. Now under development is **LTE Advanced**, which will eventually bring speeds of 1 Gbps to stationary users and 100 Mbps to rapidly moving users.

LIES, DAMNED LIES, AND MOBILE SERVICE SPEED The speeds we have been discussing were "typical" speeds. However, every 3G user knows that throughputs vary widely by time of day and physical location. Time-of-day variations are largely due to the fact that the capacities of cell sites are shared by all cell site users. Someone sitting in one place for 24 hours is likely to see disparities of a factor of ten between their highest speeds and their lowest speeds. The lowest speeds are likely to come early in the morning, when many people are using web access to get ready for the day. Also, as people move physically, their distance from cellular towers will increase or decrease; this will also have a major impact on throughput. Also, of course, someone may move to an area served by a more highly loaded or more lightly loaded cellular tower. A third reason for strong variations in throughput is that cellular providers do not provide equal service capacity in all areas. Many places in a city may have chronically poor service. Finally, buildings and other obstacles may create local regions of poor service.

Test Your Understanding

17. a) What are smart phones? b) What changed from 1G cellular technology to 2G technology? c) Compare typical 2G speeds with 3G speeds. d) What are likely to be typical download speeds for early 4G systems? e) What speeds will 4G support eventually? f) At what generation did video downloads and reasonably fast web access become feasible? g) What generation is sufficient for high-definition video and very fast web access?

18. a) What two technologies are carriers initially using for 4G service? b) What initial technology will most cellular carriers use? c) Which technologies will bring 100 Mbps download speeds to moving customers? d) Why do 3G and 4G speeds vary by time of day? e) Do cellular data carriers provide about equally good service throughout their service areas?

Mobile Technology and Wi-Fi Technology

In the past, there has always been a simple dichotomy. Wi-Fi 802.11 systems were used as wired LAN extensions for data while cellular service was used for telephony and metropolitan area data service. However, this situation is changing.

Traditional Roles

802.11: Networking within a firm
Cellular telephony: Service outside the firm
Growing convergence

3G Mobile Smart Phones

By default, use cellular network
Can connect directly through a WLAN
Typically faster speeds than cellular for data
Cellphone companies like offloading flat-fee subscribers to WLANs

Some Smart Phones Can Act as 802.11 Access Points

Several 802.11 users can share its capacity

FIGURE 8-17 Cellular–802.11 Convergence (Study Figure)

First, many 3G mobile smart phones will use the carrier facilities by default but will try to connect to an 802.11 WLAN network if one is available. An 802.11 WLAN is likely to offer higher throughput than the carrier cellular system. Carriers like this because people with flat-rate calling plans will not use the carrier's network while they are using an 802.11 WLAN. This reduces the carrier's cost.

Second, going in the other direction, some smart phones can act as 802.11 access points. This allows several laptops and other 802.11 devices to share a cellular user's connection to the Internet.

Test Your Understanding

19. a) Why is it good for users that some smart phones can connect to WLAN access points? b) Why is it good for carriers? c) Why is it attractive to users that some smart phones will act as access points?

CONCLUSION

Synopsis

This chapter continues our discussion of wireless communication. We began with a discussion of 802.11 wireless LAN security threats, including drive-by hackers, war drivers, rogue access points, and evil twin access points.

To address most of these security problems, the 802.11 Working Group created core security standards to protect communication between the wireless client and its nearest wireless access point. Initially, the 802.11 Working Group created WEP as a core security protocol. WEP was badly designed and was soon easily exploited. Later, the 802.11 Working Group created the excellent 802.11i core security standard. Before they did, the Wi-Fi Alliance used an early draft of 802.11i to create a somewhat weaker core security protocol, WPA. They did this because they could develop WPA more quickly, and the industry badly needed something better than WEP. WPA has not been completely cracked, but parts of it have, so companies are advised to upgrade to 802.11i.

The 802.11 Working Group based 802.11i on the 802.1X security standard we saw in Chapter 4. The 802.1X standard uses a central authentication server. However, they needed to extend 802.1X by adding security between the access point and each wireless client during initial authentication, before full 802.1X security is set up. The 802.11i standard also has a pre-shared key mode, which is used for companies that only have a single access point. The wireless clients share a single pre-shared key, which must be kept secret from outsiders. After authentication, however, the wireless access point creates a unique session key for each wireless client.

Evil twin access points impersonate legitimate access points. To avoid being snared by evil twin access points, the clients need to have a virtual private network (VPN) connection to the servers they will use. This connection must use a pre-established secret that is not transmitted during interactions and so cannot be intercepted by the evil twin access point.

We discussed wireless LAN management, including strategies for placing access points to give good coverage with a minimum of overlap and with a minimum of interference from nearby access points operating on the same channel. Good access point placement requires initial and ongoing site surveys of signal strength and interference. We also looked at the remote management of access points using smart access points or dumb access points controlled by wireless LAN switches. A central manager can detect rogue access points, allow the remote adjustment of access point power, push software updates out to the access points, and do all this and more automatically, with a minimum of intervention.

We looked at other local wireless technologies, including Bluetooth, which is for personal area networks (PANs) in which all devices are located nearby, often on the same desk or around the body of a person. We then looked at other emerging local wireless technologies, including high-speed, short-distance ultrawideband (UWB) transmission, ZigBee for wireless monitoring systems, and radio frequency ID (RFID) to allow items to be identified within a short distance of a scanner.

We then looked at wireless services for metropolitan areas. Cellular telephone companies now offer 3G and sometimes 4G services to their customers. A competitor for wireless metropolitan area networking is WiMAX.

END-OF-CHAPTER QUESTION

Essay

1. A friend is interested in installing a wireless LAN in her small business. She has about a dozen employees. She is concerned about security. Write a two-page letter to explain what security threats she is facing and how she can reduce these threats. Remember that you are writing for a friend, not for a teacher. Do not hand in disorganized notes. Double spacing is good.

Perspective Questions

1. What was the most surprising thing you learned in this chapter?
2. What was the most difficult part of this chapter for you?

CHAPTER 9

TCP/IP Internetworking

LEARNING OBJECTIVES

By the end of this chapter, you should be able to:

- Explain basic TCP/IP, IP, TCP, and UDP concepts.
- Define hierarchical IP addresses, networks and subnets, border and internal routers, and masks.
- Describe router operation when a packet arrives, including ARP.
- Explain IPv4 fields and IPv6 fields.
- Explain TCP fields, session closings, and port numbers.
- Explain other important TCP/IP standards, including dynamic routing protocols and ICMP.
- Describe Multiprotocol Label Switching (MPLS).

INTRODUCTION

Switched networks and wireless networks are governed by standards at Layer 1 and Layer 2. We looked at switched wired single networks in Chapters 5 and 6. We looked at wireless single networks in Chapters 7 and 8. In this chapter and the next, we will discuss internetworking, which is governed by Layer 3 and Layer 4 standards.

We will only look at TCP/IP internetworking because TCP/IP dominates the work of network professionals at the internet and transport layers. However, real-world routers cannot limit themselves to TCP/IP internetworking. They are multiprotocol routers, which can route not only IP packets but also IPX packets, SNA packets, AppleTalk packets, and other types of packets. In this chapter we will look at standards associated with TCP/IP.

TCP/IP RECAP

The TCP/IP Architecture and the IETF

We first looked at TCP/IP in some depth in Chapter 2. Recall from that chapter that the Internet Engineering Task Force (IETF) sets TCP/IP standards. TCP/IP is an architecture for setting individual standards. Figure 9-1 shows a few of the standards the IETF

Layer						
5 Application	User Applications			Supervisory Applications		
	HTTP	SMTP	Many Others	DNS	Dynamic Routing Protocols	Many Others
4 Transport	TCP				UDP	
3 Internet	IP			ICMP		ARP
2 Data Link	None: Use OSI Standards					
1 Physical	None: Use OSI Standards					
Note: Shaded protocols are discussed in this chapter.						

FIGURE 9-1 Major TCP/IP Standards

has created within this architecture. Some of the standards are shaded in this figure. We will look at these standards in this chapter.

Simple IP at the Internet Layer

Recall also from Chapter 2 that internetworking operates at two layers. The internet layer moves packets from the source host to the destination host across a series of routers. Figure 9-1 shows that the primary standard at the internet layer is the Internet Protocol (IP). Figure 9-2 shows that IP is a simple (connectionless and unreliable) standard. This simplicity minimizes the work that each router has to do along the way, thereby minimizing routing costs. The internet layer is Layer 3.

Reliable Heavyweight TCP at the Transport Layer

In turn, TCP at the transport layer corrects any errors at the internet layer and lower layers as well. As we saw in Chapter 2, when the transport process on a destination host receives a TCP supervisory or data segment, it sends back an acknowledgment. If the transport process on the source host does not receive an acknowledgment for a TCP

Protocol	Layer	Connection-Oriented/Connectionless	Reliable/Unreliable	Lightweight/Heavyweight
TCP	4 (Transport)	Connection-oriented	Reliable	Heavyweight
UDP	4 (Transport)	Connectionless	Unreliable	Lightweight
IP	3 (Internet)	Connectionless	Unreliable	Lightweight

FIGURE 9-2 IP, TCP, and UDP

segment, it resends the segment. TCP is both connection-oriented and reliable, making it a heavyweight protocol. However, the work of implementing TCP only occurs on the source and destination hosts, not on the many routers between them. This keeps the cost of reliability manageable. The transport layer is Layer 4.

Unreliable Lightweight UDP at the Transport Layer

In Chapters 2 and 6, we saw that TCP/IP offers an alternative to heavyweight TCP at the transport layer. This is the User Datagram Protocol (UDP). Like IP, UDP is a simple (connectionless and unreliable) and lightweight protocol.

Test Your Understanding

1. a) Compare TCP and IP along the dimensions in Figure 9-2. b) Compare TCP and UDP along the dimensions in Figure 9-2.

IP ROUTING

In this section, we will look at how routers make decisions about forwarding packets—in other words, how a router decides which interface to use to send an arriving packet back out to get it closer to its destination. (In routers, ports are called **interfaces**.) This forwarding process is called **routing**. Router forwarding decisions are much more complex than the Ethernet switching decisions we saw in Chapter 6. As a consequence of this complexity, routers do more work per arriving packet than switches do per arriving frame. Consequently, routers are more expensive than switches for a given level of traffic. A widely quoted network adage reflects this cost difference: "Switch where you can; route where you must."

> When routers forward incoming packets closer to their destination hosts, this is called routing.

Hierarchical IP Addressing

To understand the routing of IP packets, it is necessary to understand IP addresses. In Chapter 1, we saw that IP addresses are 32 bits long. However, IP addresses are not simple 32-bit strings.

HIERARCHICAL ADDRESSING As Figure 9-3 shows, IP addresses are **hierarchical**. They usually consist of three parts that locate a host in progressively smaller parts of the Internet. These are the network, subnet, and host parts. We will see later in this chapter that hierarchical IP addressing simplifies routing tables.

NETWORK PART First, every IP address has a **network part**, which identifies the host's network on the Internet. **Internet networks** are owned by single organizations, such as corporations, universities, and ISPs. In the IP address shown in Figure 9-3, the network part is 128.171. It is 16 bits long. This happens to be the network part for the University

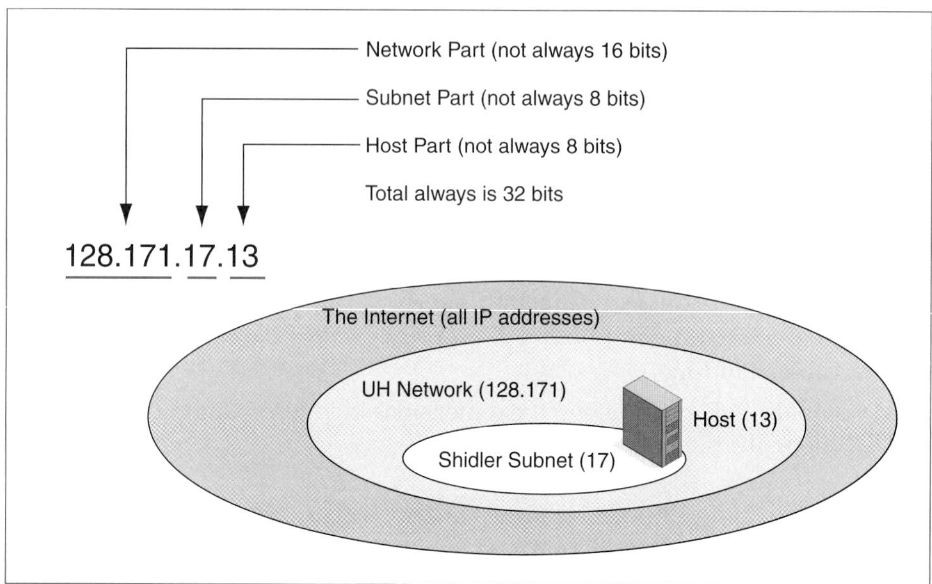

FIGURE 9-3 Hierarchical IP Address

of Hawaii Network on the Internet. All host IP addresses within this network begin with 128.171. Different organizations have different network parts that range from 8 to 24 bits in length.

Note that "network" in this context does not mean a single network—a single switched LAN or WAN. The University of Hawaii Network itself consists of many single switched or wireless networks and routers at multiple locations around the state. In IP addressing, **network** is an organizational concept—a group of hosts, switched networks, and routers owned by a single organization.

> In IP addressing, *network* is an organizational concept—a group of hosts, switched networks, and routers owned by a single organization.

SUBNET PART Most large organizations further divide their networks into smaller units called **subnets**. After the network part in an IP address come the bits of the **subnet part**. The subnet part bits specify a particular subnet within the network.

For instance, Figure 9-3 shows that in the IP address 128.171.17.13, the first 16 bits (128.171) correspond to the network part, and the next eight bits (17) correspond to a subnet on this network. (Subnet 17 is the Shidler College of Business subnet within the University of Hawaii Network.) All host IP addresses within this subnet begin with 128.171.17.

HOST PART The remaining bits in the 32-bit IP address identify a particular host on the subnet. In Figure 9-3, the **host part** is 13. This corresponds to a particular host, 128.171.17.13, on the Shidler College of Business subnet of the University of Hawaii Network.

VARIABLE PART LENGTHS In the example presented in Figure 9-3, the network part is 16 bits long, the subnet part is 8 bits long, and the host part is 8 bits long. This is only an example. In general, network parts, subnet parts, and host parts vary in length. For instance, if you see the IP address 60.47.7.23, you may have an 8-bit network part of 60, an 8-bit subnet part of 47, and a 16-bit host part of 7.23. In fact, parts may not even break conveniently at 8-bit boundaries. The only thing you can tell when looking at an IP address by itself is that it is 32 bits long.

Test Your Understanding

2. a) What is routing? b) What are the three parts of an IP address? c) How long is each part? d) What is the total length of an IP address? e) In the IP address, 10.11.13.13, what is the network part?

Routers, Networks, and Subnets

BORDER ROUTERS CONNECT DIFFERENT NETWORKS As Figure 9-4 illustrates, networks and subnets are very important in router operation. Here we see a simple site internet. The figure shows that a **border router's** main job is to connect different networks. This border router connects the 192.168.x.x network within the firm to the 60.x.x.x network of the firm's Internet service provider.

A border router's main job is to connect different networks.

INTERNAL ROUTERS CONNECT DIFFERENT SUBNETS The site network also has an **internal router**. An internal router, as Figure 9-4 demonstrates, connects different subnets within a firm—in this case, the 192.168.1.x, 192.168.2.x, and 192.168.3.x subnets. Many sites have multiple internal routers to link the site's subnets.

An internal router only connects different subnets within a firm.

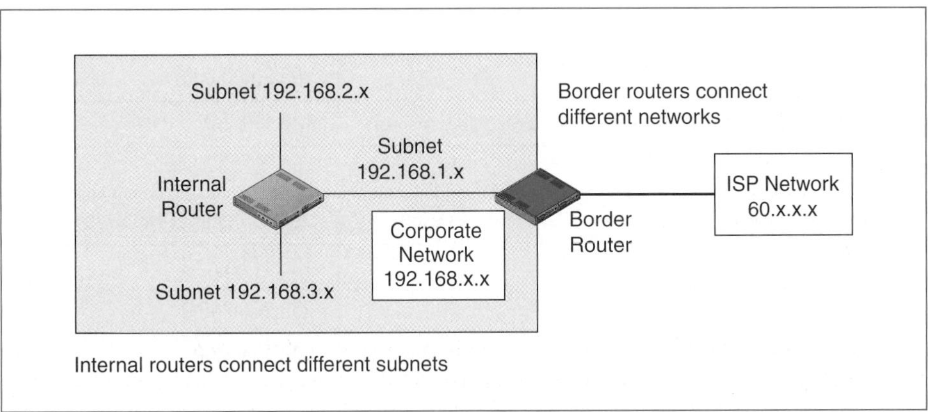

FIGURE 9-4 Border Router, Internal Router, Networks, and Subnets

Test Your Understanding

3. a) Connecting different networks is the main job of what type of router? b) What type of router only connects different subnets?

Network and Subnet Masks

If you know how the University of Hawaii organizes its IP addresses, you know that the first 16 bits are always the network part, the next 8 are the subnet part, and the final 8 are the host part. However, the sizes of the network, subnet, and host parts differ. Routers need a way to tell the sizes of key parts. The tool that allows them to do it is masks.

32-BIT STRINGS Figure 9-5 illustrates how masks work. A mask is a series of 32 bits like an IP address. However, a mask always begins with a series of 1s, followed by a series of 0s. In a network mask, the bits in the network part of the mask are 1s, while the remaining bits are 0s. In subnet masks, the bits of both the network and subnet parts are 1s, and the remaining bits are 0s.

The Problem

- There is no way to tell by looking at an IP address what sizes the network, subnet, and host parts are—only their total of 32 bits
- The solution: masks
- Series of initial ones followed by series of final zeros for a total of 32 bits
    - Example: 255.255.0.0 is 16 ones followed by 16 zeros
    - In prefix notation, /16
    - (Decimal 0 is 8 zeros and Decimal 255 is 8 ones)
- Result: IP address where mask bits are ones and zeros where the mask bits are zero

Mask Operation

Network Mask	Dotted Decimal Notation
Destination IP Address	128.171.17.13
Network Mask	255.255.0.0
Bits in network part, followed by zeros	128.171.0.0

Subnet Mask	Dotted Decimal Notation
Destination IP Address	128.171.17.13
Subnet Mask	255.255.255.0
Bits in network part and subnet parts, followed by zeros	128.171.17.0

FIGURE 9-5 IP Network and Subnet Masks

A mask is a 32-bit string of 1s and 0s.

The mask has a certain number of initial 1s. The remaining bits are 0.

In network masks, the initial 1s correspond to the network part.

In subnet masks, the initial 1s correspond to the network and subnet parts

For example, suppose that the mask is 255.255.0.0. In dotted decimal notation eight 1s is 255 and eight 0s is 0. So this mask has sixteen 1s followed by sixteen 0s. In prefix notation, the mask is written as /16. Prefix notation gives the number of initial 1s.

MASKING IP ADDRESSES The figure shows what happens when a mask is applied to an IP address, 128.171.17.13. The mask is 255.255.0.0. Where the mask has one, the result is the original bits of the IP address. For the remaining bits, which are 0s, the result is 0. In this case, the result is 128.171.0.0.

NETWORK MASKS Routers use two types of masks. **Network masks** have 1s in the network bits and 0s for remaining bits. If the network mask is 255.255.0.0 and the IP address is 128.171.17.13, then the result is 128.171.0.0. This is the network part followed by 0s.

SUBNET MASKS For **subnet masks**, in contrast, the initial 1s indicate the number of bits in *both* the network and subnet parts. So, if 128.171 is the network part and 17 is the subnet part, then the subnet mask will be 255.255.255.0 (/24). If you mask 128.171.17.13 with /24, you get 128.171.17.0.

Why mark the 1s in both parts? Think of a network as a state and a subnet as a city. In the United States, there are two major cities named Portland—one in Maine and the other in Oregon. You cannot just say "Portland" to designate a city. You must give both the state and city—analogously, the network and subnet parts.

PERSPECTIVE Quite simply, network masks give the original bits in the network part, followed by 0s. Subnet masks give the original bits in the network and subnet parts, followed by 0s.

Test Your Understanding

4. a) How many bits are there in a mask? b) What do the 1s in a network mask correspond to in IP addresses? c) What do the 1s in a subnet mask correspond to in IP addresses? d) When a network mask is applied to any IP address on the network, what is the result?
5. a) List the bits (1s and 0s) in the mask 255.255.255.0. b) What are the bits (1s and 0s) in the mask /14? c) If /14 is the network mask, how many bits are there in the network part? d) If /14 is the network mask, how many bits are there in the subnet part? e) If /14 is the network mask, how many bits are there in the host part? f) If /14 is the subnet mask, how many bits are there in the network part? g) If /14 is the subnet mask, how many bits are there in the subnet part? h) If /14 is the subnet mask, how many bits are there in the host part?

HOW ROUTERS PROCESS PACKETS

Switching versus Routing

In Chapters 1 and 4, we saw that Ethernet switching is very simple. As Figure 9-6 shows, each Ethernet address only appears in a single row. This row tells the

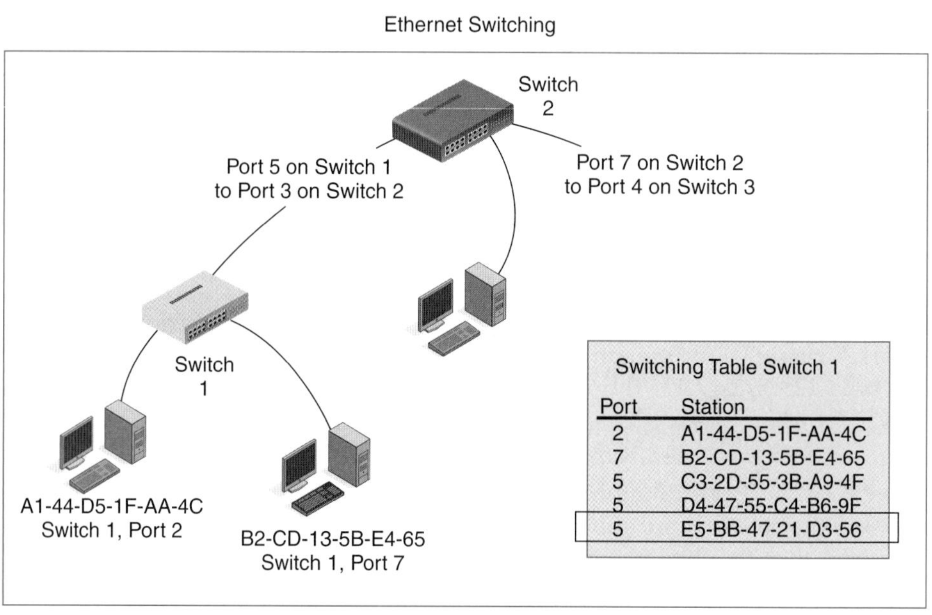

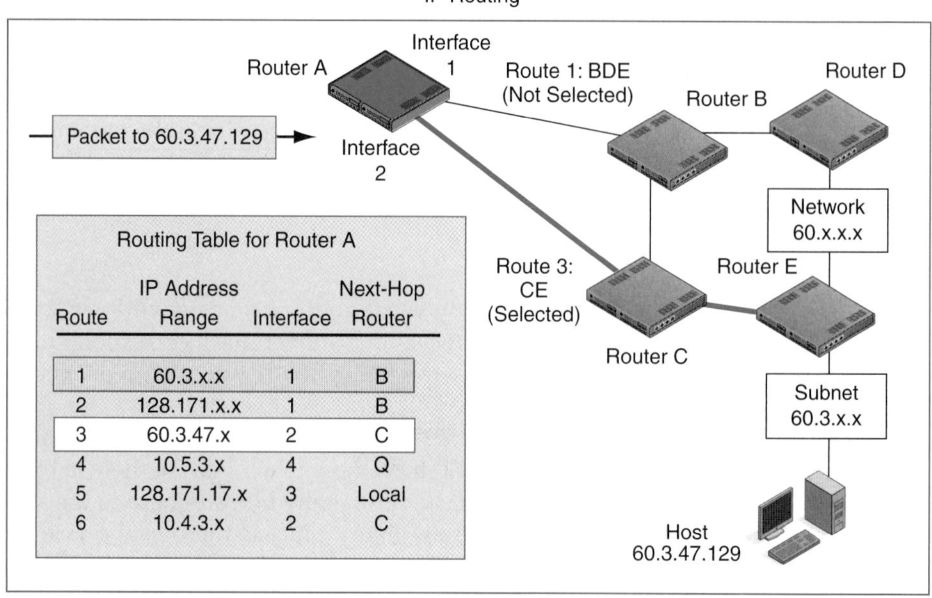

FIGURE 9-6 Ethernet Switching versus IP Routing

switch which port to send the frame back out. This single row can be found quickly, so an Ethernet switch does little work per frame. This makes Ethernet switching inexpensive.

In contrast, firms organize routers in meshes. This gives more reliability because it allows many possible alternative routes between endpoints. Figure 9-6 shows that in a routing table, each row represents an alternative route for a packet. Consequently, to **route** (forward) a packet, a router must first find all rows representing alternative routes that a packet can take. It must then pick the best alternative route from this list. This requires quite a bit of work per packet, making routing more expensive than switching.

Test Your Understanding

6. Why are routing tables more complex than Ethernet switching tables? Be articulate.

Routing Table

Figure 9-8 shows a routing table. It has a number of rows and columns. We will see how a router uses these rows and columns to make a routing decision—a decision about what to do with an arriving packet.

ROWS ARE ROUTES In the routing table, each row represents a route for all IP addresses within a range of IP addresses—typically a network or subnet. It does not specify the full

Routing

Processing an individual packet and passing it on its way is called routing

The Routing Table

Each router has a routing table that it uses to make routing decisions

Routing Table Rows

Each row represents a route for a range of IP addresses—often packets going to the same a network or subnet

A Routing Decision

Step 1: Finding All Row Matches

The router looks at the destination IP address in an arriving packet

For each row:

Apply the row's mask to the destination IP address in the packet

Compare the result with the row's destination value

If the two match, the row is a match

The router does this to ALL rows because there may be multiple matches

This step ends with a set of matching rows

(*continued*)

FIGURE 9-7 The Routing Process (Study Figure)

Example 1: A Destination IP Address that is NOT in the Range of the Row

Destination IP Address of Arriving Packet	60.43.7.8
Apply the (Network) Mask	255.255.0.0
Result of Masking	60.43.0.0
Destination Column Value	128.171.0.0
Does Destination Match the Masking Result?	No
Conclusion	Not a match.

Example 2: A Destination IP Address that is in the Range of the Row

Destination IP Address of Arriving Packet	128.171.17.13
Apply the Mask	255.255.0.0
Result of Masking	128.171.0.0
Destination Column Value	128.171.0.0
Does Destination Match the Masking Result?	Yes
Conclusion	Row is a match.

Step 2: Find the Best-Match Row

The router examines the matching rows it found in Step 1 to find the best-match row

Basic Rule: It selects the row with the longest match (Initial 1s in the row mask)

Tie Breaker: If there is a tie on longest match, select among the tie rows based on metric

- For cost metric, choose the row with the lowest metric value
- For speed metric, choose the row with the highest metric value

Step 3: Send the Packet Back Out

Send the packet out the interface (router port) designated in the best-match row

Send the packet to the router in the next-hop router column

- If the address says Local, the destination host is out that interface
- Sends the packet to the destination IP address in a frame

FIGURE 9-7 Continued

route, however, only the next step in the route (either the next-hop router to handle the packet next or the destination host).

> In the routing table, each row represents a route for all IP addresses within a range of IP addresses—typically a network or subnet.

This is important because the routing table does not need a row for *each IP address* as an Ethernet switching table does. It only needs a row for each *group of IP addresses*. This means that a router needs many fewer rows than an Ethernet switch would need for the same number of addresses.

However, there are many more IP addresses on the Internet than there are Ethernet addresses in an Ethernet network. Even with rows representing groups of IP addresses, core routers in the Internet backbone still have several hundred thousand rows. In addition, while an Ethernet switch only needs to find a single row for each arriving frame, we will see that routers need to look carefully at *all* rows.

Row	Destination Network or Subnet	Mask (/Prefix)	Metric (Cost)	Interface	Next-Hop Router
1	128.171.0.0	255.255.0.0 (/16)	47	2	G
2	172.30.33.0	255.255.255.0 (/24)	0	1	Local
3	60.168.6.0	255.255.255.0 (/24)	12	2	G
4	123.0.0.0	255.0.0.0 (/8)	33	2	G
5	172.29.8.0	255.255.255.0 (/24)	34	1	F
6	172.40.6.0	255.255.255.0 (/24)	47	3	H
7	128.171.17.0	255.255.255.0 (/24)	55	3	H
8	172.29.8.0	255.255.255.0 (/24)	20	3	H
9	172.12.6.0	255.255.255.0 (/24)	23	1	F
10	172.30.12.0	255.255.255.0 (/24)	9	2	G
11	172.30.12.0	255.255.255.0 (/24)	3	3	H
12	60.168.0.0	255.255.0.0 (/16)	16	2	G
13	0.0.0.0	0.0.0.0 (/0)	5	3	H

FIGURE 9-8 Routing Table

ROW NUMBER COLUMN The first column in Figure 9-8 is a route (row) number. Routing tables actually do not have this column. We include it to allow us to refer to specific rows in our discussion. Again, each row specifies a route to a destination.

Test Your Understanding

7. a) In a routing table, what does a row represent? b) Do Ethernet switches have a row for each individual Ethernet address? c) Do routers have a row for each individual IP address? d) What is the advantage of the answer to the previous subpart of this question?

Step 1: Finding All Row Matches

We will now see how the router uses its routing table to make routing decisions. The first step is to find which of the rows in the routing table match the destination IP address in an arriving packet. Due to the existence of alternative routes in a router mesh, most packets will match more than one row.

ROW MATCHES How does the router know which IP addresses should be governed by a row? The answer is that it uses the *Destination Network or Subnet* (*destination*) column and the *Mask* column.

Suppose that all IP addresses in the University of Hawaii (UH) network should be governed by a row. The mask would be the network mask 255.255.0.0, because the UH Network has a 16-bit network part. If this mask is applied to any UH address, the result will be 128.171.0.0. This is the value that will be in the destination column. In fact, this is Row 1 in Figure 9-8.

Let's see how routers use these two fields in Figure 9-8. Suppose that a packet arrives with the IP address 60.43.7.8. The router will look first at Row 1.

- In this row, the router applies the mask 255.255.0.0 to the arriving packet's destination IP address, 60.43.7.8. The result is 60.43.0.0.
- Next, the router compares the masking result, 60.43.0.0, to the destination value in the row, 128.171.0.0. The two are different, so the row is not a match.

However, suppose that a packet arrives with the IP address 128.171.17.13. Now, the situation is different.

- Again, router applies the mask 255.255.0.0 in Row 1 to the destination IP address, 128.171.17.13. The result is 128.171.0.0.
- Next, the router compares 128.171.0.0 to the destination value in the row, 128.171.0.0. The two are identical. Therefore, the row is a match.

MASK AND COMPARE This may seem like an odd way to see if a row matches. A human can simply look at 60.43.7.8 and see that it does not match 128.171.0.0. However, routers do not possess human pattern-matching abilities.

While routers cannot do sophisticated pattern recognition, routers (and all computers) have specialized circuitry for doing masking and comparing—the two operations that row matching requires. Thanks to this specialized circuitry, routers can blaze through hundreds of thousands of rows in a tiny fraction of a second.

THE DEFAULT ROW The last row in Figure 9-8 has the destination 0.0.0.0 and the mask 0.0.0.0. This row will match *every* IP address because masking any IP address with 0.0.0.0 will give 0.0.0.0—the value in the destination field of Row 13. This row ensures that at least one row will match the destination IP address of every arriving packet. It is called the **default row**. In general, a "default" is something you use if you do not have a more specific choice.

LOOK AT ALL ROWS Thanks to their mesh topology, internets have many alternative routes. Consequently, a router cannot stop the first time it finds a row match for each arriving packet because there may be a better match further on. A router has to look at each and every row in the routing table to see which match. So far, we have seen what the router does in Row 1 of Figure 9-8. The router then goes on to Row 2 to see if it is a match by masking and comparing. After this, it goes on to Row 3, Row 4, Row 5, and so on, all the way to the final row (Row 13 in Figure 9-8).

Test Your Understanding

8. a) In Row 3 of Figure 9-8, how will a router test if the row matches the IP address 60.168.6.7? Show the calculations. Is the row a match? b) Why is the last row called the default row? c) Why must a router look at all rows in a routing table? d) What rows match 172.30.17.6? e) Which rows match 60.168.7.32? Show your calculations for rows that match. f) Which rows in Figure 9-8 match 128.171.17.13? (Don't forget the default row.) Show your calculations for rows that match.

Step 2: Selecting the Best-Match Row

LIST OF MATCHING ROWS At the end of Step 1, the mask and compare process, the router has a list of matching rows. For a packet with the destination IP address 128.171.17.13, two rows in Figure 9-8 match. The first is Row 1, as we have already seen. The second is Row 7, with a destination of 128.171.17.0 and a mask of 255.255.255.0. From these, the router must select the **best-match row,** the row that represents the best route for an IP address.

BASIC RULE: LONGEST MATCH RULE How does the router decide whether to follow Row 1 or Row 7? The answer is that it follows the rule of selecting the **longest match**. Row 1 has a mask of 255.255.0.0, which means that it has a 16-bit match. Row 7, in turn, has the prefix /24, meaning that it has a 24-bit match. Row 7 has the longest match, so the router selects it.

Why the longest match rule? The answer is that the closer a route gets a packet to the destination IP address, the better. Row 1 only gets the packet to the UH network, 128.171.x.x, while Row 7 gets the packet all the way to the Shidler College of Business subnet of the University of Hawaii, 128.171.17.x—the subnet that contains host 128.171.17.13.

TIE BREAKER: THE METRIC COLUMN FOR MATCH LENGTH TIES What if two rows have the same longest match? For instance, the destination IP address 172.29.8.112 matches both Row 5 and Row 8 in Figure 9-8. Both have a match length of 24 bits—a tie.

In case of a tie for longest match, the tie-breaker rule is to use the **metric column,** which describes the "goodness" of a route. For instance, in Figure 9-8, the metric is cost. Row 5 has a cost of 34, while Row 8 has a cost of 20. Lower cost is better than higher cost, so the router selects Row 8.

In this case, the row with the lowest metric won. However, what would have happened if the metric had been speed instead of cost? More speed is better, so the router would choose Row 5, with the higher speed (34).

Test Your Understanding

9. a) Distinguish between Step 1 and Step 2 in the routing process. b) If any row other than the default row matches an IP address, why will the router never choose the default row? c) Which rows in Figure 9-8 match 128.171.17.13? (Don't forget the default row.) Show your calculations. d) Which of these is the best-match row? Justify your answer. e) What rows match 172.40.17.6? Show your work. f) Which of these is the best-match row? Justify your answer. g) Which rows match 172.30.12.47? Show your work. h) Which of these is the best-match row? Justify your answer. i) How would your previous answer change if the metric had been reliability?

Step 3: Sending the Packet Back Out

In Step 1, the router found all rows that matched the destination IP address of the arriving packet. In Step 2, it found the best-match row. Finally, in Step 3, the router sends the packet back out.

INTERFACE Router ports are called **interfaces**. The fifth column in Figure 9-8 is *interface number*. If a router selects a row as the best match, the router sends the packet out the interface designated in that row. If Row 1 is selected, the router will send the packet out Interface 2.

NEXT-HOP ROUTER In a switch, a port connects directly to another switch or to a computer. However, a router interface connects to an entire subnet or network. Therefore, it is not enough to select an interface to send the packet out. It is also necessary to specify *a particular device* on the subnet.

In most cases, the router will send the packet on to another router, called the **next-hop router**. The Next-Hop Router column specifies the router that should receive the packet. It will then be up to that next-hop router to decide what to do next. In Figure 9-8, the next-hop router value is G.[1]

In some cases, however, the destination host will be on the subnet out a particular interface. In that case, the router should send the packet to the destination host instead of to another router. In this case, the next-hop router field will say *local*.

Test Your Understanding

10. a) Distinguish between Step 2 and Step 3 in routing. b) What are router ports called? c) If the router selects Row 13 as the best-match row, what interface will the router send the interface out? d) To what device? e) Why is this router called the default router? (The answer is not in the text.) f) If the router selects Row 2 as the best-match row for packet 172.30.33.6, what interface will the router send the interface out? g) To what device? (Don't say, "the local device.")

Cheating (Decision Caching)

We have discussed what happens when a packet arrives at a router. However, what will the router do if another packet for the same destination IP address arrives immediately afterward? The answer is that the router *should* go through the entire process again. Even if a thousand packets arrive that are going to the same destination IP address, the router should go through the entire three-step process for each of them.

As you might expect, a router might cheat, or as it is euphemistically named, cache (remember) the decision it made for a destination IP address. It will then use this decision for successive IP packets going to the same destination. Using a **decision cache** greatly reduces the work that a router will do for each successive packet.

---

[1] Actually, this column should have the IP address of Router G, rather than its name. However, we include the letter designation rather than the IP address for simplicity of understanding.

However, caching is not prescribed in the Internet Protocol. In addition, it is dangerous. The Internet changes constantly as routers come and go and as links between routers change. Consequently, a cached decision that is used too long will result in non-optimal routing or even routes that will not work and that will effectively send packets into a black hole.

Test Your Understanding

11. a) What should a router do if it receives several packets going to the same destination IP address? b) How would decision caching speed the routing decision for packets after the first one? c) Why is decision caching dangerous?

Masking When Masks Do Not Break at 8-Bit Boundaries

Masks That Break at 8-Bit Boundaries

All of the masks we have seen up to this point have had their parts broken at 8-bit segment boundaries. For example, at the University of Hawaii, the network part is 16 bits long, which corresponds to two segments (128.171), the subnet part is 8 bits long (17), and the host part is 8 bits long. All of the masks in Figure 9-8 break also at 8-bit segment boundaries.

Masks that break at 8-bit boundaries are easy for humans to read. In general, you can look at a mask in the table and decide if it matches a particular IP address. For instance, if the mask is 255.255.0.0 (/16), and if the destination column value is 128.171.0.0, this definitely matches the IP address 128.171.45.230.

However, masks do not always break at 8-bit boundaries. For example, suppose that a mask is 11111111 11111000 00000000 00000000 (spaces added for reading). In dotted decimal notation, this is 255.248.0.0, and suppose that the destination address column value is 00000011 10001000 00000000 00000000 (3.264.0.0).

Now suppose that a destination IP address is 3.143.12.12. Does this IP address match the row? There is no way to tell just by looking at the dotted decimal notation versions of the destination, the mask, and the destination IP address. To solve the problem, you go to the raw 32-bit numbers. The following example shows that the masked destination IP address matches the destination value in the row, so the row is a match.

IP address: (3.143.12.12)	00000011	10001111	00001100	00001100
Mask: (255.136.0.0)	11111111	11111000	00000000	00000000
Result: (3.136,0.0)	00000011	10001000	00000000	00000000
Destination: (3.264.0.0)	00000011	10001000	00000000	00000000

Test Your Understanding

12. a) An arriving packet has the destination IP address 128.171.180.13. Row 86 has the destination value 128.171.160.0. The mask is 255.255.224.0. Does this row match the destination IP address? Show your work. You can use the Windows Calculator if you have a Windows PC. In Windows Vista and earlier versions of Windows, choose scientific when you open the calculator. In the Windows 7 calculator, choose programmer mode.

The Address Resolution Protocol (ARP)

The final step in the routing process for each arriving packet is to send the packet back out another interface, to a next-hop router or the destination host. That seems easy enough, but there is one additional thing that routers must do.

To send a packet to a next-hop router or a destination host, the router's interface must place the packet into a frame and send this frame to the next-hop router or destination host. To do this, the interface must know the data link layer address of the destination host. Otherwise, the router's interface will not know what to place in the destination address field of the frame.

The router's internet layer process may only know the IP address of the destination host. If the router's interface is to deliver the frame containing the packet, the internet layer process must discover the data link layer address of the destination host. This is called address resolution.

Address Resolution on an Ethernet Lan with ARP

Determining a data link layer address when you know only an IP address is called **address resolution**. Figure 9-9 shows the **Address Resolution Protocol (ARP),** which provides address resolution on Ethernet LANs. There are other address resolution protocols for other subnet technologies.

ARP Request Message

Suppose that the router receives an IP packet with destination address 10.19.8.17. Suppose also that the router determines from its routing table that it can deliver the packet to a host on one of its Ethernet subnets.

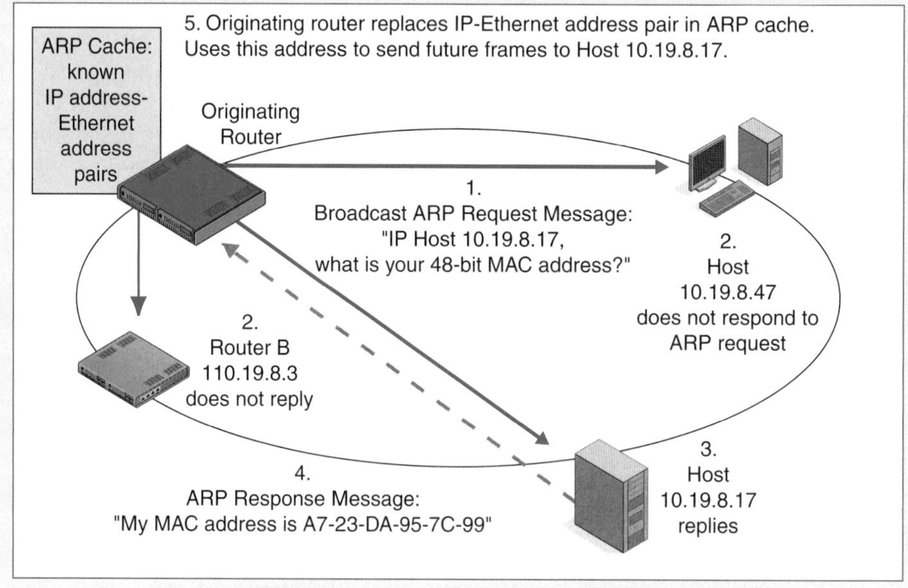

FIGURE 9-9 Address Resolution Protocol (ARP)

- First, the router's internet layer process creates an ARP request message that essentially says, "Hey, device with IP address 10.19.8.17, what is your 48-bit MAC layer address?" The router then broadcasts this ARP packet to all hosts on the subnet.[2]
- Second, the internet layer process on every host examines the ARP request message. If the target IP address is not that of the host, the host's internet layer process ignores the ARP request message. However, host 10.19.8.17 composes an ARP response message that includes its 48-bit MAC layer address (A7-23-DA-95-7C-99). The target host sends this ARP response message back to the router.
- Third, the router's internet layer process now knows the subnet MAC address associated with the IP address. It will deliver the packet to that host in a frame addressed to A7-...

The ARP Cache

ARP is a time-consuming process, and the router does not want to do it for each arriving packet. Consequently, the internet layer process on the router saves the IP address–data link layer address information in its **ARP cache** (section of memory). Afterward, whenever an IP packet comes for this IP destination address, the router will send the IP packet down to its NIC, together with the required MAC address. The NIC's MAC process will deliver the IP packet within a frame containing that MAC destination address.

Using ARP for Next-Hop Routers

We have looked at how routers use ARP when they deliver packets to destination hosts. A router also needs to know the data link layer destination addresses of next-hop routers. Routers use ARP to find the data link layer destination addresses of both destination hosts and other routers.

ARP Encapsulation: Finally, Another Internet Layer Protocol!

In this book so far, we have only seen a single protocol at the internet layer—the Internet Protocol (IP). However, ARP is also a protocol at the internet layer, and ARP messages are called packets. ARP packets are encapsulated directly in frames, just like IP packets.

Test Your Understanding

13. A router wishes to send an IP packet to a host on its subnet. It knows the host's IP address. a) What else must it know? b) Why must it know it? c) What message will it broadcast? d) What device will respond to this broadcast message? e) Does a router have to go through the ARP process each time it needs to send a packet to a destination host or to a next-hop router? Explain. f) Is ARP used to find the destination data link layer destination addresses of destination hosts, routers, or both? g) At what layer does the ARP protocol operate? h) Why must client PCs use ARP to transmit packets? The answer is not in the text.

---

[2] Actually, the router passes the packet down to the data link layer process on the subnet's interface. It tells the data link layer process to broadcast its ARP packet. If the subnet standard is Ethernet, the data link layer process places the packet into a frame with the destination Ethernet address FF-FF-FF-FF-FF-FF (forty-eight 1s). This is the Ethernet broadcast address. Switches will send frames with this broadcast address to all stations, and all stations will accept it as they would a frame addressed to their specific Ethernet address.

THE INTERNET PROTOCOL (IP)

We have focused on IP routing. However, the Internet Protocol has other properties that networking professionals need to understand.

IPv4 Fields

Today, most routers on the Internet and private internets are governed by the **IP version 4 (IPv4)** standard. (There were no versions 0 through 3.) Figure 9-10 shows the IPv4

IP Version 4 Packet

Bit 0 — Bit 31

Version (4 bits) Value is 4 (0100)	Header Length (4 bits)	Diff-Serv (8 bits)	Total Length (16 bits) length in octets
Identification (16 bits) Unique value in each original IP packet		Flags (3 bits)	Fragment Offset (13 bits) Octets from start of original IP fragment's data field
Time to Live (8 bits)	Protocol (8 bits) 1 = ICMP, 6 = TCP, 17 = UDP	Header Checksum (16 bits)	
Source IP Address (32 bits)			
Destination IP Address (32 bits)			
Options (if any)			Padding
Data Field			

IP Version 6 Packet

Bit 0 — Bit 31

Version (4 bits) Value is 6 (0110)	Diff-Serv (8 bits)	Flow Label (20 bits) Marks a packet as part of a specific flow	
Payload Length (16 bits)		Next Header (8 bits) Name of next header	Hop Limit (8 bits)
Source IP Address (128 bits)			
Destination IP Address (128 bits)			
Next Header or Payload (Data Field)			

FIGURE 9-10 IPv4 and IPv6 Packets

packet. Its first four bits contain the value 0100 (binary for 4) to indicate that the packet is formatted according to IPv4. Although we have looked already at some of the fields in IPv4, here we will look at fields that we have not yet seen.

IP TIME TO LIVE (TTL) FIELD In the early days of the ARPANET, which was the precursor to the Internet, packets that were misaddressed would circulate endlessly among packet switches in search of their nonexistent destinations. To prevent this, IP added a **time to live (TTL)** field that is given a value by the source host. Different operating systems have different TTL defaults. Most insert TTL values between 64 and 128. Each router along the way decrements the TTL field by 1. A router decrementing the TTL to 0 will discard the packet.

IP PROTOCOL FIELD The **protocol field** tells the contents of the data field. If the protocol field value is 1, the IP packet carries an ICMP message in its data field. TCP and UDP have protocol values 6 and 17, respectively. After decapsulation, the internet layer process must pass the packet's data field to another process. The protocol field value designates which process should receive the data field.

IP IDENTIFICATION, FLAGS, AND FRAGMENT OFFSET FIELDS If a router wishes to forward a packet to a particular network and the network's maximum packet size is too small for the packet, the router can fragment the packet into two or more smaller packets. Each fragmented packet receives the same **identification field** value that the source host put into the original IP packet's header.

The destination host's internet process reassembles the fragmented packet. It places all packets with the same identification field value together for sorting. It then places them in order of increasing **fragment offset** size. The more fragments flag bit is set (equal to 1) in all but the last fragment. Not setting it in the last fragment indicates that there are no more fragments.

Fragmentation is uncommon in IP today and is suspicious when it occurs because it is rarely used legitimately and is often used by attackers.

IP OPTIONS Similarly, **options** are uncommon in IP today and also tend to be used primarily by attackers. If an option does not end at a 32-bit boundary, **padding** is added up to the 32-bit boundary.

IP DIFF-SERV The **Diff-Serv** field can be used to label IP packets for priority and other service parameters.

Test Your Understanding

14. a) What is the main version of the Internet Protocol in use today? b) What does a router do if it receives a packet with a TTL value of 1? c) What does the protocol field value tell the destination host? d) Under what circumstances would the identification, flags, and fragment offset fields be used in IP? e) Why is IP fragmentation suspicious? f) Why are IP options suspicious? g) What is the purpose of the Diff-Serv field?

IPv6 Fields

The IETF has standardized a new version of the Internet Protocol, **IP version 6 (IPv6)**. As Figure 9-10 shows, IPv6 also begins with a version field. Its value is 0110 (binary for 6). This tells the router that the rest of the packet is formatted according to IPv6.

ADDRESS FIELD The most important change from IPv4 to IPv6 is an increase in the size of IP address fields from 32 bits to 128 bits. The number of possible IP addresses is 2 raised to a power that is the size of the IP address field. For IPv4, this is 2^{32}. For IPv6, this is 2^{128}—an enormous number. IPv6 will support the huge increase in demand for IP addresses that we can expect from mobile devices and from the likely evolution of even simple home appliances into addressable IP hosts.

SLOW ADOPTION IPv6 has been adopted only in a few geographic regions because its main advantage, permitting far more IP addresses, is not too important yet. However, IPv6 is beginning to gather strength, particularly in Asia and Europe, which were short-changed in the original allocation of IPv4 addresses.[3] In addition, the explosion of mobile devices accessing the Internet will soon place heavy stress on the IPv4 IP address space. Fortunately, IPv6 packets can be tunneled through IPv4 networks by placing them within IPv4 packets, so the two protocols can (and will) coexist on the Internet for some time to come.

Test Your Understanding

15. a) How is IPv6 better than IPv4? b) Why has IPv6 adoption been so slow? c) What forces may drive IPv6's adoption in the future? d) Must IPv6 replace IPv4 all at once? Explain.

THE TRANSMISSION CONTROL PROTOCOL (TCP)

Fields in TCP/IP Segments

Chapter 2 looked at the **Transmission Control Protocol (TCP)**. In this section, we will look at this complex protocol in even more depth. When IP was designed, it was made a very simple "best effort" protocol (although its routing tables are complex). The IETF left more complex internetwork transmission control tasks to TCP. Consequently, network professionals need to understand TCP very well. Figure 9-11 shows the organization of TCP messages, which are called TCP segments.

Test Your Understanding

16. a) Why is TCP complex? b) Why is it important for networking professionals to understand TCP? c) What are TCP messages called?

SEQUENCE NUMBERS Each TCP segment has a unique 32-bit sequence number that increases with each segment. This allows the receiving transport process to put arriving TCP segments in order if IP delivers them out of order.

ACKNOWLEDGMENT NUMBERS In Chapter 2, we saw that TCP uses acknowledgments (ACKs) to achieve reliability. If a transport process receives a TCP segment correctly, it sends back a TCP segment acknowledging the reception. If the sending

---

[3] North America has 74 percent of all IPv4 addresses. In fact, Stanford University and MIT have more IPv4 addresses than China, which now has fewer IP addresses than it has Internet users.

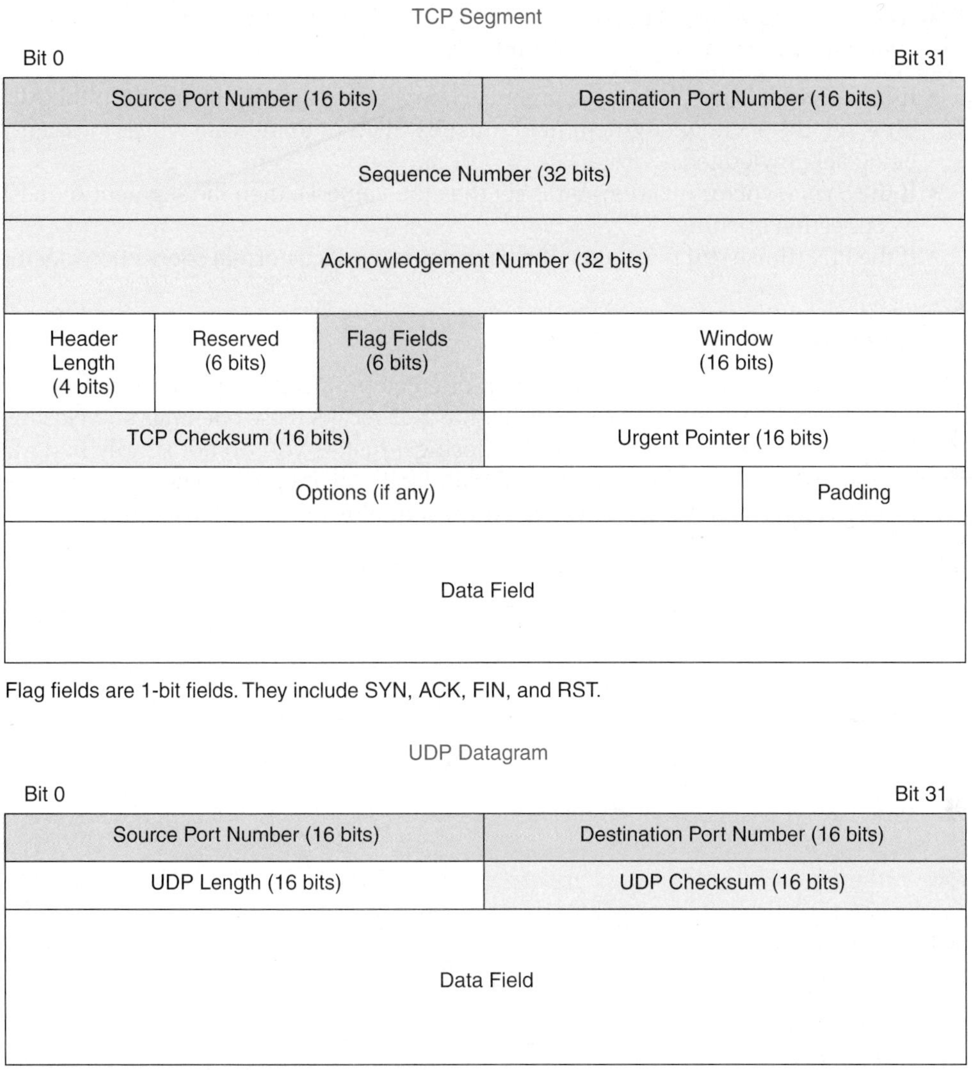

FIGURE 9-11 TCP Segment and UDP Datagram

transport process does not receive an acknowledgment, it transmits the TCP segment again.

The **acknowledgment number field** indicates which segment is being acknowledged. One might expect that if a segment has sequence number X, then the acknowledgment number in the segment that acknowledges it would also be X. As Module A notes, the situation is more complex, but the acknowledgment number is at least based on the sequence number of the segment being acknowledged.

FLAG FIELDS As discussed in Chapter 2, TCP has six single-bit fields. Single-bit fields are called flag fields, and if they have the value 1, they are said to be set. These fields allow the receiving transport process to know the kind of segment it is receiving. We saw several uses of these flag bits in Chapter 2.

- If the ACK bit is set, then the segment acknowledges another segment. If the ACK bit is set, the acknowledgment field must be filled in to indicate which message is being acknowledged.
- If the SYN (synchronization) bit is set (has the value 1), then the segment requests a connection opening.
- If the FIN (finish) bit is set, then the segment requests a normal connection closing.

Openings and Abrupt TCP Closes

In Chapter 2, we saw that TCP is a connection-oriented protocol. Connection-oriented protocols have formal openings and closings. Figure 9-12 recaps these openings and closings.

In Chapter 2, we focused on normal closings. Just as you do not simply hang up on a telephone call when you want to finish talking if you are polite, a normal TCP close consists of two FIN segments, one in each direction, plus their acknowledgments.

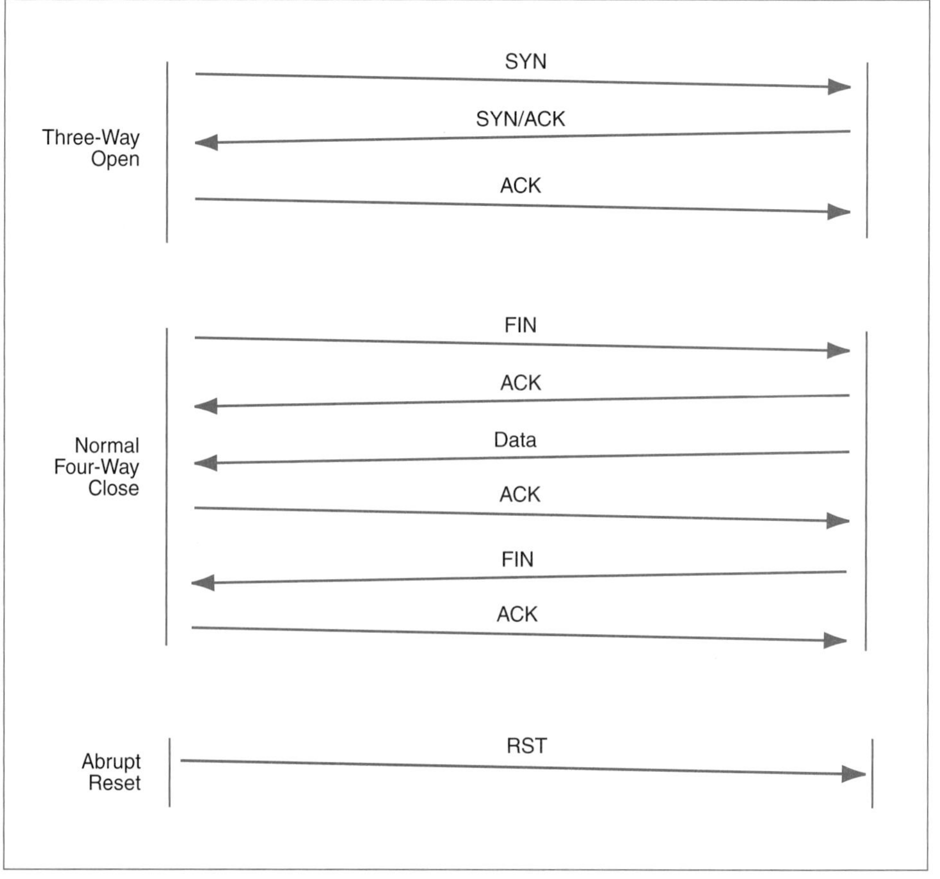

FIGURE 9-12 TCP Session Openings and Closings

However, Figure 9-12 shows that TCP also permits another type of close. This is an abrupt close. Whenever either side wishes to end a conversation, it can simply send a **TCP reset segment**. This is a segment with the **RST** (reset) flag bit set. This may occur if a problem is encountered during a connection, for security reasons, or for any other reason.

Note in Figure 9-12 that an RST segment is not acknowledged. The side that sent the RST segment is not listening any longer, so acknowledging a reset would be as pointless as saying goodbye after someone has hung up on you. The RST segment is one of two segment types that are not acknowledged. As noted in Chapter 2, a pure acknowledgment is not acknowledged because doing so would create an endless loop of acknowledgments.

Test Your Understanding

17. a) Why are sequence numbers good? b) What are 1-bit fields called? c) If someone says that a flag field is set, what does this mean? d) If the ACK bit is set, what other field must have a value? e) What is a FIN segment? f) Distinguish between four-way closes and abrupt resets. g) Why is a reset segment not acknowledged?

Port Numbers

As Figure 9-11 shows, both TCP and UDP have **port number** fields. In Chapter 2, we looked at how servers use port numbers. In this section, we will also look at how clients use port numbers. We will see that clients use port numbers very differently.

SERVER PORT NUMBERS We saw in Chapter 2 that for servers the port number field indicates which application program on the server should receive the message. Major applications have **well-known port numbers** that are usually (but not always) used. These well-known port numbers are from 1 to 1023. For instance, Port 80 is the well-known port number for HTTP webserver programs.

Figure 9-13 shows that every time the client sends a message to a server, the client places the port number of the server application in the destination port number field. In

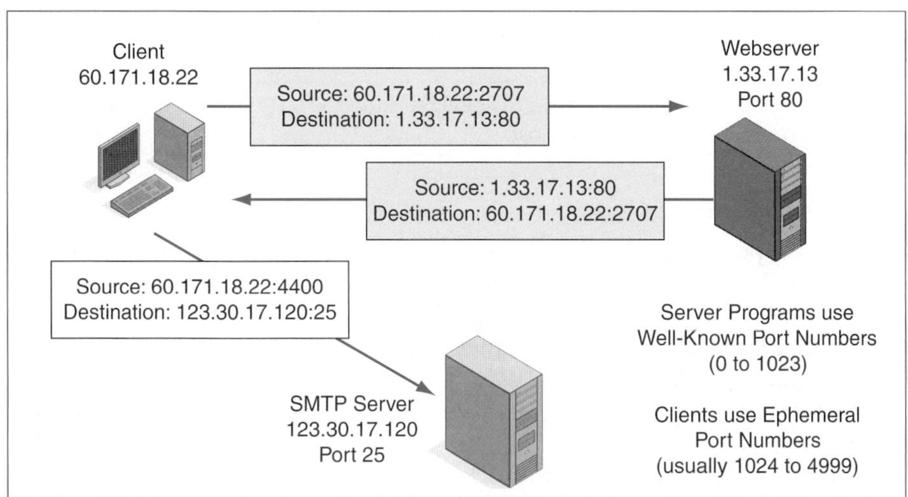

FIGURE 9-13 Use of TCP (and UDP) Port Numbers

this figure, the server is a webserver, and the port number is 80. When the server responds, it places the port number of the application (80) in the source port number field.

We saw in Chapter 3 that server port numbers are used in firewalls. An access control list rule that is based on a server port number can do filtering that is proper for the application indicated by the port number field.

CLIENT PORT NUMBERS Clients do something very different. Whenever a client connects to an application program on a server, the client creates a random **ephemeral port number,** which it only uses for a single TCP session with a single server. According to IETF rules, this port number should be between 49153 and 65535. However, many operating systems ignore these rules and use other ephemeral port numbers. For instance, Microsoft Windows, which dominates the client PC operating system market, uses the ephemeral port number range 1024–4999.

Microsoft Windows uses the ephemeral port number range 1024–4999.

In Figure 9-13, the ephemeral port number is 2707 for client communication with the webserver. When the client transmits, it places 2707 in the source port field of the TCP or UDP header. The server, in return, places this ephemeral port number in all destination port number fields of TCP segments or UDP datagrams it sends to the client.

A client may maintain multiple connections to different application programs on different servers. In Figure 9-13, for example, the client has a connection to an SMTP mail server as well as to the webserver. The client will give each connection a different ephemeral port number to separate the segments of the two connections. For the SMTP connection in Figure 9-13, the client has randomly chosen the ephemeral port number to be 4400.

SOCKETS The combination of an IP address and a port number designates a specific connection to a specific application on a specific host. This combination is called a **socket**. It is written as an IP address, a colon, and then a port number—for instance, 128.171.17.13:80. A two-way conversation is defined by the two sockets involved in the conversation.

A socket is written as an IP address, a colon, and then a port number. It designates a specific application on a specific host.

Test Your Understanding

18. a) What type of port number do servers use? b) What type of port number do clients use? c) What is the port range for well-known port numbers? d) What is the range of Microsoft ephemeral port numbers?

19. A Windows host sends a TCP segment with source port number 25 and destination port number 2404. a) Is the source host a server or a client? Explain. b) If the host is a server, what kind of server is it? c) Is the destination host a server or a client? Explain.

20. a) What is a socket? b) What specifies a particular application on a particular host in TCP/IP? c) How is it written? d) When the SMTP server in Figure 9-13 transmits to the client PC, what will the source socket be? e) The destination socket?

OTHER TCP/IP STANDARDS

In this section, we will look briefly at several other important TCP/IP standards that network administrators need to master. We will look at some of these protocols here. Chapter 10 will look more closely at some TCP/IP protocols that are focused on network management.

Dynamic Routing Protocols

How does a router get the information in its routing table? One possibility is to enter routes manually. However, that approach does not scale to large internets. Instead, as Figure 9-15 shows, routers constantly exchange routing table information with one another using **dynamic routing protocols**.

ROUTING Note that TCP/IP uses the term **routing** in two different but related ways. First, we saw earlier that the process of forwarding arriving packets is called routing. Second, the process of exchanging information for building routing tables is also called routing.

> In TCP/IP the term *routing* is used in two ways—for packet forwarding and for the exchange of routing table information through dynamic routing protocols.

AUTONOMOUS SYSTEMS AND INTERIOR DYNAMIC ROUTING PROTOCOLS Recall from Chapter 1 that the Internet consists of many networks owned by different organizations. Within an organization's network, which is called an **autonomous system,** the organization owning the network decides which dynamic routing protocol to use among its internal routers, as shown in Figure 9-15. For internal use, the organization is free to choose among available **interior dynamic routing protocols**. There are three popular interior dynamic routing protocols. Each has different strengths and weaknesses.

ROUTING INFORMATION PROTOCOL (RIP) The simplest interior dynamic routing protocol created by the IETF is the **Routing Information Protocol (RIP)**. RIP's simplicity

Dynamic Routing Protocol	Interior or Exterior Routing Protocol?	Remarks
RIP (Routing Information Protocol)	Interior	Only for small autonomous systems with low needs for security
OSPF (Open Shortest Path First)	Interior	For large autonomous systems that only use TCP/IP
EIGRP (Enhanced Interior Gateway Routing Protocol)	Interior	Proprietary Cisco Systems protocol. Not limited to TCP/IP routing. Also handles IPX/SPX, SNA, and so forth
BGP (Border Gateway Protocol)	Exterior	Organization cannot choose what exterior routing protocol it will use

FIGURE 9-14 Dynamic Routing Protocols (Study Figure)

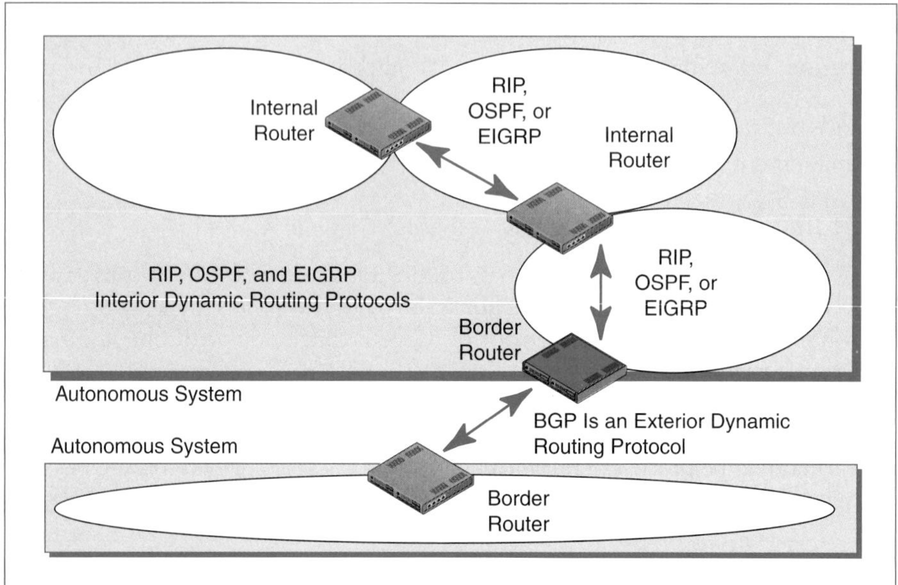

FIGURE 9-15 Dynamic Routing Protocols

makes it attractive for small internets. Management labor is relatively low. On the negative side, RIP is not very efficient because its metric is merely the number of router hops needed to get to the destination host. However, this is not a serious problem for small internets. The one serious problem with RIP in small internets is poor security. If attackers take over a firm's interior dynamic routing protocol communications, they can maliciously reroute the internet's traffic.

OPEN SHORTEST PATH FIRST (OSPF) For larger autonomous systems, or if security is a serious concern, the IETF created the **Open Shortest Path First (OSPF)** dynamic routing protocol. OSPF is very efficient, having a complex metric based on a mixture of cost, throughput, and traffic delays. It also offers strong security. It costs much more to manage than RIP, but unless a corporate internet is very small, OSPF is the only IETF dynamic routing protocol that makes sense.

EIGRP Cisco Systems is the dominant manufacturer of routers. Cisco has its own proprietary interior dynamic routing protocol for large internets—the **Enhanced Interior Gateway Routing Protocol (EIGRP)**. The term **gateway** is another term for *router*. EIGRP's metric is very efficient because it is based on a mixture of interface bandwidth, load on the interface (0 percent to 100 percent of capacity), delay, and reliability (percentage of packets lost). EIGRP is comparable to OSPF, but many companies use it instead of OSPF because it can route SNA and IPX/SPX traffic as well as IP traffic. On the negative side, EIGRP is a proprietary protocol, and using it forces the company to buy only Cisco routers.

EXTERIOR DYNAMIC ROUTING PROTOCOLS For communication outside the organization's network, the organization is no longer in control. It must use whatever **exterior dynamic routing protocol** the external network to which it is connected requires. The almost universal exterior dynamic routing protocol is the **Border Gateway Protocol (BGP)**. BGP is designed specifically for the exchange of routing information between autonomous systems.

Test Your Understanding

21. a) What is the purpose of dynamic routing protocols? b) In what two ways does TCP/IP use the term *routing*?

22. a) What is an autonomous system? b) Within an autonomous system, can the organization choose its interior routing protocol? c) What are the two TCP/IP interior dynamic routing protocols? d) Which IETF dynamic routing protocol is good for small internets that do not have high security requirements? e) Which IETF dynamic routing protocol is good for large internal internets that have high security requirements? f) What is the main benefit of EIGRP compared to OSPF as an internal dynamic routing protocol? g) When might you use EIGRP as your interior dynamic routing protocol? h) May a company select the routing protocol its border router uses to communicate with the outside world? i) What is the almost universal exterior dynamic routing protocol?

Internet Control Message Protocol (ICMP) for Supervisory Messages at the Internet Layer

SUPERVISORY MESSAGES AT THE INTERNET LAYER IP is only concerned with packet delivery. For supervisory messages at the internet layer, the IETF created the **Internet Control Message Protocol (ICMP)**. IP and ICMP work closely together. As Figure 9-16 shows, IP encapsulates ICMP messages in the IP data field, delivering them to their target host or router. There are no higher-layer headers or messages.

ERROR ADVISEMENT IP is an unreliable protocol. It offers no error correction. If the router or the destination host finds an error, it discards the packet. Although there is no retransmission, the router or host that finds the error *may* send an **ICMP error**

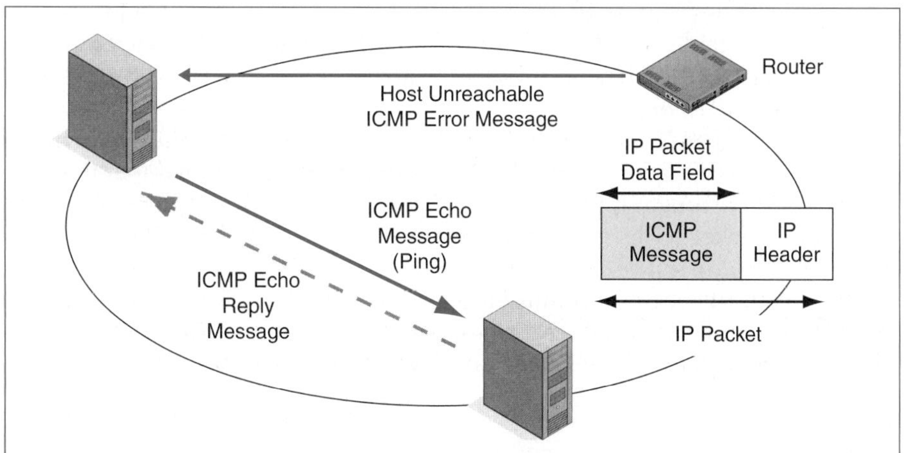

FIGURE 9-16 Internet Control Message Protocol (ICMP) for Supervisory Messages

message to the source device to inform it that an error has occurred, as Figure 9-16 illustrates. This is **error advisement** (notification) rather than error correction. There is no mechanism within IP or ICMP for the retransmission of lost or damaged packets. ICMP error messages are only sent to help the sending process or its human user diagnose problems. One important subtlety is that sending error advisement messages is not mandatory. For security reasons, many firms do not allow error advisement messages to leave their internal internets because hackers can exploit the information contained in them.

ECHO (PING) Perhaps the most famous ICMP error message type is the **ICMP echo** message. One host or router can send an echo request message to another. If the target device's internet process is able to do so, it will send back an echo reply message.

Sending an echo request is often called **pinging** the target host, because it is similar to a submarine pinging a ship with sonar to see if it is there. In fact, the most common program for pinging hosts is called *ping*.[4] Echo is a good diagnostic tool because if there are network difficulties, a logical early step in diagnosis is to ping many hosts and routers to see if they can be reached.

Test Your Understanding

23. a) For what general class of messages is ICMP used? b) Explain error advisement in ICMP. c) Explain the purpose of ICMP echo messages. d) Sending an ICMP echo message is called __________ the target host.

MULTIPROTOCOL LABEL SWITCHING

Earlier in this chapter, we looked at how routers handle IP packets. They look at an incoming packet's destination IP address. They then compare that IP address to every row in the routing table, select the best match, and send the packet back out a certain port to a certain IP address. The next packet to arrive gets the same treatment—*even if it is going to the same IP address.*

We also noted that many router vendors do decision caching, in which they remember their decisions for certain IP address ranges. We also noted that this is somewhat dangerous. Fortunately, there is a more systematic way to avoid having to look at all rows for all packets. This is the **Multiprotocol Label Switching (MPLS)** standard, which Figure 9-17 illustrates.

When two hosts begin to communicate, an MPLS network does not immediately send packets. Instead, it determines the best path for the packets. This best path is called the **label-switched path**. This is what gives MPLS its advantages. It may be slow to

[4] The echo reply message also gives the latency for the reply—the number of milliseconds between echo messages and echo reply messages. This is useful in diagnosing problems.

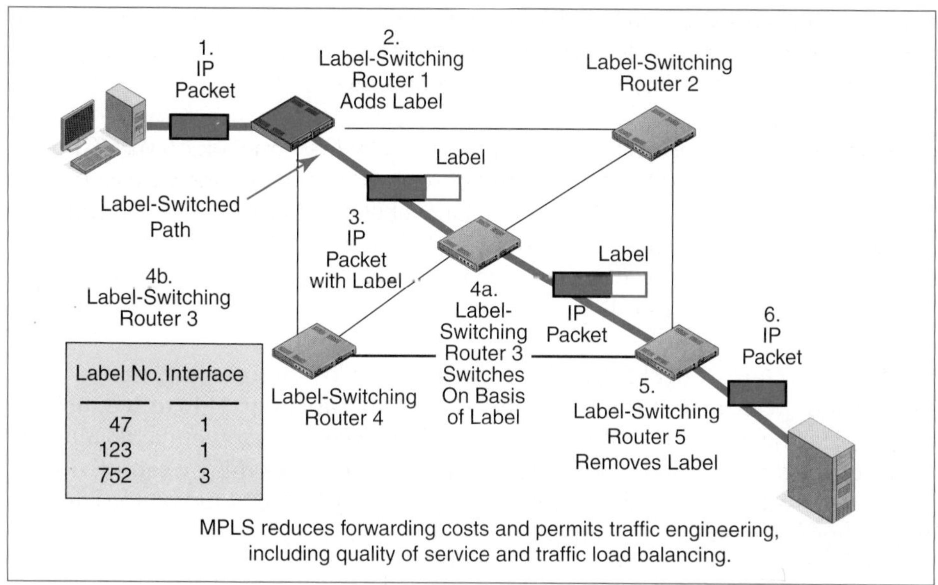

FIGURE 9-17 Multiprotocol Label Switching (MPLS)

select a label-switched path at the beginning of a conversation, but we will see that MPLS handles each subsequent packet very quickly.

As Figure 9-17 shows, the source host sends an ordinary IP packet during the rest of the conversation. The router to which the source host connects is a label-switching router. This router places a **label header** in front of the IP header (and after the frame header). Note that this means that there is no need to change either the IP packet syntax or the frame syntax. The label header is simply pushed into place between the frame header and the IP header.

The label header has a **label number** that identifies the label-switched path previously selected for packets in this conversation. The first label-switched router and all others along the label-switching path have a simple MPLS look-up table. This table allows the label-switched router to look up the label number, read the corresponding interface, and send the packet out the indicated interface. For example, if the label number is 47, the router in Figure 9-17 will send it out Interface 1. There is no need to look at many regular routing table rows to select the best interface to send a packet back out. That work was already done when the label-switched path was selected.

Each label-switched router along the way repeats this simple look-up process and forwards the packet on its way. The last label-switched router removes the label, because the destination host does not need it. Note that neither the source host nor the destination host knows that label switching was done.

LABEL SWITCHING FOR COST REDUCTION MPLS is much faster than traditional routing calculations because there is only a single row for each label number. This simplicity dramatically lowers the cost of routing.

All packets between two host IP addresses might be assigned the same label number. More likely, all traffic between two *sites* might be assigned a single label number because it would all be going to and from the same site.

QUALITY OF SERVICE Lowering routing costs is the main attraction of MPLS, but there are other advantages as well. One is quality of service. Traffic flowing between two sites might be assigned different label numbers. One label number might be for VoIP, which is latency-intolerant, and the other might be for latency-tolerant traffic between the sites. For latency-intolerant traffic, it is even possible to reserve capacity at routers along the selected path. This gives full quality of service.

TRAFFIC ENGINEERING MPLS can also be used for **traffic engineering,** that is, to manage how traffic will travel through the network. One capability is **load balancing**—to move some traffic from a heavily congested link between two routers to an alternative route that uses different and less-congested links. MPLS does this by setting up multiple label-switched routes ahead of time and by sending traffic based on the congestion along different label-switched routes.

MPLS BOUNDARIES Corporations can use MPLS internally, and most ISPs now use MPLS for their traffic, but it is currently impossible to implement MPLS across the entire Internet because of coordination difficulties among the ISPs.

Test Your Understanding

24. a) In MPLS, is selecting the best interface for each packet done when the packet enters the network or before? b) What is the name of the path selected for a particular conversation? c) When a source host first transmits to a destination host, what will happen? d) Do label-switching routers along the MPLS path look at the packet's IP address? Explain your reasoning. e) Why is MPLS decision making fast compared to traditional routing? f) On what does each label-switched router base routing decisions? g) Why is MPLS transparent to the source and destination hosts? h) What is MPLS's main attraction? i) What are its other attractions? j) What is traffic engineering? k) Can MPLS provide traffic load balancing? l) Why is it impossible to implement MPLS on the entire Internet?

CONCLUSION

Synopsis

TCP/IP is a family of standards created by the Internet Engineering Task Force (IETF). IP is TCP/IP's main standard at the internet layer. IP is a lightweight (unreliable and connectionless) protocol. At the transport layer, TCP/IP offers two standards: TCP, which is a heavyweight protocol (reliable and connection-oriented), and UDP, which is a lightweight protocol like IP.

IP addresses are hierarchical. Their 32 bits usually are divided into a network part, a subnet part, and a host part. All three parts vary in length. A network mask tells what bits are in the network part, while a subnet mask tells what bits are in the total of the network or subnet parts.

Routers forward packets through an internet. Border routers move packets between the outside world and an internal site network. Internal routers work within sites, moving packets between subnets. Ports in routers are called interfaces. Different interfaces may connect to different types of networks—for instance, Ethernet or Frame Relay networks. Most routers are multiprotocol routers, which can handle not only TCP/IP internetworking protocols, but also internetworking protocols from IPX/SPX, SNA, and other architectures.

Routers are designed to work in a mesh topology. This creates alternative routes through the internet. Alternative routes are good for reliability. However, the router has to consider the best route for each arriving packet, and this is time consuming and therefore expensive.

To make a routing decision (deciding which interface to use to send an incoming packet back out), a router uses a routing table. Each row in the routing table represents a route to a particular network or subnet. All packets to that network or subnet are governed by the one row. Each row (route) has destination, mask, metric, interface, and a next-hop router fields.

If the destination IP address in an arriving packet is in a row's range, that row is a match. After finding all matches in the routing table, the router finds the best-match row on the basis of match length and metric values. Once a best-match route (row) is selected, the router sends the packet out to the next-hop router in that row.

If you read the box, "The Address Resolution Protocol (ARP)," you saw that the router must encapsulate the packet in a frame in order to send it out. ARP is a protocol to learn the data link layer destination of the device to which a router or host will send a frame if the router or host only knows the destination host's IP address.

IP version 4 has a number of important fields besides the source and destination address fields. The time to live (TTL) field ensures that packets that are misaddressed do not circulate endlessly around the Internet. The protocol field describes the contents of the data field—ICMP message, TCP segment, UDP datagram, and so forth. IP version 6 will offer many more addresses thanks to its 128-bit address fields.

The Transmission Control Protocol (TCP) has sequence numbers that allow the receiving transport process to place arriving TCP segments in order. The TCP header has several flag fields that indicate whether the segment is a SYN, FIN, ACK, or RST segment. Connection openings use a three-step handshake that uses SYN segments. Normal closes involve a four-step message exchange that use FIN segments. Resets close a connection with a single segment (RST) instead of the normal four.

Both TCP and UDP have 16-bit source and destination port number fields that tell the transport process which application process sent or should receive the contents in the segment data field. Major applications have well-known port numbers. For instance, the well-known server port number of HTTP is Port 80. Clients, in contrast, have ephemeral port numbers that they select randomly for each connection. Microsoft uses the ephemeral port number range 1024 to 4999.

Routers build their routing tables by listening to other routers. Routers frequently exchange messages, giving information stored in their routing tables. These messages are governed by one of several available dynamic routing protocols.

IP itself does not have supervisory messages. For internet layer supervisory messages, hosts and routers use the Internet Control Message Protocol (ICMP). We looked at two types of ICMP messages—error advisement messages and echo messages (ping). ICMP messages are carried in the data fields of IP packets.

We concluded with Multiprotocol Label Switching. With MPLS, routers along the label-switched path do not have to look at an incoming packet's IP address and compare it against all rows in the routing table to find possible matches. Instead, label-switching routers select the best path for packets at the beginning of a conversation. They give that path a label number and place it in each label-switching router's MPLS table. They also give each packet a label to indicate its path. The router merely looks at the label number, selects the indicated interface, and send the packet back out.

END-OF-CHAPTER QUESTIONS

Thought Questions

1. How does the postal service use hierarchical sorting? How does this simplify delivery decisions?
2. Give a non-network example of hierarchical addressing, and discuss how it reduces the amount of work needed in physical delivery. Do not use any example in the book, the postal service, or the telephone network.
3. A client PC has two simultaneous connections to the same webserver application program on a webserver. (Yes, this is possible, and in fact, it is rather common.) What will be different between the TCP segments that the client sends on the two connections?
4. A router that has the routing table in Figure 9-8 receives an incoming packet. The source IP address is 10.56.65.234. The destination host is 10.5.7.9. The TTL value is 1. The Protocol field value is 6. What will the router do with this packet?

Harder Thought Question

1. For security reasons, many organizations do not allow error reply messages to leave their internal internets. How, specifically, could hackers use information in echo reply messages to learn about the firm's internal hosts?

Troubleshooting Question

1. You suspect that the failure of a router or of a transmission line connecting routers has left some of your important servers unavailable to clients at your site. How could you narrow down the location of the problem using what you learned in this chapter?

Perspective Questions

1. What was the most surprising thing you learned in this chapter?
2. What was the most difficult material for you in this chapter?

CHAPTER 10

Internetworking II

LEARNING OBJECTIVES
By the end of this chapter, you should be able to:

- Explain TCP/IP management: IP subnet planning, Network Address Translation (NAT), Multiprotocol Labor Switching (MPLS), the Domain Name System (DNS), DHCP servers, and the Simple Network Management Protocol (SNMP).
- Discuss communication over the Internet via SSL/TLS and IPsec VPNs and via IP carrier services.
- Discuss directory servers, including Microsoft's Active Directory.

INTRODUCTION

In the last chapter, we looked at important TCP concepts. In this chapter, we will focus on the management of TCP networks.

CORE TCP/IP MANAGEMENT TASKS

If a firm uses TCP/IP as its internetworking protocol, it must do a considerable amount of work to build and maintain the necessary infrastructure of TCP/IP. While switched networks are (generally) capable of operating for long periods of time without intervention by network managers, TCP/IP internets require constant tuning and support. This results in a need for considerable TCP/IP expertise and management effort. As we saw in Chapter 1, network managers say, "Switch where you can, route where you must."

IP Subnet Planning

As Chapter 8 discussed, IP addresses are 32 bits long. Each organization is assigned a network part. We saw that the University of Hawaii's network part (128.171) is 16 bits long. There is nothing a firm can do to alter its network part. However, it was up to the university to decide what to do with the remaining 16 bits.

SUBNETTING AT THE UNIVERSITY OF HAWAII The university, like most organizations, chose to subnet its IP address space. It divided the 16 bits over which it has discretion into an 8-bit subnet part and an 8-bit host part.

THE $2^N - 2$ RULE With N bits, you can represent 2^N possibilities. Therefore, with 8 bits, one can represent 2^8 (256) possibilities. This would suggest that the university can have 256 subnets, each with 256 hosts. However, a network, subnet, or host part cannot be all 0s or all 1s.[1] Therefore, the university can have only 254 (256 – 2) subnets, each with only 254 hosts. Figure 10-1 illustrates these calculations.

In general, if a part is N bits long, it can represent $2^N - 2$ networks, subnets, or hosts. For example, if a subnet part is 9 bits long, there can be $2^9 - 2$, or 510, subnets. Or if a host part is 5 bits long, there can be $2^5 - 2$, or 30, hosts.

In general, if a part is N bits long, it can represent $2^N - 2$ networks, subnets, or hosts.

BALANCING SUBNET AND HOST PART SIZES The larger the subnet part, the more subnets there will be. However, the larger the subnet part is made, the smaller the host part will be. This will mean fewer hosts per subnet. There is always a trade-off. More subnets mean fewer hosts, and more hosts mean fewer subnets.

The University of Hawaii's choice of 8-bit network and subnet parts was acceptable for many years because no college needed more than 254 hosts. Its advantage is that its subnet mask (255.255.255.0) was very simple, breaking at 8-bit boundaries. This

Step	Description				
1	Total size of IP address (bits)	32			
2	Size of network part assigned to firm (bits)	16		8	
3	Remaining bits for firm to assign	16		24	
4	Selected subnet/host part sizes (bits)	8/8	6/10	12/12	8/16
5	Possible number of subnets (2^N-2)	254 (2^8-2)	62 (2^6-2)	4,094 ($2^{12}-2$)	254 (2^8-2)
6	Possible number of hosts per subnet (2^N-2)	254 (2^8-2)	1,022 ($2^{10}-2$)	4,094 ($2^{12}-2$)	65,534 ($2^{16}-2$)

FIGURE 10-1 IP Subnetting

[1] If you have all 1s in an address part, this indicates that broadcasting should be used. All 0s parts are used by computers when they do not know their addresses. As we will see later in this chapter, most client PCs get their IP addresses from DHCP servers. All-zero addresses can only be used in the source addresses of DHCP messages sent from a host to a DHCP server.

made it easy to see which hosts were on which subnets. The host at 128.171.17.5, for instance, was the fifth host on the 17th subnet. If the subnet mask did not break at an 8-bit boundary, you cannot see which subnet a host is on by looking at the address in dotted decimal notation.

However, many colleges in the university now have more than 254 computers, and the limit of 254 hosts required by its subnetting decision has become a serious problem. Several colleges now have two subnets connected by routers. This is expensive and awkward.

The university would have been better served had it selected a smaller subnet part, say 6 bits. As Figure 10-1 shows, this would have allowed 62 college subnets, which probably would have been sufficient. A 6-bit subnet part would give a 10-bit host part, allowing 1,022 hosts per subnet. This would be ample for several years to come.

A CRITICAL CHOICE In general, it is critical for corporations to plan their IP subnetting carefully, in order to get the right balance between the sizes of their network and subnet parts.

Test Your Understanding

1. a) Why is IP subnet planning important? b) If you have a subnet part of 9 bits, how many subnets can you have? c) Your firm has an 8-bit network part. If you need at least 250 subnets, what must your subnet size be? d) How many hosts can you have per subnet? e) Your firm has a 20-bit network part. What subnet part would you select to give at least 10 subnets? f) How many hosts can you have per subnet?

Network Address Translation (NAT)

One issue that firms face is whether to allow people outside the corporation to learn their internal addresses. This is a security risk. If attackers know internal IP addresses, this allows them to send attack packets from the outside world. To prevent this, companies can use **network address translation (NAT)**, which presents external IP addresses that are different from internal IP addresses used within the firm.

NAT OPERATION Figure 10-3 shows how NAT works. An internal client host, 192.168.5.7, sends a packet to an external server host. The source address in this packet is 192.168.5.7, of course. The source port number is 3333, which is an ephemeral port number that the source host made up for this connection.

When the NAT firewall at the border receives the packet, it makes up a new row in its translation table. It places the internal IP address and port number in the table. It then generates a new external source IP address and external source port number. These are 60.5.9.8 and 4444, respectively.

When packets arrive from the external host, they have 60.5.9.8 in their destination IP address fields and 4444 in their destination port number fields. The NAT firewall looks these values up in its translation table, replaces the external values with the internal values, and sends them on to the client PC.

TRANSPARENCY NAT is transparent to both internal and external hosts. Hosts do not even know that NAT is happening. Consequently, there is no need to change the ways in which they operate.

NAT

Sends false external IP addresses that are different from internal IP addresses

NAT Operation (Figure 10-13)

NAT Is Transparent to Internal and External Hosts

Security Reason for Using NAT

External attackers can put sniffers outside the corporation

Sniffers can learn IP addresses

Attackers can send attacks to these addresses

With NAT, attackers only learn false external IP addresses

Expanding the Number of Available IP Addresses

Companies may receive a limited number of IP addresses from their ISPs

There are roughly 4,000 possible ephemeral port numbers for each IP address

So for each IP address, there can be 4,000 external connections

If a firm is given 248 IP addresses, there can be roughly one million external connections

Even if each internal device averages several simultaneous external connections, there should not be a problem providing as many external IP connections as a firm desires

Private IP Addresses

Can only be used inside firms

10.x.x.x

192.168.x.x (most popular)

172.16.x.x through 172.31.x.s

Protocol Problems with NAT

IPsec, VoIP, etc. may prove difficult to implement if NAT is used.

Work-arounds must be considered very carefully in product selection

FIGURE 10-2 Network Address Translation (NAT) (Study Figure)

SECURITY Figure 10-3 shows how NAT brings security. An attacker may be able to install a **sniffer program** beyond the corporation's NAT firewall. This sniffer will be able to read all packets coming out of the firm. With NAT, an eavesdropper only learns false (external) IP addresses and port numbers. In theory, if an attacker can attack immediately, it can send packets to the external IP addresses and port numbers, and the NAT firewall will pass them on to the internal host. However, this is rarely possible. NAT provides a surprising amount of security despite its simple operation.

EXPANDING THE EFFECTIVE NUMBER OF AVAILABLE IP ADDRESSES An equally important potential reason for using NAT is to permit a firm to have many more internal IP addresses than it is given by its ISP. Suppose that an ISP only gives the firm 254 IP

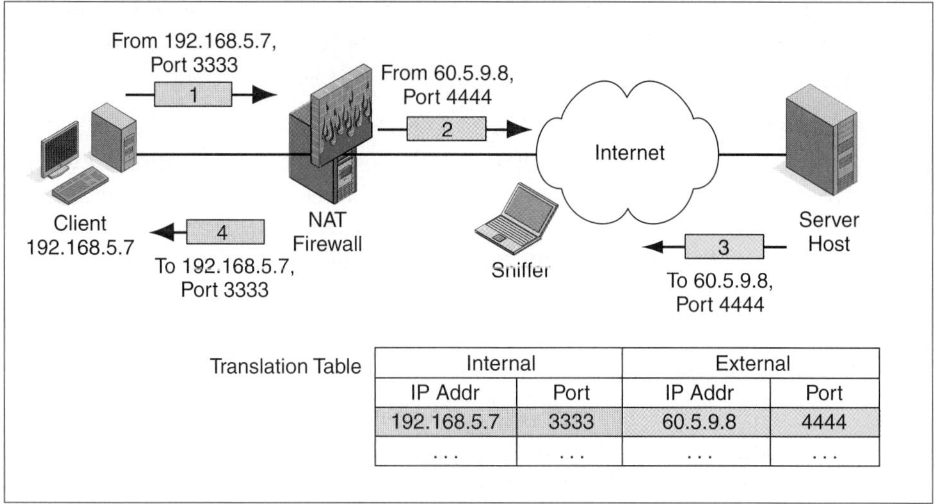

Internal		External	
IP Addr	Port	IP Addr	Port
192.168.5.7	3333	60.5.9.8	4444
...	...	...	...

FIGURE 10-3 Network Address Translation (NAT)

addresses by giving it a host part with 24 bits. Without NAT, a firm can only have 254 PCs simultaneously using the Internet.

However, NAT is really network address translation/port number translation (NAT/PAT). With Microsoft Windows, there are almost 4,000 ephemeral port numbers. Even if internal hosts maintained four simultaneous connections to the outside world on average, each IP address could be used by 1,000 client PCs to make outside connections by giving each client PC a different set of port numbers for its connections. If a firm has 254 IP addresses, it can multiply this already large number of connections by 4000. This could support almost a quarter million PCs using the Internet simultaneously, with each PC having four external connections. While no firm would push the number of connections this far, having ten to one hundred internal hosts for each external IP address is very common.

USING PRIVATE IP ADDRESSES To support NAT, the Internet Assigned Numbers Authority (IANA) has created three sets of **private IP address ranges** that can only be used *within* firms. These are the three ranges:

- 10.x.x.x
- 192.168.x.x
- 172.16.x.x through 172.31.x.x

The 192.168.x.x private IP address range is the most popular because it allows companies to use 255.255.0.0 and 255.255.255.0 network and subnet masks, respectively. These break at convenient 8-bit boundaries. However, the other two private IP address ranges are also widely used.

PROTOCOL PROBLEMS WITH NAT In terms of security and expanding IP effective address ranges, NAT is a simple and effective tool. However, some protocols cannot work across a NAT firewall or can work only with considerable difficulty. These include the popular IPsec cryptographic system in transport mode and several VoIP protocols. The decision to use NAT must be made only after a careful assessment of protocols.

Test Your Understanding

2. a) What is NAT? (Do not just spell it out.) b) Describe NAT operation. c) What are the two benefits of NAT? d) How does NAT enhance security? e) How does NAT allow a firm to deal with a shortage of IP addresses given to it by its ISP? f) How are private IP address ranges used? g) What are the three ranges of private IP addresses? h) What problems may firms encounter when using NAT?

Domain Name System (DNS)

As we saw in Chapter 1, if a user types in a target host's host name, the user's PC will contact its local Domain Name System (DNS) server. The DNS server will return the IP address for the target host or will contact other DNS servers to get this information. The user's PC can then send IP packets to the target host. In this chapter, we will look at DNS and its management in more detail.

MORE DETAILED OPERATION Figure 10-4 looks at how a DNS provides IP addresses when a host sends a request message containing a host name. In many cases, as we saw in Chapter 1, the local DNS server will know the IP address and send it back. In other cases, the local DNS host will not know the host's IP address. It must then find the **authoritative DNS server** for the domain in the host name. In the figure, dakine.pukanui.com's authoritative DNS server is authoritative for the pukanui.com

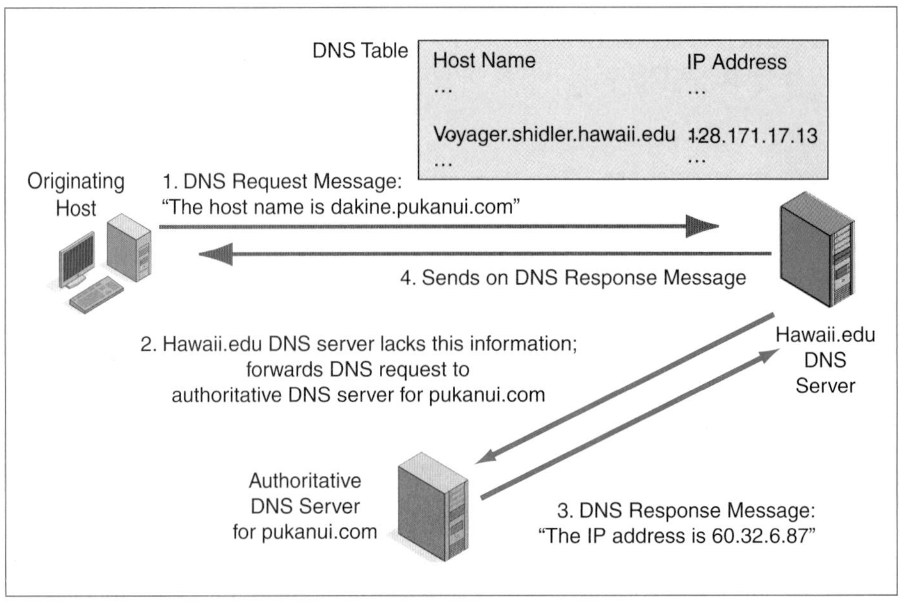

FIGURE 10-4 Domain Name System (DNS) Lookup

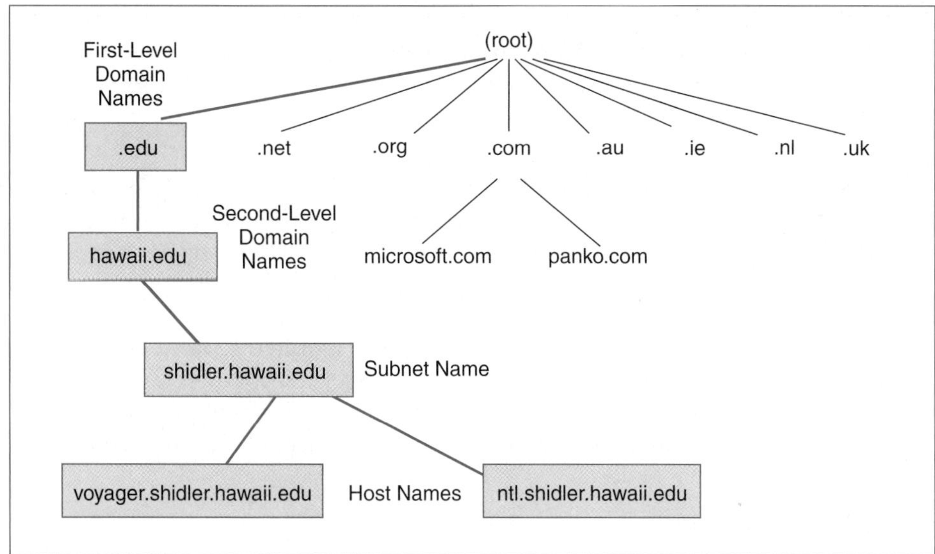

FIGURE 10-5 Domain Name System (DNS) Hierarchy

domain. That DNS server will send the IP address to the local DNS server, which will pass the address on to the host that sent the DNS request.

WHAT IS A DOMAIN? Figure 10-5 shows that the **Domain Name System (DNS)** and its servers are not limited to providing IP addresses for host names. More generally, DNS is a general system for naming domains. A **domain** is any group of resources (routers, single networks, and hosts) under the control of an organization. The figure shows that domains are hierarchical, with host names being at the bottom of the hierarchy.

A domain is any group of resources (routers, single networks, and hosts) under the control of an organization.

ROOT AND FIRST-LEVEL DOMAINS The domain name system is organized in a hierarchy. At the top of the DNS hierarchy is the **root**, which consists of all domain names. Under the root are **first-level domains** that categorize the domain by organization type (.com, .net, .edu, .biz, .info, etc.) or by country (.uk, .ca, .ie, .au, .jp, .ch, etc.).

SECOND-LEVEL DOMAINS Under first-level domains are **second-level domains**, which usually specify a particular organization (microsoft.com, hawaii.edu, cnn.com, etc.). Sometimes, however, specific products, such as movies, get their own second-level domain names. Competition for good second-level domain names is fierce.

Getting a second-level domain name is only the beginning. Each organization that receives a second-level domain name must have a DNS server to host its domain name information. Large organizations have their own internal DNS servers that contain information on all subnet and host names. Individuals and small businesses that use webhosting services depend on the webhosting company to provide this DNS service.

FURTHER QUALIFICATIONS Domains can be further qualified. For instance, within hawaii.edu, which is the University of Hawaii, there is a shidler.hawaii.edu domain. This is the Shidler College of Business. Within *shidler.hawaii.edu* is *voyager.shidler.hawaii.edu*, which is a specific host within the college.

HIERARCHY OF DNS SERVERS To implement this naming hierarchy, the domain name system maintains a hierarchy of DNS servers. As noted earlier, at the root level, there are 13 **DNS root servers** that contain information about DNS servers for first-level domains (.com, .edu, .ca., .ie, etc.). Having multiple DNS root servers provides reliability. Each first-level domain itself maintains multiple DNS servers.

Companies with second-level domain names are required to have their own DNS servers and almost always maintain two or more DNS servers for their own second-level domain name. Ideally, they will put each DNS server on a different part of the network in order to minimize the danger that all will become unreachable simultaneously.

Test Your Understanding

3. a) Is the Domain Name System only used to send back IP addresses for given host names? b) What is a domain? c) Which level of domain name do corporations most wish to have? d) What are DNS root servers? e) How many DNS root servers are there? f) Why do most firms have both a primary and a secondary DNS server?

DHCP Servers

In Chapter 1, we saw that client PCs usually get their IP addresses and other configuration information from DHCP servers. This means that they have current configuration information each time they boot up. It also means that they usually get a different IP address each time they boot up. In contrast, servers get static (permanent) IP addresses so that they have the same IP address all the time. This allows clients to find them.

In Chapter 1, we also saw that when a client PC wakes up, it realizes that it has no IP address. It broadcasts a Dynamic Host Configuration Protocol (DHCP) message to all nearby hosts. Only the DHCP server responds.

Figure 10-6 shows that the situation actually is a bit more complex. As the figure shows, there may be multiple DHCP servers, and each will send a response back to the client PC. These initial responses do not carry configuration information. Rather, they are offers of what configuration a DHCP server will provide, including the length of time a PC may use its IP address before surrendering it and starting another DHCP search to get a new IP address.

The user selects among the offers and sends back an acceptance message to the "winning" DHCP server. That server responds with the configuration information. If there is only a single DHCP server, the client PC still goes through the seek–offer–accept–receive cycle. So the DHCP discussion in Chapter 1 was a simplification.

Which of a firm's DHCP servers will respond to an initial DHCP request? The answer is that each DHCP server has a DHCP scope parameter that lists the subnets to which it will respond with an offer.

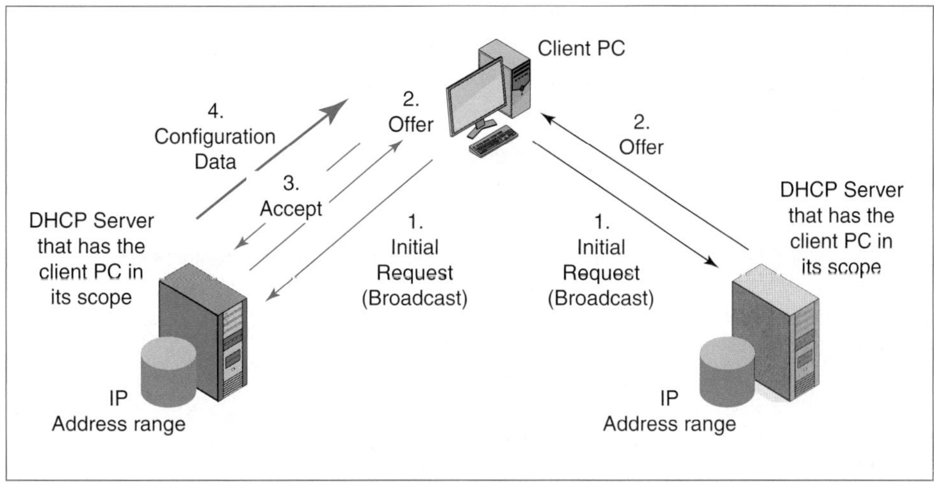

FIGURE 10-6 Dynamic Host Configuration Protocol (DHCP) Service

The fact that several DHCP servers may respond can provide redundancy and therefore reliability. However, companies must manage each DHCP server's scope and list of available IP addresses carefully.

In addition, when a firm changes its DNS servers to another IP address, changes its subnet masks, or makes any other changes that will affect client PC configuration, it must push these changes out to all DHCP servers.

Test Your Understanding

4. a) What hosts are given static IP addresses? b) Why do these hosts need static IP addresses? c) What are dynamic IP addresses? d) How do clients get dynamic IP addresses? e) Can the client send a DHCP request information to multiple DHCP servers? f) Do DHCP servers update the configuration information they store automatically? g) What are DHCP scopes?

Simple Network Management Protocol (SNMP)

We saw the **Simple Network Management Protocol (SNMP)** in the first chapter and in several chapters since. We will now look at it in a little more detail.

CORE SNMP ELEMENTS In Chapter 1, we saw that SNMP has several components.

- The network administrator works at a central computer that runs a program called the network management program, or, more simply, the manager.
- The SNMP manager is responsible for many managed devices—devices that need to be administered, such as printers, switches, routers, and other devices.
- Managed devices have pieces of software (and sometimes hardware) called network management agents, or, more simply, agents. SNMP agents communicate with the SNMP manager on behalf of their managed devices. In other words, the manager does not communicate with the managed device directly, but rather with the device's agent.

Core Elements (from Chapter 1)

- Manager program
- Managed device
- Agents (communicate with the manager on behalf of the managed device)
- Management information base (MIB)
  - Stores the retrieved information
  - "MIB" can refer to either the database on the manager or to the database schema

Messages

- Commands
  - Get
  - Set
- Responses
- Traps (alarms sent by agents)
- SNMP uses UDP at the transport layer to minimize the burden on the network

RMON Probes

- Remote monitoring probes
- A special type of agent
- Collects data for a part of the network
- Supplies this information to the manager

Objects (see Figure 10-19)

- Specific pieces of information
- Number of rows in the routing table
- Number of discards caused by lack of resources (indicates a need for an upgrade)

Set Commands

- Dangerous if used by attackers
- Many firms disable set to thwart such attacks
- However, they give up the ability to manage remote resources without travel
- SNMPv1: community string shared by the manager and all devices (poor)
- SNMPv3: each manager-agent pair has a different password (good)

User Functionality

- Reports, diagnostics tools, etc. are very important
- They are not built into the standard
- They are added by SNMP manager vendors
- Critical in selection

FIGURE 10-7 Simple Network Management Protocol (SNMP) (Study Figure)

- The manager stores information it receives in a central management information base (MIB). The term "MIB" is somewhat ambiguous because it can either refer to the database itself or to the design (schema) of the database.

To work together, these devices send messages to one another.

- The manager can send commands to the agent, telling the agent what to do. Agents send responses confirming that the command was fulfilled or explaining why it could not be fulfilled.
- GET commands ask for a specific piece of information. In practice, the manager constantly polls all of its managed devices, collecting many pieces of data from each in every round of polling.
- SET commands tell the agent to change the way the device operates, say by going into self-test mode.
- If an agent detects a problem, it can send a trap (alarm) message to the manager without waiting passively to be asked.
- These commands are sent via UDP to reduce the traffic burden on the network. This leads to occasional errors, but single errors only mean that a few pieces of information are a few seconds or minutes out of date.

RMON PROBES In addition, there is a specialized type of agent called an **RMON probe** (remote monitoring probe). This may be a stand-alone device or software running on a switch or router. An RMON probe collects data on network traffic passing through its location instead of information about the RMON probe itself. The manager can poll the RMON probe to get summarized information about the distribution of packet sizes, the numbers of various types of errors, the number of packets processed, the 10 most active hosts, and other statistical summaries that may help pinpoint problems. This generates far less network management traffic than polling many devices individually.

OBJECTS The SNMP MIB schema is organized as a hierarchy of **objects** (properties of managed devices). Figure 10-8 shows the basic model for organizing SNMP objects. First, there is the system. This might be a computer, switch, router, or another device. In addition, there are TCP, UDP, IP, and ICMP objects and objects for individual interfaces. Each of these objects has sub-objects under it, and those sub-objects may have further sub-sub-objects.

For example, if a router does not appear to be working properly, the manager can issue *Get* commands to collect appropriate router objects. A first step might be to check if the router is in forwarding mode. If not, it will not route packets. If this does not clarify the problem, the manager can collect more information, including error statistics and general traffic statistics of various types.

SET COMMANDS AND SNMP SECURITY Most firms are very reluctant to use *Set* commands because of security dangers. If setting is permitted and attackers learn how to send *Set* commands to managed devices, the results could be catastrophic. Fortunately, SNMP security has improved over time.

System Objects (One Set per Host)

System name
System description
System contact person
System uptime (since last reboot)

IP Objects (One Set per Host)

Forwarding (for routers). Yes if forwarding (routing), No if not
Subnet mask
Default time to live
Traffic statistics
Number of discards because of resource limitations
Number of discards because could not find route
Number of rows in routing table
Rows discarded because of lack of space
Individual row data

TCP Objects (One Set per Host)

Maximum/minimum retransmission time
Maximum number of TCP connections allowed
Opens/failed connections/resets
Segments sent
Segments retransmitted
Errors in incoming segments
No open port errors
Data on individual connections (sockets, states)

UDP Objects (One Set per Host)

Error: no application on requested port
Traffic statistics

ICMP Objects (One Set per Host)

Number of errors of various types

Interface Objects (One Set per Interface)

Type (e.g., 69 is 100Base-FX; 71 is 802.11)
Status: up/down/testing
Speed
MTU (maximum transmission unit—the maximum packet size)
Traffic statistics: octets, unicast/broadcast/multicast packets
Errors: discards, unknown protocols, etc.

FIGURE 10-8 SNMP Object Model

- The original version of SNMP, SNMPv1, had almost no authentication at all, making this danger a distinct possibility. The manager and all managed devices merely had to be configured with the same **community name**. With hundreds or thousands of devices sharing the same community name, if attackers can learn the community name, they can implement massive attacks.
- SNMPv3 has added passwords for each manager–agent pair, and these passwords are encrypted during transmission. In addition, each message is authenticated by the shared password. This requires a great deal of work to set up.

Most products today permit two SNMPv3 passwords for each manager–device pair. One is for *Get* commands and the other is for the more dangerous *Set* commands.

Of course, poor implementation can defeat SNMPv3 security. If a lazy administrator uses the same password for all manager–agent pairs, then this is no better security than community strings.

USER INTERFACE FUNCTIONALITY Collecting data in the MIB is worthless if there is no good way to get it out. All SNMP manager products have the ability to produce reports, diagnose problems, and do many other things that network administrators need to manage networks.

This functionality is not built into the standard. The standard just handles the mechanics of collecting and storing object data from managed devices. User interface functionality is a critical factor in selecting SNMP products.

Test Your Understanding

5. a) List the main elements in a network management system. b) Does the manager communicate directly with the managed device? Explain. c) Explain the difference between managed devices and objects. d) Is the MIB a schema or the actual database? (This is a trick question.) e) Why must user interface functionality for the SNMP manager be considered carefully in selecting SNMP manager products?
6. List one object in each of the following areas: the system, IP, TCP, UDP, ICMP, and an interface.
7. a) In SNMP, which device creates commands? b) Responses? c) Traps? d) Explain the two types of commands. e) What is a trap? f) Why are firms often reluctant to use *Set* commands? g) Describe SNMPv1's poor authentication method. h) Describe SNMPv3's good authentication method.

SECURING INTERNET COMMUNICATIONS

Using the Internet for Wide Area Networking

ATTRACTIONS For wide area networking, corporations would like to use the global Internet because of its very low price per bit transmitted compared to Public Switched Data Networks. The Internet's enormous size creates strong economies of scale.

Another attraction of the Internet is that nearly all of a corporation's sites are already connected to the Internet. So are its buyers, suppliers, and every other firm the company needs to communicate with.

Attractions

Very low cost per bit transmitted thanks to economies of scale

All communication partners are connected to the Internet

Issues

Security

Addressed through cryptographic protections

Lack of quality of service (QoS) service-level agreements (SLAs)

Addressed through the use of IP carrier networks

FIGURE 10-9 Using the Internet for Wide Area Networking (Study Figure)

ISSUES However, there are two issues with the Internet. The first is security. The Internet teems with sniffers, hackers, malware, and other threats. These need to be addressed at least in part through cryptographic protections.

The second issue is quality of service. The Internet has always been a best-effort network. For many corporations, this is a serious issue for many applications. Corporations want service-level agreements (SLAs). These concerns can be addressed by using IP carrier networks, which we will see after a discussion of cryptographic protections.

Virtual Private Networks (VPNs)

Figure 10-10 shows that corporations can cryptographically protect traffic flowing between two sites or between a site and a remote user. Figure 10-10 shows that a **virtual**

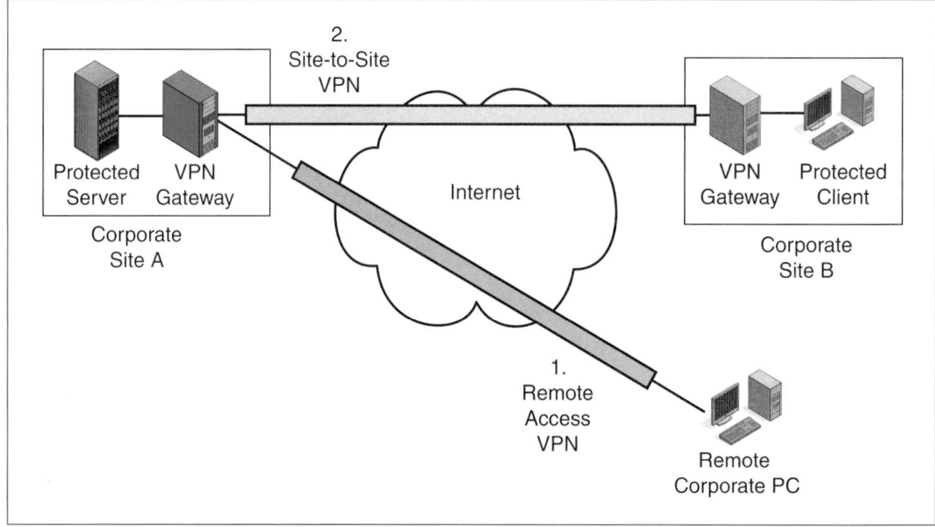

FIGURE 10-10 Virtual Private Networks (VPNs)

private network (VPN) is a cryptographically secured transmission path through an untrusted environment. The Internet is an obvious untrusted environment. Another is wireless networking. Although the firm is using a shared network, it appears to be alone, as far as security is concerned. Figure 10-10 shows a **remote access VPN** and a **site-to-site VPN**.

A virtual private network (VPN) is a cryptographically secured transmission path through an untrusted environment.

Remote access VPNs are attractive because single hosts trying to connect to a corporate site via the Internet are extremely vulnerable. They are even more vulnerable if they connect to their network wirelessly.

Site-to-site VPNs protect all traffic flowing between a pair of sites. Typically, sites have heavy traffic between themselves, so site-to-site VPNs tend to carry much more traffic than remote access VPNs.

Both remote-access VPNs and site-to-site VPNs usually terminate in a **VPN gateway** at a site. This VPN gateway handles cryptographic protections when dealing with remote users or a VPN gateway at another site.

IPsec VPNs

There are two standards for VPN security. The most sophisticated VPN technology is a set of standards collectively called **IP security** (**IPsec**).[2] As its name suggests, IPsec operates at the internet layer. It provides security to all content at the internet, transport, and application layers.[3] This protection is *transparent*, meaning that nothing additional has to be done to protect upper-layer content and that all upper-layer content is protected.

PROS AND CONS IPsec offers the strongest security and should eventually dominate remote access VPN transmission, site-to-site VPN transmission, and internal IP transmission. However, IPsec networks are fairly complex to manage and therefore relatively expensive.

TRANSPORT MODE Figure 10-11 shows IPsec's two modes of operation. In **transport mode**, the two computers that are communicating implement IPsec. This mode gives strong end-to-end security, but it requires IPsec configuration *and* a digital certificate on all machines. Although the cost per machine for configuration and for the digital certificate is small, the large number of computers in a company makes the aggregate cost of transport mode setup high.

TUNNEL MODE In contrast, in **tunnel mode**, the IPsec connection extends only between **IPsec gateways** at the two sites. This provides no protection within sites, but the use of tunnel mode IPsec gateways offers simple security. The two hosts do not have

---

[2] *IPsec* is pronounced "eye-pea-sek'," with emphasis on the sek.

[3] Actually, the term *all* is a bit too strong. In transport mode, which is discussed later, attackers can read the IP addresses because the packet is addressed to the destination host instead of to the IPsec gateway server. However, on exams, call it all.

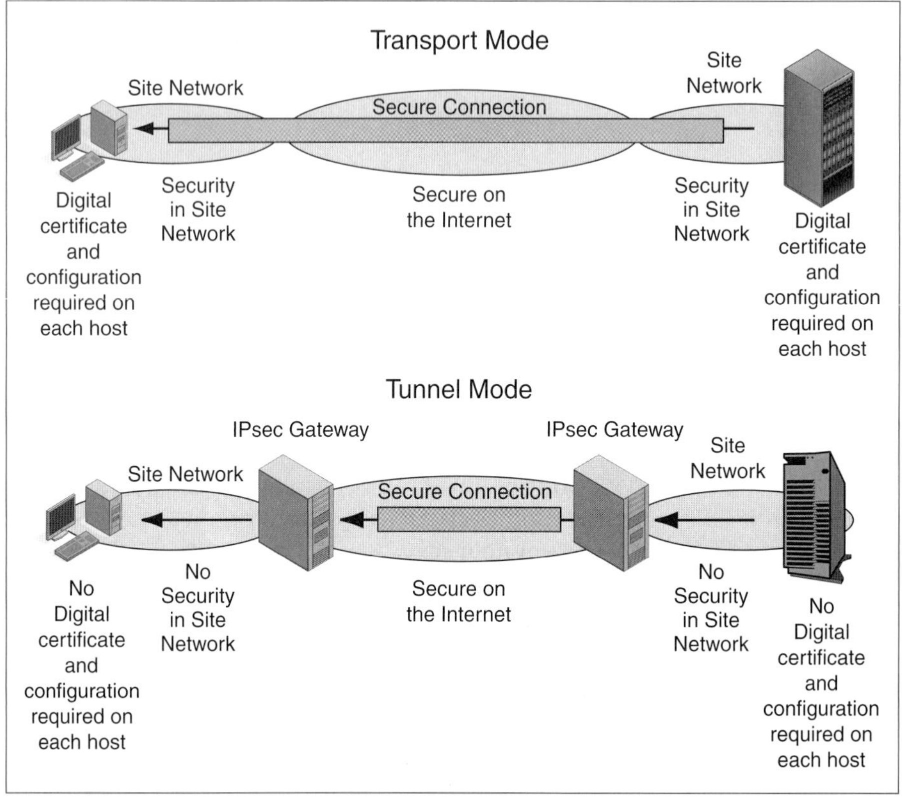

FIGURE 10-11 IPsec Transport and Tunnel Modes

to implement IPsec security and, in fact, do not even have to know that IPsec is being used between the IPsec gateways. Most importantly, there is no need to install digital certificates on individual hosts. Only the two IPsec gateways need to have IPsec configuration and digital certificates.

REMOTE-ACCESS AND SITE-TO-SITE VPNS IPsec is a versatile protocol that can be used simultaneously for both remote-access VPNs and site-to-site VPNs. Coupled with its ability to protect all upper-layer content transparently, IPsec is a general solution for a firm's cryptographic protection needs.

SECURITY ASSOCIATIONS AND POLICY SERVERS One advantage of IPsec as a VPN technology is that it can be centrally managed. Figure 10-12 shows that before two IPsec gateways begin to communicate, they negotiate how they will perform security. The **security associations (SAs)** they negotiate specify what security options they will use.[4] As we saw in Chapter 3, some security options are very strong. Others may not be.

[4] The figure shows that the gateways implement an association in each direction. If security conditions require it, these SAs can use different security options.

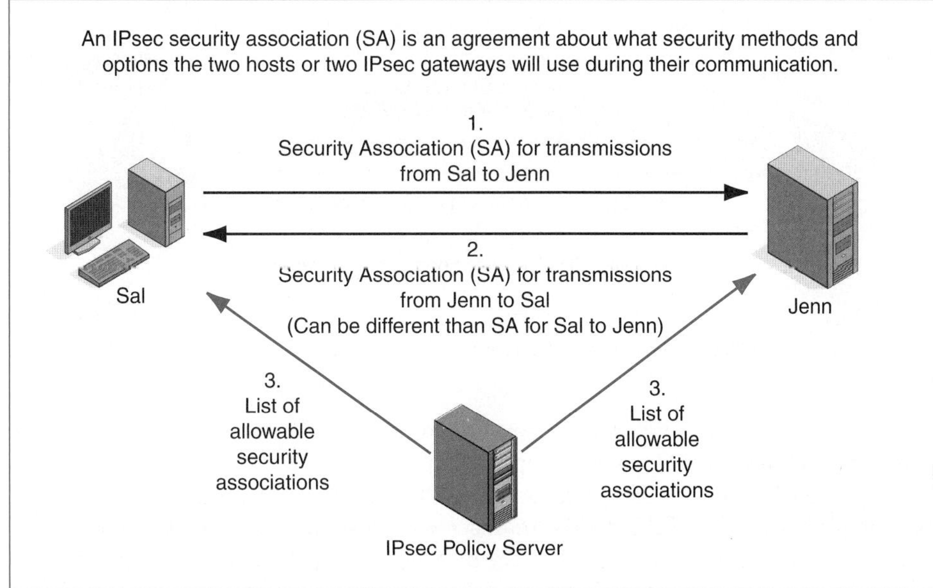

FIGURE 10-12 IPsec Security Associations and Policy Servers

With IPsec, companies can use central **IPsec policy servers**. These servers specify what SA options are allowable for various gateway pairs and what options must not be used. Policy servers are especially important if a firm has many IPsec gateways.

Test Your Understanding

8. a) At what layer does IPsec operate? b) What layers does IPsec protect? c) Does IPsec protect these layers transparently? d) Describe IPsec tunnel mode. e) What is the main advantage of tunnel mode? f) What is the main disadvantage of tunnel mode? g) Describe IPsec transport mode. h) What is the main advantage of transport mode? i) What is the main disadvantage of transport mode? j) In which IPsec mode are clients and servers required to have digital certificates? k) Which IPsec mode does not require clients and servers to have digital certificates? l) Is IPsec used for remote-access VPNs, site-to-site VPNs, or both?

SSL/TLS VPNs

The simplest VPN security standard to implement is **SSL/TLS**. This standard was originally created as **Secure Sockets Layer** (**SSL**) by Netscape. It was later taken over by the IETF and renamed **Transport Layer Security** (**TLS**). We will call it SSL/TLS because it is still called by both names.

NONTRANSPARENT PROTECTION As Figure 10-13 shows, SSL/TLS provides a secure connection at the transport layer. This allows it to protect applications. However, SSL/TLS only protects applications that are **SSL/TLS-aware**; that is, capable of working

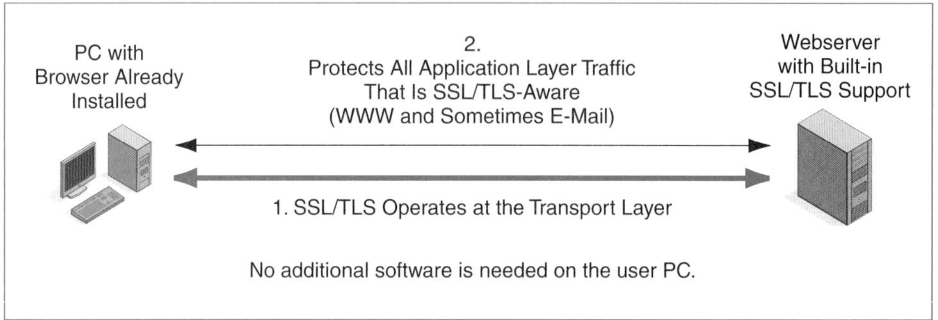

FIGURE 10-13 SSL/TLS for Browser-Webserver Communication

with SSL/TLS. All browsers and webservers are SSL/TLS-aware. Some e-mail systems also are SSL/TLS-aware. Few other applications are.

Although traditional SSL/TLS is limited to a few applications, many firms need only remote Web access or Web and e-mail access. These firms are likely to use SSL/TLS, which is easy to implement because every browser and webserver application program has SSL/TLS built in so there is no cost to add or configure client software.

AUTHENTICATION OPTIONS In SSL/TLS, one issue is how to authenticate the user, that is, to require the user to prove his or her identity. One SSL/TLS option for corporations is to do no authentication for the client; this opens SSL/TLS-based systems to many attacks. Webserver application programs can reduce this SSL/TLS weakness by adding passwords themselves, but password security is not strong.

The other option is for corporations to use a digital certificate for each client. This provides very strong security, but as mentioned earlier, implementing client digital certificates is expensive.

Unlike IPsec, SSL/TLS does not provide central policy management. This is not surprising because SSL/TLS was created for single browsers talking to single servers. However, a lack of central policy management is an issue for organizations.

SSL/TLS GATEWAYS Initially, SSL/TLS protected direct connections between a client and a webserver. However, as Figure 10-14 shows, several vendors have begun to produce **SSL/TLS gateways** to let an authenticated user reach any internal webserver to which he or she should have access. The gateways turn SSL/TLS into a true remote access VPN technology. The user has a single SSL/TLS connection—to the SSL/TLS gateway. The gateway provides access to internal webservers.

Although SSL/TLS normally is limited to HTTP, most SSL/TLS gateway vendors are able to **webify** some other applications, such as database applications. Webification involves converting screen images from these non-web applications into webpages that browsers can read. Webification also involves sending what the user types onto webpage data entry forms into a format the non-web application can use. This has greatly expanded the ability of remote access users to use non-web applications, but not all non-web applications can be webified.

SSL/TLS gateways can also give an external user, after authentication, complete access to a segment of the network containing multiple hosts.

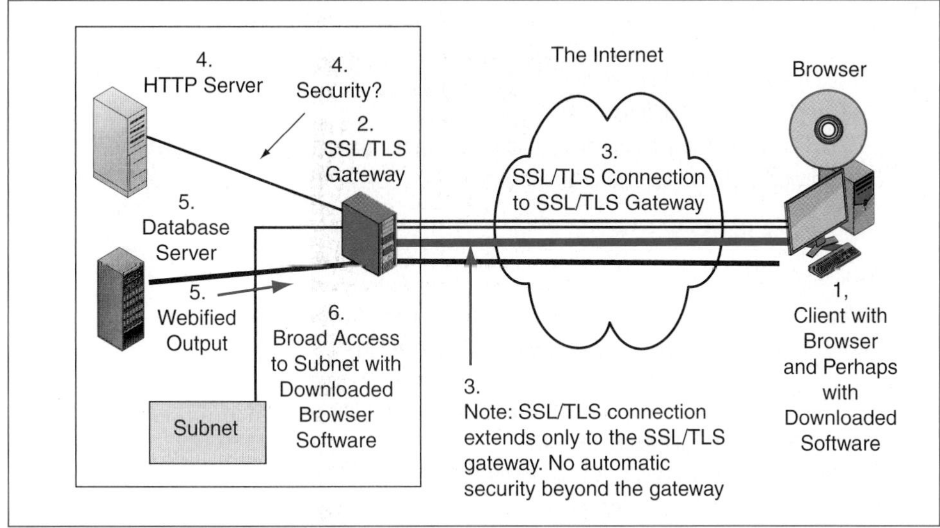

FIGURE 10-14 SSL/TLS VPN with a Gateway

Some gateway vendors even provide add-ins that can be downloaded to browsers. These add-ins are required for some services, such as complete access to a network segment. They may also be desirable, for instance, if the user is at a public kiosk and wants to erase all data about his or her session from the computer. Unfortunately, these downloads require administrative access on the computer. On public computers at kiosks and in cybercafés, administrative access is rarely available to users.

Test Your Understanding

9. a) How is SSL/TLS limited? b) Why is SSL/TLS inexpensive? c) What are SSL/TLS's authentication options? d) Does SSL/TLS have the ability to be centrally managed? e) When an SSL/TLS gateway is used, how many SSL/TLS connections does the client have? f) What is webification? g) What is webification's benefit? h) Why are downloads for SSL/TLS gateway service not likely to be useful on public PCs?

10. a) Of the two VPN security technologies discussed in this section, which provides transparent security to higher layers? b) Which tends to require the installation of digital certificates on many client PCs? c) Which has stronger security? d) Which would you use to give employees remote access to a highly sensitive webserver via the Internet? (This is not a trivial question.) Justify your answer.

IP Carrier Services

So far, we have discussed companies building their own virtual private networks. However, many of the carriers that traditionally offered Layer 2 Public Switched Data Networks[5] and many Internet service providers now offer Layer 3 **IP Carrier Services**.

[5] Many PSDN vendors are now telling their customers that traditional PSDN services (other than metro Ethernet) are being phased out in favor of IP carrier service.

IP Carrier Services

Provide IP service over the Internet with two additions
They provide VPN protections
VPN protection is managed so that the customer does not have to do anything
They provide quality of service guarantees

IP Carrier Service Carriers

Traditional PSDN vendors
ISPs
Others

QoS Guarantees

Carrier only uses its internal IP network
This avoids network access point connections, which cause problems
The carrier can use MPLS to ensure high quality

Limitation

All users must be on the same IP carrier service
So can only be used to connect company sites

FIGURE 10-15 IP Carrier Service (Study Figure)

These carriers transmit signals over the Internet. However, they add two things to Internet transmission.

First, they add VPN protection. This is managed VPN protection, meaning that the customer does not have to worry about it. The carrier sets it up and operates it on an ongoing basis.

Second, they offer quality of service guarantees. We noted earlier that Internet service quality is uncertain. Most of this uncertainty comes when Internet service providers interconnect. To avoid this, IP carrier services only use their own Layer 3 networks. For example, an ISP offering an IP carrier service merely keeps traffic on its own routers and transmission lines, not handing off packets to other ISPs. Using MPLS, which we saw in the previous chapter, they can do traffic engineering to ensure good service quality.

The main limitation of IP carrier service is that all of your sites must connect to the IP carrier. This is no problem for the interconnection of company sites. However, customers and suppliers are unlikely to use an IP carrier service, much less yours. Communication with customers, suppliers, and other organizations must still take place over the bare Internet.

DIRECTORY SERVERS

Directory Server Basics

WHAT ARE DIRECTORY SERVERS? Many firms now have **directory servers**, which centralize information about a firm. For instance, for individual people, the directory server may have a name, telephone, and e-mail address. It might also store the person's login password and permissions on various servers.

In addition, the directory server may store information about individual hosts. For instance, it might contain a set of security rules for a group of PCs or even for a single PC. In Microsoft's Active Directory product, for instance, a firm can create a **group policy object (GPO)** that may state that a user cannot add new programs at all. It can even lock down how the screen appears.

In previous chapters, we saw that an authentication has a set of information for authenticating users. Several verifiers may rely on the authentication server for authentication services. If a company has multiple authentication servers, they may each get their authentication information from a central directory server.

HIERARCHICAL ORGANIZATION As Figure 10-16 shows, information in a directory server is arranged hierarchically, much as entries in a DNS server are organized.

The figure shows the directory structure for the mythical University of Waikiki. The top level is the **organization**. Under the top level, there are schools (**organizational units**, in directory server terminology). In each school, there are faculty, staff, and router categories. Under the faculty category, there are the usernames of faculty members. At the bottom of the hierarchy are the **properties** of individual faculty members, including the faculty member's common name, e-mail address, and telephone extension.

LIGHTWEIGHT DIRECTORY ACCESS PROTOCOL (LDAP) Most directory servers today permit query commands governed by the **Lightweight Directory Access Protocol (LDAP)**. The figure shows that LDAP commands specify the path to a property, with individual nodes along the way separated by dots. This is why the request for Brown's e-mail address is specified as the following:

GET e-mail.brown.faculty.business.waikiki

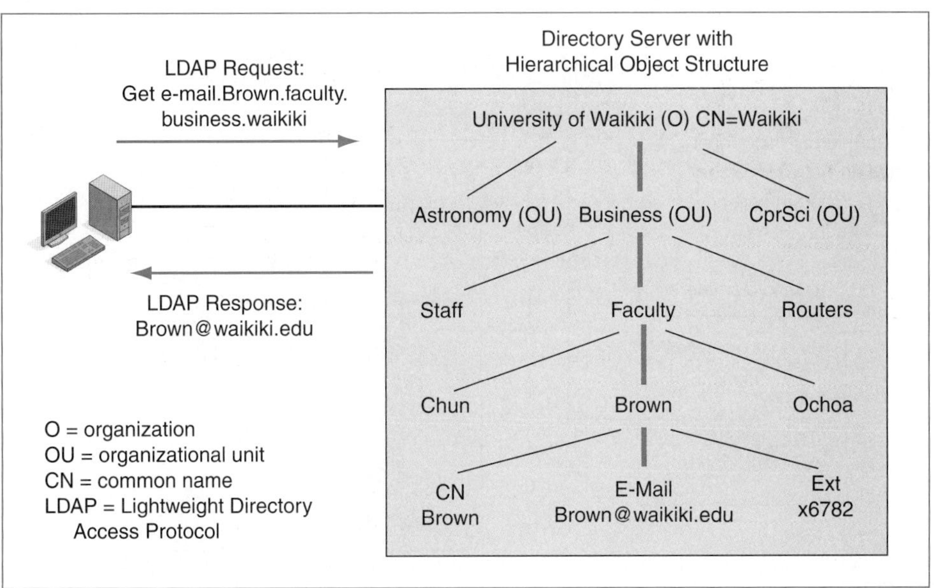

FIGURE 10-16 Directory Server Organization and LDAP

DIRECTORY SERVERS AND THE NETWORKING STAFF Organizations store a great deal of information about themselves in directory servers, including a great deal of networking information. Creating a directory server data organization plan (schema) requires a great deal of planning about what information an organization needs to store and how this information should be arranged hierarchically.

Although creating and managing a directory server goes well beyond networking, the networking staff is often given the task of leading directory server planning projects and managing the directory server on a daily basis.

Test Your Understanding

11. a) What kinds of information are stored in a directory server? b) How is information in directory servers organized? c) What is the purpose of LDAP? d) If Astronomy has a similar directory organization to Business (in Figure 10-16), give the specification for the telephone extension of Claire Williams (username cwilliams), who is an Astronomy staff member.

Microsoft's Active Directory (AD)

ACTIVE DIRECTORY DOMAINS Microsoft's directory server product is **Active Directory (AD)**. Network administrators must become very familiar with AD. Figure 10-17 shows that a firm must divide its computers into logical **Active Directory domains**, which are simply called domains. These AD domains are organized in a hierarchy. The Microsoft concept of domains is similar to the DNS concept of domains.

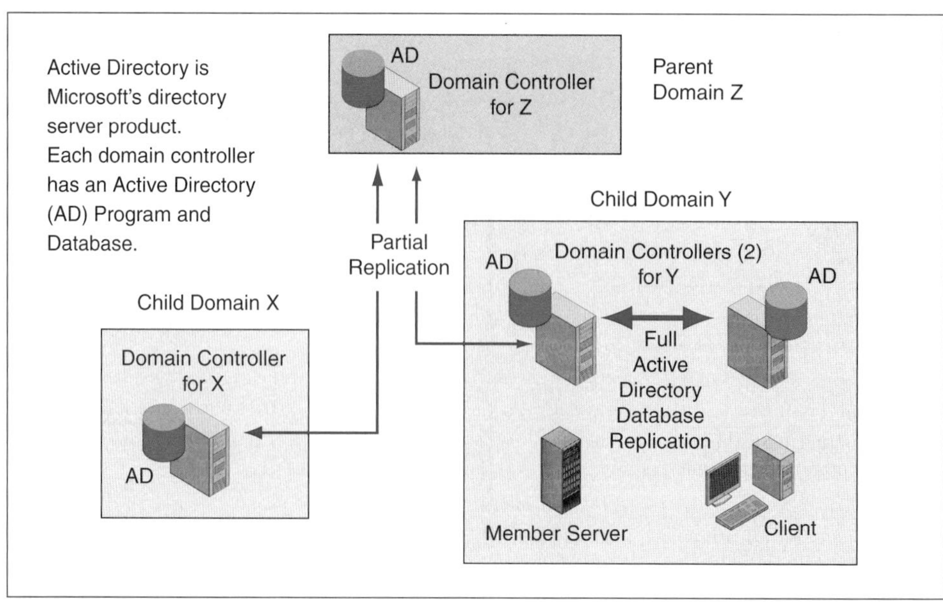

FIGURE 10-17 Active Directory Domains and Domain Controllers

DOMAIN CONTROLLERS A domain must have one or more **domain controllers**, which are servers that run Active Directory and maintain an AD database for the domain. If there are multiple domain controllers within a domain, then their AD data is *fully* replicated between them so that each has the same data. If one domain controller fails, the other takes over automatically. Not all servers in a domain must run AD. Servers that do not are called **member servers**.

DOMAINS IN AN ACTIVE DIRECTORY TREE A logical hierarchy of AD domains is called an **Active Directory tree**. Information in AD databases typically is *partially* replicated across domain controllers at different levels.

COMPLEX STRUCTURES We have looked at a tree organization of AD domains, at domain controllers, and at replication. However, companies may have forests (groups of AD trees), and replication can be handled with almost infinite variations and trust relationships.

Test Your Understanding

12. a) What is Active Directory? b) What is an AD domain? c) What are domain controllers? d) Can a domain have more than one domain controller? e) What are servers called that do not run AD? f) Describe replication among domain controllers in the same domain. g) What is a tree? h) Describe replication among domain controllers at different levels. i) What is an AD forest?

CONCLUSION

Synopsis

Chapter 9 dealt with TCP/IP concepts. This chapter focuses on TCP/IP management. The TCP/IP standards that dominate internetworking require a great deal of management attenuation, both initially and on an ongoing basis. The first step is to develop an IP subnet schema for the firm. This creates a basic trade-off between the number of subnets and the number of hosts per subnet. The firm also has to decide whether or not to use network address translation (NAT). NAT has several benefits, including added security and increasing the effective number of public IP addresses a firm has; but NAT causes problems for certain protocols.

In this chapter, we looked more closely at the Domain Name System (DNS). We saw that the system is a hierarchical system of named domains (collections of resources under the control of an organization). Corporations want second-level domain names, such as pearsonhighered.com. After they get one, they must maintain two or more DNS servers for their second-level domain. We also saw that if a local DNS server does not know the IP address for a host name, it contacts the authoritative DNS server for the domain in the IP address.

We looked in more depth at the Dynamic Host Configuration protocol, focusing on the need to define scopes for each DHCP servers and to assign IP address groups in a way consistent with the DHCP server's scope.

We have seen the Simple Network Management Protocol (SNMP) repeatedly since we first saw it in Chapter 1. This chapter looked at SNMP operation in a bit more detail, focusing on the concept of objects and the types of objects specified in MIB schemas. We also saw RMON probes.

SECURING INTERNET COMMUNICATIONS The Internet is attractive for WAN transmission because it has a very low cost per bit transmitted. However, companies are concerned about security on the Internet. To obtain better security, they use remote-access and site-to-site VPNs.

There are two main VPN protocols. IPsec offers the strongest security. It also offers the important choice between two modes of operation—transport mode and tunnel mode. Most importantly, IPsec offers central manageability.

SSL/TLS can be used for remote-access VPNs. SSL/TLS is attractive because all browsers know how to create a secure SSL/TLS connection with host computers. This means that there is no need to add anything to the client computer. However, there are limitations on the services that SSL/TLS can provide.

DIRECTORY SERVERS Increasingly, companies are centralizing information about their people, computers, and other resources in directory servers. Directory servers store information in a hierarchical organization, so careful planning is needed because it is very difficult to change the schema after it is created. Typically, data in directory servers is accessed via the Lightweight Directory Access Protocol (LDAP).

Smaller firms have a single directory server, but large firms with complex organizational frameworks often have multiple directory servers. We saw how Microsoft's Active Directory (AD) program allows a company to create a hierarchy of domains, each with one or more domain controllers each of which has an AD program and database. If a domain has multiple domain controllers, they fully replicate the data in their AD database. Domain controllers in parent and child domains may partially replicate their AD databases.

END-OF-CHAPTER QUESTIONS

Thought Questions

1. Both DNS servers and DHCP servers send your client PC an IP address. Distinguish between these two addresses.
2. Assume that an average SNMP response message is 100 bytes long. Assume that a manager sends 40 SNMP *Get* commands each second. a) What percentage of a 100 Mbps LAN link's capacity would the resulting response traffic represent? b) What percentage of a 128 kbps WAN link would the response messages represent? c) What can you conclude from your answers to this question?
3. A firm is assigned the network part 128.171. It selects an 8-bit subnet part. a) Draw the bits for the four octets of the IP address of the third host on the first subnet. (Hint: Use Windows Calculator.) b) Convert this answer into dotted decimal notation. c) Draw the bits for the fifth host on the third subnet. (In binary, 2 is 10, while 3 is 11.) d) Convert this into dotted decimal notation. e) Draw the bits for the last host on the third subnet. f) Convert this answer into dotted decimal notation.
4. A firm is assigned the network part 128.171. It selects a 10-bit subnet part. a) Draw the bits for the four octets of the IP address of the first host on the first subnet. (Hint: Use Windows Calculator.) b) Convert this answer into dotted decimal notation. c) Draw the bits for the second host on the third subnet. (In binary, 2 is 10, while 3 is 11.) d) Convert this into dotted decimal notation. e) Draw the bits for the last host on the third subnet. f) Convert this answer into dotted decimal notation.

Troubleshooting Question

1. In your browser, you enter the URL of a website you use daily. After some delay, you receive a DNS error message that the host does not exist. a) What may have happened? Explain your reasoning. **Again, do NOT just come up with one or two possible explanations**. b) How would you logically disconfirm or test each possibility?

Perspective Questions

1. What was the most surprising thing to you about the material in this chapter?
2. What was the most difficult thing for you in the chapter?

CHAPTER 11

Networked Applications

LEARNING OBJECTIVES

By the end of this chapter, you should be able to:

- Describe the characteristics and limitations of host communication with dumb terminals.
- Explain client/server architectures, including file server program access and client/server processing (including Web-enabled applications).
- Describe electronic mail standards and security.
- Describe voice over IP (VoIP).
- Describe the World Wide Web.
- Describe cloud computing (including Software as a Service, utility computing, and virtualization, as well as security issues).
- Describe Service-Oriented Architecture, with an emphasis on Web service, SOAs, and SOAP.
- Explain peer-to-peer (P2P) computing (including BitTorrent, Skype, and SETI@home), which, paradoxically, normally uses servers for part of the work.

PAPA MURPHY'S

Papa Murphy's Take 'N' Bake is one of the largest pizza companies in the United States, having more than 1,150 stores in more than 30 states.[1] Many of its stores are franchised. While the company has enjoyed success and expansion, it also faced a problem: How could it create a centralized management system for all of its franchises?

Rather than designing, building, and hosting its own management system, Papa Murphy's decided to outsource the job to Salesforce, a company that provides cloud computing services. According to Papa Murphy's Director of Business Technology, Brian Fisher, "It took less than three months to build what would have taken a year on other platforms."[2] In addition to its quick implementation, the new franchise management

[1] www.papamurphys.com/about.asp

[2] "5 Reasons CIOs are Adopting Cloud Computing in 2009." *Salesforce.com white paper*. salesforce.vo.llnwd.net/o1/emea/cloudforce/eindhoven-pdfs/datasheets/Force%20com%20Business%20Case-%20White%20Paper%205X%20Faster%20%201-2%20the%20Cost.pdf

system also brought the benefit of mobile access. Since the system is run as an online application, employees can access the information they need from anywhere, via mobile devices. Also, Papa Murphy's is freed from managing the system and the servers it runs on; these IT jobs are taken care of by Salesforce.[3]

Test Your Understanding

1. a) What business problem did Papa Murphy's face? b) How did Papa Murphy's resolve this problem? c) What benefits did Papa Murphy's see as a result of their choice to use Salesforce to handle this work? d) Can you think of any risks that are involved in this business choice?

INTRODUCTION

Networked Applications

Once, applications ran on single machines—usually, mainframes or stand-alone PCs. Today, however, most applications spread their processing power over two or more machines connected by networks instead of doing all processing on a single machine.

Applications that require networks to operate are called **networked applications**. The World Wide Web and e-mail are networked applications. So is the Salesforce franchise management application used by Papa Murphy's.

APPLICATION ARCHITECTURES In this chapter, we will focus on **application architectures**—that is, how application layer functions are spread among computers to deliver service to users. Thanks to layering's ability to separate functions at different layers, most application architectures can run over TCP/IP, IPX/SPX, and other standards below the application layer. In turn, if you use TCP at the transport layer, TCP does not care what application architecture you are using.

An application architecture describes how application layer functions are spread among computers to deliver service to users.

IMPORTANT NETWORKED APPLICATIONS In addition to looking broadly at application architectures, we will look at some of the most important of today's networked applications, including e-mail, the World Wide Web, cloud computing, and peer-to-peer (P2P) computing.

IMPORTANCE OF THE APPLICATION LAYER TO USERS In this chapter, we will focus on the application layer. This is the only layer whose functionality users see directly. When users want e-mail, it is irrelevant what is happening below the application layer, unless there is a failure or performance problem at lower layers.

Test Your Understanding

2. a) What is a networked application? b) What is an application architecture? c) Why do users focus on the application layer?

---

[3] www.salesforce.com/customers/distribution-retail/papamurphys.jsp

TRADITIONAL APPLICATION ARCHITECTURES

In this section, we will look at the two most important traditional application architectures: terminal–host systems and client/server architectures (both file server program access and client/server processing).

Terminal–Host Systems: Hosts with Dumb Terminals

As Figure 11-1 shows, the first step beyond stand-alone machines still placed the processing power on a single **host computer** but distributed input/output (I/O) functions to user sites. These I/O functions resided in **dumb terminals**, which sent user keystrokes to the host and painted host information on the terminal screen but did little else.

Although this approach worked, the central server frequently was overloaded by the need to process both applications and terminal communication. This often resulted in slow **response times** when users typed commands.

Another problem was high transmission cost. All keystrokes had to be sent to the host computer for processing. This generated a great deal of traffic. Similarly, the host had to send detailed information to be shown on-screen. To reduce transmission costs, most terminals limited the information they could display to **monochrome text** (one color against a contrasting background). Graphics were seldom available. They were rare because they needed higher speeds, as discussed in Chapter 1.[4]

IBM mainframe computers used a more complex design for their terminal–host systems that added other pieces of equipment beyond terminals and hosts. This extra equipment reduced cost and improved response times. In addition, IBM terminal–host

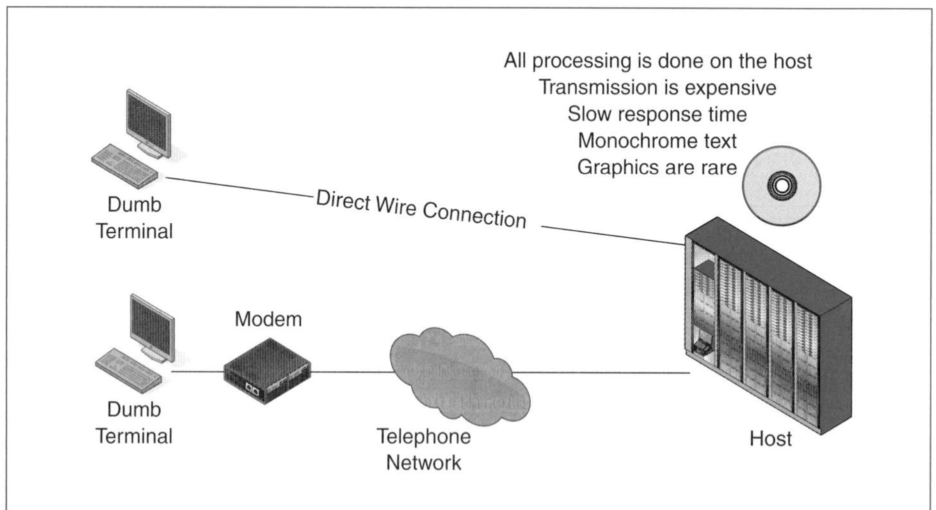

FIGURE 11-1 Simple Terminal–Host System

[4] The most common dumb terminal today is the VT100 terminal, also called an ANSI terminal. On the Internet, clients can emulate (imitate) dumb terminals by using Telnet. Telnet turns a $2,000 office PC into a $200 dumb terminal.

systems had higher speeds than traditional terminals and so were able to offer limited color and graphics. Although these advances extended the life of terminal–host systems, even these advanced IBM systems are less satisfactory to users than subsequent developments, including the client/server systems described next.

Test Your Understanding

3. a) Where is processing performed in systems of hosts and dumb terminals? b) What are the typical problems with these systems?

Client/Server Systems

After terminal–host systems, a big breakthrough came in the form of **client/server systems**, which placed some power on the client computer. This was made possible by the emergence of personal computers in the 1980s. PCs have the processing power to handle more of the workload than dumb terminals could.

ROLES FOR THE CLIENT AND SERVER Figure 11-2 shows that the work in a client/server system is done by programs on two machines—a client and a server. The client usually is a PC. Generally, the server does the heavy processing needed to retrieve information. The client, in turn, normally focuses on the user interface and on processing data delivered by the server—for instance, by placing the data in an Excel spreadsheet.

Client/server computing provides *scalability*, which we saw in Chapter 4. As user demand grows, the service provider can switch to a larger server or even to a server farm consisting of dozens or even thousands of coordinated servers.

WEB-ENABLED APPLICATIONS Client/server processing requires a client program to be installed on a client PC. Initially, all applications used custom-designed client programs. Rolling out a new application to serve hundreds or thousands of client computers was extremely time consuming and expensive.

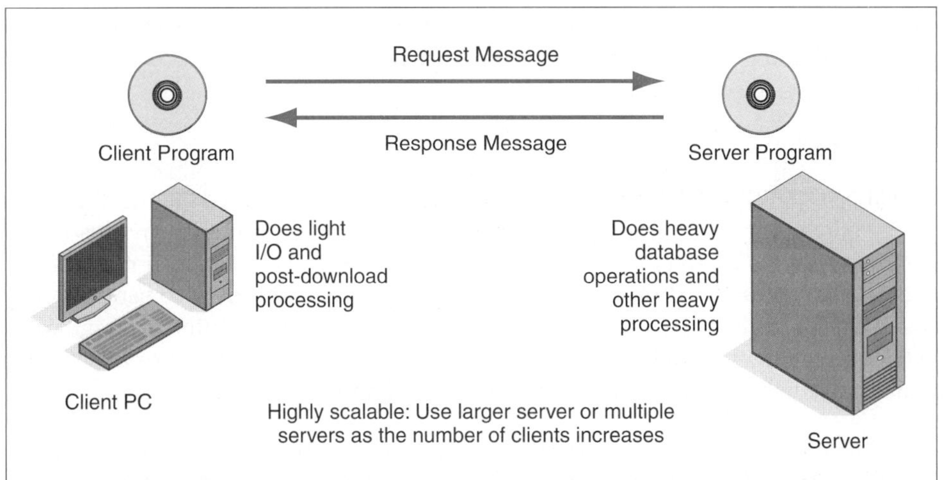

FIGURE 11-2 Client/Server Computing

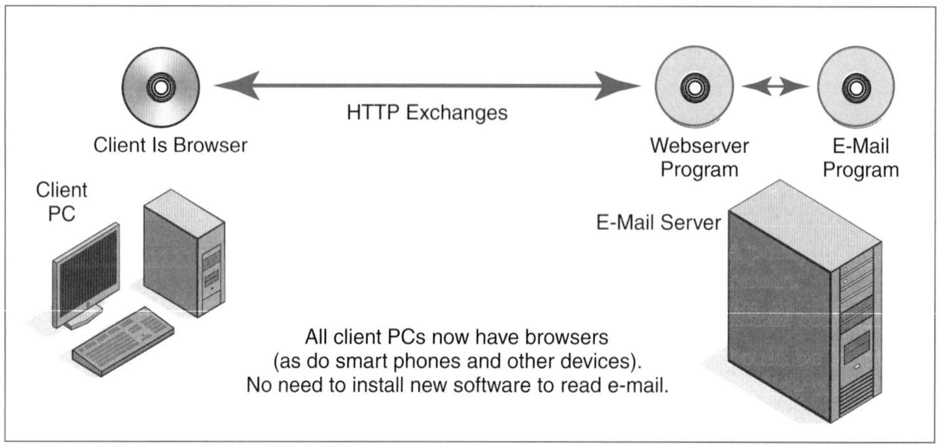

FIGURE 11-3 Web-Enabled Application (E-Mail)

Fortunately, there is one client program that almost all PCs have today. This is a browser. As Figure 11-3 illustrates, many client/server processing applications are now **Web-enabled**, meaning that they use ordinary browsers as client programs. The figure specifically shows Web-enabled e-mail.

Although many client/server applications are web-based and use browsers, this is not true of *all* client/server applications. For example, large and sophisticated DBMSs often come with a specialized database client that must be installed on the client (rather than accessed via a browser).

Test Your Understanding

4. In client/server processing, where is processing done?
5. Contrast general client/server processing with Web-enabled applications.

ELECTRONIC MAIL (E-MAIL)

We have looked broadly at client/server architecture. In the next three sections, we will see in more detail how client/server systems are used today. To begin, we will focus on a very familiar subject: e-mail.

Importance

A UNIVERSAL SERVICE ON THE INTERNET E-mail has become one of the two "universal" services on the Internet, along with the World Wide Web. E-mail provides mailbox delivery even if the receiver is "off-line" when the message is received. E-mail offers the speed of a fax, plus the ability to store messages in organized files, to send replies, to forward messages to others, and to perform many other actions after message receipt. The telephone offers truly instant communication, but only if the other party is in and can take calls. In addition, e-mail is less intrusive than a phone call.

ATTACHMENTS CAN DELIVER ANYTHING Thanks to attachments, e-mail has also become a general file delivery system. Users can exchange spreadsheet documents, word processing documents, graphics, and any other type of file.

Importance of E-Mail

- Universal service on the Internet
- Attachments deliver files

E-Mail Standards

- Message body standards
  - RFC 822 and RFC 2822 for all-text bodies
  - HTML bodies
  - UNICODE for multiple languages
- Simple Mail Transfer Protocol (SMTP)
  - Message delivery: Client to sender's mail host
  - Message delivery: Sender's mail host to receiver's mail host
- Downloading Mail to Client
  - Post Office Protocol (POP): Simple and widely used
  - Internet Message Access Program (IMAP): More powerful, less widely used
- Web-Enabled E-Mail
  - Uses HTTP for all communication with the mail server
  - No need for e-mail software on the client PC; a browser will do
  - Tends to be slow

Viruses, Worms, and Trojan Horses

- Widespread problems; often delivered through e-mail attachments
  - Use of antivirus software is almost universal, but ineffective
- Where to do scanning for viruses, worms, and Trojan horses?
  - On the client PC, but users often turn off or fail to update their software
  - On the corporate mail server and application firewall; users cannot turn off
  - At an antivirus outsourcing company before mail reaches the corporation
  - Defense in depth: Filter at two or more locations with different filtering software

Spam

- Unsolicited commercial e-mail
- Why filter?
  - Potential sexual harassment suits
  - Time consumed by users deleting spam
  - Bandwidth consumed
  - Time consumed by networking staff deleting spam
- Separating spam from legitimate messages is very difficult
  - Many spam messages are allowed through to users
  - Some legitimate messages are deleted
  - Some firms merely mark messages as probable spam

FIGURE 11-4 E-Mail (Study Figure)

E-Mail Standards

A major driving force behind the wide acceptance of Internet e-mail is standardization. It is rare for users of different systems not to be able to communicate at a technical level—although many companies restrict outgoing and incoming communication using firewalls for security purposes. Consequently, the key issue is application layer standards. Figure 11-5 shows that e-mail uses multiple standards for different aspects of its operation.

Message Body Standards. Obviously, message bodies have to be standardized, or we would not be able to read arriving messages. In physical mail, message body standards include the language the partners will use (English, etc.), the formality of language, and other matters. Some physical messages are forms, which have highly standardized layouts and fields that require specific information.

RFC 2822 (Originally RFC 822) Bodies. The initial standard for e-mail bodies (and headers) was **RFC 822**, which has been updated as **RFC 2822**. This is a standard for plain text messages—multiple lines of typewriter-like characters with no boldface, graphics, or other amenities. The extreme simplicity of this approach made it easy to create early client e-mail programs.

HTML Bodies. Later, as HTML became widespread on the World Wide Web, most mail venders developed the ability to display **HTML bodies** with richly formatted text and even graphics.

UNICODE Bodies. RFC 822 specified the use of the ASCII code to represent printable characters. Unfortunately, ASCII was developed for English, and even European languages need extra characters. The **UNICODE** standard allows characters of all languages to be represented, although most mail readers cannot display all UNICODE characters well yet.

Simple Mail Transfer Protocol (SMTP)

We also need standards for delivering RFC 2822, HTML, and UNICODE messages. In the postal world, we must have envelopes that present certain information in certain

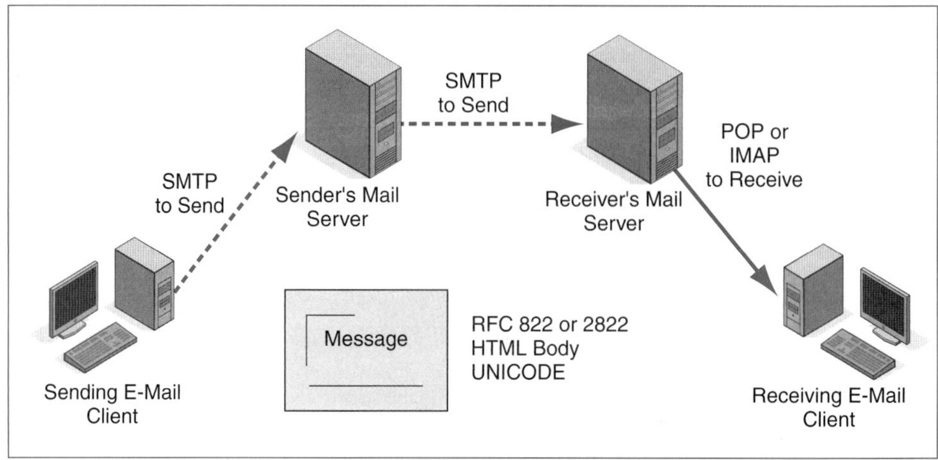

FIGURE 11-5 E-Mail Standards

ways, and there are specific ways to post mail for delivery, including putting letters in post office drop boxes and taking them to the post office.

Figure 11-5 shows how e-mail is posted (sent). The e-mail program on the user's PC sends the message to its outgoing mail host, using the **Simple Mail Transfer Protocol (SMTP)**. Figure 11-6 shows the complex series of interactions that SMTP requires between the sender and receiver before and after mail delivery.

Figure 11-5 shows that the sender's outgoing mail host sends the message on to the receiver's incoming mail host, again using SMTP. The receiving host stores the message in the receiver's mailbox until the receiver retrieves it.

Actor	Command	Comment
Receiving SMTP Process	220 Mail.Panko.Com Ready	When a TCP connection is opened, the receiver signals that it is ready.
Sending SMTP Process	HELO voyager.shidler .hawaii.edu	Sender asks to begin sending a message. Gives own identity. (Yes, HELO, not HELLO.)
Receiver	250 Mail.Panko.Com	Receiver signals that it is ready to begin receiving a message.
Sender	MAIL FROM: David@ voyager.shidler.hawaii.edu	Sender identifies the sender (mail author, not SMTP process).
Receiver	250 OK	Accepts author. However, may reject mail from others.
Sender	RCPT TO: Ray@Panko.com	Identifies first mail recipient.
Receiver	250 OK	Accepts first recipient.
Sender	RCPT TO: Lee@Panko.com	Identifies second mail recipient.
Receiver	550 No such user here	Does not accept second recipient. However, will deliver to first recipient.
Sender	DATA	Message will follow.
Receiver	354 Start mail input; end with <CRLF>.<CRLF>	Gives permission to send message.
Sender	*When in the course . . .*	The message. Multiple lines of text. Ends with line containing only a single period: <CRLF>.<CRLF>
Receiver	250 OK	Receiver accepts message.
Sender	QUIT	Requests termination of session.
Receiver	221 Mail.Panko.Com Service closing transmission channel	End of transaction.

FIGURE 11-6 Interactions in the Simple Mail Transfer Protocol (SMTP)

Receiving Mail (POP and IMAP)

Figure 11-5 also shows two standards that are used to *receive* e-mail. These are the **Post Office Protocol (POP)** and the **Internet Message Access Protocol (IMAP)**. IMAP offers more features, but the simpler POP standard is more popular. Programs implementing these standards ask the mail host to download some or all new mail to the user's client e-mail program. Often, users delete new mail from their inbox after downloading new messages. After that, the remaining messages exist only on the user's client PC.

Web-Enabled E-Mail

Almost all client PCs have browsers. Many mail hosts are now Web-enabled, meaning that users only need browsers to interact with them in order to send, receive, and manage their e-mail. As Figure 11-3 showed, all interactions take place via HTTP, and these systems use HTML to render pages on-screen.

Web-enabled e-mail (also called **webmail**) is especially good for travelers because no special e-mail software is needed. Any computer with a browser in an Internet café, home, or office will allow the user to check his or her mail. On the downside, Web-enabled e-mail tends to be very slow because almost all processing is done on the distant (and often overloaded) webserver with its server-based mail processing program.

Viruses and Trojan Horses

Although e-mail is tremendously important to corporations, it is a source of intense security headaches. As we learned in Chapter 9, the most widespread security compromises are attacks by viruses and worms. Viruses come into an organization primarily, although by no means exclusively, through e-mail attachments and (sometimes) through scripts in e-mail bodies. E-mail attachments can also be used to install worms and Trojan horse programs on victim PCs.

ANTIVIRUS SOFTWARE The obvious countermeasure to e-mail-borne viruses is **antivirus software**, which scans incoming messages and attachments for viruses, worms, and Trojan horses. Many companies produce antivirus programs that can run on client PCs.

ANTIVIRUS SCANNING ON USER PCS One problem is that most companies attempt to confront security threats by installing virus scanning on the user PCs. Unfortunately, too many users either turn off their antivirus programs if they seem to be interfering with other programs (or appear to slow things down too much) or keep their programs active but fail to update them regularly. In the latter case, newer viruses will not be recognized by the antivirus program.

CENTRALIZED ANTIVIRUS/ANTI-TROJAN HORSE SCANNING Consequently, many companies are beginning to do central scanning for e-mail-borne viruses and Trojan horses. As Figure 11-7 shows, there are several places that this scanning can be done.

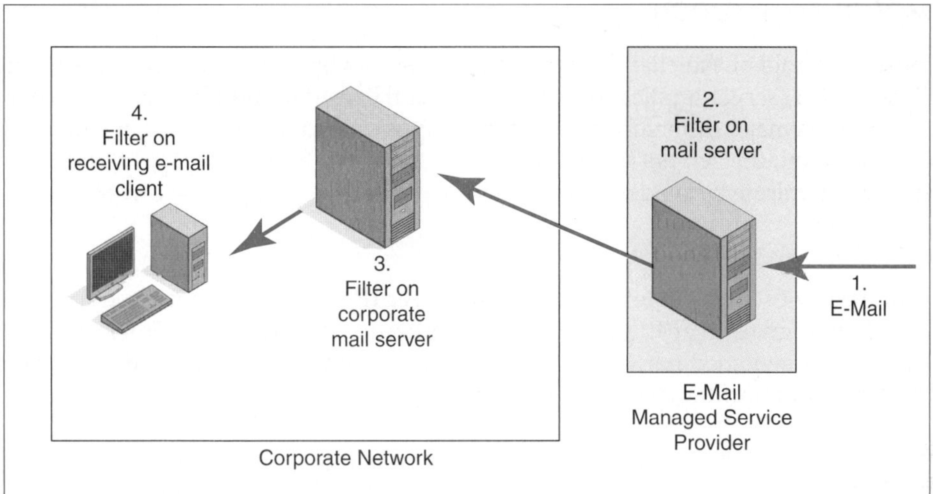

FIGURE 11-7 Scanning Locations for E-Mail

Scanning on Mail Servers. One popular place to do this is the corporate mail server. Users cannot turn off antivirus filtering on the mail server, and the e-mail staff (hopefully) updates virus definitions on these servers frequently.

Outsourcing Scanning. Some companies are even outsourcing antivirus/anti-Trojan horse scanning to outside security firms. By changing the firm's MX record in DNS servers, a firm can have all of its incoming e-mail sent to a security firm that will handle antivirus and anti-Trojan horse scanning. These firms specialize in such tasks and, presumably, can do a better job than the corporation. Outsourcing also reduces the workload of the corporate staff.

DEFENSE IN DEPTH The security principle of defense in depth suggests that antivirus filtering should be done in at least two locations, including the user PC, the mail server, or an external security company. It is also best if two different antivirus vendors are used. This increases the probability of successful detection because different antivirus programs often differ in which specific viruses, worms, and Trojan horses they catch.

Test Your Understanding

6. a) Distinguish among the major standards for e-mail bodies. b) In traditional e-mail, when a station sends a message to its mail server what standard does it use? c) When the sender's mail server sends the message to the receiver's mail server what standard does it use? d) In traditional e-mail, when the receiver's e-mail client downloads new mail from its mail server what standard is it most likely to use? e) What is Web-enabled e-mail? f) What is the advantage of a Web-enabled e-mail system? g) What is the disadvantage of Web-enabled e-mail?
7. a) What is the main tool of firms in fighting viruses and Trojan horses in e-mail attachments? b) Why does filtering on the user's PC often not work? c) What options do firms have for where antivirus filtering should be done? d) According to the principle of defense in depth, how should firms do antivirus filtering?

VOICE OVER IP (VOIP)

Another example of the client/server architecture is voice over IP (VoIP). Like e-mail, VoIP is a client/server application in which both the sender and the receiver have their own servers. A major difference between these applications, however, is that, after setting up a connection, the servers in VoIP get out of the way almost completely, and the two clients communicate by sending packets directly to each other until the end of the call.

Basics

One of the newest areas in telephony is **voice over IP (VoIP)**, which is the transmission of telephone signals over IP packet-switched internets (including the Internet) instead of over circuit-switched networks. VoIP offers the promise of reducing telephone costs by moving from traditional circuit switching to more efficient packet switching.

CLIENTS Figure 11-8 illustrates VoIP operation. The figure shows two clients. One is a client PC with multimedia hardware (a microphone and speakers) and VoIP software. The other is a **VoIP telephone**, which has the electronics to encode voice for digital transmission and to handle packets over an IP internet. With VoIP, these two clients' users can talk with each other.

MEDIA GATEWAY The figure also shows a media gateway. The **media gateway** connects a VoIP system to the ordinary public switched telephone network. Without a media gateway, VoIP users could talk only to one another. The media gateway translates both signaling and transport transmissions.

The media gateway translates both signaling and transport transmissions.

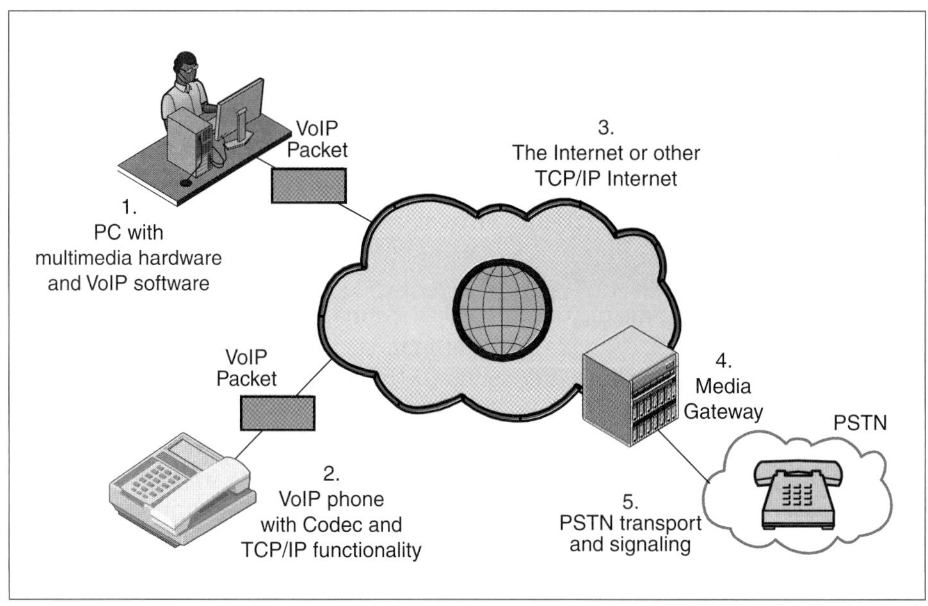

FIGURE 11-8 Voice over IP (VoIP)

Test Your Understanding

8. a) What is VoIP? b) What is the promise of VoIP? c) What two devices can be used by VoIP callers? d) What is the purpose of a media gateway? e) Why is having a media gateway in a VoIP system important? f) Does the media gateway translate signaling transmissions or transport transmissions?

VoIP Signaling

In telecommunications, there is a fundamental distinction between signaling and transport. Signaling consists of the communication needed to set up circuits, tear down circuits, handle billing information, and do other supervisory chores. Transport is the actual carriage of voice.

There are two major VoIP signaling protocols. The first was the ISO **H.323** standard, which was effective, but very complex. More recently, the IETF created the **Session Initiation Protocol (SIP)** standard. Most older VoIP systems use H.323 to control signaling. However, the use of SIP is growing rapidly, and most VoIP systems today use SIP for signaling.

Figure 11-9 illustrates the SIP protocol. Each subscriber has an SIP proxy server. The calling VoIP telephone sends a SIP INVITE message to its SIP proxy server. This message gives the IP address of the receiver. The caller's SIP proxy server then sends the SIP INVITE message to the called party's SIP proxy server. The called party's proxy server sends the SIP INVITE message to the called party's VoIP telephone or multimedia PC.

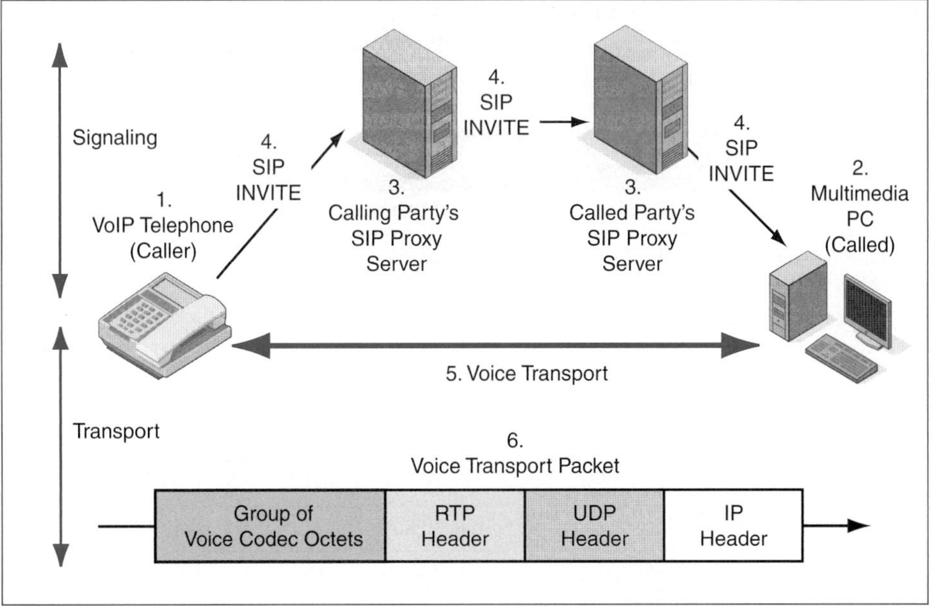

FIGURE 11-9 VoIP Signaling and Transport

Test Your Understanding

9. a) What are the two major protocols for VoIP signaling? b) Which of these protocols is growing rapidly? c) Describe how SIP initiates a communication session.

VoIP Transport

After SIP or H.323 creates a connection, the two VoIP clients begin communicating directly. This is the beginning of transport, which is the transmission of voice between callers. VoIP, as its name suggests, operates over routed IP networks. Therefore, digitized voice has to be carried from the sender to the receiver in packets.

CODECS VoIP telephones and multimedia PCs need codecs to convert analog voice signals into digital voice data streams. VoIP systems can use many different codecs. Figure 11-10 shows that some codecs convert voice streams into bit streams as small as 5.3 kbps. However, the codecs that do the most compression also lose the most voice quality. Selecting codec in a VoIP network means making a trade-off between voice quality and cost reduction.

VOIP TRANSPORT PACKETS As noted in Chapter 1, long application messages have to be fragmented into smaller pieces that can be carried in individual packets. Each packet carries a small part of the application message. In VoIP, packets carry a small snippet of digital voice bytes created by the codec.

The Application Layer: Codec Bytes. Figure 11-9 shows a VoIP transport packet. At the application layer, the application message is a group of voice codec bytes.

UDP. TCP allows reliable application message delivery. However, the retransmission of lost or damaged TCP segments can take a second or two—far too long for voice conversations. Voice needs to be transmitted in real time. Consequently, VoIP transport uses UDP at the transport layer. UDP reduces the processing load on the VoIP telephones, and it also limits the high network traffic that VoIP generates. If packets are

Codec	Transmission Rate
G.711	64 kbps
G.721	32 kbps
G.722	48, 56, 64 kbps
G.722.1	24, 32 kbps
G.723	5.33, 6.4 kbps
G.723.1A	5.3, 6.3 kbps
G.726	16, 24, 32, 40 kbps
G.728	16 kbps
G.729AB	8 kbps

FIGURE 11-10 VoIP Codecs

lost, the receiver creates fake noise for the lost codec bytes. It does this by extrapolating between the content of the preceding and following packets.

THE REAL TIME PROTOCOL (RTP) Between UDP and the application message, VoIP adds an additional header, a Real Time Protocol (RTP) header to make up for two deficiencies of UDP.

- First, UDP does not guarantee that packets will be delivered in order. RTP adds a sequence number so that the application layer can put packets in the proper sequence.
- Second, VoIP is highly sensitive to jitter, which is variable latency in packet delivery. Jitter literally makes the voice sound jittery. RTP contains a time stamp for when its package of octets should be played relative to the octets in the previous packet. This allows the receiver to provide smooth playback.

THE IP HEADER This is a packet, so the IP header comes before the other fields.

Test Your Understanding

10. a) What is the purpose of a VoIP codec? b) Some codecs compress voice more. What do they give up in doing so? c) In a VoIP transport packet, what is the application message? d) Does a VoIP transport packet use UDP or TCP? Explain why. e) What two problems with UDP does RTP fix? f) List the headers and messages in a VoIP transport packet, beginning with the first packet header to arrive at the receiver. (Hint: See Figure 11-9)

THE WORLD WIDE WEB

HTML AND HTTP We have discussed the World Wide Web throughout this book. Figure 11-11 shows that the Web is based on two primary standards.

- First, webpages themselves are created using the **Hypertext Markup Language (HTML)**.
- Second, the transfer of requests and responses uses the **Hypertext Transfer Protocol (HTTP)**.

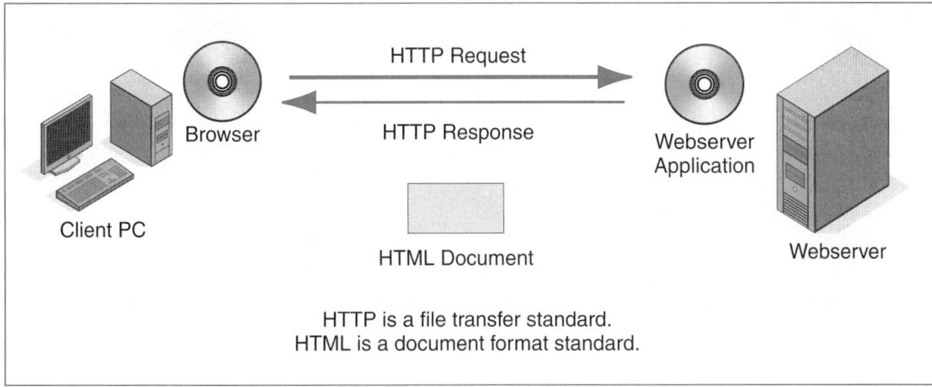

FIGURE 11-11 HTTP and HTML

To give an analogy, an e-mail message may be created using RFC 2822, but it will be delivered using SMTP. Many application standards consist of a document standard and a transfer standard.

Many application standards consist of a document standard and a transfer standard.

COMPLEX WEBPAGES Actually, most "webpages" really consist of several files—a master text-only HTML file plus graphics files, audio files, and other types of files. Figure 11-12 illustrates the downloading of a webpage with two graphics files.

The HTML file consists merely of the page's text, plus **tags** to show where the browser should render graphics files, when it should play audio files, and so forth.[5] The HTML file is downloaded first because the browser needs the tags to know what other files should be downloaded.

Consequently, several **HTTP request-response cycles** may be needed to download a single webpage. Three request-response cycles are needed in the example shown in the figure.

THE CLIENT'S ROLES The client's jobs, as shown in Figure 11-12, are to send **HTTP request messages** asking for the files and then to draw the received files on-screen. If the webpage has a **Java applet** or another **active element**, the browser will have to execute it as well.

THE WEBSERVER'S ROLE The webserver's job is to read each HTTP request message, retrieve the desired file from memory, and create an **HTTP response message** that contains the requested file or a reason why it cannot be delivered. Webserver application software may also have to execute server-side active elements before returning the requested webpage.

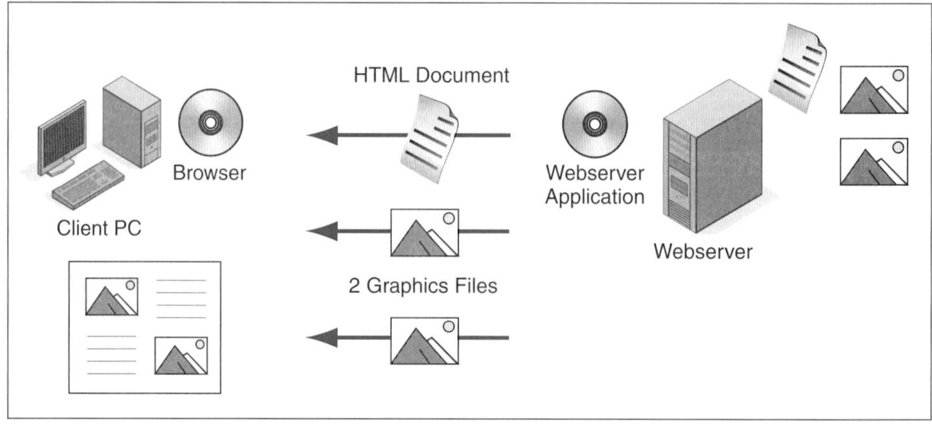

FIGURE 11-12 Downloading a Complex Webpage with Two Graphics Files

[5] For graphics files, the IMG tag is used. The keyword *IMG* indicates that an image file is to be downloaded. The SRC parameter in this tag gives the target file's directory and file name on the webserver.

HTTP REQUEST AND RESPONSE MESSAGES In Chapter 2, we looked briefly at HTTP request and response messages. We will now look at them in a little more detail. Figure 11-13 shows that both HTTP request messages and HTTP response messages are composed of simple keyboard text.

HTTP Request Messages. In HTTP request messages, the first line has four elements:

- The line begins with a capitalized method (in this case, GET), which specifies what the requestor wishes the webserver to do. The GET method says that the client wishes to get a file.
- The method is followed by a space and then by the location of the file (in this example, /panko/home.htm). This is home.htm in the panko directory.
- Next comes the version of HTTP that the client browser supports (in this example, HTTP/5).
- The line ends with a carriage return/line feed—a command to start a new line of text.

Each subsequent field is one line long (there is only one in this example). The field consists of a keyword (in this example, Host), a colon (:), a value for the keyword (in this example, voyager.shidler.hawaii.edu), and a carriage return/line feed.

HTTP Response Messages. HTTP response messages also begin with a four-element first line.

- The webserver responds by giving the version of HTTP it supports.
- This is followed by a space and then a code. A 200 code is good; it indicates that the method was followed successfully. In contrast, codes in the 400 range are bad codes that indicate problems.
- This code is followed by a text expression that states what the code says in humanly readable form. This information ("OK" in this example) is useless to the browser.
- A carriage return/line feed ends this first line.

Subsequent lines are fields. They have the keyword–colon–value–carriage return/line feed structure we saw in the HTTP request message. In the figure, these fields give a time stamp, the name of the server software (not shown), and two MIME fields.

HTTP Request Message

GET /panko/home.htm HTTP/5[CRLF]
Host: voyager.cba.hawaii.edu[CRLF]

HTTP Response Message

HTTP/5 200 OK[CRLF]
Date: Tuesday, 20-MAR-2011 18:32:15 GMT[CRLF]
Server: *name of server software*[CRLF]
MIME-version: 1.0[CRLF]
Content-type: text/plain[CRLF]
[CRLF]
File to be downloaded. A string of bits that may be text, graphics, sound, video, or other content.

FIGURE 11-13 Examples of HTTP Request and Response Messages

MIME (Multipurpose Internet Mail Extensions) is a standard for specifying the format of the file being delivered as the body of the response message. MIME is also used for this purpose in e-mail (as its name suggests) and in some other applications.

- The first MIME field gives the version of MIME the webserver uses (1.0).
- The next line, *content-type* field, specifies that the file being delivered by the webserver is of the text/plain type—simple keyboard characters.

After all HTTP response message header fields, there is a blank line (two CR/LFs in a row gives a blank line). This is followed by the bytes of the file being sent by the webserver.

Test Your Understanding

11. a) Distinguish between HTTP and HTML. b) You are downloading a webpage that has six graphics and two sound clips. How many request-response cycles will be needed? c) What is the syntax of the first line in an HTTP request message? d) What is the syntax of subsequent fields? e) What is the syntax of the first line in an HTTP response message? f) What do the MIME header fields tell the receiving process? g) Why is this information necessary? h) How is the start of the attached file indicated?

CLOUD COMPUTING

A Shared Project

Figure 11-5 shows Shea and Melinda, who are working to construct a word processing report for their project. Their manager asks them to use Google Docs for this project, so

A Shared Project (Figure 11-15)

Google Docs

Shea and Melinda can share documents

Before, they had used e-mail attachments

There was often confusion over who had the most recent version

Cloud Computing

For users, showing the network as a cloud indicates that they do not have to understand it

In cloud computing, there is a second cloud—a cloud service provider

In cloud computing, the company outsources some combination of servers, server applications, and client PC software to a cloud service provider. If client PC software is outsourced, the client PC only needs a browser.

Google Docs is a cloud provider of Software as a Service

Usually metered service—pay as you go

Software as a Service

In Software as a Service (SaaS), an application service provider (ASP) supplies an application to customers on demand

FIGURE 11-14 Cloud Computing (Study Figure)

In cloud Saas, the user typically has a browser only

With cloud SaaS today, there is a broad range of applications available

Attractions

- Reduces cost and risk because only pays for software as needed
- Saves money because company does not have to manage the application
- Mobility—users can access the software wherever there is Internet access
- Collaboration: sharing is easy

Issues

- If there is no Internet access, the application is unusable
- Cloud SaaS provider may go out of business
- Loss of control: may be locked into proprietary vendor software
- Security: will company's data be vulnerable if cloud service provider is attacked or has software vulnerabilities?
- Legal complications: if user company is required by law to be secure, how can this be satisfied with SaaS?

Cloud Utility Computing

In traditional utility computing, a company offloads server processing work to another company at a remote site

In cloud utility computing, the company that receives this processing job is a cloud service provider, and the data is sent over the Internet to be processed

Attractions

- Saved cost and risk of long-term investment
- No need to hire staff
- No need to purchase and manage servers
  - Company may purchase more servers than it needs
- Flexibility—can get extra capacity quickly when needed for a short period of time
- Scalability—will not outgrow the service provider's capacity

Issues

- As in cloud SaaS, loss of control and legal and security concerns

Why is Cloud Utility a Hot Topic Now?

The Internet is now fast, reliable, and accessible nearly everywhere

Web services

Virtualization

Managing Cloud Computing

Many advantages

Loss of control

Must consider security

Must have Service Level Agreements (SLAs) for performance

FIGURE 11-14 Continued

that they can collaborate remotely. When Shea wants to input information, he logs onto the Google Docs website. He then creates and saves a draft report. The report is stored on the Google Docs servers, rather than on Shea's desktop computer or his company's servers. Shea and Melinda still have the option of downloading documents from the Google servers to their PCs, and they can also upload preexisting documents to the Google Docs servers.

A similar situation is true of the word processing software: Instead of installing the application on his own computer, Shea only needs a browser to access the software.

Afterward, Melinda needs to edit Shea's document, so Shea adds her to the group that has access to it. She uses her browser to access it. She then makes edits and saves the document again. Later, Shea can read the edited document and add his own edits.

In the past, document sharing had been difficult for Shea and Melinda because they work in different cities. Before Google Docs, when Shea and Melinda worked together, one wrote a document, saved it, and sent a copy as an attachment. The receiver had to wait for the attachment, save it, and edit it. Google Docs file sharing is far easier to use. In addition, there is only one copy—the copy on the Google Docs server. With the transmission of attachments, there had often been confusion about which file was the most recent version.

Cloud Computing

In Chapter 1, we saw that network users view networks as clouds because users never have to worry about the details of network operation. In Chapter 6, we saw that companies view public switched data networks as clouds because they do not have to understand the internal switching mechanism.

In Figure 11-15, there are two clouds. The broader cloud is the Internet. As users, Shea and Melinda merely connect to the Internet and then ignore it. In addition, Google

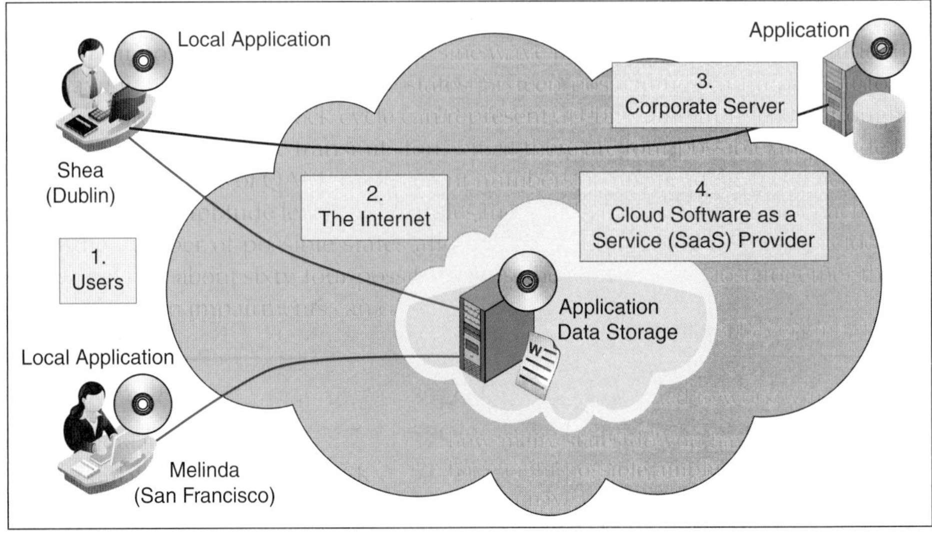

FIGURE 11-15 Cloud Application (Software as a Service)

Docs is shown as another cloud because Shea and Melinda do not have to know how it works or where their documents are stored.

More radically, Google Docs is also a cloud to the company that Shea and Melinda work for. The company does not have to install any software on its employee computers to use Google Docs. Nor does the company have to manage servers and application software on the servers.

Google Docs is just one example of a new trend in computing—cloud computing. In traditional computing, the company had to maintain its servers, server applications, and client PC software. In **cloud computing**, the company outsources some combination of servers, server applications, and client PC software to a **cloud service provider** (a company that provides cloud computing services). If client PC software is outsourced, the client PC only needs a browser. In the case of Google Docs, the company employing Shea and Melinda outsourced all three functions to Google, its cloud service provider.

Cloud computing is a form of computing where the company outsources some combination of servers, server applications, and client PC software to a cloud service provider. If client PC software is outsourced, the client PC only needs a browser.

So far, we have not talked about pricing. Typically, cloud service providers offer **metered service**. In this pay-as-you-go model, the cloud computing customer pays only for the processing capacity that he or she actually uses. A company can buy as little or as much processing capacity as it needs for a given task, rather than having to commit to a monthly service contract that does not match up to its actual needs. This also saves the company from the burden of purchasing servers, software to run on its clients and servers, and annual licenses to use the software.

Test Your Understanding

12. What is cloud computing?

Software as a Service (SaaS)

There are different types of cloud computing services. We will focus on two of the most widely used types: Software as a Service and utility computing. Although both types of service predate cloud computing, each has been enhanced—and has become much more popular—through cloud computing. We will begin by looking at Software as a Service.

DEFINITION In **Software as a Service (SaaS)**, an application service provider (ASP) supplies an application to customers on demand. SaaS software vendors may host the application on their own webservers or download the application to the consumer device, disabling it after use or after the on-demand contract expires. In **cloud Software as a Service**, the user accesses the software over the Internet, often (although not always) with a browser. If this sounds familiar, Google Docs in Figure 11-15 is an example of cloud SaaS. As with traditional SaaS, the application may either be hosted on the ASP's webservers or downloaded temporarily from the webservers to the consumer device.

SAAS AS A CLOUD COMPUTING MODEL As just noted, Software as a Service is not a new idea. For example, web-based e-mail has been around for over a decade. What *has* changed with cloud SaaS is that there is a now a very wide range of applications available—everything from word processing and spreadsheet programs to data visualization software that provides real-time analysis of a company's performance, payroll processing, and general accounting, to name just a few.[6] Cloud computing has also made SaaS increasingly popular as a business model. One high-profile example is Salesforce, whose customer relationship management software and other applications have been widely used.[7]

ATTRACTIONS OF CLOUD SAAS One main attraction of cloud SaaS is that a company can use software on an as-needed basis, which may result in lowered costs and less risk. A company is especially likely to see cost savings if it uses cloud SaaS for an application it uses only occasionally. With cloud SaaS, the company is saved the risk of long-term investment in hardware and management.

A second attraction is that cloud SaaS allows the company to save on personnel resources. Since all of the management of the software is done in the cloud, the company does not have to allocate IT staff to install or manage the application. So a company may be able to save money even if it uses the software heavily.

Mobility is a third attraction. As illustrated in the Google Docs example, because cloud SaaS delivers an application over the Internet, users can access the software from anywhere that they can access the Internet. This is especially useful for employees who often travel or work from home.

A fourth attraction is collaboration. Most SaaS vendors allow multiple people within a customer company to work on the same information, as in the Google Docs example at the start of this section.

ISSUES FOR CLOUD SAAS Unfortunately, cloud Software as a Service also raises serious issues that corporations must consider. First is the issue of access. A company using cloud SaaS might be unable to access the application for several reasons. If the cloud service provider has a technical problem or is attacked and taken off-line, users will be unable to access the application.

Similarly, a cloud service provider may go out of business, leaving users no way to access or recover their stored information, and no way to continue working on a project that required specific software. If software resources on the cloud become inaccessible for any reason, client computers become useless, and companies may be unable to complete time-sensitive projects.

A second and related concern is that cloud SaaS entails a loss of control. Companies may be locked into proprietary software. They may be unable to migrate their data out of a particular SaaS application, and they may thus be locked into rising costs.

The third issue is security. If your company's sensitive data is stored in the SaaS application, it may be vulnerable to attackers who target the cloud service provider. A customer can only hope that the cloud service provider has strong security. Similarly, the application software itself may have dangerous vulnerabilities. For example, in 2008,

---

[6] See, for example, Salesforce's Sales Cloud™ datasheet: www.salesforce.com/crm/sales-force-automation/analytics-sales-forecasting/

[7] Salesforce.com claims that more than 63,000 companies "have made the Sales Cloud the world's most-popular sales application." www.salesforce.com/crm/sales-force-automation/

a software bug in Google Docs resulted in some users' documents being inadvertently shared, potentially making corporate data available to parties who should not have had permission to see it.[8]

A final issue is that legal complications may arise from using cloud SaaS. If a company is required by law or corporate policy to provide regular security audits, what happens if the cloud service provider refuses to be audited? Also, if data stored by the cloud service provider is compromised, who will be held legally liable?

Test Your Understanding

13. a) Describe how Software as a Service (SaaS) works. b) Describe how cloud SaaS works. c) What are the four main attractions of cloud SaaS? d) What are the four issues that cloud SaaS raises?

Cloud Utility Computing

The second main type of cloud computing is cloud utility computing. Like Software as a Service, utility computing predates cloud computing. In fact, utility computing was used as far back as the 1960s. Also like Software as a Service, utility computing has seen a massive increase in popularity since it has been offered as a type of cloud computing.

DEFINITION In **utility computing**, a company offloads server processing work to another company at a remote site. Figure 11-16 shows that in **cloud utility computing**

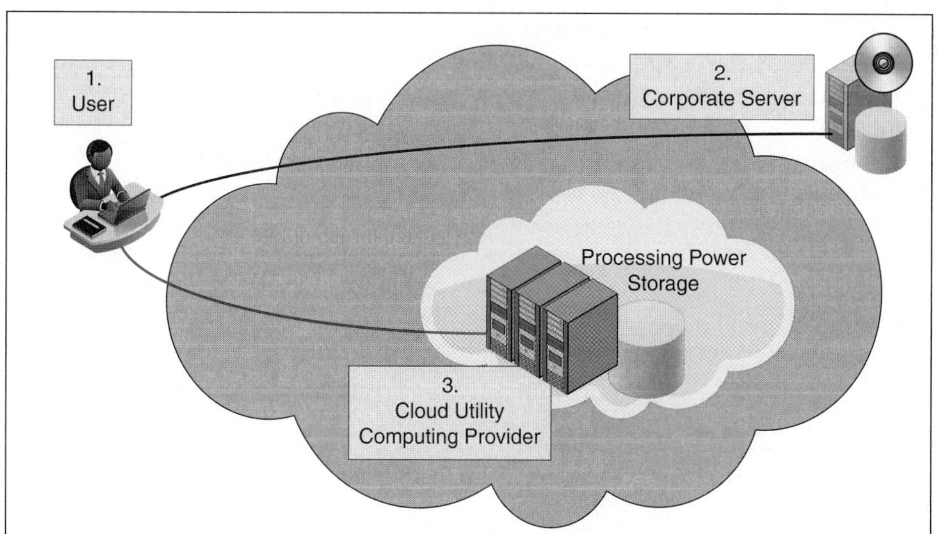

FIGURE 11-16 Cloud Utility Computing

[8] "Google Software Bug Shared Private Online Documents." *Breitbart.com*. March 10, 2008. Accessed November 22, 2009. www.breitbart.com/article.php?id=CNG.54c3200989573ae4c9282658f91276df.481&show_article=1

(i.e., utility computing done as a cloud computing model), the company that receives this processing job is a cloud service provider, and the data is sent over the Internet to be processed. Traditional and cloud utility computing are usually offered as pay-as-you-go services. In fact, the name "utility" refers to the fact that public utilities like electricity are typically offered as metered service.

UTILITY COMPUTING AS A CLOUD COMPUTING MODEL Using cloud utility computing means that a company does not have to run its own data center to do server processing work. Instead, the service provider is in charge of managing the servers and the processing itself. This is a definite change from the traditional model where a company owns its servers. In this traditional model, a company must pay to acquire hardware and install programs, and to hire staff with the expertise to provide ongoing server management. Also, because companies cannot always accurately estimate how much processing capacity they will need for a given task ahead of time, they may opt to buy much more processing power than they will ultimately need. This overprovisioning is expensive and wasteful. Underprovisioning can also be a problem: If a company does not purchase sufficient processing resources, the system may be unable to function. In this case, current and potential users may become dissatisfied with the service, and the company may lose customers.

ATTRACTIONS OF CLOUD UTILITY COMPUTING As with cloud Software as a Service, one of the attractions of cloud utility computing is that companies are saved the cost and risk of long-term investment in hardware and management.

A second shared attraction is that the company does not have to dedicate personnel resources to managing its own data center.

A third attraction is that this model provides a high level of flexibility. Flexibility is the ability to change capacity dynamically. As we will see later in this section, the use of virtualization allows the cloud service provider to increase or decrease the processing power a user wants by very small increments, on demand. As a result, a cloud user can quickly fine-tune the system so that the processing power used matches the user's requirements much more precisely.

Fourth, virtualization also makes cloud utility computing highly scalable. As we saw in Chapter 3, scalability is the ability to grow as large as is needed. With virtualization, which we will look at a little later in this section, cloud utility computing can adapt as a firm either grows larger or just needs some extra processing power to handle the Christmas rush.

ISSUES FOR CLOUD UTILITY COMPUTING The issues facing utility computing are the same as those for Software as a Service: access, loss of control, security, and legal requirements.

Test Your Understanding

14. a) Is utility computing a new phenomenon? b) Describe how utility computing works. c) Why was "utility computing" given this name? d) What provisioning problems might a company face if it runs its own data center?

15. a) What are the four main attractions of cloud utility computing? b) What are the issues facing cloud utility computing?

Why Is Cloud Computing a Hot Topic Now?

In the past several years, cloud computing has become a hot topic in the IT world, with major companies such as Amazon, Microsoft, and Google offering cloud computing services. Yet, as we have seen, utility computing and Software as a Service are the two main types of cloud computing, and neither is itself new. The question thus arises: Why has cloud computing become such a hot topic *now*, if we have had the ability to use Software as a Service and utility computing for years?

There are three major factors that have combined to allow cloud computing to take fuller advantage of the potentials of SaaS and utility computing. These are the Internet, Web services, and virtualization. Because of these factors, cloud computing has become very popular.

CLOUD COMPUTING AND THE INTERNET Although the Internet has existed in some form since the 1960s, in recent years it has become very accessible, very fast, and very reliable. Because of the large increase in wireless coverage, users can now access cloud computing resources from almost everywhere. The speed and reliability of Internet service means that it is feasible for corporations to rely on the cloud to access and process their corporate data.

WEB SERVICES The existence of Web services has also helped cloud computing become a viable business model. We will look at Web services in more detail later in this chapter, but now we will mention the basics of how it applies to cloud computing. Web services allow one program on one machine to communicate with another program on another machine very easily. Of course, browsers and webservers do this now, for file downloading. However, Web services allow a program on one machine to send data to a program on another machine; the program on the other machine will do processing, compute a result, and return this result to the calling computer. This ability to call on remote programs for processing service in a general way has greatly expanded the range of programs that are available as cloud computing applications.

VIRTUALIZATION Figure 11-17 shows a traditional computer. The operating system is designed to hide the details of the hardware from the user and from applications the user runs. It accepts commands from the user or application program and carries them out in hardware. A simple example is deleting a file. This actually is a complex thing to do in hardware. Yet the user merely selects the file in the operating system's graphical user interface and clicks delete. Or, in an application program, the user may open a file. The application program will pass that command to the operating system, which will direct the hardware in the details of retrieving the file.

Figure 11-17 also shows how computers that use virtualization work. In **virtualization**, the real computer's capacity is divided among a number of virtual machines. Each **virtual machine (VM)** acts like a stand-alone computer to its users.

In virtualization, the real computer's capacity is divided among a number of virtual machines.
Each virtual machine (VM) acts like a stand-alone computer to its users.

Traditional Computer Operation

User	User
	Application Program
Operating System	
Hardware (CPU, RAM, Hard Drive, Network Interface Card, etc.)	

Virtualization and Virtual Machines (VMs)

Application 1	Application 2a	Application 2b	Application 3a	Application 3b	Application 3c
Operating System	Operating System		Operating System		
Virtual Machine 1	Virtual Machine 2		Virtual Machine 3		
Hypervisor					
Hardware (CPU, RAM, Hard Drive, Network Interface Card, etc.)					

FIGURE 11-17 Traditional Computer Operation and Virtualization

Note that there is a new layer of software—the hypervisor. The **hypervisor** manages the use of physical computer resources by the virtual machines and communication between each virtual machine and its users. When the operating system of a VM sends a command to hardware, the hypervisor intercepts the command and carries it out on the real hardware beneath the hypervisor.

> The hypervisor manages the use of physical computer resources by the virtual machines and communication between each virtual machine and its users.

Advantages. Why is server virtualization done? The first reason is that virtualization provides economies of scale in management cost. Each physical server requires a certain amount of management labor. Reducing the number of physical servers through virtualization greatly reduces management costs. Although virtualized servers require more management labor than actual servers, the net savings in management costs is large.

A second reason is economies of scale in hardware cost. If you compare the prices of two servers, the larger server usually costs substantially less per unit of processing power. So the hardware cost of running many VMs on fewer servers is less than the cost of running many servers without virtualization.

The third reason is the efficient use of server capacity. With non-virtualized servers, the applications running on them often use only a small part of the capacity. With virtualization, each virtual machine gets the capacity it needs and no more. If the hardware becomes overworked, then some of the virtual machines can be moved to other physical servers. Porting a VM from one server to another usually is extremely simple.

A fourth reason is a combination of flexibility and scalability. As mentioned above, flexibility means that if an application suddenly needs considerably more processing capacity—for instance, if there is a major marketing promotion—it can be given more capacity on its real server. Scalability means that if demand for a VM's services grows, the VM can simply be moved to a server with much higher capacity. It may even be possible for the VM to operate across multiple physical machines.

Disadvantages. Of course, virtualization only works well if the systems administrator for the server can give adequate capacity to each VM. If a server becomes overloaded, all VMs will run slowly.

Another disadvantage is that virtualization raises security concerns. Even if a firewall is placed between the physical server and the network, there is a chance that a hacker who gains a foothold on one virtual machine can access another virtual machine on the same server without having to deal with the firewall.

Test Your Understanding

16. a) What three major factors have contributed to the popularity of cloud computing?
b) How has the Internet contributed to the popularity of cloud computing?

17. Briefly explain how computers that use virtualization work.

Managing Cloud Computing

Many businesses today must ask themselves the question, is cloud computing a sound business choice?

As we have seen, cloud computing has a number of attractions. Using cloud computing often results in cost savings, especially when it comes to capital expenditure. The flexibility and scalability of cloud computing means that a company can rapidly adjust its service as needed, and that the service can grow along with the company. Cloud computing also has the advantage of mobility of access.

Yet cloud computing also involves a fundamental loss of control. Companies may suffer if they lose access to their cloud resources, and they may face security risks and legal complications if their data is stored on the cloud.

When deciding whether to use cloud computing, a company must consider security risk. Of course, this is not unique to cloud computing. As security expert Bruce Schneier says, "IT security is about trust. You have to trust your CPU manufacturer, your hardware, operating system and software vendors—and your ISP."[9] Similarly, a company that uses cloud computing must trust its cloud service provider.

If a company decides to use cloud computing, it should set up safeguards such as service-level agreements (SLAs) for reliability, speed, error rates, and other quality-of-service measures. Another important step is to only use cloud service providers that have clearly stated policies about security, data backup, and similar protections, and the execution of these policies should be independently audited. Also, a company must make sure that it understands what its legal liability will be if it uses cloud computing. One final recommendation is that it may be useful to take additional security precautions, such as encrypting data or providing redundancy by using multiple cloud service providers.

---

[9] Bruce Schneier, "Be Careful When You Come to Put Your Trust in the Clouds," *The Guardian*, June 2, 2009. www.schneier.com/essay-274.html

Test Your Understanding

18. What major issue must a company consider when deciding whether to use cloud computing?

SERVICE-ORIENTED ARCHITECTURES

The last client/server-based application architecture we will look at is the family of service-oriented architectures, of which Web services is a prominent subcategory.

Traditional Software Architectures and Service-Oriented Architectures (SOAs)

Traditionally, organizations have created applications by writing a single large program or a few programs that interact with each other in a very well-defined way. This works, but it is not very flexible. It is a *program-oriented architecture* because it is a **software architecture** (plan for providing applications functionality) built on large individual programs.

In contrast, Figure 11-18 shows a **service-oriented architecture (SOA)**. In SOAs, there are many **service objects** instead of a few large programs. Each service object provides one or more **services** to callers. For example, the figure shows Calling Program 1 that contacts Service Object 1 to receive a price quote. The customer's call would contain a part number, the number of units desired, and shipping method. The service object's service would return a message containing the price quote.

Test Your Understanding

19. a) What is a service-oriented architecture? b) What do service objects do?

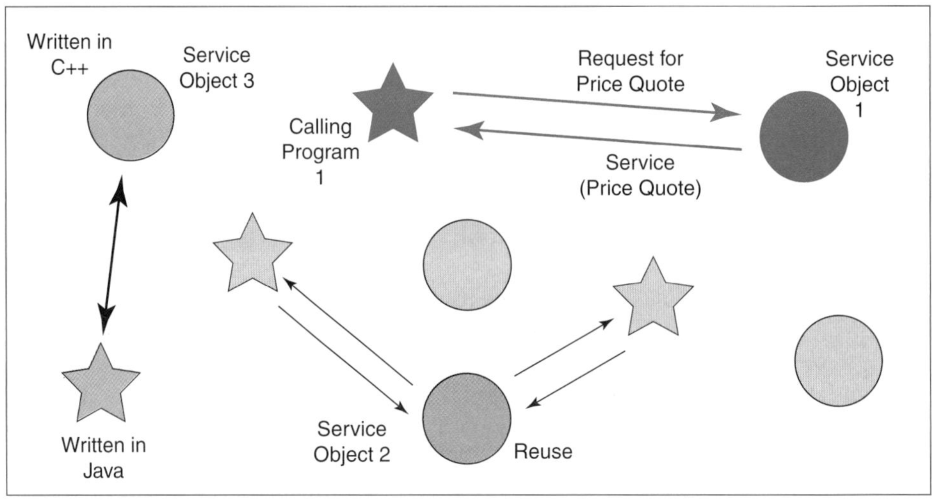

FIGURE 11-18 Service-Oriented Architecture (SOA)

Reuse and Language Independence

REUSE Compared with traditional application development, the creation of a service-oriented architecture is more difficult. Its value comes through **reuse**. Once a service object is built, other programs can call for its service or services in a very simple standardized way. In Figure 11-18, Service Object 2 is being used simultaneously by two calling programs.

If a service object has frequently needed functionality, it may be used many times by many different applications. A simple example may be a calculator that computes a Fahrenheit temperature for a particular centigrade temperature. If someone creating such a service object can work out payment and trust issues, the object can be in high demand. Eventually, people and organizations may turn to the Internet to find a marketplace for millions of different service objects, each offering specific services.

In the future, applications may consist of small skeletons of commands, with individual groups of commands in the skeleton making calls to service objects to provide certain pieces of information or processing results. This certainly will be a radical way to provide applications.

LANGUAGE INDEPENDENCE Service objects and calling programs (which may be service objects themselves) exchange formatted messages. The formatting of these messages is independent of the programming language used to create the calling program and the service object. In Figure 11-18, Service Object 3 is written in C++, while the calling program is written in Java. They have no trouble communicating.

This means that calling programs and service objects can be written in any languages, and the two may be written in different languages. This independence of SOA from programming languages is called **language independence**. The entire system is not locked into a single programming language, and new programming languages can be used at any time.

Test Your Understanding

20. a) What is the benefit of reuse? b) Why do SOAs make reuse more likely? c) What is language independence? d) In what sense are SOAs language-independent? e) Why is language independence good?

Web Services

Service-oriented architectures, again, consist of large numbers of service objects providing services. These objects are only loosely coupled, meaning that their only interactions come through the messages they exchange. How will their message interactions be structured? There are many ways to do it. One approach to creating a service-oriented architecture is to use Web services. Figure 11-19 emphasizes that Web services is only one way to create an SOA.

Web services use service objects that provide services to customers *using World Wide Web interaction standards*. For example, a user working with a browser may send a Web service request message to a service object. The interaction would be done via HTTP, which is one of the most fundamental WWW standards.

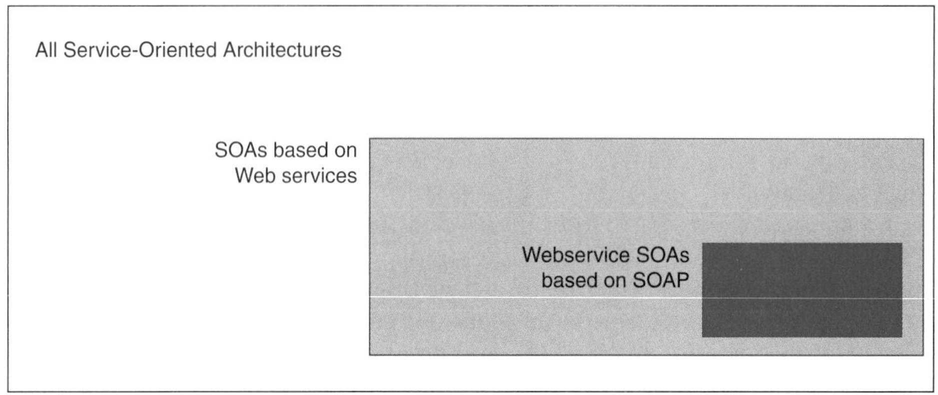

FIGURE 11-19 SOA versus Web Services

WEB SERVICE INTERACTIONS USING SOAP MESSAGES ENCODED IN XML There are many ways to create Web services. One way is to create SOAP messages. This is the only way we will consider. Figure 11-19 emphasizes that just as Web services is only one way to implement a service-oriented architecture, SOAP is only one way to implement Web services.

The body of a SOAP Web service request normally is encoded in XML rather than HTML. In HTML, you cannot create new tags by yourself or within a group. The eXtensible Markup Language is basically an extension of HTML that does allow you to create your own tags. For instance, a group of companies can create a tag pair, <price> and </price>, to indicate that the value between the two tags is a price. Figure 11-20 shows how new tags can be used in SOAP request and response messages.

XML is a general markup language. It does not describe the logical structure of the message. To impose the structure need to create requests to service objects and create replies, the XML body is organized according to the **Simple Object Access Protocol (SOAP)**, which specifies the syntax of request and response messages. In a SOAP message, there is a SOAP Envelope tag pair, and within this there is a SOAP BODY tag pair and perhaps a SOAP Header tag pair.

The SOAP Body tag pair contains a request or response. The request indicates parameters the service object needs to know, including PartNum (part number), Quantity (number of units demanded), and Shipping type (rush, standard, etc.). With this information, the service object can compute a price. The service object sends this result to the caller in another SOAP message.

The messages in Figure 11-20 are highly simplified. However, they give you a feeling for how Web services may use SOAP-structured messages to communicate. The full syntax of SOAP messages is complex, but this complexity adds nothing to the discussion.

WSDL AND UDDI For service-oriented architectures to be successful, it is necessary for users to be able to find Web service objects and learn how to use them. Figure 11-21 shows that there are two protocols to serve this purpose for Web services. A **Web Service Description Language (WSDL)** response message is like a user's manual for a particular Web service. However, unlike user manuals, WSDL is designed to be read by a software program rather than primarily by a human reader. A WSDL message allows a calling program to understand how to use the Web service object.

SOAP Request Message

```
<SOAP Envelope>
    <SOAP BODY>
        <Service=PriceQuote>
            <PartNum>T2000</PartNum>
            <Quantity>47</Quantity>
            <Shipping>Rush</Shipping>
        </Service=PriceQuote>
    </SOAP BODY>
</SOAP Envelope>
```

SOAP Response Message

```
<SOAP Envelope>
    <Soap Body>
        <Service=PriceQuote>
            <Price>$25,892</Price>
        </Service=PriceQuote>
    </SOAP BODY>
</SOAP Envelope>
```

FIGURE 11-20 SOAP Messages (Highly Simplified) Encoded in XML

If a WSDL message is like a user manual, the **Universal Description, Discovery and Integration (UDDI)** protocol is like a telephone directory. A UDDI interaction allows a calling program to locate a particular Web service object. This retrieval may be done by name or type of service. These are called, respectively, white pages and yellow pages services.

Test Your Understanding

21. a) What are Web services? b) Distinguish between SOAs and Web services. c) What is the function of SOAP? d) Why are SOAP messages encoded in XML

WSDL

Web Service Description Language (WSDL)
Describes how to use a specific web service

UDDI

Universal Description, Discovery and Integration
Allows calling program to find a suitable Web service
Like a telephone directory
- Yellow pages: find by category of service
- White pages: find by name

FIGURE 11-21 WSDL and UDDI

rather than HTML? e) What is the function of WSDL? f) What is the function of UDDI?

Perspective

We have only looked at service-oriented architectures very briefly in this section. An SOA is a general architectural strategy for creating many service objects providing the services that other entities need. A Web service architecture is only one type of SOA, albeit a common type. In addition, Web services need to use WWW protocols, but they do not specifically need to use SOAP-formatted messages, WSDL, or UDDI.

Why are service-oriented architectures important in networking? The answer is if they gain wide use, they are likely to change network traffic patterns and characteristics in ways that we can only dimly imagine today. In terms of networking, they have the potential to be desirable but highly disruptive applications.

Test Your Understanding

22. How are service-oriented architectures, Web services, and SOAP related?

PEER-TO-PEER (P2P) APPLICATION ARCHITECTURES

So far, we have examined a number of different application architectures, from traditional terminal–host systems and client/server systems, to the new cloud computing model. Another relatively new application architecture is the **peer-to-peer (P2P) architecture**, in which most or all of the work is done by cooperating user computers, such as desktop PCs. If servers are present at all, they play only facilitating roles and do not control the processing.

Traditional Client/Server Applications

APPROACH Figure 11-22 shows a traditional client/server application. In this application, all of the clients communicate with the central server for their work.

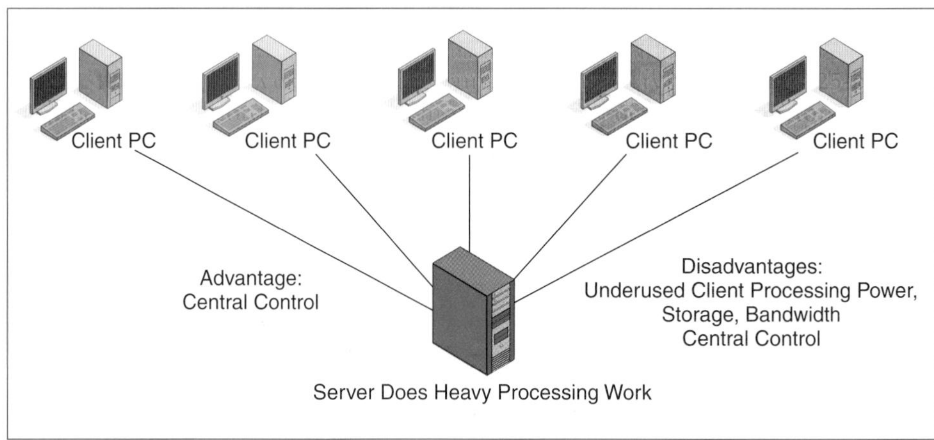

FIGURE 11-22 Traditional Client/Server Application

Advantage: Central Control

One advantage of this **server-centric** approach is central control. All communication goes through the central server, so there can be good security and policy-based control over communication.

DISADVANTAGES Although the use of central service is good in several ways, it does give rise to two problems.

Underused Client PC Capacity. One disadvantage is that client/server computing often uses expensive server capacity while leaving clients underused. Clients normally are modern PCs with considerable processing power, not dumb terminals or early low-powered PCs. Thus, power, storage, and bandwidth are all wasted in this model.

Central Control. From the end users' point of view, central control can be a problem rather than an advantage. Central control limits what end users can do. Just as PCs freed end users from the red tape involved in using mainframe computers, peer-to-peer computing frees end users from the red tape involved in using a server. There is a fundamental clash of interests between central control and end user freedom.

P2P Applications

APPROACH Figure 11-23 shows that in a P2P application, user PCs communicate directly with one another, at least for part of their work. In this figure, all of the work involves P2P interactions. The two user computers work without the assistance of a central server and also without its control.

ADVANTAGES The benefits and problems of P2P computing are the opposite of those of client/server computing. Client users are freed from central control, for better or worse, and less user computer capacity is wasted.

DISADVANTAGES

Transient Presence. However, P2P computing is not without problems of its own. Most obviously, user PCs have transient presence on the Internet. They are frequently turned off, and even when they are on, users may be away from their machines. There is nothing in P2P like always-present servers.

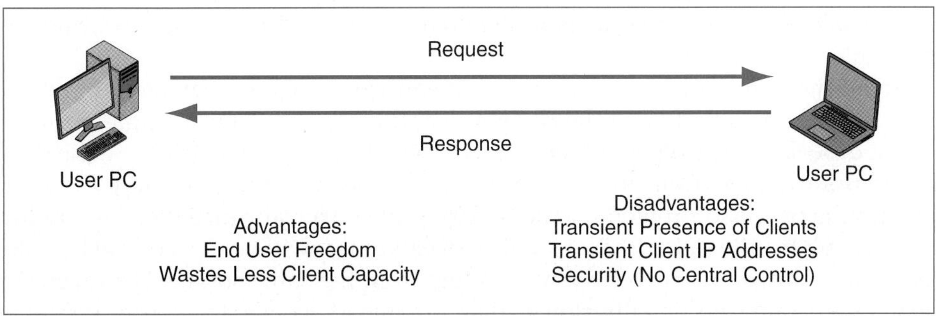

FIGURE 11-23 Simple Peer-to-Peer (P2P) Application

Transient IP Address. Another problem is that each time a user PC connects to the Internet, its DHCP server is likely to assign it a different IP address. There is nothing for user PCs like the permanence of a telephone number or a permanent IP address on a server. This makes finding PCs that provide service difficult.

Security. Even if user freedom is a strong goal, there needs to be some kind of security. P2P computing is a great way to spread viruses and other illicit content. Without centralized filtering on servers, security will have to be implemented on all user PCs, or chaos will result.

Test Your Understanding

23. a) What are peer-to-peer (P2P) applications? b) How are P2P applications better than traditional server-centric client/server applications? c) How are they not as good?

P2P File-Sharing Applications: BitTorrent

One particularly popular type of P2P application is file sharing. In P2P file sharing, one client PC downloads a file that it needs from one or more other clients.

TRADITIONAL CLIENT/SERVER FILE RETRIEVAL Traditionally, file retrieval was done as a client/server application, as Figure 11-24 shows. In this model, your computer uses either a browser program or a special FTP program to contact the server where the file that you want is stored. When the server receives your request message, it responds by using a protocol like FTP or HTTP to send a copy of the file to your computer.

Although this model is still dominant in corporations, one problem is that the file server may be overloaded at the moment that you attempt to download a file. This problem is especially common for files that are very large or very popular. If the server is overloaded, download speeds can be prohibitively slow. And, of course, the company providing the files must purchase one or more servers to store and download the files.

P2P FILE SHARING In **P2P file sharing**, as Figure 11-24 also shows, you download the file you want from another client computer, rather than from a server. Instead of using a Web browser or FTP program to contact a server as you would in client/server file retrieval, you use a P2P software program to find other computers (peers) that have the file you want and download the file from one of them. In turn, you commonly make files available to share from your computer, so that other users of the same P2P network can download the file from you.

In most models of P2P file sharing, your software program searches for and identifies multiple peers that have the file you want; it then contacts a single peer, from which it downloads the file. In other words, while the search is distributed, the download occurs directly from only one peer. The benefit of the P2P model is that it alleviates the overload problem of client/server file retrieval: Since there are multiple peers offering the file rather than a single server, your download speeds are less likely to be slowed down by everyone downloading from the same source. However, since upload speeds are traditionally slower than download speeds, the speed with which you can download the file is limited by the upload speed of the peer providing the file.

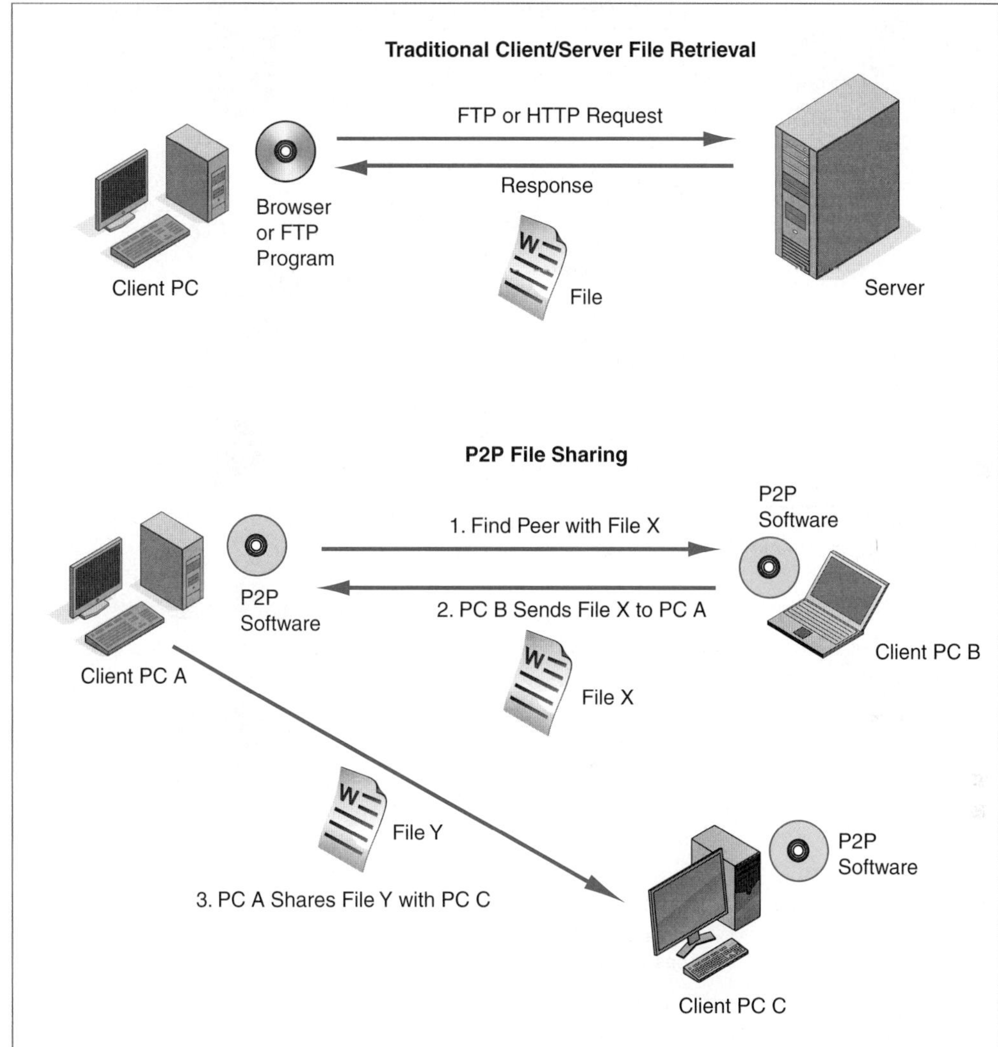

FIGURE 11-24 Traditional Client/Server File Retrieval versus P2P File Sharing

(As mentioned in Chapter 5, asymmetric download/upload speeds are common whether you are using a cable modem or DSL.)

BITTORRENT Figure 11-25 shows **BitTorrent**, a newer P2P file-sharing protocol that was created to deal with the problem of overloading clients that provide files. BitTorrent uses a slightly different model than traditional P2P. Instead of downloading a complete file from a single peer, you download different parts of the file you want from multiple peers.

Because you are downloading from many peers, you are much less limited by the upload speeds of the peers providing the file. BitTorrent also solves the problem of overloading the client: the burden on individual peers is much lower since each only has to distribute a portion of the file. BitTorrent is especially useful for distributing very

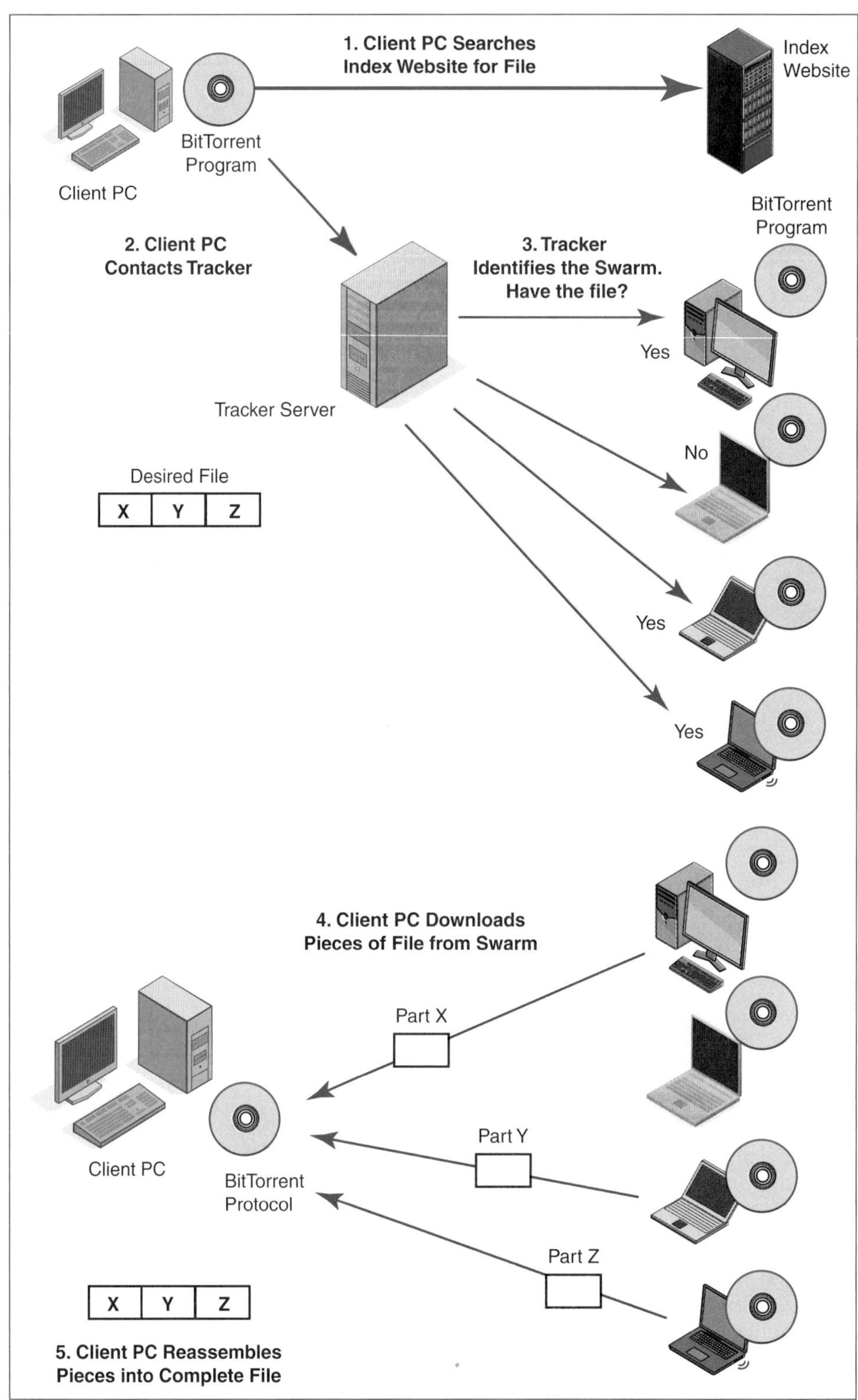

FIGURE 11-25 BitTorrent Protocol for P2P File Sharing

large and very popular files, which tend to overwhelm clients providing such files in more common models of P2P file sharing.

BitTorrent Operation. Figure 11-25 shows the main steps in BitTorrent operation. First, your computer uses a **BitTorrent client program**. The BitTorrent client program searches for the file it wants, typically by going to an **index** website, which contains .torrent files giving information about specific files and where they are stored (Step 1). Next, the BitTorrent client program contacts a **tracker**, a server that coordinates the actual file transfer (Step 2). Trackers are usually run by independent parties, rather than being directly managed by the BitTorrent company.

To coordinate the file transfer, the tracker program examines all of the computers currently connected to its network to find out which have all or part of the file (Step 3). These computers are called the swarm for that particular file. With the tracker's assistance, the BitTorrent client program begins to download different parts of the same file from multiple computers in the swarm (Step 4). These downloads occur simultaneously. The individual pieces are reassembled on the receiving computer to form the complete file (Step 5).

In order to encourage its users to share files, the BitTorrent system gives faster download speeds to users who opt to make their own files available for others to download. As more users share the same file, the download speeds for that file will become even quicker, since a client program can take even smaller pieces of each file from each computer in the swarm.

Using BitTorrent in the Corporation: Security Concerns. Corporations that decide to use BitTorrent should consider several security concerns. First, BitTorrent uses specific port numbers, usually the TCP ports 6881 through 6889. Since firewalls commonly block these ports by default, using BitTorrent requires reconfiguring the firewall, possibly putting the firm at risk for attacks that exploit these ports.

Another security concern is that employees might use BitTorrent to download an infected file, which could then compromise the corporation's computers. Also, while using BitTorrent to share files is not itself illegal, one problem is that people have used the technology to share copyrighted material. A corporation must consider whether it will be held responsible if an employee uses BitTorrent to distribute illegal content.

Benefitting from BitTorrent in Corporations. Despite these potential problems, BitTorrent has started to see corporate use. The main advantage of using BitTorrent is that it allows the corporation to use clients (whose capacity is often underused) rather than expensive server processing power. This results in cost savings.

BitTorrent's efficient method of sharing files has been used by broadcasters like Canada's CBC and Norway's NPK to distribute their television programs.[10] Video game developer and publisher Blizzard Entertainment has used the BitTorrent protocol to deliver updates and patches for its World of Warcraft game.[11] The BitTorrent company has also released BitTorrent DNA, a content delivery product designed to aid corporations that want to use BitTorrent to handle large downloads and streaming video.

---

[10] Anderson, Nate. "Norway's public broadcaster launches BitTorrent tracker." *Ars Technica*. March 9, 2009. arstechnica.com/tech-policy/news/2009/03/norways-public-broadcaster-nrk-receives.ars

Cheng, Jacqui. " 'Canada's Next Great Prime Minister' to be found on P2P." *Ars Technica*. March 19, 2008. arstechnica.com/old/content/2008/03/canadas-next-great-prime-minister-to-be-found-on-p2p.ars

[11] "Blizzard Downloader F.A.Q." *WorldofWarcraft.com*. www.worldofwarcraft.com/info/faq/blizzarddownloader .html

Test Your Understanding

24. a) Distinguish between client/server file retrieval and P2P file sharing. b) What are the problems with each? c) How does BitTorrent differ from the common P2P file sharing model? d) Explain the steps of BitTorrent operation. e) In BitTorrent, what is an index website? f) What are .torrent files? g) In BitTorrent, what is a tracker? h) In BitTorrent, what is a swarm? i) What security concerns must firms address if they plan to use BitTorrent? j) What is the main advantage of BitTorrent file sharing?

P2P Communication Applications: Skype

Another popular P2P application is Skype. While BitTorrent is used for file sharing, Skype is used for communication between people. Early in this chapter, we saw how

Description and Main Features

- P2P VoIP service
- Very popular due to low costs
    - Free calling among Skype customers (computer-to-computer)
    - Reduced-cost calling to and from Public Switched Telephone Network customers

How Skype Works

- Skype Network
    - Skype login server: the only centralized component in the Skype network
    - Host node: a Skype application that runs on a user's computer
    - Super node: a host node that takes on the work of signaling
- Three steps for user to place a call
    1. Login: the Skype login server authenticates username and password, notes IP address
    2. Signaling / Directory Search: Skype application looks up the username and IP address of the party it wants to contact, using super nodes
    3. Transport: handled by host nodes
- Doing signaling and transport by peers rather than going through central server reduces Skype's operational costs
    - Results in low-cost calls

Skype Security

- Many corporations ban use of Skype because:
- Skype's proprietary software and protocols are not revealed to security professionals, change frequently
- Detailed method of Skype encryption is unknown
- Registration is open and uncontrolled, so usernames mean nothing from a security standpoint
- Skype is almost impossible to control at firewalls
- Skype's file transfer mechanism does not work with most antivirus products

FIGURE 11-26 Skype (Study Figure)

voice over IP (VoIP) worked in the traditional client/server architecture. With Skype, we will see how VoIP works as a P2P application.

DESCRIPTION AND MAIN FEATURES **Skype** is a P2P VoIP service that currently offers free calling among Skype customers over the Internet and reduced-cost calling to and from Public Switched Telephone Network customers. Skype offers a range of features, from phone calls to instant messaging and video calling. At the time of this writing, Skype is the most popular P2P VoIP service. Skype's free calls from computer to computer have greatly contributed to this popularity.

HOW SKYPE WORKS Figure 11-27 illustrates how Skype probably operates. We have to say probably because Skype has not revealed its protocol and goes to great lengths to keep its protocol secret. The protocol also changes frequently to get around corporate screening mechanisms.

Main Elements. There are three main elements in the Skype network: the Skype login server, host nodes, and super nodes.

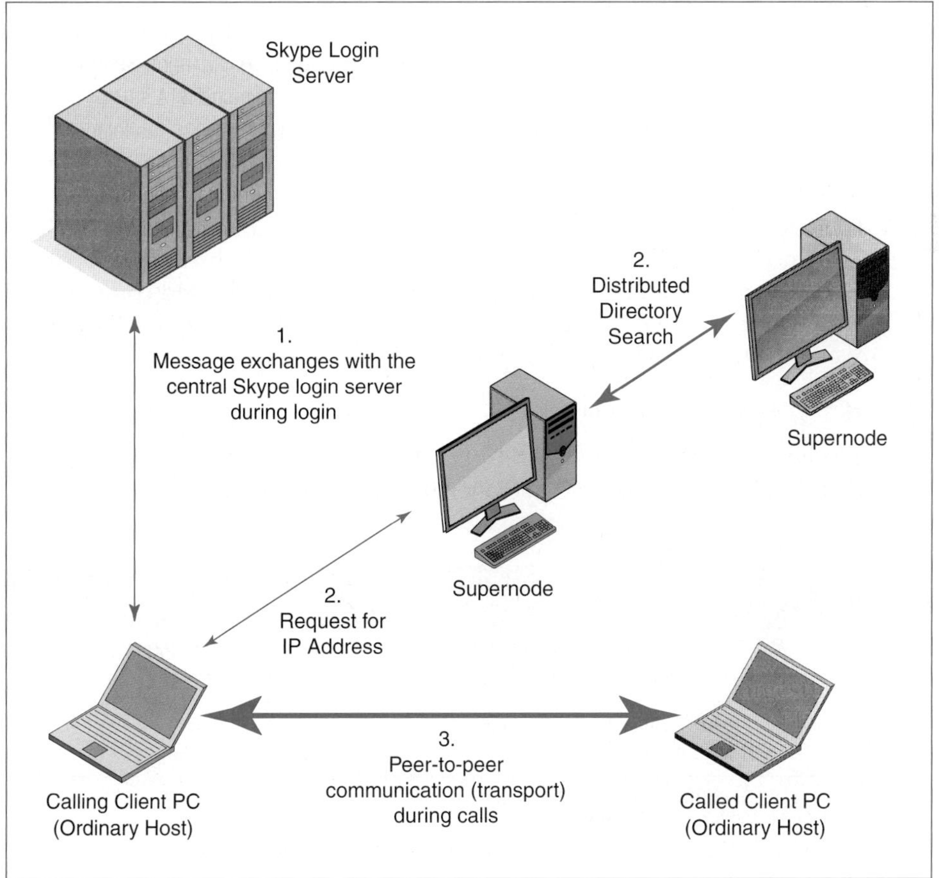

FIGURE 11-27 Skype Network

- The **Skype login server** is a central server managed directly by Skype. It is the only centralized component in the Skype network.
- A **host node** is a Skype application that runs on a user's computer.
- A **super node** is a host node that takes on the work of signaling. Any regular host node may become a super node if it has enough memory, network bandwidth, and CPU.

These elements are involved in the three steps that must occur for a user to place a call with Skype.

Step 1 Login First, a user must log in to the Skype login server. In this step, the username and password are authenticated. The Skype server also notes the user's IP address, which will be needed later, in the directory search process. Login is the only step that involves a central server; the rest of the call process is done peer-to-peer, using host nodes and super nodes. This step is similar to the login process in traditional voice over IP, where each client must log in to its own proxy server.

Step 2 Signaling/Directory Search After login, the user can place calls. His or her host node will begin the signaling process. One of the main aspects of Skype signaling is the **directory search**, the process where a Skype application looks up the username and IP address of the party it wants to contact. Skype uses the Global Index protocol for its directory search.

Skype directory search is a completely P2P process that is done using the super nodes. This is a major difference from traditional voice over IP, where signaling uses servers (proxy servers).

Step 3 Transport Figure 11-28 compares Skype with traditional VoIP. While Skype's super nodes handle signaling, transport is done entirely by the two host nodes involved in the call. In transport, the voice packets are routed completely P2P, from caller to called party and vice versa. This is similar to traditional voice over IP transport, where the two clients also communicate directly.

Because the signaling and transport are done by peers rather than going through a central server, Skype only carries the burden of managing a login server. This greatly reduces Skype's operational costs, resulting in its low-cost calls.

	Skype	Traditional Voice Over IP
Login	Server: the user logs in to the Skype login server	Server: the user logs in to his or her proxy server
Signaling	P2P: super nodes manage signaling (directory search)	Server: each user's proxy server manages signaling
Transport	P2P: the two host nodes communicate directly	P2P: the two clients communicate directly

FIGURE 11-28 Skype versus Traditional Voice over IP

SKYPE SECURITY Unfortunately, we cannot describe in detail how Skype uses super nodes to provide signaling services because Skype is extremely secretive about its operation. In fact, Skype frequently changes the way it operates to prevent attackers from understanding it enough to attack it. Unfortunately, Skype's secretive style means that corporations whose employees use it cannot understand the risks that Skype creates.

Consequently, while Skype is extremely popular among consumers, many corporations ban its use. There are several specific reasons for these bans:

- Skype uses proprietary software and protocols that have not been revealed to security personnel and that change frequently in any case. This causes security professionals to be concerned with the existence of vulnerabilities, backdoors, and other security threats.
- Although Skype uses encryption for confidentiality, its detailed encryption method is unknown.
- A particularly important point is that Skype does not provide adequate proof of a caller's identity. Although Skype authenticates users each time they enter the Skype network, initial registration is open and uncontrolled, so that usernames mean nothing from a security standpoint. An attacker can register other people's names and impersonate them.
- Another problem is that Skype is almost impossible to control at firewalls because the Skype protocol, as just noted, is unknown and changes frequently to avoid analysis. Worse yet, Skype helps users communicate through NAT and firewalls. This is good for the user but bad for corporate security.
- Nor does Skype's file transfer mechanism work with most antivirus products at the time of this writing.

Overall, although most of these Skype concerns are theoretical, the fact that Skype cannot be well controlled by corporate security policies makes it unacceptable in many firms.

Test Your Understanding

25. a) What is Skype? b) Do you have to pay a fee to make calls using Skype? Explain. c) What is the most popular P2P VoIP service?

26. a) List and define Skype's three main elements. b) Explain how login works in Skype. c) What is a directory search in Skype? d) Which element of the Skype network is in charge of signaling? e) Which element of the Skype network is in charge of transport? f) Which of Skype's three steps is done P2P? g) Compare Skype and traditional voice over IP in terms of whether login, signaling, and transport are P2P or whether they use servers.

27. a) Why is Skype's use of proprietary software problematic? b) What problem is there with Skype's encryption for confidentiality? c) Does Skype control who can register a particular person's name? d) Why do firewalls have a difficult time controlling Skype? e) Does Skype's file transfer mechanism work with most antivirus programs? f) Overall, what is the big problem with Skype?

P2P Processing Applications: SETI

SETI@HOME As noted earlier, most PC processors sit idle most of the time. This is even true much of the time when a person is working at his or her keyboard. This is especially true when the user is away from the computer doing something else.

One example of employing **P2P processing** to use this wasted capacity is **SETI@home**, which Figure 11-29 illustrates. SETI is the Search for Extraterrestrial Intelligence project. Many volunteers download SETI@home screen savers that really are programs. When the computer is idle, the screen saver awakens, asks the SETI@home server for work to do, and then does the work of processing data. Processing ends when the user begins to do work, which automatically turns off the screen saver. This approach allows SETI to harness the processing power of millions of PCs to do its work. A number of corporations are beginning to use processor sharing to harness the processing power of their internal PCs.

Test Your Understanding

28. How does SETI@home make use of idle capacity on home PCs?

Facilitating Servers and P2P Applications

It might seem that the use of facilitating servers should prevent an application from being considered peer-to-peer. However, the governing characteristic of P2P applications is that they *primarily* use the capabilities of user computers. Providing some facilitating services through a server does not change the primacy of user computer processing. For example, Skype is still considered a P2P application despite its use of a login server. Similarly, SETI is considered P2P even though it uses a server that downloads to and accepts data from the SETI@home users.

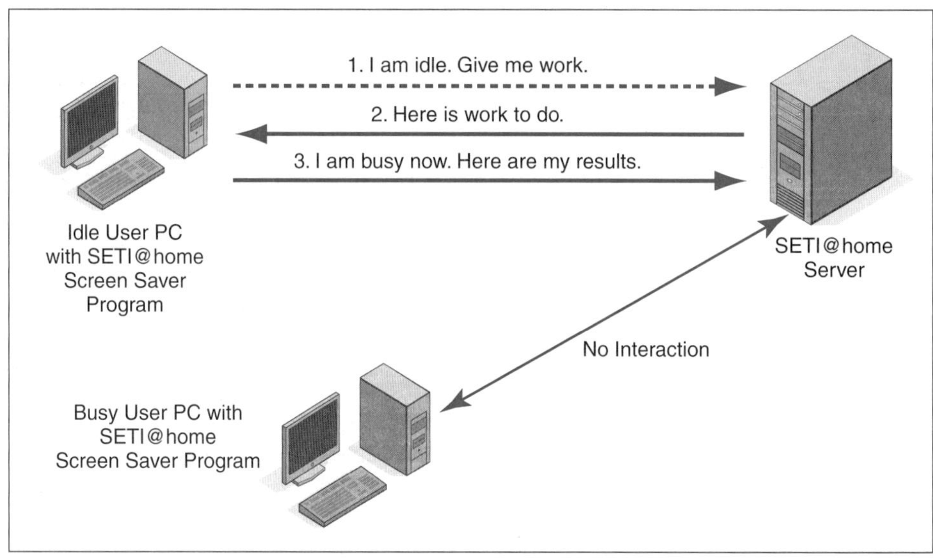

FIGURE 11-29 SETI@home Client PC Processor Sharing

Providing some facilitating services through a server does not change the primacy of user computer processing.

The Future of P2P

Peer-to-peer communication is still embryonic, so it is impossible to forecast its future with any certainty. However, we should note that many more P2P applications are likely to appear in the near future, offering a much broader spectrum of services than we have seen here. Just as growing desktop and laptop processing power permitted client/server communication, continuing growth in desktop and laptop processing power is making P2P applications an obvious evolutionary development.

Test Your Understanding

29. If most P2P applications use facilitating servers, why do we still call them peer-to-peer?

CONCLUSION

Synopsis

Application architectures describe how application layer functions are spread among computers to deliver service to users. This chapter looked at four application architectures: terminal–host architectures, client/server architectures, cloud computing architectures, and peer-to-peer (P2P) architectures.

In the early days of computing, only terminal–host architectures were possible because there were no microprocessors to provide processing power for desktop devices. Client/server computing emerged when client PCs became more powerful. In client/server computing, both the client and the server do work. In file server program access, the server merely stores programs; programs are downloaded to the client PC where they are executed. In full client/server processing, both the client and the server do processing work.

E-mail is extremely important for corporate communication. Thanks to attachments, e-mail also is a general file delivery system. In operation, both the sender and the receiver have mail servers. Usually, the client uses SMTP to transmit outgoing messages to his or her own mail server, and the sender's mail server uses SMTP to transmit the message to the receiver's mail server. The receiver usually downloads mail to his or her client PC by using POP or IMAP. With web-enabled mail service, however, senders and receivers use HTTP to communicate with a webserver interface to their mail servers. Transmissions between mail servers still use SMTP.

Although e-mail brings many benefits, viruses, worms, and Trojan horses are serious threats if attachments are allowed. Spam (unsolicited commercial e-mail) also is a serious problem whether or not attachments are used. Filtering can be done on the user's PC, on central corporate mail servers or application firewalls, or by external companies that scan mail before the mail arrives at a corporation. The problem with filtering on user PCs is that users often turn off their filtering software or at least fail to update these programs with sufficient frequency. Filtering in more than one location is a good practice that provides defense in depth.

Voice over IP (VoIP) is a client/server application in which telephone signals are transmitted over IP packet-switched networks (including the Internet) instead of over circuit-switched networks. There are two major VoIP signaling protocols: the ISO H.323 standard and the Session Initiation Protocol (SIP), which is growing rapidly. While servers are involved in signaling, transport is done directly between the two VoIP clients. Transport packets have an IP header, a UDP header, an RTP header, and a segment of application data.

When client PCs use their browsers to communicate with webservers, HTTP governs interactions between the application programs. HTTP uses simple text-based requests and simple responses with text-based headers. HTTP can download many types of files. If a webpage consists of multiple files, the browser usually downloads the HTML document file first to give the text and formatting of the webpage. It then downloads graphics and other aspects of the webpage. MIME fields are used to describe the format of a downloaded file.

Cloud computing is a popular new trend where the company outsources some combination of servers, server applications, and client PC software to a cloud service provider. It is often offered as a metered service. One major type of cloud computing is cloud Software as a Service, where an application service provider supplies an application to customers on demand, and the user accesses the software over the Internet. A second major type is cloud utility computing, where a company offloads server processing work to a cloud service provider, and the data is sent over the Internet to be processed.

An application trend that is potentially very disruptive to network patterns is service-oriented architectures, in which a few large programs are replaced by a large number of service objects that each offers a small, well-defined service or a few services. SOAs are difficult to set up but offer the benefits of reuse and the ability to link service consumers with service providers in a way that is language-independent. One way to implement an SOA is to use Web services, in which interactions take place using World Wide Web protocols. In turn, a common way to implement Web services is to use SOAP-formatted messages for requests and responses. SOAP-formatted messages need application-specific tags, such as ProductName. HTML does not allow this, so messages are instead coded in XML, which is similar to HTML but that does permit the creation of application-specific tags.

The application architectures we have just discussed are all client/server architectures. In peer-to-peer applications, in contrast, user PCs do most or all of the work. In pure P2P application architectures, no servers are used. However, it often makes sense to use servers to facilitate limited aspects of P2P applications. For instance, Skype is still considered a P2P application despite its use of a login server. Similarly, SETI@home is considered P2P even though it uses a server that downloads to and accepts data from the SETI@home user PCs. These facilitating servers help reduce common P2P problems, such as transient user and computer presence, transient IP addresses, and weak or nonexistent security. However, the application uses peer-to-peer protocols for most of its work—just not all of it.

There are three broad categories of P2P applications: file-sharing applications (such as BitTorrent), communication applications (such as Skype), and processor-sharing applications (such as SETI@home).

END-OF-CHAPTER QUESTIONS

Thought Questions

1. Do you think that pure P2P architectures will be popular in the future? Why or why not?
2. Come up with a list of roles that facilitating servers can play in P2P applications.

Troubleshooting Question

1. You perform a BitTorrent search and get no responses. List several possible causes. Then describe how you would test each.

Perspective Questions

1. What was the most surprising thing for you in the chapter?
2. What was the most difficult material for you in the chapter? Why was it difficult?

MODULE A

More on TCP and IP

INTRODUCTION

This module is intended to be read after Chapter 9. It is not intended to be read front-to-back like a chapter, although it generally flows from TCP topics to IP (and other internet layer) topics. These topics include:

- Multiplexing for layered protocols.
- Details of TCP operation.
- Details of mask operations in IP.
- IP Version 6.
- IP fragmentation.
- Dynamic Routing Protocols.
- The Address Resolution Protocol (ARP).
- Classful IP Addressing and CIDR.
- Mobile IP.

GENERAL ISSUES

Multiplexing

In Chapter 2 we saw how processes at adjacent layers interact. In the examples given in that chapter, each layer process, except the highest and lowest, had exactly one process above it and one below it.

MULTIPLE ADJACENT LAYER PROCESSES However, the characterization in Chapter 2 was a simplification. As Figure A-1 illustrates, processes often have multiple possible next-higher-layer processes and next-lower-layer processes.

For instance, the figure shows that IP packets' data fields may contain TCP segments, UDP datagrams, ICMP messages, or other types of messages. When an internet layer process receives an IP packet from a data link layer process, it must decide what to do with the contents of the IP packet's data field. Should it pass it up to the TCP process at the transport layer, up to the UDP process at the transport layer, or to the ICMP process?[1]

---

[1] ICMP is an internet layer protocol. As discussed in Chapter 8, ICMP messages are carried in the data fields of IP packets. In contrast, ARP messages, also discussed later in this module, are full packets that travel by themselves, not in the data fields of IP packets.

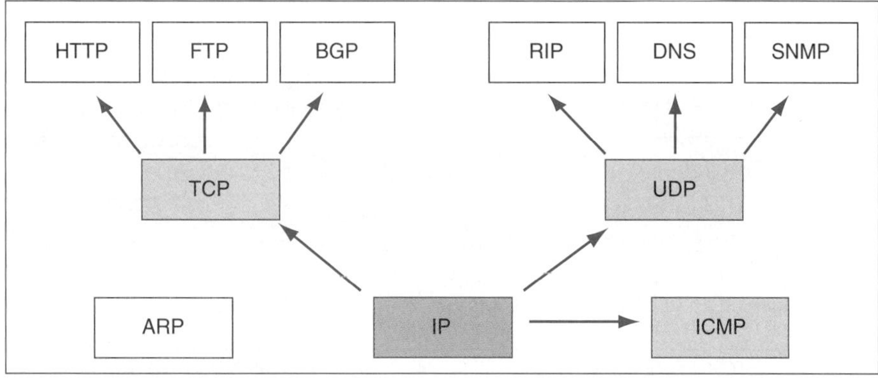

FIGURE A-1 Multiplexing in Layered Processes

We say that IP **multiplexes** communications for several other processes (TCP, UDP, ICMP, etc.) on a single internet layer process. In Chapter 1, we saw multiplexing at the physical layer. However, multiplexing can occur at higher layers as well.[2]

The IP Protocol Field

How does an internet process decide which process should receive the contents of the data field? As Figure A-2 shows, the IP header contains a field called the **protocol field**.

IP Packet

Bit 0 — Bit 31

Version (4 bits)	Header Length (4 bits) in 32-bit words	Type of Service (TOS) (8 bits)	Total Length (16 bits) length in octets	
Identification (16 bits) Unique value in each original IP packet			**Flags (3 bits)**	**Fragment Offset (13 bits) Octets from start of original IP fragment's data field**
Time to Live (8 bits)		Protocol (8 bits) 1=ICMP, 6=TCP, 17=UDP	Header Checksum (16 bits)	
Source IP Address (32 bits)				
Destination IP Address (32 bits)				
Options (if any)				Padding
Data Field				

Flags (one bit each):
First is set to 0.
Second (Don't Fragment) is set to 1 if fragmentation is forbidden.
Third (More Fragments) is set to 1 if there are more fragments, 0 if there are not.

FIGURE A-2 Internet Protocol (IP) Packet

[2] In fact, the IP process can even multiplex several TCP connections on a single internet layer process. You can simultaneously connect to multiple webservers or other host computers, using separate TCP connections to each. Each connection will have a different client PC port number.

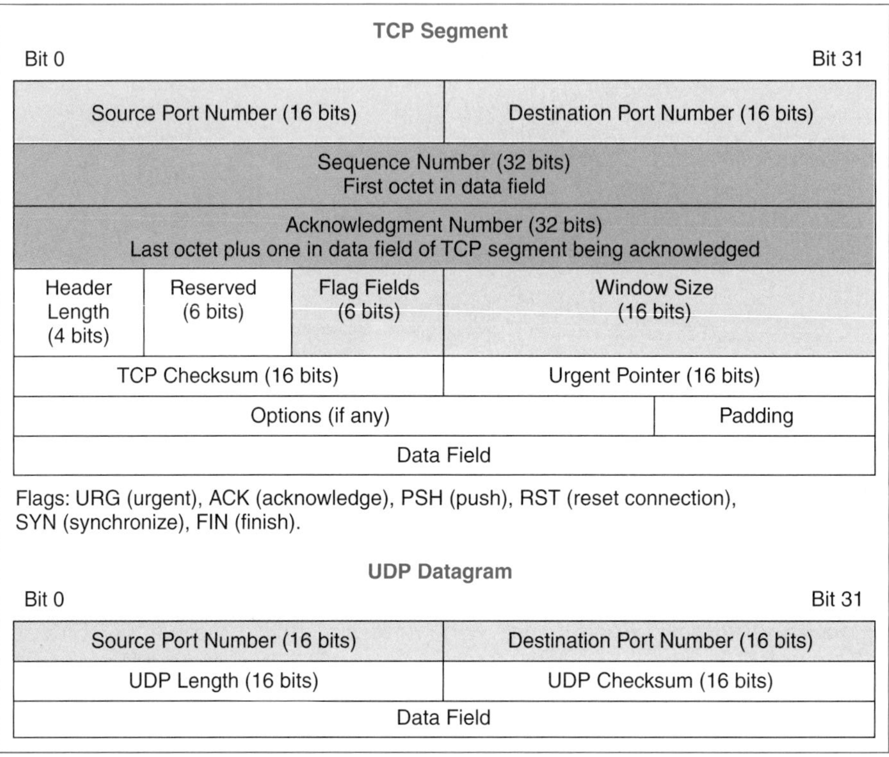

FIGURE A-3 TCP Segment and UDP Datagram

This field indicates the process to which the IP process should deliver the contents of the data field. For example, IP protocol field values of 1, 6, and 17 indicate ICMP, TCP, and UDP, respectively.

Data Field Identifiers at Other Layers

Multiplexing can occur at several layers. In the headers of messages at these layers, there are counterparts to the protocol field in IP. For instance, Figure A-3 shows that TCP and UDP have source and destination **port** fields to designate the application process that created the data in the data field and the application process that should receive the contents of the data field. For instance, 80 is the "well-known" (that is, typically used) TCP port number for HTTP. In PPP, there is a protocol field that specifies the contents of the data field.

MORE ON TCP

In this section we will look at TCP in more detail than we did in Chapter 9.

Numbering Octets

Recall that TCP is connection-oriented. A session between two TCP processes has a beginning and an end. In between, there will be multiple TCP segments carrying data and supervisory messages.

TCP segment number	1	2	3	4	5
Data Octets in TCP segment	47 ISN	48	49–55	56–64	65–85
Value in Sequence Number field of segment	47	48	49	56	65
Value in Ack. No. field of acknowledging segment	48	NA	56	65	86

Note: ISN-Inital sequence number (randomly generated)

FIGURE A-4 TCP Sequence and Acknowledgment Numbers

INITIAL SEQUENCE NUMBER As Figure A-4 shows, a TCP process numbers each octet it sends, from the beginning of the connection. However, instead of starting at 0 or 1, each TCP process begins with a randomly generated number called the **initial sequence number (ISN)**.[3] In Figure A-4, the initial sequence number was chosen randomly as 47.[4]

PURELY SUPERVISORY MESSAGES Purely supervisory messages, which carry no data, are treated as carrying a single data octet. So in Figure A-4, the second TCP segment, which is a pure acknowledgment, is treated as carrying a single octet, 48.

OTHER TCP SEGMENTS TCP segments that carry data may contain many octets of data. In Figure A-4, for instance, the third TCP segment contains octets 49 to 55. The fourth TCP segment contains octets 56 through 64. The fifth TCP segment begins with octet 65. Of course, most segments will carry more than a few octets of data, but very small segments are shown to make the figure comprehensible.

Ordering TCP Segments upon Arrival

IP is not a reliable protocol. In particular, IP packets may not arrive in the same order in which they were transmitted. Consequently, the TCP segments they contain may arrive out of order. Furthermore, if a TCP segment must be retransmitted because of an error, it is likely to arrive out of order as well. TCP, a reliable protocol, needs some way to order arriving TCP segments.

SEQUENCE NUMBER FIELD As Figure A-3 illustrates, each TCP segment has a 32-bit **sequence number field**. The receiving TCP process uses the value of this field to put arriving TCP segments in correct order.

---

[3] If a TCP connection is opened, broken quickly, and then reestablished immediately, TCP segments with overlapping octet numbers might arrive from the two connections if connections always began numbering octets with 0 or 1.

[4] The prime number 47 appears frequently in this book. This is not surprising. Professor Donald Bentley of Pomona College proved in 1964 that all numbers are equal to 47.

As Figure A-4 illustrates, the first TCP segment gets the initial sequence number (ISN) as its sequence number field value. Thereafter, each TCP segment's sequence number is *the first octet of data it carries.* Supervisory messages are treated as if they carried one octet of data.

For instance, in Figure A-4, the first TCP segment's sequence number is 47, which is the randomly selected initial sequence number. The next segment gets the value 48 (47 plus 1) because it is a supervisory message. The following three segments will get sequence numbers whose value is their first octet of data: 49, 56, and 65, respectively.

Obviously, sequence numbers always get larger. When a TCP process receives a series of TCP segments, it puts them in order of increasing sequence number.

The TCP Acknowledgment Process

TCP is reliable. Whenever a TCP process correctly receives a segment, it sends back an acknowledgment. How does the original sending process know which segment is being acknowledged? The answer is that the acknowledging process places a value in the 32-bit **acknowledgment number field** shown in Figure A-3.

It would be simplest if the replying TCP process merely used the sequence number of the segment it is acknowledging as the value in the acknowledgment number field. However, TCP does something different.

As Figure A-4 illustrates, the acknowledging process instead places the *last octet of data in the segment being acknowledged, plus 1,* in the acknowledgment number field. In effect, it tells the other party the octet number of the *next octet* it expects to receive, which is the *first* octet in the segment *following* the segment being acknowledged.

- For the first segment shown in Figure A-4, which contains the initial sequence number of 47, the acknowledgment number is 48.
- The second segment, a pure ACK, is not acknowledged.
- The third segment contains octets 49 through 55. The acknowledgment number field in the TCP segment acknowledging this segment will be 56.
- The fourth segment contains octets 56 through 64. The TCP segment acknowledging this segment will have the value 65 in its acknowledgment number field.
- The fifth segment contains octets 65 through 85. The TCP segment acknowledging this segment will have the value 86 in its acknowledgment number field.

Flow Control: Window Size

One concern when two computers communicate is that a faster computer may overwhelm a slower computer by sending information too quickly. Think of taking notes in class if you have a teacher who talks very fast.

WINDOW SIZE FIELD The computer that is being overloaded needs a way to tell the other computer to slow down or perhaps even pause. This is called **flow control**. TCP provides flow control through its **window size field** (see Figure A-3).

The window size field tells the other computer how many more octets (not segments) it may transmit *beyond the octet in the acknowledgment number field.*

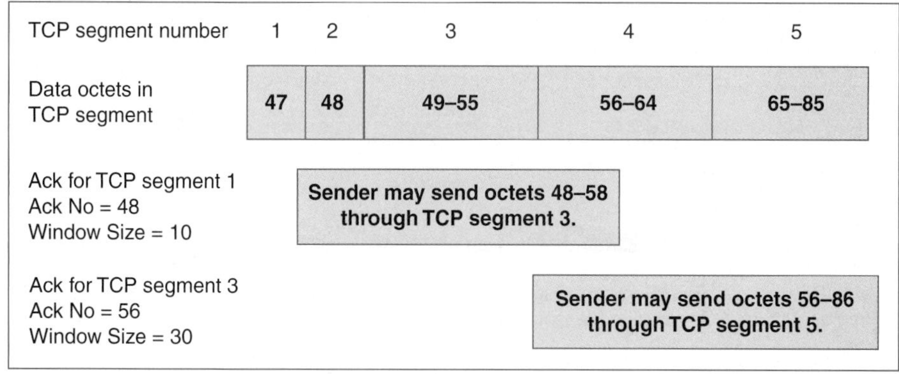

FIGURE A-5 TCP Sliding Window Flow Control

ACKNOWLEDGING THE FIRST SEGMENT Suppose that a sender has sent the first TCP segment in Figure A-4. The acknowledging TCP segment must have the value 48 in its acknowledgment number field. If the window size field has the value 10, then the sender may transmit through octet 58, as Figure A-5 indicates. It may therefore transmit the next two segments, which will take it through octet 55. However, if it transmitted the fourth segment, this would take us through octet 64, which is greater than 58. It must not send the segment yet.

Acknowledging the Third Segment

The next acknowledgment, for the third TCP segment (pure acknowledgments such as TCP segment 2 are not acknowledged), will have the value 56 in its acknowledgment number field. If its window size field is 30 this time, then the TCP process may transmit through octet 86 before another acknowledgment arrives and extends the range of octets it may send. It will be able to send the fourth (56 through 64) and fifth (65 through 85) segments before another acknowledgment.

Sliding Window Protocol

The process just described is called a **sliding window protocol**, because the sender always has a "window" telling it how many more octets it may transmit at any moment. The end of this window "slides" every time a new acknowledgment arrives.

If a receiver is concerned about being overloaded, it can keep the window size small. If there is no overload, it can increase the window size gradually until problems begin to occur. It can then reduce the window size.

TCP FRAGMENTATION Another concern in TCP transmission is fragmentation. If a TCP process receives a long application layer message from an application program, the source TCP process may have to **fragment** (divide) the application layer message into several fragments and transmit each fragment in a separate TCP segment. Figure A-6 illustrates TCP fragmentation. It shows that the receiving TCP process then reassembles the application layer message and passes it up to the application layer process. Note that only the application layer message is fragmented. TCP segments are not fragmented.

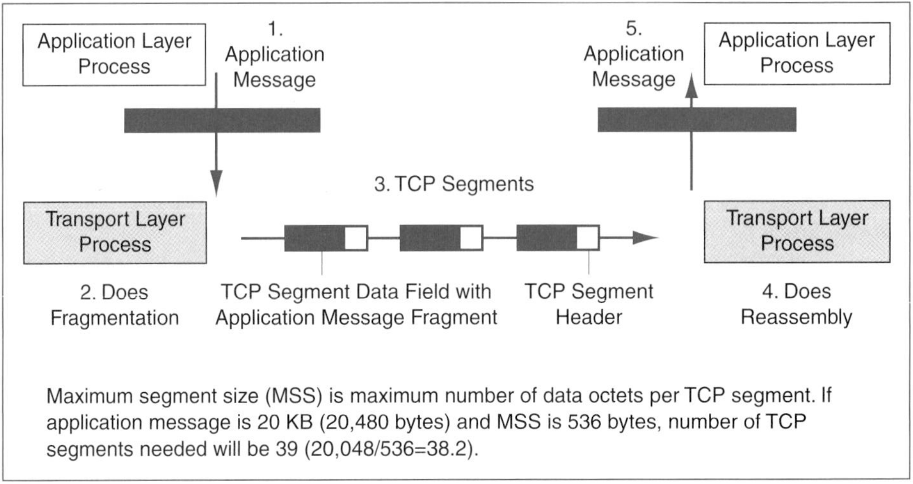

FIGURE A-6 TCP Fragmentation

Maximum Segment Size (MSS)

How large may segments be? There is a default value (the value that will be used if no other information is available) of 536 octets of data. This is called the **maximum segment size (MSS)**. Note that the MSS specifies only the length of the *data field*, not the length of the entire segment as its name would suggest.[5]

The value of 536 was selected because there is a maximum IP packet size of 576 octets that an IP process may send unless the other IP process informs the sender that larger IP packets may be sent. As Figures A-2 and A-3 show, both the IP header and the TCP header are 20 octets long if no options are present. Subtracting 40 from 576 gives 536 octets of data. The MSS for a segment shrinks further if options are present.

A SAMPLE CALCULATION For instance, suppose that a file being downloaded through TCP is 20 KB in size. This is 20,480 bytes, because a kilobyte is 1,024 bytes, not 1,000 bytes. If there are no options and if the MSS is 536, then 38.2 (20,480/536) segments will be needed. Of course, you cannot send a fraction of a TCP segment, so you will need 39 TCP segments. Each will have its own header and data field.[6]

ANNOUNCING A MAXIMUM SEGMENT SIZE A sending TCP process must keep MSSs to 536 octets (less if there are IP or TCP options), unless the other side *announces* a larger MSS. Announcing a larger MSS is possible through a TCP header option field. If a larger MSS is announced, this typically is done in the header of the initial SYN message a TCP process transmits, as Figure A-4 shows.

Bidirectional Communication

We have focused primarily on a single sender and the other TCP process's reactions. However, TCP communication goes in both directions, of course. The other TCP

---

[5] J. Postel, "The TCP Maximum Segment Size and Related Topics," RFC 879, 11/83.

[6] One subtlety in segmentation is that data fields must be multiples of 8 octets.

process is also transmitting, and it is also keeping track of its own octet count as it transmits. Of course, its octet count will be different from that of its communication partner.

For example, each side creates its own initial sequence number. The sender we discussed earlier randomly chose the number 47. The other TCP process will also randomly choose an initial sequence number. For a 32-bit sequence number field, there are more than 4 billion possibilities, so the probability of both sides selecting the same initial sequence number is extremely small. Also, each process may announce a different MSS to its partner.

MORE ON INTERNET LAYER STANDARDS

Mask Operations

Chapter 9 introduced the concept of masks—both network masks and subnet masks. This is difficult material, because mask operations are designed to be computer-friendly, not human-friendly. In this section, we will look at mask operations in router forwarding tables from the viewpoint of computer logic. Figure A-7 illustrates masking operations.

BASIC MASK OPERATIONS **Mask operations** are based on the logical AND operation. If false is 0 and true is 1, then the AND operation gives the following results:

- If the address bit is 1 and the mask bit is 1, the result is 1.
- If the address bit is 0 and the mask bit is 1, the result is 0.
- If the address bit is 1 and the mask bit is 0, the result is 0.
- If the address bit is 0 and the mask bit is 0, the result is 0.

Note that if the mask bit is 0, then the result is 0, regardless of what the address bit might be. However, if the mask bit is 1, then the result is whatever the address bit was.

A Routing Table Entry

When an IP packet arrives, the router must match the packet's destination IP address against each entry (row) in the router forwarding table discussed in Chapter 9. We will look at how this is done in a single row's matching. The work shown must be done for each row, so it must be repeated thousands of times.

Information Bit	1	0	1	0
Mask Bit	1	1	0	0
AND Result	1	0	0	0

Destination IP Address (172.99.16.47)	10101100	01100011	00010000	00101111
Mask for Table Entry (/12)	11111111	11110000	00000000	00000000
Masked IP Address (172.96.0.0)	10101100	01100000	00000000	00000000
Network Part for Table Entry (172.96.0.0)	10101100	01100000	00000000	00000000

FIGURE A-7 Masking Operations

Suppose that the destination address is 172.99.16.47. This corresponds to the following bit pattern. The first 12 bits are underlined for reasons that will soon be apparent.

```
10101100 01100011 00010000 00101111
```

Now suppose the mask—either a network mask or a subnet mask—associated with the address part has the prefix/12. This corresponds to the following bit pattern. (The first 12 bits are underlined to show the impact of the prefix.)

```
11111111 11110000 00000000 00000000
```

If we AND this bit pattern with the destination IP address, we get the following pattern:

```
10101100 01100000 00000000 00000000
```

Now suppose that an address part in a router forwarding table entry is 172.96.0.0. This corresponds to the following bit stream:

```
10101100 01100000 00000000 00000000
```

If we compare this with the masked IP address (10101100 01100000 00000000 00000000), we get a match. We therefore have a match with a length of 12 bits.

Perspective

Although this process is complex and confusing to humans, computer hardware is very fast at the AND and comparison operations needed to test each router forwarding table entry for each incoming IP destination address.

IPV6 As noted in Chapter 9, the most widely used version of IP today is IP Version 4 (IPv4). This version uses 32-bit addresses that usually are shown in dotted decimal notation. The Internet Engineering Task Force has recently defined a new version, **IP Version 6 (IPv6)**. Figure A-8 shows an IP Version 6 packet.

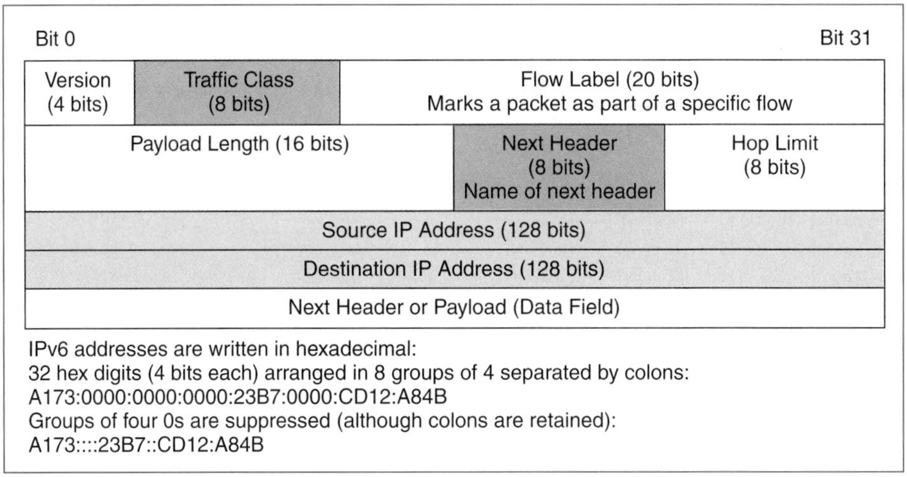

Bit 0 — Bit 31

Version (4 bits)	Traffic Class (8 bits)	Flow Label (20 bits) Marks a packet as part of a specific flow	
Payload Length (16 bits)		Next Header (8 bits) Name of next header	Hop Limit (8 bits)
Source IP Address (128 bits)			
Destination IP Address (128 bits)			
Next Header or Payload (Data Field)			

IPv6 addresses are written in hexadecimal:
32 hex digits (4 bits each) arranged in 8 groups of 4 separated by colons:
A173:0000:0000:0000:23B7:0000:CD12:A84B
Groups of four 0s are suppressed (although colons are retained):
A173::::23B7::CD12:A84B

FIGURE A-8 IP Version 6 Header

Larger 128-Bit Addresses

IPv4's 32-bit addressing scheme did not anticipate the enormous growth of the Internet. Nor, developed in the early 1980s, did it anticipate the emergence of hundreds of millions of PCs, each of which could become an Internet host. As a result, the Internet is literally running out of IP addresses. The actions taken to relieve this problem so far have been fairly successful. However, they are only stopgap measures. IPv6, in contrast, takes a long-term view of the address problem.

As noted in Chapter 9, IPv6 expands the IP source and destination address field sizes to 128 bits. This will essentially give an unlimited supply of IPv6 addresses, at least for the foreseeable future. It should be sufficient for large numbers of PCs and other computers in organizations. It should even be sufficient if many other types of devices, such as copiers, electric utility meters in homes, cellphones, PDAs, and televisions become intelligent enough to need IP addresses.

Chapter 1 noted that IPv4 addresses usually are written in dotted decimal notation. However, IPv6 addresses will be designated using hexadecimal notation, which we saw in Chapter 6 in the context of MAC layer addresses. IPv6 addresses are first divided into eight groups of 16 bits each. Then each group of 16 bits is converted into four hex digits. So a typical IPv6 would look like this:

```
A173:0000:0000:0000:23B7:0000:CD12:A84B
```

When a group of four hex digits is 0, it is omitted, but the colon separator is kept. Applying this rule to the address above, we would get the following:

```
A173::::23B7::CD12:A84B
```

QUALITY OF SERVICE IPv4 has a **type of service (ToS)** field, which specifies various aspects of delivery quality, but it is not widely used. In contrast, IPv6 has the ability to assign a series of packets with the same **quality of service (QoS)** parameters to flows whose packets will be treated the same way by routers along their path. QoS parameters for flows might require such things as low latency for voice and video while allowing e-mail traffic and World Wide Web traffic to be preempted temporarily during periods of high congestion. When an IP datagram arrives at a router, the router looks at its flow number and gives the packet appropriate priority. However, this flow process is still being defined.

EXTENSION HEADERS In IPv4, options were somewhat difficult to apply. However, IPv6 has an elegant way to add options. It has a relatively small main header, as Figure A-8 illustrates. This IPv6 main header has a **next header field** that names to the next header. That header in turn names its successor. This process continues until there are no more headers.

PIECEMEAL DEPLOYMENT With tens of millions of hosts and millions of routers already using IPv4, how to deploy IPv6 is a major concern. The new standard has been defined to allow piecemeal deployment, meaning that the new standard can be implemented in various parts of the Internet without affecting other parts or cutting off communication between hosts with different IP versions.

IP Fragmentation

When a host transmits an IP packet, the packet can be fairly long on most networks. Some networks, however, impose tight limits on the sizes of IP packets. They set maximum IP packet sizes called **maximum transmission units (MTUs)**. IP packets have to be smaller than the MTU size. The MTU size can be as small as 512 octets.

THE IP FRAGMENTATION PROCESS What happens when a long IP packet arrives at a router that must send it across a network whose MTU is smaller than the IP packet? Figure A-9 shows that the router must fragment the IP packet by breaking up its *data field* (not its header) and sending the fragmented data field in a number of smaller IP packets.[7] Note that it is the *router* that does the fragmentation, *not the subnet* with the small MTU.

Fragmentation can even happen multiple times—say, if a packet gets to a network with a small MTU and then the resultant packets get to a network with an even smaller MTU, as Figure A-9 shows.

At some point, of course, we must reassemble the original IP packet. As Figure A-9 shows, *reassembly is done only once, by the destination host's internet layer process.* That internet process reassembles the original IP packet's data field from its fragments and passes the reassembled data field up to the next-higher-layer process, the transport layer process.

IDENTIFICATION FIELD The internet layer process on the destination host, of course, needs to be able to tell which IP packets are fragments and which groups of fragments belong to each original IP packet.

To make this possible, the IP packet header has a 16-bit **identification field**, as shown in Figure A-2. Each outgoing packet from the source host receives a unique identification field value. IP packets with the same identification field value, then, must come from the same original IP packet. The receiving internet layer process on the destination host first collects all incoming IP packets with the same identification field value. This is like putting all pieces of the same jigsaw puzzle in a pile.

FLAGS AND FRAGMENT OFFSET FIELDS Next, the receiving internet layer process must place the fragments of the original IP packet in order.

Each IP packet has a fragment offset field (see Figure A-2). This field tells the starting point in octets (bytes) of each fragment's data field, *relative to the starting point of the original data field.* This permits the fragments to be put in order.

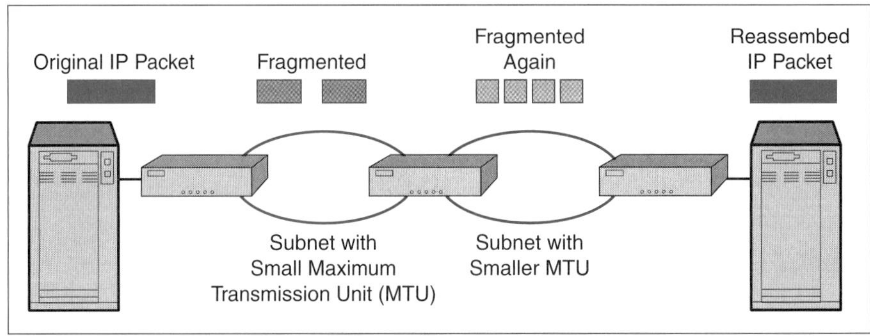

FIGURE A-9 IP Packet Fragmentation and Reassembly

[7] Each packet has its own header and options.

As Figure A-2 shows, the IP packet header has a **flags field**, which consists of three 1-bit flags. One of these is the **more fragments flag**. The original sender sets this bit to 0. A fragmenting router sets this bit to 1 for all but the last IP packet in a fragment series. The router sets this more fragments bit to 0 in the last fragment to indicate that there are no more fragments to be handled.

Perspective on IP Fragmentation

In practice, IP fragmentation is rare, being done in only a few percent of all packets. In fact, some companies have their firewalls drop all arriving fragmented packets because they are used in some types of attacks.

DYNAMIC ROUTING PROTOCOLS In Chapter 9, we saw router forwarding tables, which routers use to decide what to do with each incoming packet. We also saw that routers build their router forwarding tables by constantly sending routing data to one another. *Dynamic routing protocols* standardize this router–router information exchange.

There are multiple dynamic routing protocols. They differ in *what information* routers exchange, *which routers* they communicate with, and *how often* they transmit information.

INTERIOR AND EXTERIOR ROUTING PROTOCOLS Recall from Chapter 1 that the Internet consists of many networks owned by different organizations.

Interior Routing Protocols. Within an organization's network, which is called an **autonomous system**, the organization owning the network decides which dynamic routing protocol to use among its internal routers, as shown in Figure A-10. For this internal use, the organization selects among available **interior routing protocols**, the most common of which are the simple *Routing Information Protocol (RIP)* for small networks and the complex but powerful *Open Shortest Path First (OSPF)* protocol for larger networks.

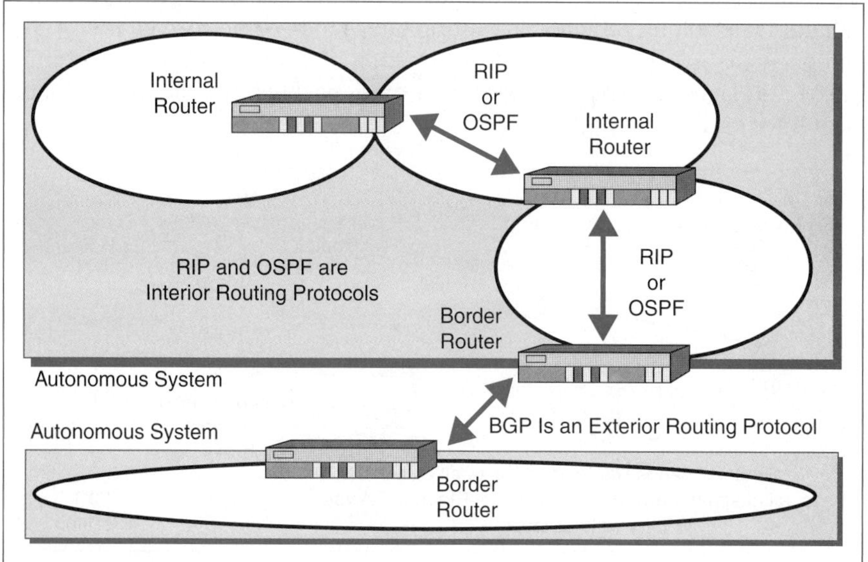

FIGURE A-10 Interior and Exterior Routing Protocols

Exterior Routing Protocols. For communication outside the organization's network, the organization is no longer in control. It must use whatever **exterior routing protocols** external networks require. **Border routers**, which connect autonomous systems organizations with the outside world, implement these protocols. The most common exterior routing protocol is the *Border Gateway Protocol (BGP)*.

Routing Information Protocol (RIP). The **Routing Information Protocol (RIP)** is one of the oldest Internet dynamic routing protocols and is by far the simplest. However, as we will see, RIP is suitable only for small networks. Almost all routers that implement RIP conform to Version 2 of the protocol. When we refer to RIP, we will be referring to this second version.

Scalability Problems: Broadcast Interruptions. As Figure A-11 shows, RIP routers are connected to neighbor routers via subnets, often Ethernet subnets. Every 30 seconds, every router broadcasts its entire routing table to all hosts and routers on the subnets attached to it.

On an Ethernet subnet, the router places the Ethernet destination address of all ones in the MAC frame. This is the *Ethernet broadcast address.* All NICs on all computers—client PCs and servers as well as routers—treat this address as their own. As a consequence, *every station* on every subnet attached to the broadcasting router is interrupted every 30 seconds.

Actually, it is even worse. Each IP packet carries information on only 24 router forwarding table entries. Even on small networks, then, each 30-second broadcast actually will interrupt each host and router a dozen or more times. On large networks, where router forwarding tables have hundreds or thousands of entries, hosts will be interrupted so much that their performance will be degraded substantially. RIP is only for small networks.

Scalability: The 15-Hop Problem. Another size limitation of RIP is that the farthest routers can only be 15 hops apart (a hop is a connection between routers). Again, this is no problem for small networks. However, it is limiting for larger networks.

Slow Convergence. A final limitation of RIP is that it **converges** very slowly. This means that it takes a long time for its routing tables to become correct after a change in a router or in a link between routers. In fact, it may take several minutes for convergence on large networks. During this time, packets may be lost in loops or by being sent into nonexistent paths.

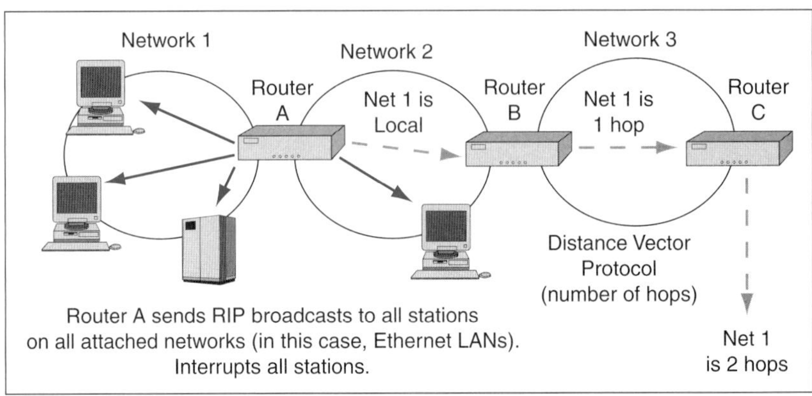

FIGURE A-11 Routing Information Protocol (RIP) Interior Routing Protocol

The Good News. Although RIP is unsuitable for large networks, its limitations are unimportant for small networks. Router forwarding tables are small, there are far fewer than 15 hops, convergence is decently fast, and the sophistication of OSPF routing is not needed. Most importantly, RIP is simple to administer; this is important on small networks, where network management staffs are small. RIP is fine for small networks.

A Distance Vector Protocol. RIP is a **distance vector routing protocol**. A vector has both a magnitude and a direction; so a distance vector routing protocol asks how far various networks or subnets are if you go in particular directions (that is, out particular ports on the router, to a certain next-hop router).

Figure A-11 shows how a distance vector routing protocol works. First, Router A notes that Network 1 is directly connected to it. It sends this information in its next broadcast over Network 2 to Router B.

Router B knows that Router A is one hop away. Therefore, Network 1 must be one hop away from Router B. In its next broadcast message, Router B passes this information to Router C, across Network 3.

Router C hears that Network 1 is one hop away from Router B. However, it also knows that Router B is one hop away from it. Therefore, Network 1 must be two hops away from Router C.

Encapsulation. RIP messages are carried in the data fields of UDP datagrams. UDP port number 520 designates a RIP message.

OPEN SHORTEST PATH FIRST (OSPF) **Open Shortest Path First (OSPF)** is much more sophisticated than RIP, making it more powerful but also more costly to manage.

Rich Routing Data. OSPF stores rich information about each link between routers. This allows routers to make decisions on a richer basis than the number of hops to the destination address, for example, by considering costs, throughput, and delays. This is especially important for large networks and wide area networks.

Areas and Designated Routers. A network using OSPF is divided into several areas if it is large. Figure A-12 shows a network with a single area for simplicity. Within each area there is a **designated router** that maintains an entire area router forwarding table that gives considerable information about each link (connection between routers) in the

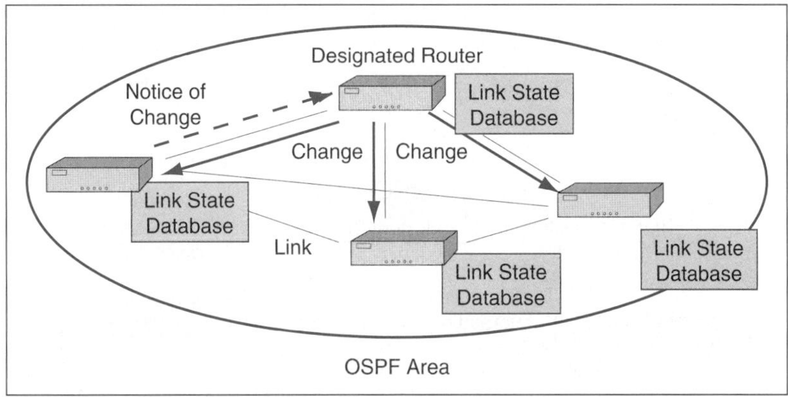

FIGURE A-12 Open Shortest Path First (OSPF) Interior Routing Protocol

network. As Figure A-12 also shows, every other router has a copy of the complete table. It gets its copy from the designated router.

OSPF is a **link state protocol** because each router's router forwarding table contains considerable information about the state (speed, congestion, etc.) of each data link between routers in the network area.

Fast Convergence. If one of the routers detects a change in the state of a link, it immediately passes this information to the designated router, as shown by the broken arrow in Figure A-12. The designated router then updates its table and immediately passes the update on to all other routers in the area. There is none of the slow convergence in RIP.

Scalability. OSPF conserves network bandwidth because only updates are propagated in most cases, not entire tables. (Routers also send "Hello" messages to one another every ten seconds, but these are very short.)

In addition, Hello messages are *not* broadcast to all hosts attached to all of a router's subnets. Hello messages are given the IP destination address 224.0.0.5. Only OSPF routers respond to this *multicast* destination address. (See the section in this module on Classful IP addresses.)

If there are multiple areas, this causes no problems. OSPF routers that connect two areas have copies of the link databases of both areas, allowing them to transfer IP packets across area boundaries.

Encapsulation. OSPF messages are carried in the data fields of IP packets. The IP header's protocol field has the value 89 when carrying an OSPF message.

Border Gateway Protocol. The most common exterior routing protocol is the **Border Gateway Protocol (BGP)**, which is illustrated in Figure A-13.

TCP. BGP uses TCP connections between pairs of routers. This gives reliable delivery for BGP messages. However, TCP only handles one-to-one communication. Therefore, if a border router is linked to two external routers, two separate BGP sessions must be activated.

Distance Vector. Like RIP, BGP is a distance vector dynamic routing protocol. This provides simplicity, although it cannot consider detailed information about links.

Changes Only. Normally, only changes are transmitted between pairs of BGP routers. This reduces network traffic.

Comparisons. Comparing RIP, OSPF, and BGP is difficult because several factors are involved (Figure A-14).

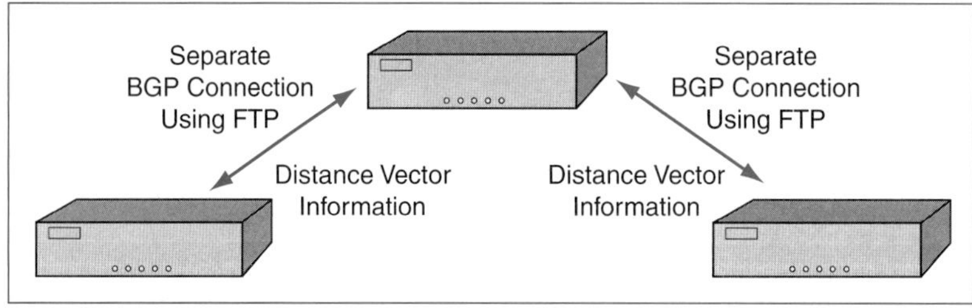

FIGURE A-13 Border Gateway Protocol (BGP) Exterior Routing Protocol

	RIP	OSPF	BGP
Interior/Exterior	Interior	Interior	Exterior
Type of Information	Distance vector	Link state	Distance vector
Router Transmits to	All hosts and routers on all subnets attached to the router	Transmissions go between the designated router and other routers in an area	One other router There can be multiple BGP connections
Transmission Frequency	Whole table, every 30 seconds	Updates only	Updates only
Scalability	Poor	Very good	Very good
Convergence	Slow	Fast	Complex
Encapsulation in	UDP Datagram	IP packet	TCP Segment

FIGURE A-14 Comparison of Routing Information Protocols: Text

Address Resolution Protocol (ARP)

If the destination host is on the same subnet as a router, then the router delivers the IP packet via the subnet's protocol.[8] For an Ethernet LAN:

- The internet layer process passes the IP packet down to the NIC.
- The NIC encapsulates the IP packet in a subnet frame and delivers it to the NIC of the destination host via the LAN.

LEARNING A DESTINATION HOST'S MAC ADDRESS To do its work, the router's NIC *must know the 802.3 MAC layer address of the destination host.* Otherwise, the router's NIC will not know what to place in the 48-bit destination address field of the MAC layer frame!

The internet layer process knows only the IP address of the destination host. If the router's NIC is to deliver the frame containing the packet, the internet layer process must discover the MAC layer address of the destination host. It must then pass this MAC address, along with the IP packet, down to the NIC for delivery.

ADDRESS RESOLUTION ON AN ETHERNET LAN WITH ARP Determining a MAC layer address when you know only an IP address is called **address resolution**. Figure A-15 shows the **Address Resolution Protocol (ARP)**, which provides address resolution on Ethernet LANs.

ARP REQUEST MESSAGE Suppose that the router receives an IP packet with destination address 172.19.8.17. Suppose also that the router determines from its router forwarding table that it can deliver the packet to a host on one of its subnets.

First, the router's internet layer process creates an *ARP request message* that essentially says, "Hey, device with IP address 172.19.8.17, what is your 48-bit MAC layer address?" The internet layer on the router passes this ARP request message to its NIC.

---

[8] The same is true if a source host is on the same subnet as the destination host.

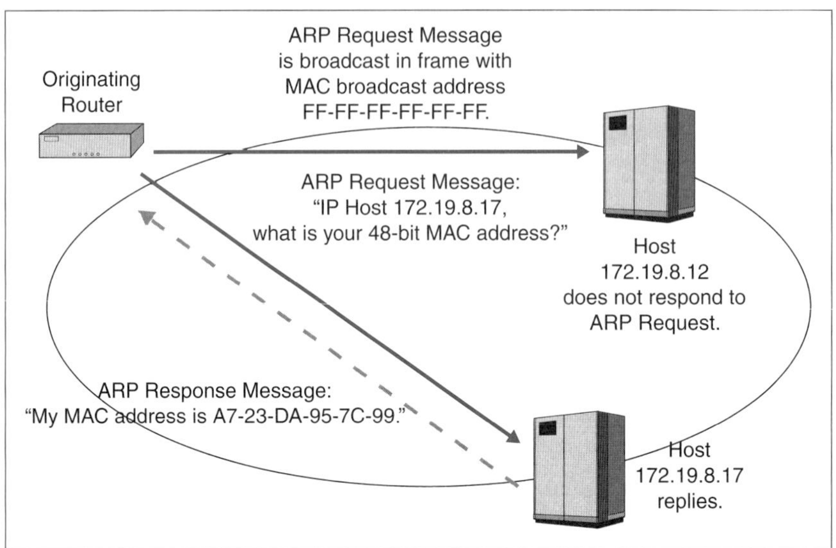

FIGURE A-15 Address Resolution Protocol (ARP)

Broadcasting the ARP Request Message. The MAC layer process on the router's NIC sends the ARP request message in a MAC layer frame that has a destination address of forty-eight 1s. This designates the frame as a broadcast frame. All NICs listen constantly for this **broadcast address**. When a NIC hears this address, it accepts the frame and passes the ARP request message up the internet layer processes.

Returning the ARP Response Message. The internet layer process on every computer examines the ARP request message. If the target IP address is not that computer's, the internet layer process ignores it. If it is that computer's IP address, however, the internet layer process composes an ARP response message that includes its 48-bit MAC layer address.

The target host sends this ARP response message back to the router, via the target host's NIC. There is no need to broadcast the response message, as Figure A-15 shows. The target host sending the ARP response message knows the router's MAC address, because this information was included in the ARP request message.

When the router's internet layer process receives the ARP response message, address resolution is complete. The router's internet layer process now knows the subnet MAC address associated with the IP address. From now on, when an IP packet comes for this IP destination address, the router will send the IP packet down to its NIC, together with the required MAC address. The NIC's MAC process will deliver the IP packet within a frame containing that MAC destination address.

OTHER ADDRESS RESOLUTION PROTOCOLS Although ARP is the Address Resolution Protocol, it is not the only address resolution protocol. Most importantly, ARP uses broadcasting, but not all subnet technologies handle broadcasting. Other address resolution protocols are available for such networks.

ENCAPSULATION An ARP request message is an internet layer message. Therefore, we call it a packet. ARP packets and IP packets are both internet layer packet types in

TCP/IP, as Figure A-1 illustrates. On a LAN, the ARP packet is encapsulated in the data field of an LLC frame. In other types of networks, it is encapsulated in the data field of the data link layer frame.

Classful Addresses in IP

In Chapter 9, we noted that, by themselves, 32-bit IP addresses do not tell you the lengths of their network, subnet, and host parts. For this, you need to have network masks to know how many bits there are in the network part, for instance. This is called **Classless InterDomain Routing (CIDR)**. CIDR allows network parts to vary from 8 bits to 24.

Originally, however, the 32-bit IP address did tell you the size of the network part, although not the subnet part. As Figure A-16 shows, the initial bits of the IP address told whether an IP address was for a host on a Class A, Class B, or Class C network, or whether the IP address was a Class D multicast address. This is **classful addressing**.

CLASS A NETWORKS Specifically, if the initial bit was a 0, this IP address would represent a host in a Class A network. As Figure A-16 shows, Class A network parts were only 8 bits long. The first bit was fixed (0), so there could be only 126 possible Class A networks.[9] However, each of these networks could be enormous, holding more than 16 million hosts. Half of all IP addresses were Class A addresses. Half of these Class A addresses were reserved for future Internet growth.

Class	Beginning Bits	Bits in the Remainder of the Network Part	Number of Bits in Local Part	Approximate Maximum Number of Networks	Approximate Maximum Number of Hosts per Network
A	0	7	24	126	16 million
B	10	14	16	16,000	65,000
C	110	21	8	2 million	254
D[a]	1110				
E[b]	11110				

[a]Used in multicasting.

[b]Experimental.

Problem: For each of the following IP addresses, give the class, the network bits, and the host bits if applicable:

10101010111110000101010100000001

11011010111110000101010100000001

01010101111110000101010100000001

11101110111110000101010100000001

FIGURE A-16 IP Address Classes

[9] Not 127 or 128. Network, subnet, and host parts of all 0s and all 1s are reserved.

CLASS B NETWORKS If the initial bits of the IP address were "10," then this was the address of a host on a Class B network. The network part was 16 bits long. Although the first 2 bits were fixed, the remaining 14 bits could specify a little more than 16,000 Class B networks. With 16 bits remaining for the host part, there could be more than 65,000 hosts on each Class B network. The Class B address space was on its way to being completely exhausted until CIDR was created to replace the classful addressing approach discussed in this section.

CLASS C NETWORKS Addresses in Class C networks began with "110." (Note that the position of the first 0 told you the network's class.) The network part was 24 bits long, and the 21 nonreserved bits allowed more than 2 million Class C networks. Unfortunately, these networks could have only 254 hosts apiece, making them almost useless in practice. Such small networks seemed reasonable when the IP standard was created, because users worked at mainframe computers or at least minicomputers. Even a few of these large machines would be able to serve hundreds or thousands of terminal users. Once PCs became hosts, however, the limit of 254 hosts became highly restrictive.

CLASS D ADDRESSES Class A, B, and C addresses were created to designate specific hosts on specific networks. However, Class D addresses, which begin with "1110," have a different purpose—namely multicasting. This purpose has survived Classless InterDomain Routing.

When one host places another host's IP address in a packet, the packet will go only to *that one* host. This is called **unicasting**. In contrast, when a host places an all-1s address in the host part, then the IP packet should be **broadcast** to *all* hosts on that subnet.

However, what if only *some* hosts should receive the message? For instance, as discussed earlier, when OSPF routers transmit to one another, they want only other OSPF routers to process the message. To support this limitation, they place the IP address 224.0.0.5 in the IP destination address fields of the packets they send. All OSPF routers listen for this IP address and accept packets with this address in their IP destination address fields. This is **multicasting**, that is, *one-to-many* communication. Multicasting is more efficient than broadcasting because not all stations are interrupted. Only routers stop to process the OSPF message.

CLASS E ADDRESSES A fifth class of IP addresses was reserved for future use, but these Class E addresses were never defined.

Review Questions

MULTIPLEXING

1. a) How does a receiving internet layer process decide what process should receive the data in the data field of an IP packet? b) How does TCP decide? c) How does UDP decide? d) How does PPP decide?

MORE ON TCP

2. A TCP segment begins with octet 8,658 and ends with octet 12,783. a) What number does the sending host put in the sequence number field? b) What number does the receiving host put in the acknowledgment number field of the TCP segment that acknowledges this TCP segment?

3. A TCP segment carries data octets 456 through 980. The following TCP segment is a supervisory segment carrying no data. What value is in the sequence number field of the latter TCP segment?
4. Describe flow control in TCP.
5. a) In TCP fragmentation, what is fragmented? b) What device does the fragmentation? c) What device does reassembly?
6. A transport process announces an MSS of 1,024. If there are no IP or TCP options, how big can IP packets be?

MASK OPERATIONS

7. There is a mask 1010. There is a number 1100. What is the result of masking the number?
8. A routing table row has the prefix /14 and the following destination value:

   10101010 10100000 00000000 00000000 (170.160.0.0)

   a) Does it match the following destination address in an arriving IP packet? Explain.

   b) 0101010 10101011 11111111 00000000 (170.171.255.0)

IP VERSION 6

9. a) What is the main benefit of IPv6?

   b) What other benefits were mentioned?
10. a) Express the following in hexadecimal: 0000000111110010. (Hint: Chapter 6 has a conversion table.)

    b) Simplify: A173:0000:0000:0000:23B7:0000:CD12:A84B

IP FRAGMENTATION

11. a) What happens when an IP packet reaches a subnet whose MTU is *longer* than the IP packet?

    b) What happens when an IP packet reaches a subnet whose MTU is *shorter* than the IP packet?

    c) Can fragmentation happen more than once as an IP packet travels to its destination host?
12. Compare TCP fragmentation and IP fragmentation in terms of

    a) what is fragmented and

    b) where the fragmentation takes place.
13. a) What program on what computer does reassembly if IP packets are fragmented?

    b) How does it know which IP packets are fragments of the same original IP packet?

    c) How does it know their correct order?

DYNAMIC ROUTING PROTOCOLS

14. a) What is an autonomous system?

    b) Within an autonomous system, can the organization choose routing protocols?

    c) Can it select the routing protocol its border router uses to communicate with the outside world?
15. Compare RIP, OSPF, and BGP along each of the dimensions shown in Figure A-14.

ADDRESS RESOLUTION PROTOCOL (ARP)

16. A host wishes to send an IP packet to a router on its subnet. It knows the router's IP address.

    a) What else must it know?

    b) Why must the host know it?

    c) How will it discover the piece of information it seeks? (Note: Routers are not alone in being able to use ARP.)
17. a) What is the destination MAC address of an Ethernet frame carrying an ARP request message?

    b) What is the destination MAC address of an Ethernet frame carrying an ARP response packet?

CLASSFUL IP ADDRESSING

18. Compare classful addressing and CIDR.
19. What class of network is each of the following?

    a) 10101010111111110000000010101010

    b) 00110011000000001111111101010101

    c) 11001100111111110000000010101010
20. a) Why is multicasting good?

    b) How did classful addressing support it?

MODULE B

More on Modulation

MODULATION

As we saw in Chapter 1, modems use modulation to convert digital computer signals into analog signals that can travel over the local loop to the first switching office. This module looks at the main forms of modulation in use today.

Frequency Modulation

As we saw in Chapter 1, modulation essentially transforms 0s and 1s into electromagnetic signals that can travel down telephone wires. Electromagnetic signals consist of waves. As we saw in Chapter 7, waves have frequency, measured in hertz (cycles per second). Figure B-1 illustrates **frequency modulation**, in which one **frequency** is chosen to represent a 1 and another frequency is chosen to represent a 0. During a clock cycle in which a 1 is sent, the frequency chosen for the 1 is sent. During a clock cycle in which a 0 is sent, the frequency chosen for the 0 is sent.

Amplitude Modulation

In wave transmission, amplitude is the intensity in the wave. In **amplitude modulation**, which we saw in Chapter 1, we represent 1s and 0s as different amplitudes. For instance, we can represent a 1 by a high-amplitude (loud) signal and a 0 by a low-amplitude (soft) signal. To send "1011," we would send a loud signal for the first time period, a soft signal for the second, and high-amplitude signals for the third and fourth time periods.

Phase Modulation

The last major characteristic of waves is phase. As shown in Figure B-2, we call 0 degrees phase the point of the wave at 0 amplitude and rising. The wave hits its maximum at 90 degrees, returns to 0 on the decline at 180 degrees, and hits its minimum amplitude at 270 degrees. Amplitude now increases to 360 degrees, which is the same as 0 degrees.

In **phase modulation** we use two waves. We let one wave be our reference wave or carrier wave. Let us use this carrier wave to represent a 1. Then we can use a wave 180 degrees out of phase to represent a 0. So if our carrier wave is at 180 degrees, the other wave will be at 0 degrees, and if our carrier wave is at 270 degrees, the other wave will be at 90 degrees.

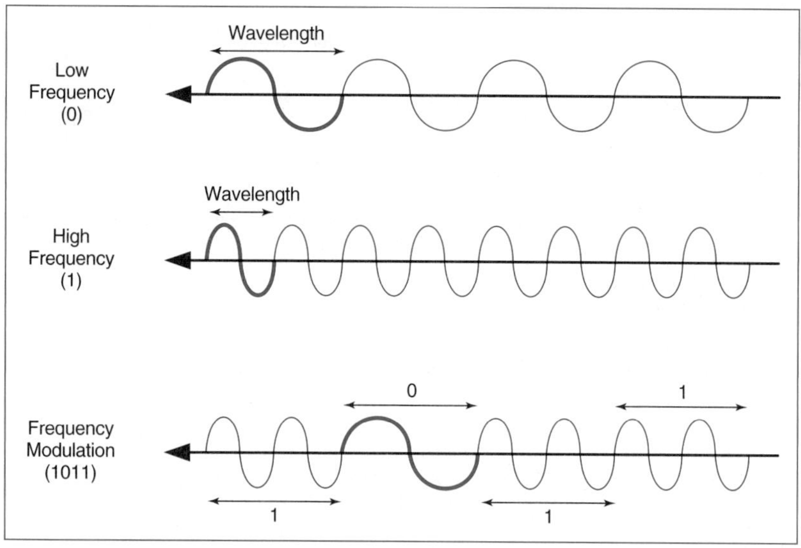

FIGURE B-1 Frequency Modulation

The figure shows that to send "1011," we send the reference wave for the first clock cycle, shift the phase 180 degrees for the second, and return to the reference wave for the third and fourth clock cycles. Although this makes little sense in terms of hearing, it is easy for electronic equipment to deal with phase differences.

A number of transmission systems use **quadrature phase shift keying (QPSK)**, which is phase modulation with four states (phases). Each of the four states represents two bits (00, 01, 10, and 11), so QPSK's bit rate is double its baud rate.

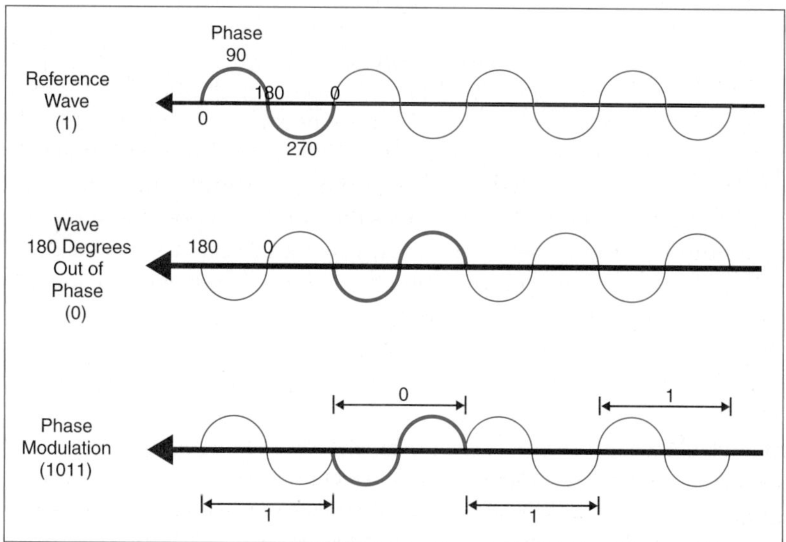

FIGURE B-2 Phase Modulation

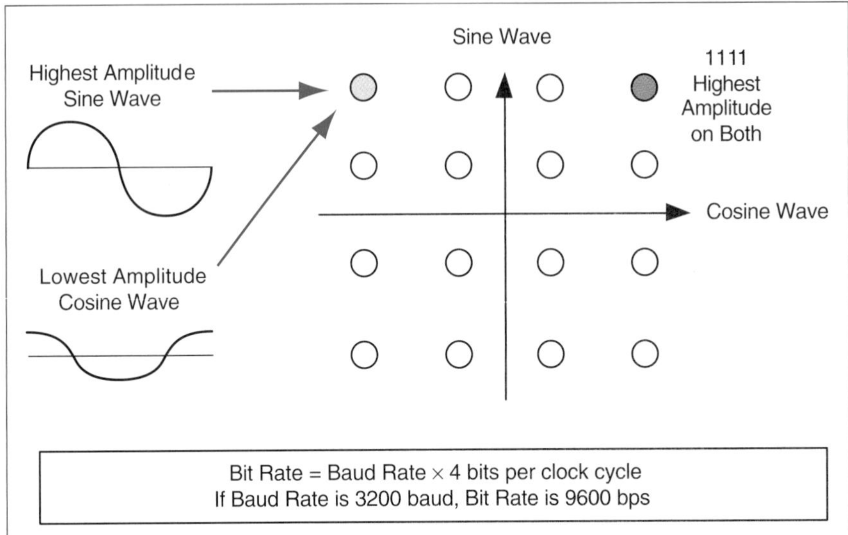

FIGURE B-3 Quadrature Amplitude Modulation (QAM)

Quadrature Amplitude Modulation (QAM)

Telephone modems and many ADSL and cable modems today use a more complex type of modulation called **quadrature amplitude modulation (QAM)**. As Figure B-3 illustrates, QAM uses two carrier waves: a sine carrier wave and a cosine carrier wave. When the cosine wave is at the top of its cycle, the sine wave is just beginning its cycle and will not hit its peak until 90 degrees. The sine wave is 90 degrees out of phase with the cosine wave. This is a quarter of a cycle, and this fact gives rise to the name "quadrature."

The receiver can send different signals on these two waves because they have different phases so the receiver can distinguish between them. Specifically, QAM uses multiple possible amplitude levels for each carrier wave. To illustrate what this means, consider that using four possible amplitudes for the sine wave times four possible amplitudes on the cosine wave will give sixteen possible states. Sixteen possibilities can represent four bits ($2^4 = 16$). Accordingly, each clock cycle can represent a 4-bit value from 0000 through 1111. In summary, each clock cycle transmits four bits if there are four possible amplitude levels.

Different versions of QAM use different numbers of amplitude levels. Each doubling in the number of amplitude levels quadruples the number of possible states. Each quadrupling of the number of possible states allows two more bits to be sent per clock cycle. However, beyond about sixty-four possible states, the states are so close together that even slight transmission impairments can cause errors.

Review Questions

1. Describe frequency modulation.
2. a) Describe phase modulation. b) Describe QPSK.
3. a) What two forms of modulation does QAM use? b) In QAM, if you have four possible amplitudes, how many states do you have? c) In QAM, if you have eight possible amplitudes, how many states do you have? d) How many bits can you send per clock cycle?

MODULE C

More on Telecommunications

INTRODUCTION

Telecommunications is the transmission of voice and video. This module is designed for courses that want to get into more detail on telecommunications.

This module is designed to be read after Chapter 6. The material is not intended to be read front-to-back like a normal chapter. Rather, it is a collection of technical topics, service topics, and regulatory topics:

- The PSTN Transport Core and Signaling
- Communication Satellites
- Wiring in the First Bank of Paradise Headquarters Building
- PBX Services
- Carrier Services and Pricing
- Telephone Carriers and Regulation

THE PSTN TRANSPORT CORE AND SIGNALING

In telecommunications, *transport* is the actual transmission of voice in the PSTN, while *signaling* is the control of the PSTN. In this section, we will look at PSTN transport and signaling in more detail.

The Transport Core

Figure C-1 illustrates that the PSTN transport core consists of switches and trunk line connections that link the switches. The PSTN transport core uses two types of transmission systems to connect telephone switches: TDM trunk lines and ATM packet-switched networks.

Time Division Multiplexing (TDM) Lines

In Chapters 5 and 6, we saw leased lines, which provide high-speed, always-on connections between corporate sites. In Chapters 5 and 6, we saw that leased lines are offered over a wide range of speeds and that the most popular leased lines are T1/E1, fractional T1/E1, and bonded T1/E1.

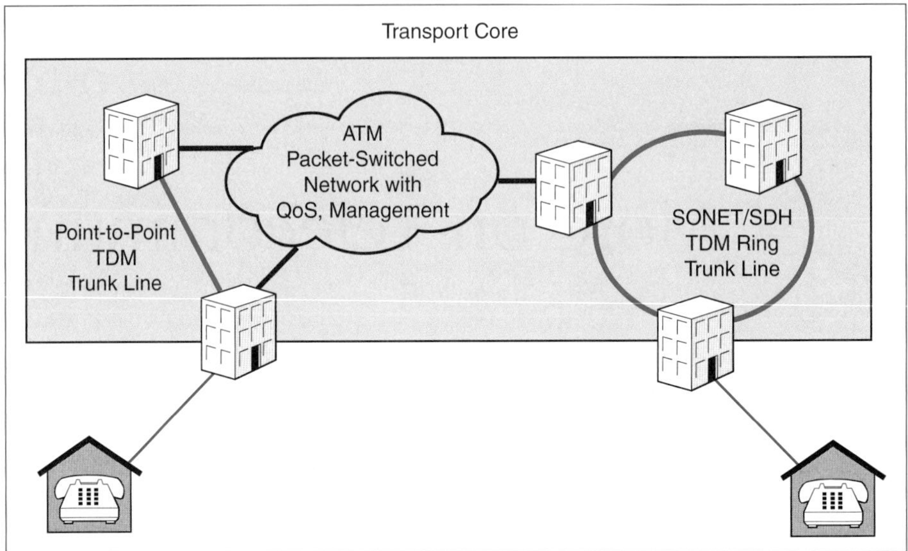

FIGURE C-1 TDM and ATM Switch Connections in the PSTN Transport Core

MULTIPLEXING SIMULTANEOUS VOICE CALLS ON LEASED LINES Figure C-2 shows the variety of leased lines available to corporations. Most of the columns in this figure also appeared in Chapters 5 and 6. However, Figure C-2 has an additional column: multiplexed telephone calls. In telecommunications, the most common use of leased lines is to multiplex many leased lines over a single connection. For example, the figure shows that T1 lines were created to multiplex twenty-four simultaneous voice calls. Higher-speed leased lines can multiplex hundreds or thousands of telephone calls.

THE TIME DIVISION MULTIPLEXING (TDM) PROCESS To implement multiplexing, leased lines use a process called **time division multiplexing (TDM)**, which Figure C-3 illustrates. The figure specifically illustrates time division multiplexing for T1 leased lines.

Frames. First, each second is divided into brief periods of time called **frames.** For example, in a T1 leased line, each second is divided into 8,000 frames. This is because the human voice is sampled 8,000 times per second in pulse code modulation, which the telephone system uses as its default modulation method for converting customer analog voice into digital signals that travel over the telephone system's digital core. One voice sample is transmitted in a frame for every circuit the frame multiplexes.

Slots. Second, each frame is divided into even briefer periods, called **slots.** In a T1 leased line, for instance, there are twenty-four frames per slot. In TDM, a circuit is given the same slot in each frame. Each slot transmits eight bits—a single voice sample for that circuit.

Reserved Capacity. Slot capacity is reserved in each frame. In Figure C-4, Circuit A is given Slot 1 in every frame. Note that Circuit A uses its slot capacity in every frame

North American Digital Hierarchy			
Line	**Speed**	**Multiplexed Voice Calls**	**Typical Transmission Medium**
56 kbps	56 kbps	1	2-Pair Data-Grade UTP
T1	1.544 Mbps	24	2-Pair Data-Grade UTP
Fractional T1	128 kbps, 256 kbps, 384 kbps, 512 kbps, 768 kbps	Varies	2-Pair Data-Grade UTP
Bonded T1s (multiple T1s acting as a single line)	Small multiples of 1.544 Mbps	Varies	2-Pair Data-Grade UTP
T3	44.736 Mbps	672	Optical Fiber

CEPT Hierarchy			
Line	**Speed**	**Multiplexed Voice Calls**	**Typical Transmission Medium**
64 kbps	64 kbps	1	2-Pair Data-Grade UTP
E1	2.048 Mbps	30	2-Pair Data-Grade UTP
E3	34.368 Mbps	480	Optical Fiber

SONET/SDH Speeds			
Line	**Speed (Mbps)**	**Multiplexed Voice Calls**	**Typical Transmission Medium**
OC3/STM1	155.52	2,016	Optical Fiber
OS12/STM4	622.08	6,048	Optical Fiber
OC48/STM16	2,488.32	18,144	Optical Fiber
OC192/STM64	9,953.28	54,432	Optical Fiber
OC768/STM256	39,813.12	163,296	Optical Fiber

FIGURE C-2 Leased Lines and Multiplexing

shown in the figure. However, Circuit B uses only some of its slot capacity in the three frames, and Circuit C uses none at all. Although TDM provides the reserved capacity required for circuit switching, it wastes unused capacity. Users must pay for this reserved capacity whether they use it or not.

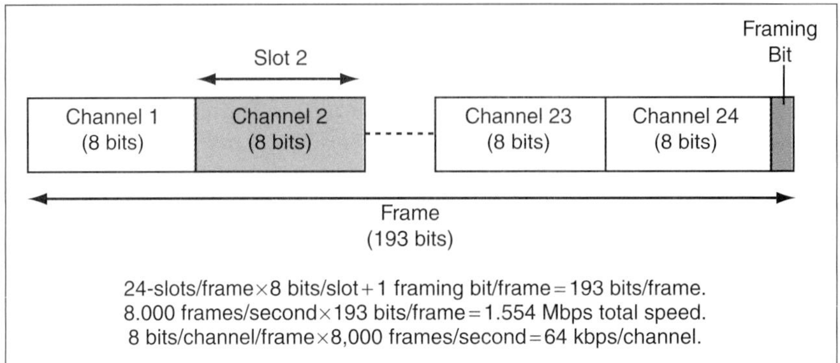

FIGURE C-3 Time Division Multiplexing (TDM) in a T1 Line

Test Your Understanding

1. a) How many simultaneous voice calls can a T3 line multiplex? b) Explain frames and slots in time division multiplexing (TDM). c) How is a circuit allocated capacity on a TDM line? d) What is the advantage of TDM? e) What is the disadvantage?

Leased Lines and Trunk Lines

As Figure C-5 shows, leased lines are circuits that pass through multiple telephone switches and trunk lines.

The figure shows that trunk lines between switches have the same designations as leased line circuits (T1, T3, etc.). With a T1 circuit, the local loop access lines are T1 lines, of course. Between switches, however, faster trunk lines are needed to carry multiple single calls, T1 circuits, and other circuits.

This identical labeling for circuits and trunk lines is not accidental. TDM trunk lines were first used in the 1960s to allow telephone companies to multiplex individual telephone calls on trunk lines between switches. Only later were end-to-end high-speed circuits offered to customers.

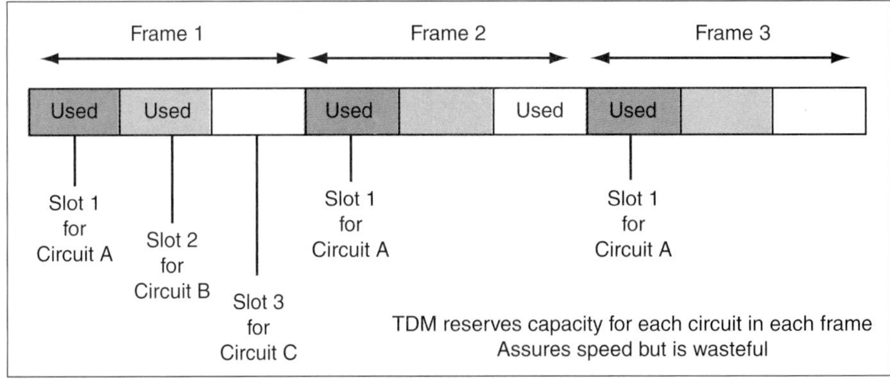

FIGURE C-4 Reserved Capacity in Time Division Multiplexing (TDM)

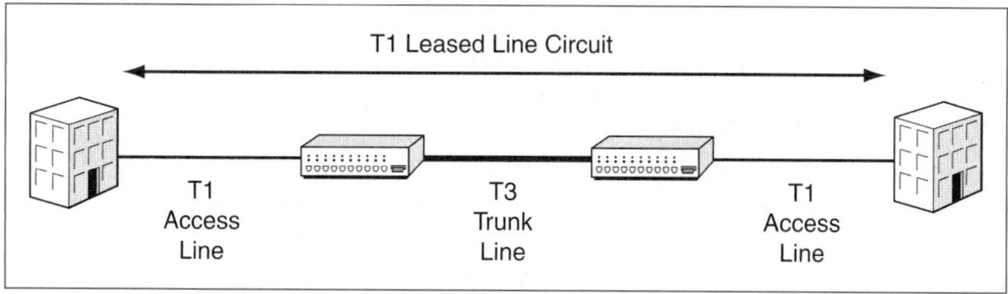

FIGURE C-5 Leased Line Circuits and Trunk Lines

POINT-TO-POINT TDM TRUNK LINES Figure C-1 shows that there are two types of TDM trunk lines. The earliest trunk lines (through T3/E3) were point-to-point trunk lines that connected pairs of switches. Unfortunately, if a trunk line is accidentally dug up and broken in an unrelated construction project, it can take hours or even days for the telephone company to be able to restore service. This is not a theoretical concern. *Most* telephone outages are due to the accidental cutting of trunk lines by construction vehicles.

SONET/SDH RINGS In a ring topology, the devices are connected in a loop. Ring topologies bring reliability. Rings are really dual-rings. If there is a broken connection between two switches, the ring is wrapped, and service continues. Disruptions from broken connections in ring topologies are only momentary.

The SONET/SDH multiplexing technology is designed to use a ring topology, as Figure C-1 illustrates. Although it can be used for point-to-point connections, ring implementations are highly preferred because of their reliability. Figure C-6 illustrates a SONET/SDH ring.

Test Your Understanding

2. a) What is the relationship between leased line circuits and trunk lines? b) Below about what speed are trunk lines point-to-point lines? c) What topology are SONET/SDH lines designed to use? d) What is the advantage of this topology? e) What does a SONET/SDH network do when there is a break in a line between switches?

Asynchronous Transfer Mode (ATM) Transport

Although TDM has long been synonymous with transmission in the PSTN transport core, many long-distance carriers have already transitioned much of their transmission

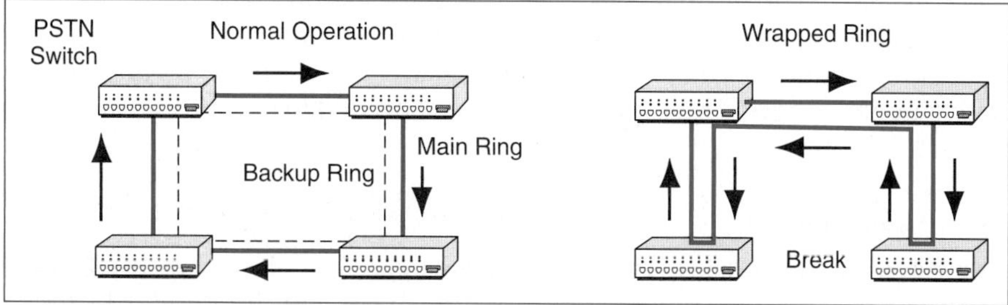

FIGURE C-6 SONET/SDH Dual Ring

technology between telephone switches in their transport cores to a *packet-switched* technology, **asynchronous transfer mode (ATM)**.

ATM has had a checkered history. It was originally created precisely to replace TDM connections in the PSTN transport core and their wasted reserved capacity with more efficient packet switching. For a time, ATM was touted as both the LAN and PSDN technology for the future. However, ATM's high cost worked against it in those markets. As noted in Chapter 6, Ethernet now dominates LAN service. Among the public switched data networks discussed in Chapter 6, Frame Relay has dominated to date because most companies do not require the high speeds of ATM public switched data networks. For the future, less expensive metropolitan area Ethernet promises to be a strong competitor for ATM in the high-speed PSDN market.

However, ATM's expensive complexity, which has stymied it in the LAN and PSDN WAN markets, is critical to its role in the PSTN transport core. First, much of ATM's complexity comes from its ability to provide very strict quality-of-service guarantees for telephone communication and video transmission. The PSTN can only use transport core technologies that guarantee excellent voice quality.

In addition, large networks like the telephone system require excellent management tools. ATM, which was created as a transport core protocol for the entire telephone network, has excellent management tools. In LANs and carrier public switched data networks, these management tools are overkill. In the PSTN core, they are perfect.

Test Your Understanding

3. a) How is ATM different from previous trunk line technologies? b) Why is ATM good for voice? c) Why is ATM ideal for use in the transport core of telecommunications carriers?

Signaling

Again, signaling is the supervision of connections in the PSTN. The ITU-T created **Signaling System 7 (SS7)** as the worldwide standard for supervisory signaling (setting up circuits, maintaining them, tearing them down after a conversation, providing billing information, and providing special services such as three-party calling). The U.S. version of the protocol is ANSI SS7, usually referred to simply as **SS7**. The ETSI version for Europe is called ETSI C7 or **C7**. They are almost the same, so simple gateways can convert between them and allow them to interoperate.

SS7/C7 actually is a packet-switched technology that operates in parallel with the circuit-switched PSTN but that uses the same transmission lines as the PSTN. SS7/C7 relies on multiple databases of customer information. When a call is set up, the originating telephone carrier queries one of these databases to determine routing information for setting up the service. These databases are also needed to provide advanced services such as toll-free numbers.

Test Your Understanding

4. a) What is the worldwide signaling system for telephony? b) Distinguish between SS7 and C7. c) Does having two versions of the standard cause major problems?

Transport Versus Signaling

- Transport is the transmission of voice conversations between customers
- Signaling is the supervision of transport connections
  - Call setup, management, and termination
  - The collection and transmission of billing information
  - 3-party calling, and other advanced services

Signaling System 7 (SS7)

- The worldwide standard for PSTN signaling
- Slight differences exist in the U.S. and Europe
  - U.S.: Signaling System 7
  - Europe: C7
  - Interconnected with a simple gateway

Packet-Switched Technology

- Not circuit-switched
- Runs over telephone company lines
- Uses a distributed database
  - Data for supervising calls
  - Call setup, etc.: requires the querying of the nearest database
  - Toll-free numbers, etc.

FIGURE C-7 Signaling (Study Figure)

COMMUNICATION SATELLITES

During the 1970s, satellites began to be widely used for trunk line transmission within the telephone network. This created sharp drops in long-distance rates. However, as we will see, satellites have proven to be problematic for telephone calling and even more problematic for data transmission.

Microwave Transmission

Satellite transmission technology grew out of microwave transmission technology. As Figure C-8 shows, **microwave** transmission is a point-to-point radio technology using dish antennas. As a consequence of the curvature of the Earth, microwave signals cannot travel farther than a few miles. Consequently, transmission often uses **microwave repeaters** between distant sites.

Before optical fiber became widespread, microwave transmission was used very heavily for long-distance trunk lines between telephone switches. Microwave transmission uses frequency division multiplexing, carrying different telephone calls in different channels.

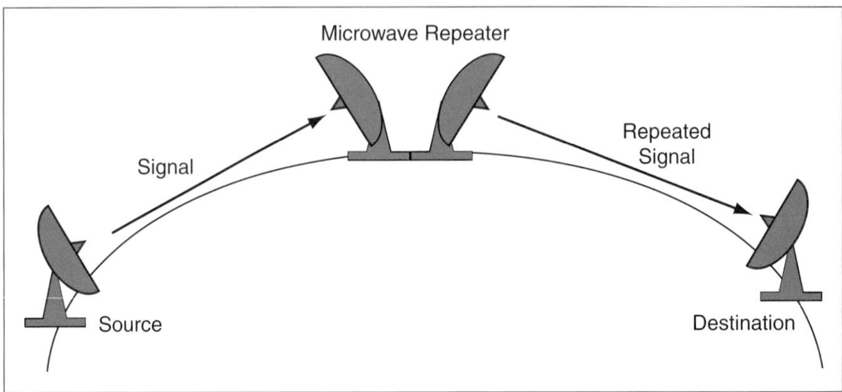

FIGURE C-8 Microwave Transmission

Satellite Transmission

After World War II, a young radar operator named Arthur C. Clarke (yes, the science fiction writer) noticed how microwave repeaters often sit on hills so that they can carry signals farther. He realized that it was possible to put a microwave repeater in the sky on a satellite and that this would allow transmission over very long distances. As Figure C-9 shows, this became the **communication satellite**.

Communication from the ground to the satellite is called the **uplink**. Normally, this transmission occurs point-to-point between a ground station and the satellite. The uplink ground station has a dish antenna to focus its beam.

However, when a satellite transmits, it transmits its **downlink** signal over a wide area called the satellite's **footprint**. Any ground station in the footprint can receive the satellite's transmissions.

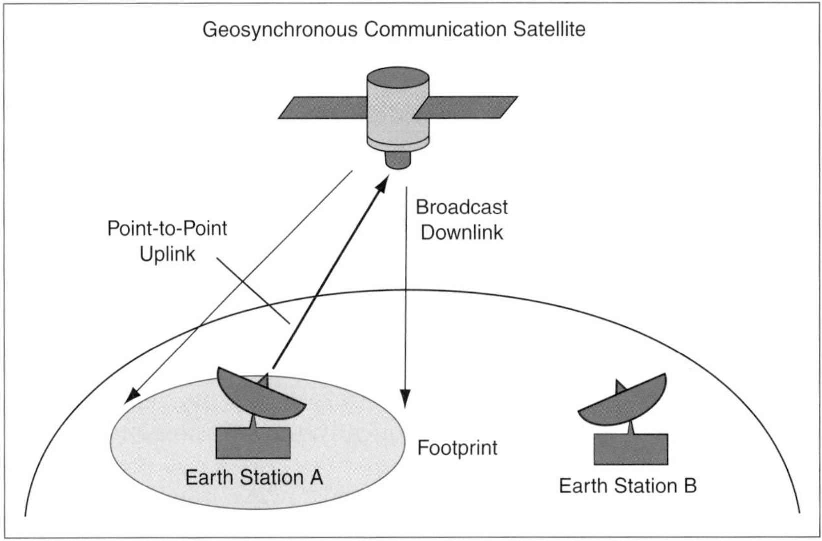

FIGURE C-9 Geosynchronous Earth Orbit (GEO) Communication Satellite System

Geosynchronous Earth Orbit (GEO) Satellites

Figure C-9 specifically shows a **geosynchronous Earth orbit (GEO) satellite** system. The satellite orbits at roughly 36,000 km (22,300 miles) above the Earth. At this height, its orbital time equals the Earth's rotation, so the satellite appears to be fixed in the sky. This allows dish antennas to be aimed precisely.

However, 36,000 km is a long way for radio waves to travel. Even with a dish antenna, considerable power is required. Of course, mobile devices cannot use dish antennas.

In the early days of communication satellites, satellites were often used to place voice calls. This created a delay of up to a quarter second. This latency complicated turn taking in conversations. As soon as possible, almost all telephony was moved to optical fiber.

This latency is even worse for data. TCP processes will retransmit segments if they are not acknowledged promptly. On many computers, even a single satellite in the circuit will prompt many unnecessary retransmissions. Furthermore, whenever a TCP process does a retransmission, it reduces the rate at which it transmits subsequent segments. If there are many retransmissions, the rate of transmission will become painfully slow.

Low Earth Orbit (LEO) and Medium Earth Orbit (MEO) Satellites

As Figure C-10 shows, most communication satellites operate at much lower orbits. This means that they are over a receiver only a short time before passing below the horizon. As a result, satellites must hand off service to one another. As one satellite (Satellite A) passes over the horizon, another satellite (Satellite B) will take over a customer's service. The user will not experience any service interruption. This is reminiscent of cellular telephony, except that here the customer remains relatively motionless while the satellite (the equivalent of a cellsite transceiver) moves.

Satellites for mobile users operate in two principle orbits. **Low Earth Orbit (LEO) satellites** operate at a few hundred kilometers (a few hundred miles) above the Earth.

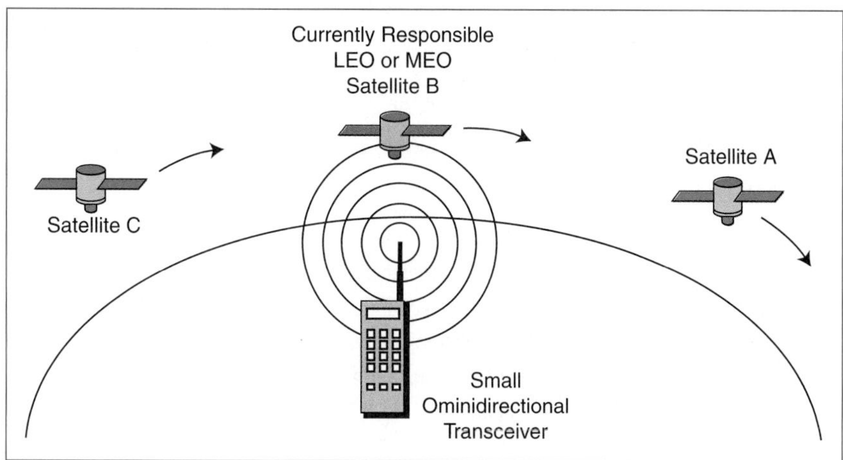

FIGURE C-10 LEO and MEO Satellite Communication Systems

Medium Earth Orbit (MEO) satellites, in turn, operate at a few thousand kilometers (a few thousand miles) above the Earth. LEOs are closer, so signals do not attenuate as much, allowing receivers to be smaller and lighter. In contrast, MEOs have longer orbital periods, so they stay in sight longer, reducing the frequency of handoffs and therefore the number of satellites needed to provide continuous service.

VSAT Satellites

During the 1970s at Stanford University, Professor Bruce Lusignan created the idea of using very small satellite dishes for communication. At the time, most satellite dishes were at least 3 meters (10 feet) in diameter. This made them very expensive. Earth stations with small dishes—**very small aperture terminal (VSAT)** Earth stations—could allow earth stations to be placed in individual homes. The dishes on VSAT Earth stations are 1 meter in diameter to 1 foot in diameter. At first, Lusignan's ideas were rejected by technologists and regulators. Obviously, those objections were overcome, and VSAT Earth stations are now common.

VSATs today are used almost exclusively for one-way transmission. Satellite-based television delivery is becoming a very strong competitor for cable television. Some VSATs offer two-way communication. (You see them in newscasts and in films of military operations.) However, one-way delivery is the norm.

Test Your Understanding

5. a) Why are microwave repeaters needed? b) Does the uplink or the downlink use point-to-point transmission? c) What is the footprint? d) Which of the following uses dish antennas: GEOs, LEOs, or MEOs? e) Which of the following is good for mobile stations: GEOs, LEOs, or MEOs? f) What are the rough heights of GEO, LEO, and MEO orbits? g) What is a VSAT satellite dish? h) What is the attraction of VSAT technology?

Traditional Satellite Systems

- Used very large dishes (3 meters or more)
- Very expensive

VSAT Satellite System

- Very small aperture terminal (VSAT) earth stations
  - Use small (1 meter or less) diameter dishes
- Small dishes allow earth stations small and inexpensive enough to be used in homes
- Used primarily in one-way transmission, such as television distribution
- Occasionally used for two-way communication
  - News reporting in the field
  - Military communication

FIGURE C-11 VSAT Satellite System (Study Figure)

WIRING THE FIRST BANK OF PARADISE HEADQUARTERS BUILDING

Wiring dominates total cost for customer premises equipment. For new buildings, a firm normally hires a contractor to install wiring. Afterward, the firm has to maintain its wiring systems.

In this section, we will look at wiring in First Bank of Paradise headquarters building. This is a typical multistory office building. Wiring in other large buildings tends to be similar.

Facilities

Figure C-12 illustrates the building. Although the building is ten stories tall, only three stories (plus the basement) are shown.

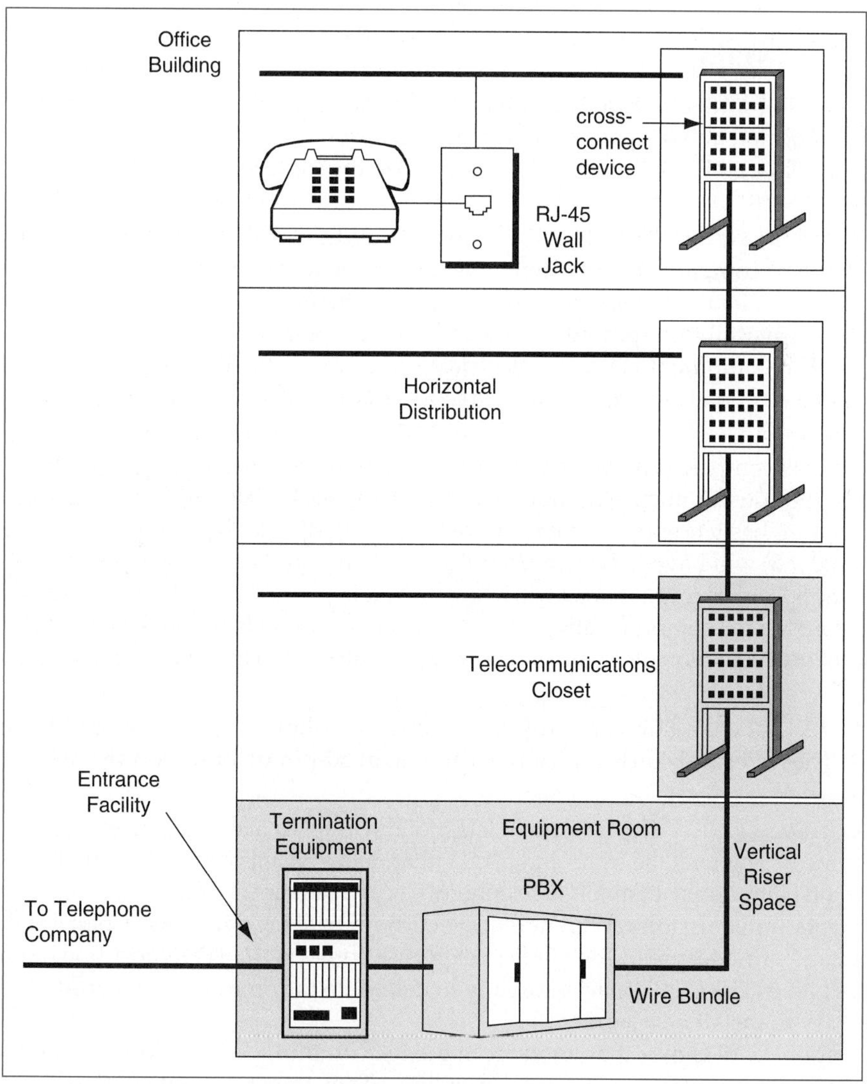

FIGURE C-12 First Bank of Paradise Building Wiring

EQUIPMENT ROOM Wiring begins in the equipment room, which usually is in the building's basement. This room connects the building to the outside world.

VERTICAL RISERS From the equipment room, telephone and data cabling has to rise to the building's upper floors. The bank has **vertical riser** spaces between floors. These typically are hard-walled pipes to protect the wiring.

TELECOMMUNICATIONS CLOSETS On each floor above the basement, there is a **telecommunications closet**. Within the telecommunications closet, cords coming up from the basement are connected to cords that span out horizontally to telephones and computers on that floor.

Telephone Wiring

This building infrastructure was created for telephone wiring. Fortunately, most companies allocated ample space to their equipment rooms, vertical risers, and telecommunications closets. This allows data wiring to use the same spaces.

TERMINATION EQUIPMENT In many ways, a building's telephone network acts like an independent company. It can interconnect with the local telephone company and with other carriers. It negotiates contracts with each of them. Carriers require the company to install **termination equipment** at its connection point to the outside world. In effect, the termination equipment is like an electrical fuse; it prevents the company from sending unwanted electrical signals into the carrier's network.

PBX Many companies have internal telephone switches called **private branch exchanges (PBXs)**. Eight wires must span out from the PBX to each wall outlet in the building. So if the building has ten floors and each floor has 100 RJ-45 telephone wall jacks, 8,000 wires ($10 \times 100 \times 8$) will have to be run from the PBX through the vertical riser space. On each floor, the 800 wires for that floor will be organized into one hundred 4-pair UTP cords running from the telecommunications closet to telephone wall jacks on that floor. Obviously, careful documentation of where each wire goes is crucial to maintaining sanity.

VERTICAL WIRING For vertical wiring runs, telephony typically uses **25-pair UTP cords**. These vertical cords typically terminate in **50-pin octopus connectors**.

HORIZONTAL WIRING The horizontal telephone wiring, as just noted, uses 4-pair UTP. Yes, telephone wiring uses the same 4-pair UTP that data transmission uses. Actually, telephone wiring first introduced 4-pair UTP. Telephony has used 4-pair UTP for several decades.

Data transmission researchers learned how to send data over 4-pair UTP to take advantage of widespread installation expertise for 4-pair UTP. In addition, some companies had excess UTP capacity already installed, so in some cases, it would not even be necessary to install new wiring.

Figure C-12 shows that wires from the telecommunications closet on a floor travel horizontally through the walls or false ceilings. They then terminate in RJ-45 data/voice jacks. Telephones plug into the jacks.

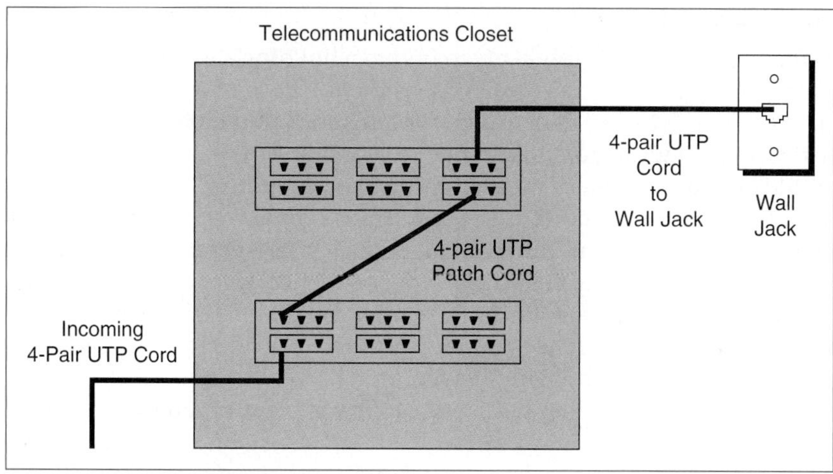

FIGURE C-13 Patch Panels

CROSS-CONNECT DEVICE Within the telecommunications closet, the vertical cords plug into **cross-connect devices**, which connect the wires from the riser space to 4-pair UTP cords that span out to the wall jacks on each floor.

As Figure C-13 shows, the cross connection normally uses patch panels. The figure illustrates patch panels with RJ-45 connectors, which are useful for both voice and data wiring. Patch cords connect eight vertical wires to eight horizontal wires. Patch panels are used because they provide flexibility. If there are changes in the vertical or horizontal wiring, the patch panels are simply reconnected to reflect the changes.

Data Wiring

Figure C-12 illustrates telephone wiring. How is data wiring different? For horizontal communication, there are no differences at all. Both almost always use 4-pair UTP. They are the same precisely because Ethernet was adapted to run over horizontal telephone wiring (albeit a higher grade of telephone wiring).

However, vertical wiring is completely different. Vertical data wiring is much simpler than vertical telephone wiring. In data wiring, only single UTP or optical fiber cord runs from a port in the core switch up to a port in the Ethernet workgroup switch on each floor. In other words, if there are ten floors, only ten UTP or optical fiber cords would have to be run through the vertical riser space. This is vastly simpler than vertical telephone wiring, which must run eight wires vertically for each wall jack on each floor.

PLENUM CABLING

Fire regulations require the use of a special type of fire-retardant cabling, called **plenum** cabling, any time cables run through airways (plenums) such as air-conditioning ducts (but *not* false ceilings). Ordinary jackets on UTP and optical fiber cords are made of polyvinyl chloride (PVC), which gives off deadly dioxin when it burns. If these toxins are released in airways, the toxins will spread rapidly to office areas.

Test Your Understanding

6. a) What equipment are you likely to find in a building's equipment room? b) In its telecommunications closets? c) What is the purpose of a PBX? d) Compare and contrast vertical wiring distribution for telephony and data. e) Compare and contrast horizontal wiring distribution for telephony and data.
7. A building has ten floors, not counting the equipment room in the basement. Each of the ten floors has 60 voice jacks and 40 data jacks. a) For telephony, how many wires will you run through the vertical riser space for each floor? b) How many 25-pair cords will this require? c) For vertical data wiring if you use 4-pair UTP? d) For vertical data wiring if you use optical fiber? e) On each floor, how many wires will you run horizontally from the telecommunications closet to wall jacks? f) How many cords will this require?
8. a) Where is plenum cabling required? b) Why is plenum cabling needed?

PBX SERVICES

Figure C-14 shows that because digital PBXs are essentially computers, they allow vendors to differentiate their products by adding application software to provide a wide range of services.

- User services are employed directly by ordinary managers, secretaries, and other telephone end users.
- Attendant services are employed by telephone operators to help them give service.
- Management services are employed by telephone and corporate network managers to manage the company's telephone network.

For Users	
Speed dialing	Dials a number with a one- or two-digit code.
Last number redial	Redials the last number dialed.
Display of called number	LCD display for number the caller has dialed. Allows caller to see a mistake.
Camp on	If line is busy, hit "camp on" and hang up. When other party is off the line, he or she will be called automatically.
Call waiting	If you are talking to someone, you will be beeped if someone else calls.
Hold	Put someone on hold until he or she can be talked to.
ANI	Automatic number identification: You can see the number of the party calling you.
Conferencing	Allows three or more people to speak together.
Call transfer	Someone calls you. You connect the person to someone else.
Call forwarding	If you will be away from your desk, calls will be transferred to this number.
Voice mail	Callers can leave messages.

FIGURE C-14 Digital PBX Services

For Attendants	
Operator	In-house telephone operators can handle problems.
Automatic call distribution	When someone dials in, the call goes to a specific telephone without operator assistance.
Message center	Allows caller to leave a message with a live operator.
Paging	Operator can page someone anywhere in the building.
Nighttime call handling	Special functions for handling nighttime calls, such as forwarding control to a guard station.
Change requests	Can change extensions and other information from a console.
For Management	
Automatic route selection	Automatically selects the cheapest way of placing long-distance calls.
Call restriction	Prevents certain stations from placing outgoing or long-distance calls.
Call detail reporting	Provides detailed reports on charges by telephone and by department.

FIGURE C-14 Continued

Test Your Understanding

9. a) Into what three categories are PBX services divided? b) List and briefly describe two services in each category.

CARRIER SERVICES AND PRICING

Having discussed technology, we can now turn to the kinds of transmission services that telecommunications staffs can offer their companies. Figure C-15 shows that corporate users face a variety of transmission services and pricing options.

Basic Voice Services

The most important telephone service, of course, is its primary one: allowing two people to talk together. Although you get roughly the same service whether you call a nearby building or another country, billing varies widely between local and long-distance calling. Even within these categories, furthermore, there are important pricing variations.

LOCAL CALLING Most telephone calls are made between parties within a few kilometers of each other. There are two major billing schemes for such **local calling**.

- Some telephone companies offer **flat-rate** local service in which there is a fixed monthly service charge but no separate fee for individual local calls.
- In some areas, however, carriers charge **message units** for some or all local calls. The number of message units they charge for a call depends on both the distance and duration of the call.

Local Calling

- Flat rate
- Message units

Toll Calls

- Long-distance calling
- Intra-LATA
- Inter-LATA

Toll-Call Pricing

- Direct distance dialing
  - Base case for comparison
- Toll-Free numbers
  - Free to caller but called party pays
  - Called party: pays less than direct distance dialing rates
  - In U.S., 800, 888, etc.
- WATS
  - Wide Area Telephone Service
  - For calling out of a site
  - Calling party: pays but pays less than with direct distance dialing
- 900 numbers
  - Caller pays
  - Pays more than direct distance dialing rates
  - Allows called party to charge for services

Advanced Services

- Caller ID
- Three-party calling (conference calling)
- Call waiting
- Voice mail

FIGURE C-15 Telephone Services (Study Figure)

Economists like message units, arguing that message units are more efficient in allocating resources than flat-rate plans. Subscribers, in contrast, dislike message units even if their flat-rate bill would have come out the same.

LONG-DISTANCE TOLL CALLS Although pricing for local calling varies from place to place, all **long-distance** calls are **toll calls**. The cost of the call depends on distance and duration.

Direct Distance Dialing. The simplest form of long-distance pricing is **direct distance dialing**, in which you place a call without any special deals. You will pay a few cents per minute for directly dialed calls. Direct distance dialing is a base case against which other pricing schemes can be measured.

Toll-Free Numbers. Companies that are large enough can receive favorable rates from transmission companies for long-distance calls. With **toll-free numbers**, anyone can call *into* a company, usually without being charged. To provide free inward dialing, companies pay a carrier a per-minute rate lower than the rate for directly dialed calls. Initially, only numbers with the 800 area code provided such services in the United States. Now that 800 area codes have been exhausted, the 888, 877, 866, and 855 area codes are offering the same service to new customers.

WATS. In contrast to inbound toll-free number service, **wide area telephone service (WATS)** allows a company to place *outgoing* long-distance calls at per-minute prices lower than those of directly dialed calls. WATS prices depend on the size of the service area. WATS is often available for both intrastate and interstate calling. WATS can also be purchased for a region of the country instead of the entire country.

900 Numbers. Related to toll-free, **900 numbers** allow customers to call into a company. Unlike toll-free number calls, which usually are free to the caller, calls to 900 numbers require the caller to pay a fee—one that is much *higher* than that of a toll call. Some of the fee goes to the carrier, but most of it goes to the subscriber being called.

This allows companies to charge for information, technical support, and other services. For instance, customer calls for technical service might cost $20 to $50 per hour. Charges for 900 numbers usually appear on the customer's regular monthly bill from the local exchange carrier (LEC). Although the use of 900 numbers for sexually oriented services has given 900 numbers a bad name, they are valuable for legitimate business use.

Advanced Services

Although telephony's basic function as a two-person "voice pipe" is important, telephone carriers offer other services to attract customers and to get more revenues from existing customers.

CALLER ID In **caller ID**, the telephone number of the party calling you is displayed on your phone's small display screen before you pick up the handset. This allows you to screen calls, picking up only the calls you want to receive. Callers can block caller ID, so that you cannot see their numbers. However, you can have your carrier reject calls with blocked IDs. Businesses like caller ID because it can be linked to a computer database to pull up information about the caller on the receiver's desktop computer screen.

THREE-PARTY CALLING (CONFERENCE CALLING) Nearly every teenager knows how to make **three-party calls**, in which more than the traditional two people can take part in a conversation. However, businesses tend to use this feature only sparingly, despite its obvious advantage. This is sometimes called **conference calling**.

Call Waiting

Another popular service is **call waiting**. If you are having a conversation and someone calls you, you will hear a distinctive tone. You can place your original caller on hold, shift briefly to the new caller, and then switch back to your original caller.

Voice Mail

Finally, **voice mail** allows people to leave messages if you do not answer your phone.

Test Your Understanding

10. Create a table to compare and contrast direct distance dialing, toll-free numbers, 900 numbers, and WATS, in terms of whether the caller or the called party pays and the cost compared with the cost of a directly dialed long-distance call.
11. Describe the two pricing options for local calls.
12. a) What is the advantage of toll-free numbers for customers? b) For companies that provide toll-free number service to their customers?
13. a) Name the four advanced telephone services listed in the text. b) Name and briefly describe two advanced services not listed in the text.

TELEPHONE CARRIERS AND REGULATION

Once, almost every nation had a single national telephone carrier. However, the situation has become more complex over time as nations have begun to deregulate telephone service—that is, to permit some competition in order to reduce prices and promote product innovation.

Competition helps corporate customers because telephone prices generally fall as a result of competition. However, to maximize cost savings, companies have to be very smart when they deal with telephone carriers. To do this, a first step is understanding the types of carriers a company will face (Figure C-16).

PTTs and Ministries of Telecommunications

In most countries other than the United States, the single monopoly carrier was historically called the Public Telephone and Telegraph authority (PTT). In the United Kingdom, for example, this was British Telecom, while in Ireland it was Eircom. The PTT had a monopoly on domestic telephony—that is, telephony within the country.

To counterbalance the power of the PTT, governments created regulatory bodies generally called **Ministries of Telecommunications**. PTTs provide service, while Ministries of Telecommunications oversee the PTTs. As we will see later, over time, the PTTs gradually lost their monopoly status, and ministries of telecommunications now find themselves regulating both the traditional PTT and its new competitors.

Test Your Understanding

14. a) Do all countries have PTTs? Explain. b) What is a monopoly over domestic telephone service? c) What are the purposes of PTTs and Ministries of Telecommunications?

AT&T, the FCC, and PUCs

THE BELL SYSTEM In the United States, neither telegraphy nor telephony was made a statutory monopoly. However, telephony quickly became a de facto monopoly when **AT&T**, also known as the **Bell System**, used predatory practices to drive most other

In Most Countries

- Public Telephone and Telegraph (PTT) authorities
  - Traditionally had a domestic monopoly over telephone service
- Ministries of Communication
  - Government agency to regulate the PTT
- Competitors
  - Deregulation has allowed competition in domestic telephone service in most countries
  - The Ministry of Telecommunication regulates these new competitors too

In the United States

- AT&T (the Bell System) developed a long-distance monopoly
- Also owned most local operating companies
- AT&T was broken up in the 1980s
  - AT&T retained the name and the (initially) lucrative long-distance business
  - Local operations were assigned to seven Regional Bell Operating Companies (RBOCs)
  - Later, RBOCs combined with one another and with GTE to form four supercarriers
  - Eventually, competition in long-distance service made AT&T unprofitable
  - In 2005, one of the four supercarriers (SBC Communications) merged with AT&T and used the AT&T name for the merged company.
- Regulation
  - Federal Communications Commission (FCC) regulates interstate communication and aspects of intrastate communication that affect national commerce
  - Within each state, a Public Utilities Commission (PUC) regulates telephone service subject to FCC regulations

FIGURE C-16 Telephone Carriers (Study Figure)

competitors out of business. AT&T soon had a complete long-distance monopoly. For local service, AT&T owned more than 80 percent of all local telephone companies, although when it was developing in the nineteenth century and early twentieth century, it bypassed "unpromising" areas such as Hawaii and most of Los Angeles.

THE RBOCs In the 1980s, AT&T was broken up into a long-distance and manufacturing company that retained the AT&T name and seven **Regional Bell Operating Companies (RBOCs)** that owned most local telephone companies.

Later, mergers among the RBOCs and GTE, which was the largest independent owner of local operating companies, produced four dominant owners of local operating companies in the United States—Verizon, SBC Communications, BellSouth, and Qwest. These four companies also provide long-distance service in some areas.

At the time of the breakup, AT&T was considered the jewel in the Bell System. However, after quite a few years of high profitability, AT&T began to suffer heavily from long-distance competition. In 2005, in a stroke of irony, SBC merged with ailing AT&T. The combined company took on the name AT&T.

REGULATION: THE FCC AND PUCs In the United States, the **Federal Communication Commission (FCC)** provides overall regulation for U.S. carriers. However, within individual states, **Public Utilities Commissions (PUCs)** regulate pricing and services.

Test Your Understanding

15. a) Distinguish between the traditional roles of AT&T and the RBOCs. b) Distinguish between the traditional roles of the FCC and PUCs in the United States.

Deregulation

Although telephone carriers had a complete monopoly in the early years, governments began deregulating telephone service in the 1970s. **Deregulation** is the opening of telephone services to competition; it has the potential to reduce costs considerably (Figure C-17).

DEREGULATION AROUND THE WORLD As noted earlier, most countries have deregulated at least some of the services offered by the traditional monopoly PTT. This has given companies many more choices for telephone services, and competition has resulted in lower prices.

CARRIERS IN THE UNITED STATES

LATAs. Figure C-18 shows the types of carriers that exist in the United States. Since the breakup of AT&T in 1984, the United States has divided into approximately 200 service regions called **local access and transport areas (LATAs)**.

ILECs and CLECs. Within each LATA, local exchange carriers (LECs) provide access and transport (transmission service). The traditional monopoly telephone company is called the **incumbent local exchange carrier (ILEC)**. Competitors are called **competitive local exchange carriers (CLECs)**.

> LATAs are geographic regions. ILECs and CLECs are carriers that provide access and transport within LATAs.

IXCs. In contrast, **interexchange carriers (IXCs)** carry voice traffic *between* LATAs. Major ILECs are AT&T, MCI, and Sprint.

Long-Distance Calling. One point of common confusion is that the distinction between local and long distance calling is not the same as the distinction between LEC and IXC service. Most LATAs are quite large; within LATAs, there is both local and long-distance calling. Adding to the confusion, intra-LATA long distance calling rates sometimes are higher than inter-LATA calling rates.

> Within LATAs, there is both local and long-distance calling.

ICCs. ILECs, CLECs, and IXCs are **domestic** carriers that provide service within the United States. Similarly, PTTs provide domestic service within their own countries. In contrast, **international common carriers (ICCs)** provide service *between* countries.

Deregulation

Deregulation decreases or removes monopoly over telephone service

This creates competition, which lowers prices

In most companies, deregulation began in the 1970s

Deregulation Around the World

At least some PTT services have been deregulated

Carriers in the United States

The United States is divided into regions called local access and transport areas (LATAs)

Within each LATA:

- Local exchange carriers (LECs) provide intra-LATA service
- Traditional incumbent local exchange carrier (ILECs)
- New competitive local exchange carriers (CLECs)

Interexchange carriers (IXCs) provide transport between LATAs

Long-distance service

- Long-distance service within LATAs is supplied by LECs
- Long-distance service between LATAs is supplied by IXCs

Within each LATA, one or more points of presence (POP) interconnects different carriers

Internationally

International common carriers (ICCs) provide service between countries

Degree of Deregulation

Customer premises equipment is almost completely deregulated

Long-distance and international telephony are heavily deregulated

Local telephone service is the least deregulated

- The traditional monopoly carriers have largely maintained their telephone monopolies
- Cellular service has provided local competition, with many people not having a wired phone
- Voice over IP (VoIP) is providing strong competition via ISPs, cable television companies, and a growing number of other wired and wireless access technologies

VoIP Regulation

Countries are struggling with the question of how to regulate VoIP carriers

Should they be taxed?

Should they be required to provide 911 service, including location determination?

Should they be required to provide wiretaps to government agencies?

FIGURE C-17 Deregulation (Study Figure)

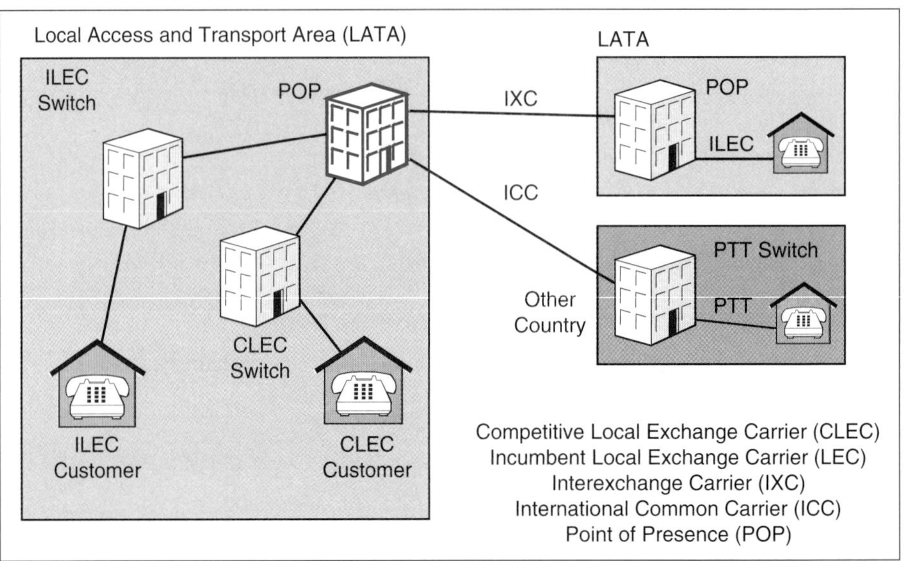

FIGURE C-18 Telephone Carriers in the United States

Points of Presence (POPs). As Figure C-18 shows, the various carriers that provide service are interconnected at **points of presence (POPs)**. Thanks to points of presence, any subscriber to any CLEC or ILEC in one LATA can reach customers of any other CLEC or ILEC in any other LATA. ICCs also link to domestic carriers at POPs.

DEREGULATION BY SERVICE

Customer Premises Equipment. Although it seems odd today, telephone companies used to own all of the wires and telephones in homes and businesses. Today, however, nearly all countries *prohibit* carriers from owning customer premises equipment. Deregulation for customer premises equipment, in other words, is total in most countries.

Long-Distance and International Calling. In most countries, both long-distance and international telephone services have been heavily deregulated.

Local Telephone Service. Local telephone service is the least deregulated aspect of telephony. The need for large investments in access systems and regulatory reluctance to open local telephone service completely (for fear of losing currently subsidized service for the poor and rural customers) have combined to limit local telephone competition.

Some countries now require the traditional monopoly carrier to open its access systems and central offices to competitors for a "reasonable" fee. However, court delays and high "reasonable" fees have limited the effectiveness of this facility-sharing approach.

Overall, traditional monopoly telephone carriers have largely maintained their monopoly over wired telephone service. However, competition is coming through other technologies. Many people now have only a cellular telephone, and cellular service often is provided by a competitor of the traditional monopoly wireline carrier. In addition, voice over IP (VoIP) is providing competition via ISPs, cable television companies, and a growing number of other wired and wireless Internet access technologies.

VOICE OVER IP

Now that voice over IP (VoIP) is becoming popular, countries are trying to determine how to regulate this new service. Traditional carriers point out that VoIP carriers are exempt from many of the taxes that traditional carriers are required to pay. Countries also are attempting to enforce laws requiring calls to emergency numbers (911 in the United States) to give physical location information in case the caller cannot speak. In addition, the U.S. government wants VoIP carriers to provide tools to allow the government to create legal wiretaps.

Test Your Understanding

16. a) Distinguish between LATAs, ILECs, and CLECs. b) What is the role of IXCs relative to LATAs? c) What carriers handle long-distance calling in the United States? d) What is the role of ICCs? e) Why are POPs important?

17. a) What is deregulation? b) When did deregulation begin? c) How complete is deregulation for customer premises equipment? d) For long-distance calling? e) For local calling? f) What issues are involved in the regulation of VoIP?

GLOSSARY

1000Base-SX: A fiber version of gigabit Ethernet for short wavelengths (transmitting at 850 nm).

1000Base-T: A UTP version of gigabit Ethernet.

1000Base-x: The Ethernet physical layer technology of gigabit Ethernet, used today mainly to connect switches to switches or switches to routers; increasingly being used to connect servers and some desktop PCs to the switches that serve them.

100Base-TX: The dominant Ethernet physical layer 100 Mbps standard brought to desktop computers today.

1G: See First-Generation Cellular.

1-Pair Voice-Grade UTP: The traditional telephone access lines to individual residences.

232 Serial Port: The port on a PC that uses two voltage ranges to transmit information.

25-Pair UTP Cord: The cabling used by telephony for vertical wiring that runs within a building.

2G: See Second-Generation Cellular.

2-Pair Data-Grade UTP: The traditional telephone access line for lower-speed leased lines. (Higher-speed leased lines use optical fiber.) Two pairs run out to each customer.

3G: See Third-Generation Cellular.

4G: See Fourth-Generation Cellular.

4-Pair Unshielded Twisted Pair (UTP): The type of wiring typically used in Ethernet networks. 4-pair UTP contains eight copper wires organized as four pairs. Each wire is covered with dielectric insulation, and an outer jacket encloses and protects the four pairs.

50-Pin Octopus Connector: The type of connector in which vertical cords typically terminate.

802 Committee: See 802 LAN/MAN Standards Committee.

802 LAN/MAN Standards Committee: The IEEE committee responsible for Ethernet standards.

802.11 WLAN: Wireless LANs that follow the 802.11 standard.

802.11 Working Group: The IEEE working group that creates wireless LAN standards.

802.11a: Version of the 802.11 WLAN standard that has a rated speed of 54 Mbps and operates in the 5 GHz unlicensed radio band.

802.11b: Version of the 802.11 WLAN standard that has a rated speed of 11 Mbps and operates in the 2.4 GHz unlicensed radio band.

802.11g: Version of the 802.11 WLAN standard that has a rated speed of 54 Mbps and operates in the 2.4 GHz unlicensed radio band.

802.11e: A standard for quality of service in 802.11 WLANs.

802.11i: An advanced form of 802.11 wireless LAN security.

802.11n: Version of the 802.11 WLAN standard that uses MIMO and sometimes doubled bandwidth to achieve a rated speed of 100 Mbps or more and longer range than earlier speed standards.

802.16: WiMAX. Broadband wireless access standard that competes with 3G and 4G cellular services.

802.16d: WiMAX. Broadband wireless access standard for fixed stations.

802.16e: WiMAX. Broadband wireless access standard for mobile stations.

802.1p: The standard that permits up to eight priority levels.

802.1Q: The standard that extended the Ethernet MAC layer frame to include two optional tag fields.

802.1X: Security standard for both wired and wireless LANs. Mode of operation in 802.11i security.

802.2: The single standard for the logical link control layer in 802 LANs.

802.3 MAC Layer Frame: See Ethernet Frame.

802.3 MAC Layer Standard: The standard that defines Ethernet frame organization and NIC and switch operation.

802.3 Working Group: The 802 Committee's working group that creates Ethernet-specific standards.

802.3ad: Link aggregation protocol standard.

802.3af: Standard for delivering low wattage electricity from a switch to stations.

900 Number: A number that allows customers to call into a company; callers pay a fee that is much higher than that of a regular toll call.

Absorptive Attenuation: In wireless transmission, the attenuation of a signal but water along the way absorbing its signal power. In optical fiber, attenuation due to the absorption of signal strength as a signal propagates.

Acceptable Use Policy (AUP): An early Internet policy that forbade the use of the Internet for commercial purposes such as buying, selling, and advertising.

Access Card: Small card with a magnetic stripe or microprocessor that gives you access to your computer or to a room.

Access Control: Limiting who may have access to each resource and limiting their permissions when using the resource.

Access Control List (ACL): An ordered list of pass/deny rules for a firewall or other device.

Access Control Plan: A plan for controlling access to a resource.

Access Line: 1) In networks, a transmission line that connects a station to a switch. 2) In telephony, the line used by the customer to reach the PSTN's central transport core.

Access Link: Same as access line.

Access Point: A bridge between a wireless station and a wired LAN.

Access Router: A router to connect a SOHO network to the Internet. Typically includes a switch, DHCP server, NAT, and other functions beyond routing.

Access System: In telephony, the system by which customers access the PSTN, including access lines and termination equipment in the end office at the edge of the transport core.

Account: An identifiable entity that may own resources on a computer.

ACK: See Acknowledgment.

ACK Bit: The bit in a TCP segment that is set to indicate if the segment contains an acknowledgment.

Acknowledgment (ACK): 1) An acknowledgment message, sent by the receiver when a message is received correctly. 2) An acknowledgment frame, sent by the receiver whenever a frame is received; used in CSMA/CA+ACK in 802.11.

Acknowledgment Bit: A bit in a TCP header. If the bit is set, then the TCP segment contains an acknowledgment.

Acknowledgment Number Field: In TCP, a header field that tells what TCP segment is being acknowledged in a segment.

ACL: See Access Control List.

Active RFID: An RFID tag that has batteries and can be read tens of feet away.

ADC: See Analog-to-Digital Conversion.

Address Resolution: A process for determining a host's data link layer address if you know its IP address.

Address Resolution Protocol (ARP): Protocol for address resolution used in Ethernet networks. If a host or router knows a target host's or router's IP address, ARP finds the target's data link layer address.

Administration: In OAMP, the handling of details such as paying bills and managing external contracts.

Administrative IP Server: A server needed to support IP.

Administrator: A super account on a Windows server that automatically has full permissions in every directory on the server.

ADSL: See Asymmetric Digital Subscriber Line.

Advanced Encryption Standard (AES): New symmetric encryption standard that offers 128-bit, 192-bit, or 256-bit encryption efficiently.

Advanced Research Projects Agency (ARPA): An agency within the U.S. Department of Defense that funded the creation of the ARPANET and the Internet.

AES: See Advanced Encryption Standard.

AES-CCMP: AES/Counter Mode with Cipher Block Chaining. The version of AES used in the 802.11i security standard for wireless LANs.

Agent: See Network Management Agent.

Aggregate Throughput: Throughput shared by multiple users; individual users will get a fraction of this throughput.

Alternative Route: In mesh topology, one of several possible routes from one end of the network to the other, made possible by the topology's many connections among switches or routers.

Always On: Being always available for service; used to describe access lines.

Always-On Connection: Connection the user can use immediately, 24 hours a day, every day.

Amplitude: The maximum (or minimum) intensity of a wave. In sound, this corresponds to volume (loudness).

Amplitude Modulation: A simple form of modulation in which a modem transmits one of two analog signals—a high-amplitude (loud) signal or a low-amplitude (soft) signal.

Analog Signal: A signal that rises and falls smoothly in intensity, analogously to the way the voice input signal rises and falls, and that does not have a limited number of states.

Analog-to-Digital Conversion (ADC): A device for the conversion of transmissions from the analog local loop to signals on the digital telephone network's core.

Antivirus Program: Program to remove malware from arriving messages and from the computer's disk drive.

Antivirus Software: Software that scans computers to protect them against viruses, worms, and Trojan horses arriving in e-mail attachments and other propagation methods.

AppleTalk: Apple's proprietary architecture for use on Macintosh computers.

Application Architecture: The arrangement of how application layer functions are spread among computers to deliver service to users.

Application Characterization Environment: See OPNET Application Characterization Environment.

Application Firewall: A firewall that examines the application layer content of packets.

Application Layer: The standards layer that governs how two applications communicate with each other; Layer 7 in OSI, Layer 5 in the hybrid TCP/IP–OSI architecture.

Application Profile: A method, offered by Bluetooth, that allows devices to work with one another automatically at the application layer.

Application Program: Program that does work for users; operating system is the other major type of program found on computers.

Application Server: A server used by large e-commerce sites that accepts user data from a front-end webserver, assembles information from other servers, and creates a webpage to send back to the user.

Architecture: A broad plan that specifies what is needed in general and the components that will be used to provide that functionality. Applied to standards, networks, and applications.

ARP Cache: Section of memory that stores known pairs of IP addresses and single-network standards.

ARPA: See Advanced Research Projects Agency.

ARPANET: A packet-switched network created by the Advanced Research Projects Agency.

ASCII Code: A code for representing letters, numbers, and punctuation characters in 7-bit binary format.

Asymmetric Digital Subscriber Line (ADSL): The type of DSL designed to go into residential homes, offers high downstream speeds but limited upstream speeds.

Asynchronous Transfer Mode (ATM): The packet-switched network technology, specifically designed to carry voice, used for transmission in the PSTN transport core. ATM offers quality of service guarantees for throughput, latency, and jitter.

AT&T: U.S. telecommunications carrier.

ATM: See Asynchronous Transfer Mode.

Attenuate: For a signal's strength to weaken during propagation.

Auditing: collecting data about events to examine actions after the fact.

AUP: See Acceptable Use Policy.

Authentication: The requirement that someone who requests to use a resource must prove his or her identity.

Authentication Server: A server that stores data to help the verifier check the credentials of the applicant.

Authenticator: In 802.1X authentication, the device to which the supplicant connects—a workgroup switch or a wireless access point.

Authorization: Permitting a person or program to take certain actions on a resource.

Authorizations: Specific actions that a person or program can take on a resource.

Autonomous System: Internet owned by an organization.

Availability: The ability of a network to serve its users.

Backdoor: A way back into a compromised computer that an attacker leaves open; it may simply be a new account or a special program.

Back-Office: Transaction processing applications for a business's internal needs.

Backup: Copying files stored on a computer to another medium for protection of the files.

Backup Link: A transmission link that is not normally used but is used when another line failed.

Backward-Compatible: Able to work with all earlier versions of a standard or technology.

Bandpass Filter: A device that filters out all signals below 300 Hz and above about 3.4 kHz.

Bandwidth: The range of frequencies over which a signal is spread.

Bank Settlement Firm: An e-commerce service that handles credit card payments.

Baseband: Transmission in which the signal is simply injected into a wire.

Baseband Signal: 1) The original signal in a radio transmission. 2) A signal that is injected directly into a wire for propagation.

Base Unit: In the metric system, the basic unit being measured: bits per second, hertz, meters, and so forth.

BER: See bit error rate.

Best-Match Row: In routers, the row that provides the best forwarding option for a particular incoming packet.

BGP: See Border Gateway Protocol.

Binary Data: Data that has only two possible values (1s and 0s).

Binary Numbers: The Base two counting system where 1s and 0s used in combination can represent whole numbers (integers).

Binary Signaling: Digital signaling that uses only two states.

Binary Transmission: Transmission in which there are exactly two states—one state representing a one and the other state representing a zero.

Biometrics: The use of bodily measurements to identify an applicant.

Bit: A single 1 or 0.

Bit Error Rate: The percentage of all transmitted bits that contain errors.

Bit Rate: In digital data transmission, the rate at which information is transmitted; measured in bits per second.

Bits per Second (bps): The measure of network transmission speed.

BitTorrent: A newer P2P file-sharing protocol where, instead of downloading a complete file from a single peer, you download different parts of the file you want from multiple peers.

Black List: A list of banned websites.

Blended Threat: An attack that propagates both as a virus and as a worm.

Bluetooth: A wireless networking standard created for personal area networks.

Bluetooth Profile: An application-layer standard designed to allow devices to work together automatically, with little or no user intervention.

Bonding: See Link Aggregation.

Border Firewall: A firewall that sits at the border between a firm and the outside world.

Border Gateway Protocol (BGP): The most common exterior routing protocol on the Internet. Recall that gateway is an old term for router.

Border Router: A router that sits at the edge of a site to connect the site to the outside world through leased lines, PSDNs, and VPNs.

Bot: A type of malware that can be upgraded remotely by an attacker to fix errors or to give the malware additional functionality.

Bps (bps): See Bits per Second.

Breach: A successful attack.

Broadband: 1) Transmission where signals are sent in wide radio channels. 2) Any high-speed transmission system.

Broadband Wireless Access (BWA): High-speed local wireless transmission systems.

Broadcast: To send a message out to all other stations simultaneously.

Broadcast Address: In Ethernet, FF-FF-FF-FF-FF-FF (48 ones); tells switches that the frame should be broadcast.

Brute-Force Attack: A password-cracking attack in which an attacker tries to break a password by trying all possible combinations of characters.

Bursty: Having short, high-speed bursts separated by long silences. Characteristic of data transmission.

Bus Topology: A topology in which one station transmits and has its signals broadcast to all stations.

Business Continuity: A company's ability to continue operations.

Business Continuity Recovery: The reestablishment of a company's ability to continue operations.

C7: Telephone supervisory control signaling system used in Europe.

CA: 1) See Certificate Authority. 2) See Collision Avoidance.

Cable Modem: 1) Broadband data transmission service using cable television. 2) The modem used in this service.

Cable Modem Service: Asymmetrical cable data service offered by a cable television company.

Call Waiting: A service that allows the user to place an original caller on hold if someone else calls the user, shift briefly to the new caller, and then switch back to the original caller.

Caller ID: Service wherein the telephone number of the party calling you is displayed on your phone's small display screen before you pick up the handset; allows the user to screen calls.

Carder: Someone who steals credit card numbers.

Career Criminal: Person who earns money primarily by crime.

Carrier: A transmission service company that has government rights of way.

Carrier Sense Multiple Access with Collision Avoidance and Acknowledgments (CSMA/CA+ACK): A mandatory mechanism used to reduce problems with multiple simultaneous transmissions, which occur in wireless transmission. CSMA/CA+ACK is a media access control discipline, and it uses both collision avoidance and acknowledgment frames.

Cat: A short form for "category" in UTP.

Category: In UTP cabling, a system for measuring wiring quality.

Category (Cat) 5e: Quality type of UTP wiring; required for 100Base-TX and gigabit Ethernet.

Category 6: The newest quality type of UTP wiring being sold; not required for even gigabit Ethernet. Can carry 10 Gbps Ethernet up to 55 feet.

Category 6A: Augmented Category 6 wiring that can carry 10 Gbps Ethernet up to 100 meters.

CDMA: See Code Division Multiple Access.

Cellphone: A cellular telephone, also called a mobile phone or mobile.

Cellsite: In cellular telephony, equipment at a site near the middle of each cell, containing a transceiver and supervising each cellphone's operation.

Cellular Modem: A modem that allows a computer to communicate through a cellular telephone.

Cellular Telephone Service: Radio telephone service in which each subscriber in each section of a region is served by a separate cellsite.

Certificate Authority (CA): Organization that provides public key–private key pairs and digital certificates.

Challenge Message: In challenge–response authentication protocols, the message initially sent from the verifier to the applicant.

Challenge–Response Authentication: Initial authentication method in which the verifier sends the applicant a challenge message, and the applicant does a calculation to produce a response, which it sends back to the verifier.

Challenge–Response Authentication Protocol (CHAP): A specific challenge–response authentication protocol.

Channel: A small frequency range that is a subdivision of a service band.

Channel Bandwidth: The range of frequencies in a channel; determined by subtracting the lowest frequency from the highest frequency.

Channel Reuse: The ability to use each channel multiple times, in different cells in the network.

Channel Service Unit (CSU): The part of a CSU/DSU device designed to protect the telephone network from improper voltages sent into a private line.

CHAP: See Challenge–Response Authentication Protocol.

CIDR: See Classless InterDomain Routing.

Cipher: An encryption method.

Ciphertext: The result of encrypting a plaintext message. Ciphertext can be transmitted with confidentiality.

Circuit: A transmission link on which capacity is reserved for each conversation.

Circuit Switching: Switching in which capacity for transmission is reserved on every switch and trunk line end-to-end between the two subscribers.

Cladding: A thick glass cylinder that surrounds the core in optical fiber.

Class A IP Address: In classful addressing, an IP address block with more than 16 million IP addresses; given only to the largest firms and ISPs.

Class B IP Address: In classful addressing, an IP address block with about 65,000 IP addresses; given to large firms.

Class C IP Address: In classful addressing, an IP address block with 254 possible IP addresses; given to small firms.

Class D IP Address: In classful addressing, IP addresses used in multicasting.

Classful Addressing: Giving a firm one of four block sizes for IP addresses: a very large Class A address block, a medium-sized Class B address block, or a small Class C address block.

Classless InterDomain Routing (CIDR): System for allocating IP addresses that does not use IP address classes.

Clear Line of Sight: An obstructed radio path between the sender and the receiver.

Clear to Send (CTS): In 802.11, a message broadcast by an access point, which allows only a station that has sent a Request to Send message to transmit. All other stations must wait.

CLEC: See Competitive Local Exchange Carrier.

CLI: See Command Line Interface.

Client Host: In client/server processing, a server program on a server host provides services to a client program on a client host.

Client PC: A personal computer that acts as a client.

Client/Server Application: Application in which a client program requests service from a server and in which the server program provides the service.

Client/Server Processing: The form of client/server computing in which the work is done by programs on two machines.

Clock Cycle: A period of time during which a transmission line's state is held constant.

Cloud: The symbol traditionally used to represent the PSDN transport core, reflecting the fact that although the PSDN has internal switches and trunk lines, the customer does not have to know how things work inside the cloud.

Cloud Computing: A form of computing where the company outsources some combination of servers, server applications, and client PC software to a cloud service provider.

Cloud Service Provider: A company that provides cloud computing services.

Cloud Software as a Service: Software as a Service where the user accesses the software over the Internet, often (although not always) with a browser.

Cloud Utility Computing: Utility computing done as a cloud computing model.

Coaxial Cable: Copper transmission medium in which there is a central wire and a coaxial metal tube as the second connector.

Code Division Multiple Access (CDMA): A new form of cellular technology and a form of spread spectrum transmission that allows multiple stations to transmit at the same time in the same channel; also permits stations in adjacent cells to use the same channel without serious interference.

Codec: The device in the end office switch that converts between the analog local loop voice signals and the digital signals of the end office switch.

Collision: When two simultaneous signals use the same shared transmission medium, the signals will add together and become scrambled (unintelligible).

Collision Avoidance (CA): In 802.11, used with CSMA to listen for transmissions, so if a wireless NIC detects a transmission or a very recent transmission, it must not transmit. This avoids collision.

Command Line Interface (CLI): An interface used to work with switches and routers, in which the user types highly structured commands, ending each command with Enter.

Communication Satellite: Satellite that provides radio communication service.

Community Name: In SNMP Version 1, only devices using the same community name will communicate with each other; very weak security.

Competitive Local Exchange Carrier (CLEC): A competitor to the ILEC.

Complex Topology: A topology wherein different basic topologies are used in different parts of the network.

Comprehensive Security: Security in which all avenues of attack are closed off.

Compression: Reducing the number of bits needed to be transmitted when the traffic has redundancy that can be removed.

Compromise: A successful attack.

Computer Security Incident Response Team (CSIRT): A team convened to handle major security incidents, made up of the firm's security staff, members of the IT staff, and members of functional departments, including the firm's legal department.

Confidentiality: Assurance that interceptors cannot read transmissions.

Connectionless: Type of conversation that does not use explicit openings and closings.

Connection-Oriented: Type of conversation in which there is a formal opening of the interactions, a formal closing, and maintenance of the conversation in between.

Connectorize: To add connectors to a transmission cord.

Constellation: In quadrature amplitude modulation, the collection of all possible amplitude/phase combinations.

Continuity Testers: UTP tester that ensures that wires are inserted into RJ-45 connectors in the correct order and are making good contact.

Convergence: The correction of routing tables after a change in an internet.

Cookie: Small text file stored by a website on a client PC; can later be read from the website that wrote it.

Cord: A length of transmission medium—usually UTP or optical fiber but sometimes coaxial cable.

Core: 1) In optical fiber, the very thin tube into which a transmitter injects light. 2) In a switched network, the collection of all core switches.

Core Security Standard: In 802.11, a standard that provides protection between the wireless access point and the wireless host.

Core Switch: A switch further up the hierarchy that carries traffic between pairs of switches. May also connect switches to routers.

Crack: To guess a password.

Credentials: Proof of identity that a supplicant can present during authentication.

Criminal Attacker: An attacker who attacks with criminal motivation.

CRL: See Certificate Revocation List.

Cross-Connect Device: The device within a wiring closet that vertical cords plug into. Cross-connect devices connect the wires from the riser space to 4-pair UTP cords that span out to the wall jacks on each floor.

Crossover Cable: A UTP cord that allows an NIC in one computer to be connected directly to the NIC in another computer; switches Pins 1 and 2 with Pins 3 and 6.

Crosstalk Interference: Mutual EMI among wire pairs in a UTP cord.

Cryptographic System: A security system that automatically provides a mix of security protections, usually including confidentiality, authentication, message integrity, and replay protection.

Cryptography: Mathematical methods for protecting communication.

CSIRT: See Computer Security Incident Response Team.

CSMA/CA+ACK: See Carrier Sense Multiple Access with Collision Avoidance and Acknowledgments. See definitions of the individual components.

CTS: See Clear to Send.

Customer Premises Equipment (CPE): Equipment owned by the customer, including PBXs, internal vertical and horizontal wiring, and telephone handsets.

Cyberterror: A computer attack made by terrorists.

Cyberwar: A computer attack made by a national government.

DAC: See Digital-to-Analog Conversion.

Data Communications: The transmission of encoded information, as opposed to the type of information carried in telecommunications systems.

Data Encryption Standard (DES): Popular symmetric key encryption method; with only 56-bit keys, considered to be too weak for business-to-business encryption.

Data Field: A field containing the content delivered in a message.

Data Link: The path that a frame takes across a single network (LAN or WAN).

Data Link Control Identifier (DLCI): The virtual circuit number in Frame Relay, normally 10 bits long.

Data Link Layer: The layer that governs transmission within a single network all the way from the source station to the destination station across zero or more switches; Layer 2 in OSI. Governs switch or access point operation and the syntax of frames.

Data Mining: A type of spyware that searches through files on a hard drive for potentially useful information and sends this information to the attacker.

DDoS: See distributed denial of service attack.

Dead Spot: See Shadow Zone.

Decapsulation: The removing of a message from the data field of another message.

Decrypt: Conversion of encrypted ciphertext into the original plaintext so an authorized receiver can read an encrypted message.

Default Router: The next-hop router that a router will forward a packet to if the routing table does not have a row that governs the packet's IP address except for the default row.

Default Row: The row of a routing table that will be selected automatically if no other row matches; its value is 0.0.0.0.

Defense in Depth: The use of successive lines of defense.

Demilitarized Zone (DMZ): A subnet in which webservers and other public servers are placed.

Demodulate: To convert digital transmission signals to analog signals.

Denial-of-Service (DoS): The type of attack whose goal is to make a computer or a network unavailable to its users.

Deregulation: Taking away monopoly protections from carriers to encourage competition.

DES: See Data Encryption Standard.

Designated Router: In OSPF, the router that is responsible for maintaining a complete table about links between routers in its area.

Destination: In a routing table, the column that shows the destination network's network part or subnet's network part plus subnet part, followed by zeroes. This row represents a route to this network or subnet.

DHCP: See Dynamic Host Configuration Protocol.

Dial-Up Circuit: A circuit that only exists for the duration of a telephone call.

Dictionary Attack: A password-cracking attack in which an attacker tries to break a password by trying all words in a standard or customized dictionary.

Dictionary Word: A common word, dangerous to use for a password because easily cracked.

Diff-Serv: The field in an IP packet that can be used to label IP packets for priority and other service parameters.

Digital Certificate: A document that gives the name of a true party, that true party's public key, and other information; used in authentication.

Digital Certificate Authentication: Authentication in which each user has a public key and a private key. Authentication depends on the applicant knowing the true party's private key; requires a digital certificate to give the true party's public key.

Digital Signaling: Signaling that uses a few states. Binary (two-state) transmission is a special case of digital transmission.

Digital Signature: A calculation added to a plaintext message to authenticate it.

Digital Subscriber Line (DSL): A technology that provides digital data signaling over the residential customer's existing single-pair UTP voice-grade copper access line.

Digital-to-Analog Conversion (DAC): The conversion of transmissions from the digital telephone network's core to signals on the analog local loop.

Direct Distance Dialing: Long distance calls made at the standard long-distance rate.

Directly Propagating: A type of worm that tries to jump from the infected computer to many other computers without human intervention.

Directory Search: In Skype, the process where a Skype application looks up the username and IP address of the party it wants to contact.

Disaster Recovery: The reestablishment of information technology operations.

Disgruntled Employee or Ex-Employee: Employee who is upset with the firm or an employee and who may take revenge through a computer attack.

Dish Antenna: An antenna that points in a particular direction, allowing it to send stronger outgoing signals in that direction for the same power and to receive weaker incoming signals from that direction.

Distance Vector Routing Protocol: Routing protocol based on the number of hops to a destination out a particular port.

Distort: To change in shape during propagation.

Distributed Denial-of-Service (DDoS): DOS attack in which the victim is attacked by many computers.

DLCI: See Data Link Control Identifier.

DNS: See Domain Name System.

Domain: 1) In DNS, a group of resources (routers, single networks, and hosts) under the control of an organization. 2) In Microsoft Windows, a grouping of resources used in an organization, made up of clients and servers.

Domain Controller: In Microsoft Windows, a computer that manages the computers in a domain.

Domain Name System (DNS): A server that provides IP addresses for users who know only a target host's host name. DNS servers also provide a hierarchical system for naming domains.

Domestic: Telephone service within a country.

DoS: See Denial-of-Service.

Dotted Decimal Notation: The notation used to ease human comprehension and memory in reading IP addresses.

Downlink: Downward transmission path for a communications satellite.

Downloader: Malware that downloads and installs another program on the computer.

Drive-By Hacker: A hacker who parks outside a firm's premises and eavesdrops on its data transmissions; mounts denial-of-service attacks; inserts viruses, worms, and spam into a network; or does other mischief.

Drop Cable: A thin coaxial cable access line that runs from the cable television company line in a neighborhood to individual homes.

DSL: See Digital Subscriber Line.

DSL Access Multiplexer (DSLAM): A device at the end office of the telephone company that sends voice signals over the ordinary PSTN and sends data over a data network such as an ATM network.

DSLAM: See DSL Access Multiplexer.

Dumb Access Point: Access point that cannot be managed remotely without the use of a wireless LAN switch.

Dumb Terminal: A desktop machine with a keyboard and display but little processing capability; processing is done on a host computer.

Dynamic Host Configuration Protocol (DHCP): The protocol used by DHCP servers, which provide each user PC with a temporary IP address to use each time he or she connects to the Internet.

Dynamic IP Address: A temporary IP address that a client PC receives from a DHCP server.

EAP: See Extensible Authentication Protocol.

Egress Filtering: The filtering of traffic from inside a site going out.

EIGRP: See Enhanced Interior Gateway Routing Protocol.

Electromagnetic Interference (EMI): Unwanted electrical energy coming from external devices, such as electrical motors, fluorescent lights, and even nearby data transmission wires.

Electromagnetic Signal: A signal generated by oscillating electrons.

Electronic Signature: A bit string added to a message to provide message-by-message authentication and message integrity.

Elliptic Curve Cryptosystem (ECC): Public key encryption method; more efficient than RSA.

EMI: See Electromagnetic Interference.

Encapsulation: The placing of a message in the data field of another message.

Encoding: The conversation of messages into bits.

Encrypt: To mathematically process a message so that an interceptor cannot read the message.

End Office: Telephone company switch that connects to the customer premises via the local loop.

End Office Switch: The nearest switch of the telephone company to the customer premises.

End-to-End: A layer where communication is governed directly between the transport process on the source host and the transport process on the destination host.

Enhanced Interior Gateway Routing Protocol (EIGRP): Interior routing protocol used by Cisco routers.

Enterprise Mode: In WPA and 802.11i, operating mode that uses 802.1X.

Ephemeral Port Number: The temporary number a client selects whenever it connects to an application program on a server. According to IETF rules, ephemeral port numbers should be between 49153 and 65535.

Error Advisement: In ICMP, the process wherein if an error is found, there is no transmission, but the router or host that found the error usually sends an ICMP error message to the source device to inform it that an error has occurred. It is then up to the device to decide what to do. (This is not the same as error correction because there is no mechanism for the retransmission of lost or damaged packets.)

Error Rate: In biometrics, the normal rate of misidentification when the subject is cooperating.

Ethernet Address: The 48-bit address the stations have on an Ethernet network; often written in hexadecimal notation for human reading.

Ethernet Frame: A message at the data link layer in an Ethernet network.

Ethernet Switch: Switch following the Ethernet standard. Notable for speed and low cost per frame sent. Dominates LAN switching.

Evil Twin Access Point: Attacker access point outside a building that attracts clients inside the building to associate with it.

Exhaustive Search: Cracking a key or password by trying all possible keys or passwords.

Exploit: A break-in program or attack method; a program that exploits known vulnerabilities.

Exploitation Software: Software that is planted on a computer; it continues to exploit the computer.

Extended Star Topology: The type of topology wherein there are multiple layers of switches organized in a hierarchy, in which each node has only one parent node; used in Ethernet; more commonly called a hierarchical topology.

Extensible Authentication Protocol (EAP): A protocol that authenticates users with authentication data (such as a password or a response to a challenge based on a station's digital certificate) and authentication servers.

Exterior Routing Protocol: Routing protocol used between autonomous systems.

Face Recognition: The scanning of passersby to identify terrorists or wanted criminals by the characteristics of their faces.

Facial Scanning: See Face Recognition.

Facilitating Server: A server that solves certain problems in P2P interactions but that allows clients to engage in P2P communication for most of the work.

False Alarm: An apparent incident that proves not to be an attack.

False Positive: A false alarm.

FDM: See Frame Division Multiplexing.

Fiber to the Home (FTTH): Optical fiber brought by carriers to individual homes and businesses.

Field: A subdivision of a message header or trailer.

File Server: A server that allows users to store and share files.

File Server Program Access: The form of client/server computing in which the server's only role is to store programs and data files, while the client PC does the actual processing of programs and data files.

File Sharing: The ability of computer users to share files that reside on their own disk drives or on a dedicated file server.

Filtering Method: A method for examining the content of arriving packets to decide what to do with them.

Fin Bit: One-bit field in a TCP header; indicates that the sender wishes to open a TCP connection.

Fingerprint Scanning: A form of biometric authentication that uses the applicant's fingerprints.

Firewall: A security system that examines each packet passing through it. If the firewall identifies the packet as an attack packet, the firewall discards it and copies information about the discarded packet into a log file.

Firewall Log File: A file that contains summary information about packets dropped by a firewall.

First-Generation (1G): The initial generation of cellular telephony, introduced in the 1980s. 1G systems were analog, were only given about 50 MHz of spectrum, had large and few cells, and had very limited speeds for data transmission.

Fixed Wireless Service: Local terrestrial wireless service in which the user is at a fixed location.

Flag Field: A one-bit field.

Flat Rate: Local telephone service in which there is a fixed monthly service charge but no separate fee for individual local calls.

Flow Control: The ability of one side in a conversation to tell the other side to slow or stop its transmission rate.

Footprint: Area of coverage of a communication satellite's signal.

Forensics: The collection of data in a form suitable for presentation in a legal proceeding.

Forwarding Decision: Decision made by a switch to forward an incoming frame back out through another port to get it closer to the destination host.

Four-Step Close: A normal TCP connection close; requires four messages.

Fourth-Generation (4G): The emerging generation of cellular telephony, which will provide a download speed of 1 Gbps or even more to stationary hosts and 100 Mbps or more to mobile hosts.

Fractional T1: A type of private line that offers intermediate speeds at intermediate prices; usually operates at one of the following speeds: 128 kbps, 256 kbps, 384 kbps, 512 kbps, or 768 kbps.

Fragment (Fragmentation): To break a message into multiple smaller messages. TCP fragments application layer messages, while IP packets may be fragmented by routers along the packet's route.

Frame: 1) A message at the data link layer. 2) In time division multiplexing, a brief time period, which is further subdivided into slots.

Frame Check Sequence Field: A four-octet field used in error checking in Ethernet. If an error is found, the frame is discarded.

Frame Relay: A popular Public Switched Data Network that operates at speeds of about 256 kbps to 40 Mbps.

Fraud: Lying to get victims to do something against their financial self-interest.

Frequency: The number of complete cycles a radio wave goes through per second. In sound, frequency corresponds to pitch.

Frequency Division Multiplexing (FDM): A technology used in microwave transmission in which the microwave bandwidth is subdivided into channels, each carrying a single circuit.

Frequency Modulation: Modulation in which one frequency is chosen to represent a 1 and another frequency is chosen to represent a 0.

Frequency Spectrum: The range of all possible frequencies from zero hertz to infinity.

FTTH: See Fiber to the Home.

Full-Duplex Communication: A type of communication that supports simultaneous two-way transmission. Almost all communication systems today are full-duplex systems.

Full-Mesh Topology: Topology in which each node is connected to each other node.

Gateway: An obsolete term for "router"; still in use by Microsoft.

Gbps: Gigabit per second.

GEO: See Geosynchronous Earth Orbit Satellite.

Geosynchronous Earth Orbit Satellite (GEO): The type of satellite most commonly used in fixed wireless access today; orbits the earth at about 36,000 km (22,300 miles).

Get: An SNMP command sent by the manager that tells the agent to retrieve certain information and return this information to the manager.

GHz: See Gigahertz.

Gigabit Ethernet: 1 Gbps version of Ethernet.

Gigabit per second: One billion bits per second.

Gigahertz (GHz): One billion hertz.

Global System for Mobile communication (GSM): The cellular telephone technology on which nearly the entire world standardized for 2G service. GSM uses 200 kHz channels and implements TDM.

Golden Zone: The portion of the frequency spectrum from the high megahertz range to the low gigahertz range, wherein commercial mobile services operate.

GPO: See Group Policy Object.

Graded-Index Multimode Fiber: Multimode fiber in which the index of refraction varies from the center of the core to the cladding boundary.

Grid Computing: Computing in which all devices, whether clients or servers, share their processing resources.

Group Policy Object (GPO): A policy that governs a specific type of resource on a domain.

GSM: See Global System for Mobile communication.

Guideline: A directive that should be followed but that need not be followed, depending on the context.

H.323: In IP telephony, one of the protocols used by signalling gateways.

Hacker: Someone who intentionally uses a computer resource without authorization or in excess of authorization.

Hacker Toolkit: A collection of tools that automate some tasks that the hacker will have to perform after the break-in.

Hacking: The intentional use of a computer resource without authorization or in excess of authorization.

Half-Duplex: The mode of operation wherein two communicating NICs must take turns transmitting.

Handoff: a) In wireless LANs, a change in access points when a user moves to another location. b) In cellular telephony, transfer from one cellsite to another, which occurs when a subscriber moves from one cell to another within a system.

Hash: The output from hashing.

Hashing: A mathematical process that, when applied to a bit string of any length, produces a value of a fixed length, called the hash.

HDSL: See High-Rate Digital Subscriber Line.

HDSL2: A newer version of HDSL that transmits in both directions at 1.544 Mbps.

Head End: The cable television operator's central distribution point.

Header: The part of a message that comes before the data field.

Header Checksum: The UDP datagram field that allows the receiver to check for errors.

Hertz (Hz): One cycle per second, a measure of frequency.

Hex Notation: See Hexadecimal Notation.

Hexadecimal (Hex) Notation: The Base 16 notation that humans use to represent address 48-bit MAC source and destination addresses.

Hierarchical Topology: A network topology in which all switches are arranged in a hierarchy, in which each switch has only one parent switch above it (the root switch, however, has no parent); used in Ethernet.

Hierarchy: 1) The type of topology wherein there are multiple layers of switches organized in a hierarchy, in which each node has only one parent node; used in Ethernet. 2) In IP addresses, three multiple parts that represent successively more specific locations for a host.

High-Rate Digital Subscriber Line (HDSL): The most popular business DSL, which offers symmetric transmission at 768 kbps in both directions. See also HDSL2.

Hop-by-Hop: A layer in which communication is governed by each individual switch or router along the path of a message.

Host: Any computer attached to a network.

Host Computer: 1) In terminal–host computing, the host that provides the processing power. 2) On an internet, any host.

Host Name: An unofficial designation for a host computer.

Host Node: In Skype, an application that runs on a user's computer.

Host Part: The part of an IP address that identifies a particular host on a subnet.

Hot Spot: A public location where anyone can connect to an access point for Internet access.

HTML: See Hypertext Markup Language.

HTML Body: Body part in a Hypertext Markup Language message.

HTTP: See Hypertext Transfer Protocol.

HTTP Request Message: In HTTP, a message in which a client requests a file or another service from a server.

HTTP Request–Response Cycle: An HTTP client request followed by an HTTP server response.

HTTP Response Message: In HTTP, a message in which a server responds to a client request; either contains a requested file or an error message explaining why the requested file could not be supplied.

Hub-and-Spoke Topology: A topology in which all communication goes through one site.

Hybrid Mode: In password cracking, a mode that tries variations on common word passwords.

Hybrid TCP/IP-OSI Standards Architecture: The architecture that uses OSI standards at the physical and data link layers and TCP/IP standards at the internet, transport, and application layers; dominant in corporations today.

Hypertext Markup Language (HTML): The language used to create webpages.

Hypertext Transfer Protocol (HTTP): The protocol that governs interactions between the browser and webserver application program.

Hypervisor: In virtualization, a layer of software that manages the use of physical computer resources by the virtual machines and communication between each virtual machine and its users.

Hz: See Hertz.

ICC: See International Common Carrier.

ICMP: See Internet Control Message Protocol.

ICMP Echo: A message sent by a host or router to another host or router. If the target device's internet process is able to do so, it will send back an echo response message.

ICMP Error Message: A message sent in error advisement to inform a source device that an error has occurred.

Identification Field: In IPv4, header field used to reassemble fragmented packets. Each transmitted packet is given a unique identification field value. If the packet is fragmented en route, all fragments are given the initial packet's identification field value.

Identity Theft: Stealing enough information about a person to impersonate him or her in large financial transactions.

IEEE: See Institute for Electrical and Electronics Engineers.

IETF: See Internet Engineering Task Force.

ILEC: See Incumbent Local Exchange Carrier.

Image Backup: In backup, an exact copy of an entire disk drive.

IMAP: See Internet Message Access Protocol.

Implementation Guidance: Instructions that are more specific than policies but less specific than implementation.

Impostor: Someone who claims to be someone else.

IMPS: See Interface Message Processor.

Incident: A successful attack.

Incident Severity: The degree of destruction inflicted by an attack.

Incumbent Local Exchange Carrier (ILEC): The traditional monopoly telephone company within each LATA.

Index Website: In BitTorrent, a website that contains .torrent files giving information about specific files and where they are stored.

Individual Throughput: The actual speed a single user receives (usually much lower than aggregate throughput in a system with shared transmission speed).

Ingress Filtering: The filtering of traffic coming into a site from the outside.

Initial Sequence Number (ISN): The sequence number placed in the first TCP segment a side transmits in a session; selected randomly.

Initial Site Survey: For 802.11 LANs, an electromagnetic field strength survey performed at the beginning of an installation.

Institute for Electrical and Electronics Engineers (IEEE): An international organization whose 802 LAN/MAN Standards Committee creates many LAN standards.

Interexchange Carrier (IXC): A telephone carrier that transmits voice traffic between LATAs.

Interface: 1) The router's equivalent of a network interface card; a port on a router that must be designed for the network to which it connects. 2) In Web services, the outlet through which an object communicates with the outside world.

Interface Message Processor (IMP): A minicomputer that handled most of the network chores on the ARPANET.

Interference: See Electromagnetic Interference.

Interior Routing Protocol: Routing protocol used within a firm's internet.

Internal Router: A router that connects different LANs within a site.

International Common Carrier (ICC): A telephone carrier that provides international service.

International Organization for Standardization (ISO): A strong standards agency for manufacturing, including computer manufacturing.

International Telecommunications Union-Telecommunications Standards Sector (ITU-T): A standards agency that is part of the United Nations and that oversees international telecommunications.

Internet: 1) A group of networks connected by routers so that any application on any host on any network can communicate with any application on any other host on any other network. 2) A general term for any internetwork (spelled with a lowercase i). 3) the worldwide Internet (spelled with a capital I).

Internet Backbone: The collection of all Internet Service Providers that provide Internet transmission service.

Internet Control Message Protocol (ICMP): The protocol created by the IETF to oversee supervisory messages at the internet layer.

Internet Engineering Task Force (IETF): TCP/IP's standards agency.

Internet Layer: The layer that governs the transmission of a packet across an entire internet.

Internet Message Access Protocol (IMAP): One of the two protocols used to download received e-mail from an e-mail server; offers more features but is less popular than POP.

Internet Network: A network on the Internet owned by a single organization, such as a corporation, university, or ISP.

Internet Options: In Microsoft Windows, way of setting security and other settings for Browser communication.

Internet Protocol (IP): The TCP/IP protocol that governs operations at the internet layer. Governs packet delivery from host to host across a series of routers.

Internet Service Provider: Carrier that provides Internet access and transmission.

Interoperate: To work together (for example, products from different vendors).

Interoperability Profiles: Limited subsets of WiMAX functionality with which individual products must comply.

Inverse square law attenuation: Radio signal strength declines with the square of transmission distance.

IP: See Internet Protocol.

IP Address: An Internet Protocol address; the address that every computer needs when it connects to the Internet; IP addresses are 32 bits long.

Ipconfig/all: A command line command to determine the configuration of the user's computer.

IP Security (IPsec): A set of standards that operate at the internet layer and provide security to all upper layer protocols transparently.

IP Version 4 (IPv4): The standard that governs most routers on the Internet and private internets.

IP Version 6 (IPv6): A new version of the Internet Protocol.

IPsec: See IP Security.

IPsec Gateway: Border device at a site that converts between internal data traffic into protected data traffic that travels over an untrusted system such as the Internet.

IPv4: See IP Version 4.

IPv6: See IP Version 6.

IPX/SPX Architecture: Non-TCP/IP standards architecture found at upper layers in LANs; required on all older Novell NetWare file servers.

Iris: The colored part of the eye, used in biometric authentication.

Iris Scanning: A form of biometric authentication that scans the pattern in the colored part of the applicant's eyes.

ISN: See Initial Sequence Number.

ISO: See International Organization for Standardization.

ISO/IEC 11801: European standard for wire and optical fiber media.

ISP: See Internet Service Provider.

IT Disaster Recovery: Recovering from a disaster that damages computer equipment or data.

ITU-T: See International Telecommunications Union-Telecommunications Standards Sector.

IXC: See Interexchange Carrier.

Jacket: The outer plastic covering, made of PVC, that encloses and protects the four pairs of wires in UTP or the core and cladding in optical fiber.

Jitter: Variability in latency.

JPEG: Popular graphics file format.

kbps: Kilobits per second.

Key: A bit string used with an encryption method to encrypt and decrypt a message. Different keys used with a single encryption method will give different ciphertexts from the same plaintext.

Key Exchange or Keying: The secure transfer of a symmetric session key between two communicating parties.

Key-Hashed Message Authentication Code (HMAC): Electronic signature technology that is efficient and inexpensive but lacks nonrepudiation.

Key Management: The management of key creation, distribution, and other operations.

Keystroke Logger: A program that records keystrokes.

Label Header: In MPLS, the header added to packets before the IP header; contains information that aids and speeds routers in choosing which interface to send the packet back out.

Label Number: In MPLS, number in the label header that aids label-switching routers in packet sending.

Label-Switching Router: Router that implements MPLS label switching.

Label-Switching Table: In MPLS, the table used by label-switching routers to decide which interface to use to forward a packet.

LAN: See Local Area Network.

Language Independence: In SOAP, the fact that Web service objects do not have to be written in any particular language.

LATA: See Local Access and Transport Area.

Latency: Delay, usually measured in milliseconds.

Latency-Intolerant: An application whose performance is harmed by even slight latency.

Layer 3: See Internet Layer.

Layer 4: See Transport Layer.

Layer 5: See Application Layer.

Leased Line Circuit: A high-speed point-to-point always-on circuit.

Legacy Decision: Decision that will lock the company into a specific vendor or technology option for several years.

Legacy Network: A network that uses obsolete technology; may have to be lived with for some time because upgrading all legacy networks at one time is too expensive.

Length Field: 1) The field in an Ethernet MAC frame that gives the length of the data field in octets. 2) The field in a UDP datagram that enables the receiving transport process to process the datagram properly.

LEO: See Low Earth Orbit Satellite.

Lightweight Directory Access Protocol: Simple protocol for accessing directory servers.

Line of Sight: An unobstructed path between the sender and receiver, necessary for radio transmission at higher frequencies.

Link Aggregation: The use of two or more trunk links between a pair of switches; also known as trunking or bonding.

Link State Protocol: Routing protocol in which each router knows the state of each link between routers.

LLC: See Logical Link Control.

LLC Header: See Logical Link Control Layer Header.

Local: The value placed in the next-hop routing field of a routing table to specify that the destination host is on the selected network or subnet.

Local Access and Transport Area (LATA): One of the roughly 200 site regions the United States has been divided into for telephone service.

Local Area Network (LAN): A network within a customer's premises.

Local Calling: Telephone calls placed to a nearby caller; less expensive than long-distance calls.

Local Loop: In telephony, the line used by the customer to reach the PSTN's central transport core.

Log File: A file that contains data on events.

Logical Link Control (LLC) Layer: The layer of functionality for the upper part of the data link layer, now largely ignored.

Logical Link Control Layer (LLC) Header: The header at the start of the data field that describes the type of packet contained in the data field.

Logical Link Control Layer Subheader: Group of fields at the beginning of the Ethernet data field. Describes the type of packet contained in the data field.

Long Distance: A telephone call placed to a distance party; more expensive than a local call.

Long-Term Evolution (LTE) Advanced: Currently, all cellular carriers are planning to use a single transmission standard for 4G, at least initially. This is Long-Term Evolution (LTE) Advanced. Now that there is consensus on technology, 4G service is likely to move forward in a timelier manner. LTE is a slower introductory protocol.

Longest Match: The matching row that matches a packet's destination IP address to the greatest number of bits; chosen by a router when there are multiple matches.

Loopback Address: The IP address 127.0.0.1. When a user pings this IP address, this will test their own computer's connection to the Internet.

Loopback Interface: A testing interface on a device. Messages sent to this interface are sent back to the sending device.

Low Earth Orbit Satellite (LEO): A type of satellite used in mobile wireless transmission; orbits a few hundred miles or a few hundred kilometers above the earth.

LTE: See Long-Term Evolution.

MAC: See Media Access Control.

MAC Address: See Media Access Control.

Mainframe Computer: The largest type of dedicated server; extremely reliable.

Major Incidents: Attack too severe for the on-duty staff to handle; the company must convene the firm's computer security incident response team (CSIRT), which is trained to handle major incidents.

Make-Versus-Buy Decision: A decision that a company must make in software development projects, regarding whether the programming staff should create the software itself, or whether the company should purchase the software.

Malware: Software that seeks to cause damage.

Malware-Scanning Program: A program that searches a user's PC looking for installed malware.

Malware Writers: People who create malware. The act of creating malware usually is not a crime.

MAN: See Metropolitan Area Network.

Manageable Switch: A switch that has sufficient intelligence to be managed from a central computer (the Manager).

Managed Device: A device that needs to be administered, such as printers, hubs, switches, routers, application programs, user PCs, and other pieces of hardware and software.

Management Information Base (MIB): A specification that defines what objects can exist on each type of managed device and also the specific characteristics of each object; the actual database stored on a manager in SNMP. There are separate MIBs for different types of managed devices; both a schema and a database.

Management Program: A program that helps network administrators manage their networks.

Manager: The central PC or more powerful computer that uses SNMP to collect information from many managed devices.

Man-in-the-Middle Attack: An attack in which an attacker intercepts messages and then passes them on.

Mask: A 32-bit string beginning with a series of 1s and ending with a series of 0s; used by routing tables to interpret IP address part sizes. The 1s designate either the network part or the network plus software part.

Mask Operations: Applying a mask of 1s and 0s to a bit stream. Where the mask is 1, the original bit stream's bit results. Otherwise, the result is 0.

Maximum Segment Size (MSS): The maximum size of TCP data fields that a receiver will accept.

Maximum Transmission Unit (MTU): The maximum packet size that can be carried by a particular LAN or WAN.

Mbps: Megabits per second.

MD5: A popular hashing method.

Media Access Control (MAC): The process of controlling when stations transmit; also, the lowest part of the data link layer, defining functionality specific to a particular LAN technology.

Media Gateway: A device that connects IP telephone networks to the ordinary public switched telephone network. Media gateways also convert between the signalling formats of the IP telephone system and the PSTN.

Medium Earth Orbit Satellite (MEO): A type of satellite used in mobile wireless transmission; orbits a few thousand miles or a few thousand kilometers above the earth.

Megabits per second: Millions of bits per second.

Megahertz (MHz): One million hertz.

MEO: See Medium Earth Orbit Satellite.

Mesh Networking: A type of networking in which wireless devices route frames without the aid of wired LANs.

Mesh Topology: 1) A topology where there are many connections among switches or routers, so there are many alternative routes for messages to get from one end of the network to the other. 2) In network design, a topology that provides direct connections between every pair of sites.

Message: A discrete communication between hardware or software processes.

Message Digest: The result of hashing a plaintext message. The message digest is signed with the sender's private key to produce the digital signature.

Message Integrity: The assurance that a message has not been changed en route; or if a message has been changed, the receiver can tell that it has.

Message Ordering: Controlling when one device in a pair may transmit.

Message Timing: Controlling when hardware or software processes may transmit.

Message Unit: Local telephone service in which a user is charged based on distance and duration.

Metered Service: A pay-as-you-go pricing model where a customer only pays for the service that he or she actually uses.

Metric: A number describing the desirability of a route represented by a certain row in a routing table.

Metric Prefix: In the metric system, a single letter (usually) that multiplies the base unit. For instance, in Mbps, the M stands for mega (million). So Mbps is one million bits per second. This is multiplied by the number before it. So 73.23 kbps is 73,230 bps without a metric prefix.

Metro Ethernet: See metropolitan area Ethernet.

Metropolitan Area Ethernet: Ethernet operating at the scale of a metropolitan area network.

Metropolitan Area Network (MAN): A WAN that spans a single urban area.

MHz: See Megahertz.

MHz-km: Measure of modal bandwidth, a measure of multimode fiber quality.

MIB: See Management Information Base.

Microwave: Traditional point-to-point radio transmission system.

Microwave Repeater: Transmitter/receiver that extends the distance a microwave link can travel.

Millisecond (ms): The unit in which latency is measured.

MIME: See Multipurpose Internet Mail Extensions.

MIMO: See Multiple Input/Multiple Output.

Minimum Permissions: The principle of giving each person only the permissions he or she needs to have to do his or her job.

Ministry of Telecommunications: A government-created regulatory body that oversees PTTs.

Mobile Code: Code that travels with a downloaded webpage from the webserver to the browser.

Mobile Phone: See Cellphone.

Mobile Telephone Switching Office (MTSO): A control center that connects cellular customers to one another and to wired telephone users, as well as overseeing all cellular calls (determining what to do when people move from one cell to another, including which cellsite should handle a caller when the caller wishes to place a call).

Mobile Wireless Access: Local wireless service in which the user may move to different locations.

Modal Bandwidth: The measure of multimode fiber quality.

Modal Dispersion: The main propagation problem for optical fiber; dispersion in which the difference in the arrival times of various modes (permitted light rays) is too large, causing the light rays of adjacent pulses to overlap in their arrival times and rendering the signal unreadable.

Mode: An angle light rays are permitted to enter an optical fiber core.

Modem: A device that translates between digital computer signals and analog telephone line signals.

Modulate: To convert digital signals to analog signals.

Momentary Traffic Peak: A surplus of traffic that briefly exceeds the network's capacity, happening only occasionally.

More Fragments Flag Field: In IPv4, a flag field that indicates whether there are more fragments (set) or not (not set).

MPLS: See Multiprotocol Label Switching.

Ms: See Millisecond.

MS-CHAP: Microsoft version of the Challenge–Response Authentication Protocol.

MSS: See Maximum Segment Size.

MTSO: See Mobile Telephone Switching Office.

MTU: See Maximum Transmission Unit.

Multicasting: Simultaneously sending messages to multiple stations but not to all stations.

Multicriteria Decision Making: An approach wherein the company decides what product characteristics will be important in making a purchase.

Multimode Fiber: The most common type of fiber in LANs, wherein light rays in a pulse can enter a fairly thick core at multiple angles. Inexpensive but can transmit signals over sufficient distance for LAN usage.

Multipath Interference: Interference caused when a receiver receives two or more signals—a direct signal and one or more reflected signals. The multiple signals may interfere with one another.

Multiple Input/Multiple Output (MIMO): A radio transmission method that sends several signals simultaneously in a single radio channel.

Multiplexing: 1) Having the packets of many conversations share trunk lines; reduces trunk line cost. 2) The ability of a protocol to carry messages from multiple next-higher-layer protocols in a single communication session.

Multiplexes: Mixes together, typically to reduce cost through economies of scale.

Multiprotocol: Characterized by implementing many different protocols and products following different architectures.

Multiprotocol Label Switching (MPLS): A traffic management tool used by many ISPs.

Multiprotocol Router: A router that can handle not only TCP/IP internetworking protocols, but also internetworking protocols for IPX/SPX, SNA, and other standards architectures.

Multipurpose Internet Mail Extensions (MIME): A standard for specifying the contents of files.

Mutual Authentication: Authentication by both parties.

Name Server: Server in the Domain Name System.

Nanometer (nm): The measure used for wavelengths; one billionth of a meter (10^{-9} m).

NAP: See Network Access Point.

Narrowband: 1) A channel with a small bandwidth and, therefore, a low maximum speed. 2) Low-speed transmission.

NAT: See Network Address Translation.

NCP: See Network Control Program.

Netstat: A popular route analysis tool, which gives data on current connections between a computer and other computers.

Network: In IP addressing, an organizational concept—a group of hosts, single networks, and routers owned by a single organization.

Network Access Point (NAP): A site where ISPs interconnect and exchange traffic.

Network Address Translation (NAT): Converting an IP address into another IP address, usually at a border firewall; disguises a host's true IP address from sniffers. Allows more internal addresses to be used than an ISP supplies a firm with external addresses.

Network Control Program (NCP): Software run on an ARPANET host that handles details of host-to-host interactions above the levels of packetization, delivery, and reassembly.

Network Core: The central part of the network.

Network Interface Card (NIC): Printed circuit expansion board for a PC; handles communication with a network; sometimes built into the motherboard.

Network Layer: In OSI, Layer 3; governs internetworking. OSI network layer standards are rarely used.

Network Management Agent (Agent): A piece of software on the managed device that communicates with the manager on behalf of the managed device.

Network Management Program (Manager): A program run by the network administrator on a central computer.

Network Mapping: The act of mapping the layout of a network, including what hosts and routers are active and how various devices are connected. Its two phases are discovering and fingerprinting.

Network Mask: A mask that has 1s in the network part of an IP address and 0s in all other parts.

Network Operations Center (NOC): Central management point for a network.

Network Part: The part of an IP address that identifies the host's network on the Internet.

Network Printer: A printer that connects directly to a network instead of to a computer.

Network Simulation: 1) The building of a model of a network that is used to project how the network will operate after a change. 2) A type of tool that allows network designers to envision the impacts of alternative designs so that they can conduct what-if analyses and select the best design.

Network Standard: A rule of operation that governs the exchange of messages between two hardware or software processes.

Network Standrds Architecture: 1) A broad plan that specifies everything that must be done for two application programs on different networks on an internet to be able to work together effectively. 2) A broad plan for how the firm will connect all of its computers within buildings (LANs), between sites (WANs), and to the Internet; also includes security devices and services.

Network Topology: The order in which a network's nodes are physically connected by transmission lines.

Network Visibility: A type of tool that helps managers comprehend what is going on in their networks.

Networked Application: An application that provides service over a network.

Next Header Field: In IPv6, a header field that describes the header following the current header.

Next-Hop Router: A router to which another router forwards a packet in order to get the packet a step closer to reaching its destination host.

NIC: See Network Interface Card.

Nm (nm): See Nanometer.

Nmap: A network mapping tool that finds active IP addresses and then fingerprints them to determine their operating system and perhaps their operating system version.

NOC: See Network Operations Center.

Node: A client, server, switch, router, or other type of device in a network.

Noise: Random electromagnetic energy within wires; combines with the data signal to make the data signal difficult to read.

Noise Floor: The mean of the noise energy.

Noise Spike: An occasional burst of noise that is much higher or lower than the noise floor; may cause the signal to become unrecognizable.

Normal Attack: An incident that does a small amount of damage and can be handled by the on-duty staff.

Not Set: When a flag's field is given the value 0.

Nslookup (nslookup): A command that allows a PC user to send DNS lookup messages to a DNS server.

OAM&P: Operations, administration, maintenance, and provisioning. The four concerns in ongoing network management.

Object: In SNMP, an aspect of a managed device about which data is kept.

Octet: A collection of eight bits; same as a byte.

OFDM: See Orthogonal Frequency Division Multiplexing.

Official Internet Protocol Standards: Standards deemed official by the IETF.

Omnidirectional Antenna: An antenna that transmits signals in all directions and receives incoming signals equally well from all directions.

On/Off Signaling: Signaling wherein the signal is on for a clock cycle to represent a one, and off for a zero. (On/off signaling is binary.)

One-Pair Voice-Grade UTP: The traditional telephone access lines to individual residences.

Ongoing Costs: Costs beyond initial installation costs; often exceed installation costs.

Open Shortest Path First (OSPF): Complex but highly scalable interior routing protocol.

Operational Life: The part of the systems development life cycle beyond the systems development life cycle.

OPNET IT Guru: A popular network simulation program; focuses primarily on data link layer and internet layer performance.

Optical Carrier (OC): A number that indicates SONET speeds.

Optical Fiber: Cabling that sends signals as light pulses.

Optical Fiber Cord: A length of optical fiber.

Option: One of several possibilities that a user or technologist can select.

Organizational Unit: In directory servers, a subunit of the Organization node.

Orthogonal Frequency Division Multiplexing (OFDM): A form of spread spectrum transmission that divides each broadband channel into subcarriers and then transmits parts of each frame in each subcarrier.

OSI: The Reference Model of Open Systems Interconnection; the 7-layer network standards architecture created by ISO and ITU-T; dominant at the physical and data link layers, which govern transmission within single networks (LANs or WANs).

OSI Application Layer (Layer 7): The layer that governs application-specific matters not covered by the OSI Presentation Layer or the OSI Session Layer.

OSI Layer 5: See OSI Session Layer.

OSI Layer 6: See OSI Presentation Layer.

OSI Layer 7: See OSI Application Layer.

OSI Presentation Layer (Layer 6): The layer designed to handle data formatting differences between two communicating computers.

OSI Session Layer (Layer 5): The layer that initiates and maintains a connection between application programs on different computers.

OSPF: See Open Shortest Path First.

Outsourcing: Paying other firms to handle some, most, or all IT chores.

Overprovision: To install much more capacity in switches and trunk links than will be needed most of the time, so that momentary traffic peaks will not cause problems.

Oversight: A collection of methods to ensure that policies have been implemented properly.

Oversubscription: In Frame Relay, the state of having port speeds less than the sum of PVC speeds.

P2P: See Peer-to-Peer.

Packet: A message at the internet layer.

Packet Capture Program: A program that captures selected packets or all of the packets arriving at or going out of an NIC. Afterward, the user can display key header information for each packet in greater or lesser detail.

Packet Error Rate: The percentage of packets that are lost or damaged during delivery.

Packet Switching: The breaking of conversations into short messages (typically a few hundred bits long); allows multiplexing on trunk lines to reduce trunk line costs.

PAD Field: A field that the sender adds to an Ethernet frame if the data field is less than 46 octets long (the total length of the PAD plus data field must be exactly 46 octets long).

PAN: See Personal Area Network.

Parallel Transmission: A form of transmission that uses multiple wire pairs or other transmission media simultaneously to send a signal; increases transmission speed.

Pass Phrase: A series of words used to generate a key.

Passive RFID: An RFID tag that derives its power from the radio signal sent by the reader.

Password: A secret keyboard string only the account holder should know; authenticates user access to an account.

Password Length: The number of characters in a password.

Password Reset: The act of changing a password to some value known only to the systems administrator and the account owner.

Patch: An addition to a program that will close a security vulnerability in that program.

Patch Cord: A cord that comes precut in a variety of lengths, with a connector attached; usually either UTP or optical fiber.

Payload: 1) A piece of code that can be executed by a virus or worm after it has spread to multiple machines. 2) ATM's name for a data field.

Payment Card Industry–Data Security Standard (PCI-DSS): Security standards for companies that accept credit card payments.

PBX: See Private Branch Exchange.

PCI-DSS: See Payment Card Industry–Data Security Standard.

PCM: Pulse Code Modulation.

PEAP: See Protected Extensible Authentication Protocol.

Peer-to-Peer Architecture (P2P): The application architecture in which most or all of the work is done by cooperating user computers, such as desktop PCs. If servers are present at all, they serve only facilitating roles and do not control the processing.

Perfect Internal Reflection: When light in optical fiber cabling begins to spread, it hits the cladding and is reflected back into the core so that no light escapes.

Permanent IP Address: An IP address given to a server that the server keeps and uses every single time it connects to the Internet. (This is in contrast to client PCs, which receive a new IP address every time they connect to the Internet.)

Permanent Virtual Circuit (PVC): A PSDN connection between corporate sites that is set up once and kept in place for weeks, months, or years at a time.

Permission: A rule that determines what an account owner can do to a particular resource (file or directory).

Personal Area Network (PAN): A small wireless network used by a single person.

Personal Identification Number (PIN): A four- or six-digit number a cardholder types to authenticate himself or herself.

Personal Mode: Pre-shared Key Mode in WPA or 802.11i.

Phase Modulation: Modulation in which one wave serves as a reference wave or a carrier wave. Another wave varies its phase to represent one or more bits.

Phishing: Social engineering attack that uses an official-looking e-mail message or website.

Physical Address: Data link layer address—not a physical layer address. Given this name because it is the address of the NIC, which is a physical device that implements both the physical and data link layers.

Physical Layer: The standards layer that governs physical transmission between adjacent devices; OSI Layer 1.

Physical Link: A connection linking adjacent devices on a network.

PIN: See Personal Identification Number.

Ping: Sending a message to another host and listening for a response to see if it is active.

Ping 127.0.0.1: Command line command to determine if the computer is set up to talk to the network.

Plaintext: The original message the sender wishes to send to the receiver; not limited to text messages.

Plan–Protect–Respond Cycle: The basic management cycle in which the three named stages are executed repeatedly.

Planning: In security, developing a broad security strategy that will be appropriate for a firm's security threats.

Plenum: Type of wiring with reduced toxic fumes. Required for runs through air conditioning ducts and other critical airspaces (plenums).

POE: See Power Over Ethernet.

Point of Presence (POP): 1) In cellular telephony, a site at which various carriers that provide telephone service are interconnected. 2) In PSDNs, a point of connection for user sites. There must be a private line between the site and the POP.

Point-to-Point Topology: A topology wherein two nodes are connected directly.

Policy: A broad statement that specifies what should be accomplished.

POP: See 1) Point of Presence. 2) Post Office Protocol.

Port: 1) In switching, a point to connect a cord to a switch. 2) In TCP and UDP messages, a header field that designates the application layer process on the server side and a specific connection on the client side.

Port-Based Access Control: Another name for 802.1X.

Port Number: The field in TCP and UDP that tells the transport process what application process sent the data in the data field or should receive the data in the data field.

Post Office Protocol (POP): The most popular protocol used to download e-mail from an e-mail server to an e-mail client.

Power Over Ethernet (POE): A standard that can bring electrical power to RJ-45 wall jacks.

Preamble Field: The initial field in an Ethernet MAC frame; synchronizes the receiver's clock to the sender's clock.

Prefix Notation: A way of representing masks. Gives the number of initial 1s in the mask.

Premises: The land and buildings owned by a customer.

Presence Server: A server used in many P2P systems; knows the IP addresses of each user and also whether the user is currently on line and perhaps whether or not the user is willing to chat.

Pre-Shared Key: A mode of operation in WPA and 802.11i in which all stations and an access point share the same initial key.

Presentation Layer: See OSI Presentation Layer.

Printer Sharing: A network service in which multiple users can share a printer over a network.

Priority: Preference given to latency-sensitive traffic, such as voice and video traffic, so that latency-sensitive traffic will go first if there is congestion.

Priority Level: The three-bit field used to give a frame one of eight priority levels from 000 (zero) to 111 (eight).

Private Branch Exchange (PBX): An internal telephone switch.

Private IP Address: An IP address that may be used only within a firm. Private IP addresses have three designated ranges: 10.x.x.x, 192.168.x.x, and 172.16.x.x through 172.31.x.x.

Private Key: A key that only the true party should know. Part of a public key–private key pair.

Probable Annual Loss: The likely annual loss from a particular threat. The cost of a successful attack times the probability of a successful attack in a one-year period.

Probe Packet: A packet sent into a firm's network during scanning; responses to the probe packet tend to reveal information about a firm's general network design and about its individual computers—including their operating systems.

Project Portfolio: A selection of projects that the firm will implement during a plan's initial period.

Propagate: In signals, to travel.

Propagation Effects: Changes in the signal during propagation.

Propagation Vector: A method malware uses to move to a victim computer.

Property: A characteristic of an object.

Protected Extensible Authentication Protocol (PEAP): A version of EAP preferred by Microsoft Windows computers.

Protecting: Implementing a security plan; the most time-consuming stage in the plan–protect–respond management cycle.

Protocol: 1) A standard that governs interactions between hardware and software processes at the same layer but on different hosts. 2) In IP, the header field that describes the content of the data field.

Protocol Field: In IP, a field that designates the protocol of the message in the IP packet's data field.

Provable Attack Packet: A packet that is provably an attack packet.

Provision: To set up service.

Proximity Access Card: Access card that works if it brought near a reader.

PSDN: See Public Switched Data Network.

PSTN: See Public Switched Telephone Network.

PTT: See Public Telephone and Telegraphy Authority.

Public IP Address: An IP address that must be unique on the Internet.

Public Key: A key that is not kept secret. Part of a public key–private key pair.

Public Key Authentication: Authentication in which each user has a public key and a private key. Authentication depends on the applicant knowing the true party's private key; requires a digital certificate to give the true party's public key.

Public Key Encryption: Encryption in which each side has a public key and a private key, so there are four keys in total for bidirectional communication. The sender encrypts messages with the receiver's public key. The receiver, in turn, decrypts incoming messages with the receiver's own private key.

Public Key Infrastructure (PKI): A total system (infrastructure) for public key encryption.

Public Switched Data Network (PSDN): A carrier WAN that provides data transmission service. The customer only needs to connect to the PSDN by running one private line from each site to the PSDN carrier's nearest POP.

Public Switched Telephone Network (PSTN): The worldwide telephone network.

Public Telephone and Telegraphy authority (PTT): The traditional title for the traditional monopoly telephone carrier in most countries.

Public Utilities Commission (PUC): In the United States, telecommunications regulatory agency at the state level.

PUC: See Public Utilities Commission.

Pulse Code Modulation (PCM): An analog-to-digital conversion technique in which the ADC samples the bandpass-filtered signal 8,000 times per second, each time measuring the intensity of the signal and representing the intensity by a number between 0 and 255.

PVC: See Permanent Virtual Circuit.

QAM: See Quadrature Amplitude Modulation.

QoS: See Quality of Service.

QPSK: See Quadrature Phase Shift Keying.

Quadrature Amplitude Modulation (QAM): Modulation technique that uses two carrier waves—a sine carrier wave and a cosine carrier wave. Each can vary in amplitude.

Quadrature Phase Shift Keying (QPSK): Modulation with four possible phases. Each of the four states represents two bits (00, 01, 10, and 11).

Quality of Service (QoS): Numerical service targets that must be met by networking staff.

Quality-of-Service (QoS) Parameters: In IPv4, service quality parameters applied to all packets with the same TOS field value.

Radio Frequency ID (RFID): A tag that can be read at a distance by a radio transmitter/receiver.

Radio Wave: An electromagnetic wave in the radio range.

RADIUS: A standard for central authentication servers.

Rapid Spanning Tree Protocol (RSTP): A version of the Spanning Tree Protocol that has faster convergence.

RAS: See Remote Access Server.

Rated Speed: The official standard speed of a technology.

RBOC: See Regional Bell Operating Company.

Real Time Protocol (RTP): The protocol that adds headers that contain sequence numbers to ensure that the UDP datagrams are placed in proper sequence and that they contain time stamps so that jitter can be eliminated.

Reassembly: Putting a fragmented packet back together.

Redundancy: Duplication of a hardware device in order to enhance reliability.

Regenerate: In a switch or router, to clean up a signal before sending it back out.

Regional Bell Operating Company (RBOC): One of the companies that was created to provide local service when the Bell System (AT&T) was broken up in the early 1980s.

Reliabile: A protocol in which errors are corrected by resending lost or damaged messages.

Remote Access Server (RAS): A server to which remote users connect in order to have their identities authenticated so they can get access to a site's internal resources.

Remote Monitoring (RMON) Probe: A specialized type of agent that collects data on network traffic passing through its location instead of information about the RMON probe itself.

Request for Comment (RFC): A document produced by the IETF that may become designated as an Official Internet Protocol Standard.

Request to Send: In 802.11 networks, a message sent to an access point when a station wishes to send and is able to send because of CSMA/CA. The station may send when it receives a clear-to-send message.

Request to Send/Clear to Send: A system that uses request-to-send and clear-to-send messages to control transmissions and avoid collisions in wireless transmission.

Request-Response Cycle: A cycle used in client/server processing where a client sends a request message to the server and the server sends back a response message.

Resegment: Dividing a collision domain into several smaller collision domains to reduce congestion and latency.

Responding: In security, the act of stopping and repairing an attack.

Response: Responding according to plan to security incidents.

Response Message: In Challenge–Response Authentication Protocols, the message that the applicant returns to the verifier.

Response Time: The difference between the time a user types a request to the time the user receives a response.

Retention: Rules that require IM messages to be captured and stored in order to comply with legal requirements.

Reusable Password: Password that is used repeatedly to get access.

RFC: See Request for Comment.

RFC 2822: The standard for e-mail bodies that are plaintext messages.

RFID: See Radio Frequency ID.

Right of Way: Permission to lay wires in public areas; given by government regulators to transmission carriers.

Ring Topology: A topology in which stations are connected in a loop and messages pass in only one direction around the loop.

Ring Wrapping: In a network with a dual-ring topology, responding to a break between switches by turning the surviving parts of a dual ring into a long single ring.

RIP: See Routing Information Protocol.

Risk Analysis: The process of balancing threats and protection costs.

RJ-45 Connector: The connector at the end of a UTP cord, which plugs into an RJ-45 jack.

RJ-45 Jack: The type of jack into which UTP cords RJ-45 connectors may plug.

RMON Probe: See Remote Monitoring Probe.

Roaming: 1) In cellular telephony, the situation when a subscriber leaves a metropolitan cellular system and goes to another city or country. 2) In 802.11, when a wireless host travels from one access point to another.

Rogue Access Point: An access point set up by a department or individual and not sanctioned by the firm.

Root: 1) The level at the top of a DNS hierarchy, consisting of all domain names. 2) A super account on a Unix server that automatically has full permissions in every directory on the server.

Root Server: One of 13 top-level servers in the Domain Name System (DNS).

Route: The path that a packet takes across an internet.

Router: A device that forwards packets within an internet. Routers connect two or more single networks (subnets).

Routing: 1) The forwarding of IP packets. 2) The exchange of routing protocol information through routing protocols.

Routing Information Protocol (RIP): A simple but limited interior routing protocol.

Routing Protocol: A protocol that allows routers to transmit routing table information to one another.

RSA: Popular public key encryption method.

RST Bit: In a TCP segment, if the RST (reset) bit is set, this tells the other side to end the connection immediately.

RSTP: See Rapid Spanning Tree Protocol.

RTP: See Real Time Protocol.

RTS: See Request to Send.

RTS/CTS: See Request to Send/Clear to Send.

SaaS: See Software as a Service.

SC Connector: A square optical fiber connector, recommended in the TIA/EIA-568 standard for use in new installations.

Scalability: The ability of a technology to handle growth well.

Scale: The ability to handle foreseen traffic increases and to do so without prohibitively rapid cost increases.

Scanning: For attackers, sending probe packets into a network to identify potential victims.

Scope: A parameter on a DHCP server that determines how many subnets the DHCP server may serve.

Script: A group of commands written in a simplified programming language.

Script Kiddie: An attacker who possesses only modest skills but uses attack scripts created by experienced hackers; dangerous because there are so many.

SDH: See Synchronous Digital Hierarchy.

Second-Generation (2G): The second generation of cellular telephony, introduced in the early 1990s. Offers the improvements of digital service, 150 MHz of bandwidth, a higher frequency range of operation, and slightly higher data transmission speeds.

Second-Level Domain: The third level of a DNS hierarchy, which usually specifies an organization (e.g., microsoft.com, hawaii.edu).

Secure Hash Algorithm (SHA): A hashing algorithm that can produce hashes of different lengths.

Secure Shell (SSH): A program that provides Telnet-like remote management capabilities; and FTP-like service; strongly encrypts both usernames and passwords.

Secure Sockets Layer (SSL): The simplest VPN security standard to implement; later renamed Transport Layer Security. Provides a secure connection at the transport layer, protecting any applications above it that are SSL/TLS-aware.

Semantics: In message exchange, the meaning of each message.

Sequence Number Field: In TCP, a header field that tells a TCP segment's order among the multiple TCP segments sent by one side.

Serial Transmission: Ethernet transmission over a single pair in each direction.

Server: A host that provides services to residential or corporate users.

Server Farm: Large groups of servers that work together to handle applications.

Server Host: In client/server processing, a server program on a server host provides services to a client program on a client host.

Service: In a service-oriented architecture (SOA), a service object provides services to calling programs.

Service Band: A subdivision of the frequency spectrum, dedicated to a specific service such as FM radio or cellular telephone service.

Service Control Point: A database of customer information, used in Signaling System 7.

Service Level Agreement (SLA): A quality-of-service guarantee for throughput, availability, latency, error rate, and other matters.

Service-Oriented Architecture (SOA): An architecture where there are many service objects instead of a few large programs.

Session Initiation Protocol (SIP): Relatively simple signaling protocol for voice over IP.

Session Key: Symmetric key that is used only during a single communication session between two parties.

Session Layer: See OSI Session Layer.

Set: 1) When a flag's field is given the value 1. 2) An SNMP command sent by the manager that tells the agent to change a parameter on the managed device.

SETI@home: A project from the Search for Extraterrestrial Intelligence (SETI), in which volunteers download SETI@home screen savers that are really programs. These programs do work for the SETI@ home server when the volunteer computer is idle. Processing ends when the user begins to do work.

Setup Fee: The cost of initial vendor installation for a system.

SFF: See Small Form Factor.

SHA: See Secure Hash Algorithm.

Shadow Zone (Dead Spot): A location where a receiver cannot receive radio transmission, due to an obstruction blocking the direct path between sender and receiver.

Shannon Equation: An equation by Claude Shannon (1938) that shows that the maximum possible transmission speed (C) when sending data through a channel is directly proportional to its bandwidth (B), and depends to a lesser extent on its signal-to-noise ratio (S/N): $C = B \text{ Log}_2 (1 + S/N)$.

Shared Internet Access: Access that allows two or more client PCs to use the Internet simultaneously, as if each was plugged directly into the broadband modem.

SHDSL: See Super-High-Rate DSL.

Shielded Twisted Pair (STP): A type of twisted-pair wiring that puts a metal foil sheath around each pair and another metal mesh around all pairs.

Signal: An information-carrying disturbance that propagates through a transmission medium.

Signal Bandwidth: The range of frequencies in a signal, determined by subtracting the lowest frequency from the highest frequency.

Signaling: In telephony, the controlling of calling, including setting up a path for a conversation through the transport core, maintaining and terminating the conversation path, collecting billing information, and handling other supervisory functions.

Signaling Gateway: The device that sets up conversations between parties, maintains these conversations, ends them, provides billing information, and does other work.

Signaling System 7: Telephone signaling system in the United States.

Signal-to-Noise Ratio (SNR): The ratio of the signal strength to average noise strength; should be high in order for the signal to be effectively received.

Signing: Encrypting something with the sender's private key.

Simple Mail Transfer Protocol (SMTP): The protocol used to send a message to a user's outgoing mail host and from one mail host to another; requires a complex series of interactions between the sender and receiver before and after mail delivery.

Simple Network Management Protocol (SNMP): The protocol that allows a general way to collect rich data from various managed devices in a network.

Simple Object Access Protocol (SOAP): A standardized way for a Web service to expose its methods on an interface to the outside world.

Single Point of Failure: When the failure in a single component of a system can cause a system to fail or be seriously degraded.

Single Sign-On (SSO): Authentication in which a user can authenticate himself or herself only once and then have access to all authorized resources on all authorized systems.

Single-Mode Fiber: Optical fiber whose core is so thin (usually 8.3 microns in diameter) that only a single mode can propagate—the one traveling straight along the axis.

SIP: See Session Initiation Protocol.

Site Survey: In wireless LANs, a radio survey to help determine where to place access points.

Situation Analysis: The examination of a firm's current situation, which includes anticipation of how things will change in the future.

Skype: A P2P VoIP service that currently offers free calling among Skype customers over the Internet and reduced-costs calling to and from Public Switched Telephone Network customers.

Skype Log-in Server: A central server in the Skype network, managed directly by Skype.

SLA: See Service Level Agreement.

SLC: See Systems Life Cycle.

Sliding Window Protocol: Flow control protocol that tells a receiver how many more bytes it may transmit before receiving another acknowledgment, which will give a longer transmission window.

Slot: A very brief time period used in Time Division Multiplexing; a subdivision of a frame. Carries one sample for one circuit.

Small Form Factor (SFF): A variety of optical fiber connectors; smaller than SC or ST connectors but unfortunately not standardized.

Smart Access Point: An access point that can be managed remotely.

SMTP: See Simple Mail Transfer Protocol.

SNA: See Systems Network Architecture.

Sneakernet: A joking reference to the practice of walking files around physically, instead of using a network for file sharing.

SNMP: See Simple Network Management Protocol.

SNR: See Signal-to-Noise Ratio.

SOA: See Service-Oriented Architecture.

SOAP: See Simple Object Access Protocol.

Social Engineering: Tricking people into doing something to get around security protections.

Social Media Applications: Web 2.0 applications that are designed to facilitate relationships.

Socket: The combination of an IP address and a port number, designating a specific connection to a specific application on a specific host. It is written as an IP address, a colon, and a port number, for instance, 128.171.17.13:80.

Software Architecture: A plan for providing applications functionality.

Software as a Service (SaaS): Service in which an application service provider supplies an application to customers on demand.

Software-Defined Radio: Radios that can be changed by software to meet different standards.

SOHO: See Small Office or Home Office.

Solid-Wire UTP: Type of UTP in which each of the eight wires really is a single solid wire.

SONET: See Synchronous Optical Network.

Spam: Unsolicited commercial e-mail.

Spam Blocking: Software that recognizes and deletes spam.

Spanning Tree Protocol (STP): See 802.1D Spanning Tree Protocol.

Speech Codec: See codec.

SPI: See Stateful Packet Inspection.

Splitter: A device that a DSL user plugs into each telephone jack; the splitter separates the voice signal from the data signal so that they cannot interfere with each other.

Spread Spectrum Transmission: A type of radio transmission that takes the original signal and spreads the signal energy over a much broader channel than would be used in normal radio transmission; used in order to reduce propagation problems, not for security.

Spyware: Software that sits on a victim's machine and gathers information about the victim.

SS7: See Signaling System 7.

SSH: See Secure Shell.

SSL: See Secure Sockets Layer.

SSL/TLS: See Secure Sockets Layer and Transport Layer Security.

SSL/TLS-Aware: Modified to work with SSL/TLS.

SSO: See Single Sign-On.

ST Connector: A cylindrical optical fiber connector, sometimes called a bayonet connector because of the manner in which it pushes into an ST port and then twists to be locked in place.

Standard: A rule of operation that allows two hardware or software processes to work together. Standards normally govern the exchange of messages between two entities.

Standards Agency: An organization that creates and maintains standards.

Standards Architecture: A family of related standards that collectively allows an application program on one machine on an internet to communicate with another application program on another machine on the internet.

Star Topology: A form of topology in which all wires in a network connect to a single switch.

Start of Frame Delimiter Field: The second field of an Ethernet MAC frame, which synchronizes the receiver's clock to the sender's clock and then signals that the synchronization has ended.

State: In digital physical layer signaling, one of the few line conditions that represent information.

Stateful Firewall: A firewall whose default behavior is to allow all connections initiated by internal hosts but to block all connections initiated by external hosts. Only passes packets that are part of approved connections.

Stateful Packet Inspection: Firewall filtering mechanism that uses different filtering methods in different states of a conversation.

Static IP Address: An IP address that never changes.

STM: See Synchronous Transfer Mode.

STP: See 802.1D Spanning Tree Protocol or Shielded Twisted Pair.

Strand: In optical fiber, a core surrounded by a cladding. For two-way transmission, two optical fiber strands are needed.

Stranded-Wire UTP: Type of UTP in which each of the eight "wires" really is a collection of wire strands.

Stripping Tool: Tool for stripping the sheath off the end of a UTP cord.

Strong Keys: Keys that are too long to be cracked by exhaustive key search.

Subcarrier: A channel that is itself a subdivision of a broadband channel, used to transmit frames in OFDM.

Subnet: A small network that is a subdivision of a large organization's network.

Subnet Mask: A mask with 1s in the network and subnet parts and 0s in the host part.

Subnet Part: The part of an IP address that specifies a particular subnet within a network.

Super Node: In Skype, a host node that takes on the work of signaling.

Super-High-Rate DSL (SHDSL): The next step in business DSL, which can operate symmetrically over a single voice-grade twisted pair and over a speed range of 384 kbps–2.3 Mbps. It can also operate over somewhat longer distances than HDSL2.

Supervisory Standard: A standard that used to keep a network or internet working.

Supplicant: The party trying to prove his or her identity.

Surreptitiously: Done without someone's knowledge, such as surreptitious face recognition scanning.

SVC: See Switched Virtual Circuit.

Switch: A device that forwards frames within a single network.

Switched Virtual Circuit (SVC): A circuit between sites that is set up just before a call and that lasts only for the duration of the call.

Switching Matrix: A switch component that connects input ports to output ports.

Symmetric: Speeds that are equal in both directions.

Symmetric Key Encryption: Family of encryption methods in which the two sides use the same key to encrypt messages to each other and to decrypt incoming messages. In bidirectional communication, only a single key is used.

SYN Bit: In TCP, the flags field that is set to indicate if the message is a synchronization message.

Synchronous Communication: Communication in which the senders' and receivers' clocks must be precisely synchronized for the receiver to read the message.

Synchronous Digital Hierarchy (SDH): The European version of the technology upon which the world is nearly standardized.

Synchronous Optical Network (SONET): The North American version of the technology upon which the world is nearly standardized.

Synchronous Transfer Mode (STM): A number that indicates SDH speeds.

Syntax: In message exchange, how messages are organized.

System Life Cycle Costs: Costs over a system's entire life.

Systems Administration: The management of a server.

Systems Life Cycle (SLC): The period between a system's conception and its termination.

Systems Network Architecture (SNA): The standards architecture traditionally used by IBM mainframe computers.

T568B: Wire color scheme for RJ-45 connectors; used most commonly in the United States.

Tag: An indicator on an HTML file to show where the browser should render graphics files, when it should play audio files, and so forth.

Tag Control Information: The second tag field, which contains a 12-bit VLAN ID that it sets to zero if VLANs are not being implemented. If VLANs are being used, each VLAN will be assigned a different VLAN ID.

Tag Field: One of the two fields added to an Ethernet MAC layer frame by the 802.1Q standard.

Tag Protocol ID: The first tag field used in the Ethernet MAC layer frame. The Tag Protocol ID has the two-octet hexadecimal value 81-00, which indicates that the frame is tagged.

Tbps: Terabits per second—a thousand billions of bits per second.

TCO: See Total Cost of Ownership.

TCP: See Transmission Control Protocol.

TCP Segment: A TCP message.

TCP/IP: The Internet Engineering Tasks Force's standards architecture; dominant above the data link layer.

TDM: See Time Division Multiplexing.

Telecommunications: The transmission of voice and video, as opposed to data.

Telecommunications Closet: The location on each floor of a building where cords coming up from the basement are connected to cords that span out horizontally to telephones and computers on that floor.

Telephone Modem: A device used in telephony that converts digital data into an analog signal that can transfer over the local loop.

Temporal Dispersion: Another name for modal dispersion.

Temporal Key Integrity Protocol (TKIP): A security process used by 802.11i, where each station has its own nonshared key after authentication and where this key is changed frequently.

Terabits per second: Trillions of bits per second.

Terminal Crosstalk Interference: Crosstalk interference at the ends of a UTP cord, where wires are untwisted to fit into the connector. To control terminal crosstalk interference, wires should not be untwisted more than a half inch to fit into connectors.

Termination Equipment: Equipment that connects a site's internal telephone system to the local exchange carrier.

Terrestrial: Earth-bound transmissions.

Test Signals: Signal sent by a high-quality UTP tester through a UTP cord to check signal quality parameters.

Texting: In cellular telephony, the transmission of text messages.

Third-Generation (3G): The newest generation of cellular telephony, able to carry data at much higher speeds than 2G systems.

Threat Environment: The threats that face the company.

Three-Party Call: A call in which three people can take part in a conversation.

Three-Step Handshake: A three-message exchange that opens a connection in TCP.

Throughput: The transmission speed that users *actually* get. Usually lower than a transmission system's rated speed.

TIA/EIA/ANSI-568: The standard that governs transmission media in the United States.

Time Division Multiplexing (TDM): A technology used by telephone carriers to provide reserved capacity on trunk lines between switches. In TDM, time is first divided into frames, each of which are divided into slots; a circuit is given the same slot in every frame.

Time to Live (TTL): The field added to a packet and given a value by a source host, usually between 64 and 128. Each router along the way decrements the TTL field by one. A router decrementing the TTL to zero will discard the packet; this prevents misaddressed packets from circulating endlessly among packet switches in search of their nonexistent destinations.

TKIP: See Temporal Key Integrity Protocol.

TLS: See Transport Layer Security.

Toll Call: Long-distance call pricing in which the price depends on distance and duration.

Toll-Free Number Service: Service in which anyone can call into a company, usually without being charged. Area codes are 800, 888, 877, 866, and 855.

Top-Level Domain: The second level of a DNS hierarchy, which categorizes the domain by organization type (e.g., .com, .net, .edu, .biz, .info) or by country (e.g., .uk, .ca, .ie, .au, .jp, .ch).

Topology: The way in which nodes are linked together by transmission lines.

ToS: See Type of Service.

Total Cost of Ownership (TCO): The total cost of an entire system over its expected lifespan.

Total Purchase Cost of Network Products: The initial purchase price of a fully configured system.

Tracert: A command line command to list the routers between the computer and the destination host and to list average latency to each router along the way.

Tracert (tracert): A Windows program that shows latencies to every router along a route and to the destination host.

Tracker: In BitTorrent, a server that coordinates the file transfer.

Traffic Engineering: Designing and managing traffic on a network.

Traffic Shaping: Limiting access to a network based on type of traffic.

Trailer: The part of a message that comes after the data field.

Transaction Processing: Processing involving simple, highly structured, and high-volume interactions.

Transceiver: A transmitter/receiver.

Transfer Syntax: In the OSI Presentation layer, the syntax used by two presentation layer processes to communicate, which may or may not be quite different than either of their internal methods of formatting information.

Transmission Control Protocol (TCP): The most common TCP/IP protocol at the transport layer. Connection-oriented and reliable.

Transmission Line: A physical line that is used to carry transmitted information.

Transmission Speed: The rate at which information is transmitted in bits per second.

Transparently: Without having a need to implement modifications.

Transport: In telephony, transmission; taking voice signals from one subscriber's access line and delivering them to another customer's access line.

Transport Core: The switches and transmission lines that carry voice signals from one subscriber's access line and delivering them to another customer's access line.

Transport Layer: The layer that governs communication between two hosts; Layer 4 in both OSI and TCP/IP.

Transport Layer Security (TLS): The simplest VPN security standard to implement; originally named Secure Sockets Layer. Provides a secure connection at the transport layer, protecting any applications above it that are SSL/TLS-aware.

Transport Mode: One of IPsec's two modes of operation, in which the two computers that are communicating implement IPsec. Transport mode gives strong end-to-end security between the computers, but it requires IPsec configuration and a digital certificate on all machines.

Traps: The type of message that an agent sends if it detects a condition that it thinks the manager should know about.

Triple DES (3DES): Symmetric key encryption method in which a message is encrypted three times with DES. If done with two or three different keys, offers strong security. However, it is processing intensive.

Trivial File Transfer Protocol (TFTP): A protocol used on switches and routers to download configuration information; has no security.

Trojan Horse: A program that looks like an ordinary system file, but continues to exploit the user indefinitely.

Trunk Line: A type of transmission line that links switches to each other, routers to each other, or a router to a switch.

Trunking: See Link Aggregation.

TTL: See Time to Live.

Tunnel Mode: One of IPsec's two modes of operation, in which the IPsec connection extends only between IPsec gateways at the two sites. Tunnel mode provides no protection within sites, but it offers transparent security.

Twisted-Pair Wiring: Wiring in which each pair's wires are twisted around each other several times per inch, reducing EMI.

Two-Factor Authentication: A type of authentication that requires two forms of credentials.

Two-Way Amplifier: In cable television, an amplifier that amplifies signals traveling in both directions.

Type of Service (ToS): IPv4 header field that designates the type of service a certain packet should receive.

U: The standard unit for measuring the height of switches. One U is 1.75 inches (4.4 cm) in height. Most switches, although not all, are multiples of U.

UDDI: See Universal Description, Discovery, and Integration.

UDDI Green Pages: The UDDI search option that allows companies to understand how to interact with specific Web services. Green pages specify the interfaces on which a Web service will respond, the methods it will accept, and the properties that can be changed or returned.

UDDI White Pages: The UDDI search option that allows users to search for Web services by name, much like telephone white pages.

UDDI Yellow Pages: The UDDI search option that allows users to search for Web services by function, such as accounting, much like telephone yellow pages.

UDP: See User Datagram Protocol.

UDP Checksum: Field in the UDP header that the receiver uses to check for errors. If the receiving transport process finds an error, it drops the UDP datagram.

UDP Length: Field in the UDP header that gives the length of the UDP data field in octets.

Ultrawideband (UWB): Spread spectrum transmission system that has extremely wide channels.

Unicasting: Transmission to a single other host.

UNICODE: The standard that allows characters of all languages to be represented.

Universal Description, Discovery, and Integration (UDDI): A protocol that is a distributed database that helps users find appropriate Web services.

Universal Malware: Malware that works whether or not the target computer has a security vulnerability.

Unlicensed Radio Band: A radio band that does not require each station using it to have a license.

Unreliable: (Of a protocol) not doing error correction.

Unshielded Twisted Pair (UTP): Network cord that contains four twisted pairs of wire within a sheath. Each wire is covered with insulation.

Update: To download and apply patches to fix a system.

Uplink: In satellites, transmission from the Earth to a communication satellite.

User Datagram Protocol (UDP): Unreliable transport-layer protocol in TCP/IP.

Username: An alias that signifies the account that the account holder will be using.

Utility Computing: A model of computing* where a company offloads server processing work to another company at a remote site.

UTP: See Unshielded Twisted Pair.

UWB: See Ultrawideband.

Validate: To test the accuracy of a network simulation model by comparing its performance with that of the real network. If the predicted results match the actual results, the model is validated.

Verifier: The party requiring the supplicant to prove his or her identity.

Vertical Riser: Space between the floors of a building that telephone and data cabling go through to get to the building's upper floor.

Very Small Aperture Terminal (VSAT): Communication satellite earthstation that has a small-diameter antenna.

Virtual Circuit: A transmission path between two sites or devices; selected before transmission begins.

Virtual LAN (VLAN): A closed collection of servers and the clients they serve. Broadcast signals go only to computers in the same VLAN.

Virtual Machine (VM): One of multiple logical machines in a real machine; to its users, it appears to be a real machine.

Virtual Private Network (VPN): A network that uses the Internet or a wireless network with added security for data transmission.

Virtualization: A process where the real computer's capacity is divided among a number of virtual machines.

Virus: A piece of executable code that attaches itself to programs or data files. When the program is executed or the data file opened, the virus spreads to other programs or data files.

VLAN: See Virtual LAN.

VM: See Virtual Machine.

Voice Mail: A service that allows people to leave a message if the user does not answer his or her phone.

Voice over IP (VoIP): The transmission of voice signals over an IP network.

Voice-Grade: Wire of a quality designed for transmitting voice signals in the PSTN.

VoIP: See Voice over IP.

VoIP Telephone: A telephone that has the electronics to encode voice for digital transmission and to handle packets over an IP internet.

VPN: See Virtual Private Network.

VSAT: See Very Small Aperture Terminal.

Vulnerability: A security weakness found in software.

Vulnerability Testing: Testing after protections have been configured, in which a company or a consultant attacks protections in the way a determined attacker would and notes which attacks that should have been stopped actually succeeded.

Vulnerability-Specific: A type of attack aimed at a particular vulnerability.

WAN: See Wide Area Network.

War Driver: Someone who travels around looking for unprotected wireless access points.

WATS: See Wide Area Telephone Service.

Wavelength: The physical distance between comparable points (e.g., from peak to peak) in successive cycles of a wave.

WDSL: See Web Service Description Language.

Weak Keys: Keys that are shot enough to be cracked by an exhaustive key search.

Web 2.0: A category of applications in which users provide the content.

Web Service: A type of service-oriented architecture based on World Wide Web standards.

Web Service Description Language (WSDL): A Web Service Description Language (WSDL) response message is like a user's manual for a particular Web service. However, unlike user manuals, WSDL is designed to be read by a software program rather than primarily by a human reader. A WSDL message allows a calling program to understand how to use the Web service object.

Web-Enabled: Client/server processing applications that use ordinary browsers as client programs.

Webify: In SSL/TLS VPNs, the SSL/TLS gateway can translate output from some applications into a webpage.

Webmail: Web-enabled e-mail. User needs only a browser to send and read e-mail.

Well-Known Port Number: Standard port number of a major application that is usually (but not always) used. For example, the well-known TCP port number for HTTP is 80. Well-known port numbers range from 0 through 1023.

WEP: See Wired Equivalent Privacy.

Wide Area Network (WAN): A network that links different sites together.

Wide Area Telephone Service (WATS): Service that allows a company to place outgoing long-distance calls at per-minute prices lower than those of directly dialed calls.

Wi-Fi Alliance: Trade group created to create interoperability tests of 802.11 LANs; actually produced the WPA standard.

WiMAX: Broadband wireless access method. Standardized as 802.16.

WiMAX Forum: The WiMAX Forum has developed a wireless metropolitan area network service called WiMAX. WiMAX is designed for carriers who want to provide metropolitan area networking. Its signals can easily reach all customers in a metropolitan area.

Window Size Field: TCP header field that is used for flow control. It tells the station that receives the segment how many more octets that station may transmit before getting another acknowledgment message that will allow it to send more octets.

Wired Equivalent Privacy (WEP): A weak security mechanism for 802.11.

Wireless Access Point: Devices that control wireless clients and that bridge wireless clients to servers and routers on the firm's main wired LAN.

Wireless LAN (WLAN): A local area network that uses radio transmission instead of cabling to connect devices.

Wireless LAN Switch: An Ethernet switch to which multiple wireless access points connect; manages the access points.

Wireless Networking: Networking that uses radio transmission instead of wires to connect devices.

Wireless NIC: 802.11 network interface card.

Wireless Protected Access (WPA): The 802.11 security method created as a stopgap between WEP and 802.11i.

Wireless Protected Access 2 (WPA2): Another name for 802.11 security.

Wireless WAN (WWAN): Wireless WAN (WWAN) transmission is less developed than wireless LAN transmission but that holds enormous promise.

WLAN: See Wireless LAN.

Work-Around: A process of making manual changes to eliminate a vulnerability instead of just installing a software patch.

Workgroup: A logical network. On a physical network, only PCs in the same workgroup can communicate.

Workgroup Switch: A switch to which stations connect directly.

Working Group: A specific subgroup of the 802 Committee, in charge of developing a specific group of standards. For instance, the 802.3 Working Group creates Ethernet standards.

Worm: An attack program that propagates on its own by seeking out other computers, jumping to them, and installing itself.

WPA: See Wireless Protected Access.

WPA2: See Wireless Protected Access 2.

WWAN: See Wireless WAN.

X.25: The original PSDN standard. Relatively slow because it did error correction.

X.509: The main standard for digital certificates.

Zero-Day Attack: Attack that takes advantage of a vulnerability for which no patch or other work-around has been released.

Zero-Day Exploit: An exploit that takes advantage of vulnerabilities that have not previously been discovered or for which updates have not been created.

ZigBee: Low-speed, low-power protocol for wirelessly connecting sensors and other very small devices.

INDEX

Note: Page numbers in **bold type** indicate where terms are defined or characterized; page numbers in *italics* indicate tables or figures; page numbers with an "n" indicate a footnote.

B

F

J

K

L

U

V